MATHEMATICS RESEARCH DEVELOPMENTS

ADVANCES IN LINEAR ALGEBRA RESEARCH

MATHEMATICS RESEARCH DEVELOPMENTS

Additional books in this series can be found on Nova's website
under the Series tab.

Additional e-books in this series can be found on Nova's website
under the e-book tab.

MATHEMATICS RESEARCH DEVELOPMENTS

ADVANCES IN LINEAR ALGEBRA RESEARCH

IVAN KYRCHEI

EDITOR

nova publishers

New York

LIBRARY OF CONGRESS CATALOGING-IN-PUBLICATION DATA

Advances in linear algebra research / Ivan Kyrchei (National Academy of Sciences of Ukraine), editor.
pages cm. -- (Mathematics research developments)
Includes bibliographical references and index.
ISBN 978-1-63463-565-3 (hardcover)
1. Algebras, Linear. I. Kyrchei, Ivan, editor.
QA184.2.A38 2015
512'.5--dc23

2014043171

Published by Nova Science Publishers, Inc. † New York

CONTENTS

PREFACE

This book presents original studies on the leading edge of linear algebra. Each chapter has been carefully selected in an attempt to present substantial research results across a broad spectrum. The main goal of Chapter One is to define and investigate the restricted generalized inverses corresponding to minimization of constrained quadratic form. As stated in Chapter Two, in systems and control theory, Linear Time Invariant (LTI) descriptor (Differential-Algebraic) systems are intimately related to the matrix pencil theory. A review of the most interesting properties of the Projective Equivalence and the Extended Hermite Equivalence classes is presented in the chapter. New determinantal representations of generalized inverse matrices based on their limit representations are introduced in Chapter Three. Using the obtained analogues of the adjoint matrix, Cramer's rules for the least squares solution with the minimum norm and for the Drazin inverse solution of singular linear systems have been obtained in the chapter. In Chapter Four, a very interesting application of linear algebra of commutative rings to systems theory, is explored. Chapter Five gives a comprehensive investigation to behaviors of a general Hermitian quadratic matrix-valued function by using ranks and inertias of matrices. In Chapter Six, the theory of triangular matrices (tables) is introduced. The main "characters" of the chapter are special triangular tables (which will be called triangular matrices) and their functions paradeterminants and parapermanents. The aim of Chapter Seven is to present the latest developments in iterative methods for solving linear matrix equations. The problems of existence of common eigenvectors and simultaneous triangularization of a pair of matrices over a principal ideal domain with quadratic minimal polynomials are investigated in Chapter Eight. Two approaches to define a noncommutative determinant (a determinant of a matrix with noncommutative elements) are considered in Chapter Nine. The last, Chapter 10, is an example of how the methods of linear algebra are used in natural sciences, particularly in chemistry. In this chapter, it is shown that in a First Order Chemical Kinetics Mechanisms matrix, all columns add to zero, all the diagonal elements are non-positive and all the other matrix entries are non-negative. As a result of this particular structure, the Gershgorin Circles Theorem can be applied to show that all the eigenvalues are negative or zero.

Minimization of a quadratic form $\langle x, Tx\rangle + \langle p, x\rangle + a$ under constraints defined by a linear system is a common optimization problem. In Chapter 1, it is assumed that the operator T is symmetric positive definite or positive semidefinite. Several extensions to different sets of linear matrix constraints are investigated. Solutions of this problem may be given using the Moore-Penrose inverse and/or the Drazin inverse. In addition, several new classes of generalized inverses are defined minimizing the seminorm defined by the quadratic forms, depending on the matrix equation that is used as a constraint.

A number of possibilities for further investigation are considered.

In systems and control theory, Linear Time Invariant (LTI) descriptor (Differential-Algebraic) systems are intimately related to the matrix pencil theory. Actually, a large number of systems are reduced to the study of differential (difference) systems $S(F, G)$ of the form:

$$S(F,G) : F\dot{x}(t) = Gx(t) \ \ (\text{or the dual } Fx = G\dot{x}(t)),$$

and

$$S(F,G) : Fx_{k+1} = Gx_k \ \ (\text{or the dual } Fx_k = Gx_{k+1}), \ \ F, G \in \mathbb{C}^{m\times n}$$

and their properties can be characterized by the homogeneous pencil $sF - \hat{s}G$. An essential problem in matrix pencil theory is the study of invariants of $sF - \hat{s}G$ under the *bilinear strict equivalence*. This problem is equivalent to the study of complete *Projective Equivalence* (PE), $\mathcal{E}_{\mathcal{P}}$, defined on the set $\mathbb{C}_r$ of complex homogeneous binary polynomials of fixed homogeneous degree r. For a $f(s, \hat{s}) \in \mathbb{C}_r$, the study of invariants of the PE class $\mathcal{E}_{\mathcal{P}}$ is reduced to a study of invariants of matrices of the set $\mathbb{C}^{k\times 2}$ (for $k \geqslant 3$ with all 2×2-minors non-zero) under the *Extended Hermite Equivalence* (EHE), $\mathcal{E}_{rh}$. In Chapter 2, the authors present a review of the most interesting properties of the PE and the EHE classes. Moreover, the appropriate projective transformation $d \in RGL(1, \mathbb{C}/\mathbb{R})$ is provided analytically ([1]).

By a generalized inverse of a given matrix, the authors mean a matrix that exists for a larger class of matrices than the nonsingular matrices, that has some of the properties of the usual inverse, and that agrees with inverse when given matrix happens to be nonsingular. In theory, there are many different generalized inverses that exist. The authors shall consider the Moore Penrose, weighted Moore-Penrose, Drazin and weighted Drazin inverses.

New determinantal representations of these generalized inverse based on their limit representations are introduced in Chapter 3. Application of this new method allows us to obtain analogues classical adjoint matrix. Using the obtained analogues of the adjoint matrix, the authors get Cramer's rules for the least squares solution with the minimum norm and for the Drazin inverse solution of singular linear systems. Cramer's rules for the minimum norm least squares solutions and the Drazin inverse solutions of the matrix equations $\mathbf{AX} = \mathbf{D}$, $\mathbf{XB} = \mathbf{D}$ and $\mathbf{AXB} = \mathbf{D}$ are also obtained, where $\mathbf{A}$, $\mathbf{B}$ can be singular matrices of appropriate size. Finally, the authors derive determinantal representations of solutions of the differential matrix equations, $\mathbf{X}' + \mathbf{AX} = \mathbf{B}$ and $\mathbf{X}' + \mathbf{XA} = \mathbf{B}$, where the matrix $\mathbf{A}$ is singular.

Many physical systems in science and engineering can be described at time t in terms of an n-dimensional state vector $x(t)$ and an m-dimensional input vector $u(t)$, governed by an evolution equation of the form $x'(t) = A \cdot x(t) + B \cdot u(t)$, if the time is continuous, or $x(t+1) = A \cdot x(t) + B \cdot u(t)$ in the discrete case. Thus, the system is completely described by the pair of matrices (A, B) of sizes $n \times n$ and $n \times m$ respectively.

In two instances feedback is used to modify the structure of a given system (A, B): first, A can be replaced by $A + BF$, with some characteristic polynomial that ensures stability of the new system $(A + BF, B)$; and second, combining changes of bases with a feedback action $A \mapsto A + BF$ one obtains an equivalent system with a simpler structure.

Given a system (A, B), let (A, B) denote the set of states reachable at finite time when starting with initial condition $x(0) = 0$ and varying $u(t)$, i.e., (A, B) is the right image of the matrix $[B|AB|A^2B|\cdots]$. Also, let $Pols(A, B)$ denote the set of characteristic polynomials of all possible matrices $A + BF$, as F varies.

Classically, (A, B) have their entries in the field of real or complex numbers, but the concept of discrete-time system is generalized to matrix pairs with coefficients in an arbitrary commutative ring R. Therefore, techniques from Linear Algebra over commutative rings are needed.

In Chapter 4, the following problems are studied and solved when R is a commutative von Neumann regular ring:

- A canonical form is obtained for the feedback equivalence of systems (combination of basis changes with a feedback action).
- Given a system (A, B), it is proved that there exist a matrix F and a vector u such that the single-input system $(A + BF, Bu)$ has the same reachable states and the same assignable polynomials as the original system, i.e. $(A + BF, Bu) = (A, B)$ and $Pols(A + BF, Bu) = Pols(A, B)$.

Chapter 5 gives a comprehensive investigation to behaviors of a general Hermitian quadratic matrix-valued function

$$\phi(X) = (\,AXB + C\,)M(\,AXB + C)^* + D$$

by using ranks and inertias of matrices. The author first establishes a group of analytical formulas for calculating the global maximal and minimal ranks and inertias of $\phi(X)$. Based on the formulas, the author derives necessary and sufficient conditions for $\phi(X)$ to be a positive definite, positive semi-definite, negative definite, negative semi-definite function, respectively, and then solves two optimization problems of finding two matrices $\widehat{X}$ or $\widetilde{X}$ such that $\phi(X) \succcurlyeq \phi(\widehat{X})$ and $\phi(X) \preccurlyeq \phi(\widetilde{X})$ hold for all X, respectively. As extensions, the author considers definiteness and optimization problems in the Löwner sense of the following two types of multiple Hermitian quadratic matrix-valued function

$$\phi(\,X_1, \ldots, X_k\,) = \left(\sum_{i=1}^{k} A_iX_iB_i + C\right)M\left(\sum_{i=1}^{k} A_iX_iB_i + C\right)^* + D,$$

$$\psi(\,X_1, \ldots, X_k\,) = \sum_{i=1}^{k}(\,A_iX_iB_i + C_i\,)M_i(\,A_iX_iB_i + C_i\,)^* + D.$$

Some open problems on algebraic properties of these matrix-valued functions are mentioned at the end of Chapter 5.

In Chapter 6, the author considers elements of linear algebra based on triangular tables with entries in some number field and their functions, analogical to the classical notions of a matrix, determinant and permanent. Some properties are investigated and applications in various areas of mathematics are given.

The aim of Chapter 7 is to present the latest developments in iterative methods for solving linear matrix equations. The iterative methods are obtained by extending the methods presented to solve the linear system $Ax = b$. Numerical examples are investigated to confirm the efficiency of the methods.

The problems of existence of common eigenvectors and simultaneous triangularization of a pair of matrices over a principal ideal domain with quadratic minimal polynomials are investigated in Chapter 8. The necessary and sufficient conditions of simultaneous triangularization of a pair of matrices with quadratic minimal polynomials are obtained. As a result, the approach offered provides the necessary and sufficient conditions of simultaneous triangularization of pairs of idempotent matrices and pairs of involutory matrices over a principal ideal domain.

Since product of quaternions is noncommutative, there is a problem how to determine a determinant of a matrix with noncommutative elements (it's called a noncommutative determinant). The authors consider two approaches to define a noncommutative determinant. Primarily, there are row – column determinants that are an extension of the classical definition of the determinant; however, the authors assume predetermined order of elements in each of the terms of the determinant. In Chapter 9, the authors extend the concept of an immanant (permanent, determinant) to a split quaternion algebra using methods of the theory of the row and column determinants.

Properties of the determinant of a Hermitian matrix are established. Based on these properties, analogs of the classical adjont matrix over a quaternion skew field have been obtained. As a result, the authors have a solution of a system of linear equations over a quaternion division algebra according to Cramer's rule by using row–column determinants.

Quasideterminants appeared from the analysis of the procedure of a matrix inversion. By using quasideterminants, solving of a system of linear equations over a quaternion division algebra is similar to the Gauss elimination method.

The common feature in definition of row and column determinants and quasideterminants is that the authors have not one determinant of a quadratic matrix of order n with noncommutative entries, but certain set (there are n^2 quasideterminants, n row determinants, and n column determinants). The authors have obtained a relation of row-column determinants with quasideterminants of a matrix over a quaternion division algebra.

First order chemical reaction mechanisms are modeled through Ordinary Differential Equations (O.D.E.) systems of the form: , being the chemical species concentrations vector, its time derivative, and the associated system matrix.

A typical example of these reactions, which involves two species, is the Mutarotation of Glucose, which has a corresponding matrix with a null eigenvalue whereas the other one is negative.

A very simple example with three chemical compoundsis grape juice, when it is converted into wine and then transformed into vinegar. A more complicated example,also involving three species, is the adsorption of Carbon Dioxide over Platinum surfaces. Although, in these examples the chemical mechanisms are very different, in both cases the O.D.E. system matrix has two negative eigenvalues and the other one is zero. Consequently, in all these cases that involve two or three chemical species, solutions show a weak stability (i.e., they are stable but not asymptotically). This fact implies that small errors due to measurements in the initial concentrations will remain bounded, but they do not tend to vanish as the reaction proceeds.

In order to know if these results can be extended or not to other chemical mechanisms, a possible general result is studied through an inverse modeling approach, like in previous papers. For this purpose, theoretical mechanisms involving two or more species are proposed and a general type of matrices - so-called First Order Chemical Kinetics Mechanisms (F.O.C.K.M.) matrices - is studied from the eigenvalues and eigenvectors view point.

Chapter 10 shows that in an F.O.C.K.M. matrix all columns add to zero, all the diagonal elements are non-positive and all the other matrix entries are non-negative. Because of this particular structure, the Gershgorin Circles Theorem can be applied to show that all the eigenvalues are negative or zero. Moreover, it can be proved that in the case of the null eigenvalues - under certain conditions - algebraic and geometric multiplicities give the same number.

As an application of these results, several conclusions about the stability of the O.D.E. solutions are obtained for these chemical reactions, and its consequences on the propagation of concentrations and/or surface concentration measurement errors, are analyzed.

In: Advances in Linear Algebra Research
Editor: Ivan Kyrchei, pp. 1-55
ISBN: 978-1-63463-565-3

Chapter 1

MINIMIZATION OF QUADRATIC FORMS AND GENERALIZED INVERSES

***Predrag S. Stanimirović*[1,*], *Dimitrios Pappas*[2,†] *and Vasilios N. Katsikis*[3,‡]**
[1]University of Niš, Faculty of Sciences and Mathematics, Niš, Serbia
[2]Athens University of Economics and Business
Department of Statistics, Athens, Greece
[3]National and Kapodistrian University of Athens, Department of Economics
Division of Mathematics and Informatics, Athens, Greece

Abstract

Minimization of a quadratic form $\langle x, Tx\rangle + \langle p, x\rangle + a$ under constraints defined by a linear system is a common optimization problem. It is assumed that the operator T is symmetric positive definite or positive semidefinite. Several extensions to different sets of linear matrix constraints are investigated. Solutions of this problem may be given using the Moore-Penrose inverse and/or the Drazin inverse. In addition, several new classes of generalized inverses are defined minimizing the seminorm defined by the quadratic forms, depending on the matrix equation that is used as a constraint.

A number of possibilities for further investigation are considered

Keywords: Quadratic functional, quadratic optimization, generalized inverse, Moore-Penrose inverse, Drazin inverse, outer inverse, system of linear equations, matrix equation, generalized inverse solution, Drazin inverse solution

AMS Subject Classification: 90C20, 15A09, 15A24, 11E04, 47N10

1. Introduction

It is necessary to mention several common and usual notations. By $\mathbb{R}^{m\times n}$ (resp. $\mathbb{C}^{m\times n}$) we denote the space of all real (resp. complex) matrices of dimension $m \times n$. If $A \in$

*E-mail address: pecko@pmf.ni.ac.rs
†E-mail address: pappdimitris@gmail.com
‡E-mail address: vaskatsikis@econ.uoa.gr

$\mathbb{R}^{m\times n}$ (resp. $\mathbb{C}^{m\times n}$), by $A^T \in \mathbb{R}^{n\times m}$ (resp. $A^* \in \mathbb{R}^{n\times m}$) is denoted the transpose (resp. conjugate and transpose) matrix of A. As it is usual, by $\mathcal{N}(A)$ we denote the null-space of A, by $\mathcal{R}(A)$ the range of A, and $\mathrm{ind}(A)$ will denote the index of the matrix A.

1.1. Quadratic Functions, Optimization and Quadratic Forms

Definition 1.1. *A square matrix $A \in \mathbb{C}^{n\times n}$ (resp. $A \in \mathbb{R}^{n\times n}$) is:*

1) *Hermitian (Symmetric) matrix if $A^* = A$ ($A^T = A$),*
2) *normal, if $A^*A = AA^*$ ($A^TA = AA^T$),*
3) *lower-triangular, if $a_{ij} = 0$ for $i < j$,*
4) *upper-triangular, if $a_{ij} = 0$ for $i > j$,*
5) *positive semi-definite, if $\mathrm{Re}\,(x^*Ax) \geq 0$ for all $x \in \mathbb{C}^{n\times 1}$. Additionally, if it holds $\mathrm{Re}\,(x^*Ax) > 0$ for all $x \in \mathbb{C}^{n\times 1} \setminus \{0\}$, then the matrix A is positive definite.*

6) *Unitary (resp. orthogonal) matrix A is a square matrix whose inverse is equal to its conjugate transpose (resp. transpose), $A^{-1} = A^*$ (resp. $A^{-1} = A^T$).*

Definition 1.2. *Let $A \in \mathbb{C}^{m\times n}$. A real or complex scalar λ which satisfies the following equation*

$$Ax = \lambda x, \quad \text{i.e.,} \quad (A - \lambda I)x = 0,$$

is called an eigenvalue of A, and x is called an eigenvector of A corresponding to λ.

The eigenvalues and eigenvectors of a matrix play a very important role in matrix theory. They represent a tool which enables to understand the structure of a matrix. For example, if a given square matrix of complex numbers is self-adjoint, then there exist basis of $\mathbb{C}^m$ and $\mathbb{C}^n$, consisting of distinct eigenvectors of A, with respect to which the matrix A can be represented as a diagonal matrix. But, in the general case, not every matrix has enough distinct eigenvectors to enable its diagonal decomposition. The following definition, given as a generalization of the previous one, is useful to resolve this problem.

Definition 1.3. *Let $A \in \mathbb{C}^{m\times n}$ and λ is an eigenvalue of A. A vector x is called generalized eigenvector of A of grade p corresponding to λ, or λ-vector of A of grade p, if it satisfies the following equation*

$$(A - \lambda I)^p x = 0.$$

Namely, for each square matrix there exists a basis composed of generalized eigenvectors with respect to which, a matrix can be represented in the Jordan form. Corresponding statement is stated in the following proposition.

Proposition 1.1. [1] *(The Jordan decomposition). Let the matrix $A \in \mathbb{C}^{n\times n}$ has p distinct eigenvalues $\{\lambda_1, \lambda_2, \ldots, \lambda_p\}$. Then A is similar to a block diagonal matrix J with Jordan blocks on its diagonal, i.e., there exists a nonsingular matrix P which satisfies*

$$AP = PJ = \begin{bmatrix} J_{k_1}(\lambda_1) & 0 & \ldots & 0 \\ 0 & J_{k_2}(\lambda_2) & \ldots & 0 \\ \vdots & \vdots & \ddots & \vdots \\ 0 & 0 & \ldots & J_k(\lambda_p) \end{bmatrix},$$

where the Jordan blocks are defined by

$$J_{k_i}(\lambda_i) = \begin{bmatrix} \lambda_i & 1 & 0 & \dots & 0 \\ 0 & \lambda_i & 1 & \dots & 0 \\ \vdots & \vdots & \ddots & \vdots & \\ 0 & 0 & \dots & \lambda_i & 1 \end{bmatrix}$$

and the matrix J is unique up to a rearrangement of its blocks.

The following Proposition 1.2 gives us an alternative way to obtain even simpler decomposition of the matrix A, than the one given with the Jordan decomposition, but with respect to a different basis of $\mathbb{C}^n$. This decomposition is known as the Singular Value Decomposition (SVD shortly) and it is based on the notion of singular values, given in Definition 1.4.

Definition 1.4. *Let $A \in \mathbb{C}^{m\times n}$ and $\{\lambda_1, \dots, \lambda_p\}$ be the nonzero eigenvalues of AA^*. The singular values of A, denoted by $\sigma_i(A)$, $i = 1, \dots, p$ are defined by*

$$\sigma_i(A) = \sqrt{\lambda_i}, \quad i = 1, \dots, p.$$

Proposition 1.2. *(Singular value decomposition)* [1] *Let $A \in \mathbb{R}^{m\times n}$ be a matrix with singular values $\{\sigma_1, \dots, \sigma_r\}$. Then there exist orthogonal matrices $U \in \mathbb{R}^{m\times m}$ and $V \in \mathbb{R}^{n\times n}$ such that*

$$A = U\Sigma V^T,$$

where Σ is a nonsquare diagonal matrix

$$\Sigma = \begin{bmatrix} \sigma_1 & & & \vdots & 0 \\ & \ddots & & \vdots & 0 \\ & & \sigma_r & \vdots & \\ \dots & \dots & \dots & \dots & \dots \\ & 0 & & \vdots & 0 \end{bmatrix}, \sigma_1 \geq \sigma_2 \geq \cdots \geq \sigma_r.$$

A square matrix T of the order n is symmetric and positive semidefinite (abbreviated SPSD and denoted by $T \succeq 0$) if

$$v^T T v \geq 0 \text{ for all } v \in \mathbb{R}^n.$$

T is symmetric and positive definite (abbreviated SPD and denoted by $T \succ 0$) if

$$v^T T v > 0 \text{ for all } v \in \mathbb{R}^n, v \neq 0.$$

Recall that a symmetric matrix T is positive definite if and only if all its eigenvalues are nonnegative. The corresponding minimization problem, stated originally in linear algebra and frequently used in many scientific areas, is to minimize the quadratic form

$$\frac{1}{2}x^T T x + p^T x + a = \frac{1}{2}\langle x, Tx\rangle + p^T x + a \tag{1.1}$$

with respect to unknown vector $x \in \mathbb{R}^n$. Here T is a square positive definite matrix of the order n, $p \in \mathbb{R}^n$ is a vector of length n and a is a real scalar. Optimization problem (1.1) is called an *unconstrained quadratic optimization problem.*

Let $x, p, a \in \mathbb{R}^n$ are real vectors and T is a symmetric $n \times n$ matrix. The linearly constrained quadratic programming problem can be formulated as follows (see, for example, [2]):
Minimize the goal function (1.1) subject to one or more inequality and/or equality constraints defined by two $n \times n$ matrices A, E and two n-dimensional vectors b, d:

$$Ax \le b$$
$$Ex = d.$$

Notice that in the general Quadratic Programming model (QP model shortly) we can always presume that T is a symmetric matrix. Indeed, because

$$x^T T x = \frac{1}{2} x^T (T + T^T) x.$$

it is possible to replace T by the symmetric matrix $\tilde{T} = \frac{1}{2}(T + T^T)$.

Proposition 1.3. *An arbitrary symmetric matrix T is diagonalizable in the general form $T = RDR^T$, where R is an orthonormal matrix, the columns of R are an orthonormal basis of eigenvectors of T, and D is a diagonal matrix of the eigenvalues of T.*

Proposition 1.4. *If $T \in \mathbb{R}^{n \times n}$ is symmetric PSD matrix, then the following statements are equivalent:*
1) $T = MM^T$, *for an appropriate matrix M of the order $M \in \mathbb{R}^{n \times k}$, $k \ge 1$.*
2) $v^T T v \ge 0$ *for all $v \in \mathbb{R}^n, v \ne 0$.*
3) *There exist vectors v_i, $i = 1, \dots, n \in \mathbb{R}^k$ (for some $k \ge 1$) such that $T_{ij} = v_i^T v_j$ for all $i, j = 1, \dots, n$. The vectors v_i, $i = 1, \dots, n$, are called a Gram representation of T.*
4) *All principal minors of T are non-negative.*

Proposition 1.5. *Let $T \in \mathbb{C}^{n \times n}$ is symmetric. Then $T \succeq 0$ and it is nonsingular if and only if $T \succ 0$.*

Quadratic forms have played a significant role in the history of mathematics in both the finite and infinite dimensional cases. A number of authors have studied problems on minimizing (or maximizing) quadratic forms under various constraints such as vectors constrained to lie within the unit simplex (see Broom [3]), and the minimization of a more general case of a quadratic form defined in a finite-dimensional real Euclidean space under linear constraints (see e.g. La Cruz [4], Manherz and Hakimi [5]), with many applications in network analysis and control theory (for more on this subject, see also [6,7]). In a classical book on optimization theory, Luenberger [8], presented similar optimization problems for both finite and infinite dimensions. Quadratic problems are very important cases in both constrained and non-constrained optimization theory, and they find application in many different areas. First of all, quadratic forms are simple to be described and analyzed, and thus by their investigation, it is convenient to explain the convergence characteristics of the iterative optimization methods. The conjugate gradient methods invariably are invented and analyzed for the purely quadratic unconstrained problem, and then extended, by approximation, to more general problems, etc.

Applicability of the quadratic forms can be observed in other practical areas such as: network analysis and control theory [4,5,9], the *Asset Pricing Theory and Arbitrage Pricing Theory* [10], etc.

1.2. Short Overview of Generalized Inverses and Underlying Results

As previously mentioned, the main idea of defining generalized inverses originates from the need to solve the problem of finding a solution of the following system

$$Ax = b, \tag{1.2}$$

where $A \in \mathbb{C}^{m \times n}$ and $b \in \mathbb{C}^m$.

Definition 1.5. *For a given matrix $A \in \mathbb{C}^{n \times n}$, the inverse of the matrix A is a square matrix A^{-1} such that it satisfies the following equalities*

$$AA^{-1} = I \quad \textit{and} \quad A^{-1}A = I.$$

Proposition 1.6. *A square matrix $A \in \mathbb{C}^{n \times n}$ has a unique inverse if and only if $\det(A) \neq 0$, in which case we say that the matrix A is nonsingular matrix.*

Remark 1.1. *In order to distinguish between generalized inverses, the inverse of a matrix defined with Definition 1.5 will be called the ordinary inverse.*

In the case when the matrix A from the system (1.2) is nonsingular, the vector

$$x = A^{-1}b,$$

provides a solution of the system (1.2). However, many problems that usually arise in practice, reduce to a problem of the type (1.2), where the matrix A is singular, and moreover, in many cases it is not even a square matrix.

1.2.1. The Moore-Penrose Inverse

Let $A \in \mathbb{C}^{m \times n}$. The matrix $X \in \mathbb{C}^{n \times m}$ satisfying the conditions

$$(1)\ \ AXA = A \qquad (2)\ \ XAX = X \qquad (3)\ \ (AX)^* = AX \qquad (4)\ \ (XA)^* = XA$$

is called the Moore-Penrose inverse of A and denoted by $A^\dagger$.

It is easy to see that $AA^\dagger$ is the orthogonal projection of $\mathcal{H}$ onto $\mathcal{R}(A)$, denoted by P_A, and that $A^\dagger A$ is the orthogonal projection of $\mathcal{H}$ onto $\mathcal{R}(A^*)$ noted by P_{A^*}. It is well known that $\mathcal{R}(A^\dagger) = \mathcal{R}(A^*)$.

The set of matrices obeying the equations defined by the numbers contained in a sequence $\mathcal{S}$ from the set $\{1, 2, 3, 4\}$ is denoted by $A\{\mathcal{S}\}$. A matrix from $A\{\mathcal{S}\}$ is called an $\mathcal{S}$-inverse of A. An arbitrary $\mathcal{S}$-inverse of A of A is denoted by $A^{(\mathcal{S})}$.

If A commutes with $A^\dagger$, then A is called an EP matrix. EP matrices constitute a wide class of matrices which includes the self adjoint matrices, the normal matrices and the invertible matrices. Since the symmetric matrices are EP, the positive matrix T in the quadratic form studied in this work is EP. It is easy to see that for EP matrices we have the following:

$$A \text{ is EP } \Leftrightarrow \mathcal{R}(A) = \mathcal{R}(A^*) \Leftrightarrow \mathcal{N}(A) = \mathcal{N}(A^*) \Leftrightarrow \mathcal{R}(A) \oplus \mathcal{N}(A) = \mathcal{H}. \tag{1.3}$$

Let A be EP. Then, A has a matrix decomposition with respect to the orthogonal decomposition $\mathcal{H} = \mathcal{R}(A) \oplus \mathcal{N}(A)$:

$$T = \begin{bmatrix} A_1 & 0 \\ 0 & 0 \end{bmatrix} : \begin{bmatrix} \mathcal{R}(A) \\ \mathcal{N}(A) \end{bmatrix} \to \begin{bmatrix} \mathcal{R}(A) \\ \mathcal{N}(A) \end{bmatrix}$$

where the operator $A_1 : \mathcal{R}(T) \to \mathcal{R}(T)$ is invertible. Moreover,

$$A^\dagger = \begin{bmatrix} A_1^{-1} & 0 \\ 0 & 0 \end{bmatrix} : \begin{bmatrix} \mathcal{R}(A) \\ \mathcal{N}(A) \end{bmatrix} \to \begin{bmatrix} \mathcal{R}(A) \\ \mathcal{N}(A) \end{bmatrix}.$$

Lemma 1.1. *Let $A \in \mathbb{C}^{m \times n}$ be an arbitrary matrix. Then the following properties are valid.*

1) $(A^\dagger)^\dagger = A$, $(A^\dagger)^* = (A^*)^\dagger$;

2) $(AA^*)^\dagger = (A^*)^\dagger A^\dagger$, $(A^*A)^\dagger = A^\dagger (A^*)^\dagger$;

3) $A^\dagger AA^* = A^* = A^*AA^\dagger$;

4) $A^\dagger = (A^*A)^\dagger A^* = A^*(AA^*)^\dagger$;

5) $\mathcal{N}(AA^\dagger) = \mathcal{N}(A^\dagger) = \mathcal{N}(A^*) = \mathcal{R}(A)$

6) $\mathcal{R}(AA^*) = \mathcal{R}(AA^{(1)}) = \mathcal{R}(A)$, $\operatorname{rank}(AA^{(1)}) = \operatorname{rank}(A^{(1)}A) = \operatorname{rank}(A)$;

7) $AA^\dagger = P_{\mathcal{R}(A^*),\mathcal{N}(A)}$ *and* $A^\dagger A = P_{\mathcal{R}(A),\mathcal{N}(A^*)}$.

Lemma 1.2. *Let $A \in \mathbb{C}^{m \times n}$ be an arbitrary matrix. Then the matrix A can be written in the following way:*

$$A \sim \begin{bmatrix} A_1 & 0 \\ 0 & 0 \end{bmatrix} : \begin{bmatrix} \mathcal{R}(A^*) \\ \mathcal{N}(A) \end{bmatrix} \to \begin{bmatrix} \mathcal{R}(A) \\ \mathcal{N}(A^*) \end{bmatrix}, \tag{1.4}$$

where A_1 is invertible. Hence,

$$A^\dagger \sim \begin{bmatrix} A_1^{-1} & 0 \\ 0 & 0 \end{bmatrix} : \begin{bmatrix} \mathcal{R}(A) \\ \mathcal{N}(A^*) \end{bmatrix} \to \begin{bmatrix} \mathcal{R}(A^*) \\ \mathcal{N}(A) \end{bmatrix}.$$

Let us consider the equation $Ax = b$, $A \in B(\mathcal{H})$, where A is singular. If $b \notin R(A)$, then the equation has no solution. Therefore, instead of trying to solve the system of linear equations $\|Ax - b\| = 0$, we may look for a vector u that minimizes the norm $\|Ax - b\|$. Note that the vector u is unique. In this case we consider the equation $Tx = P_{R(A)}b$, where $P_{R(A)}$ is the orthogonal projection on $\mathcal{R}(A)$.

If the system (1.2) is such that $b \notin \mathcal{R}(A)$, then we search for an approximate solution of the system (1.2) by trying to find a vector x for which the norm of the vector $Ax - b$ is minimal.

Definition 1.6. *Let $A \in \mathbb{C}^{m\times n}$ and $b \in \mathbb{C}^m$. A vector $\hat{x}$, which satisfies the equality*

$$\|A\hat{x} - b\|^2 = \min_{x\in\mathbb{C}^n} \|Ax - b\|^2. \tag{1.5}$$

is called a least-squares solution of the system (1.2).

The next lemma gives a characterization of all least-squares solutions of the system (1.2).

Lemma 1.3. *The vector x is a least-squares solution of the system* (1.2) *if and only if x is a solution of the normal equation, defined by*

$$A^*Ax = A^*b. \tag{1.6}$$

The following proposition from [11] shows that $\|Ax - b\|$ is minimized by choosing $x = A^{(1,3)}b$, thus establishing a relation between the $\{1,3\}$-inverses and the least-squares solutions of the system (1.2).

Proposition 1.7. [11] *Let $A \in \mathbb{C}^{m\times n}$, $b \in \mathbb{C}^m$. Then $\|Ax - b\|$ is smallest when $x = A^{(1,3)}b$, where $A^{(1,3)} \in A\{1,3\}$. Conversely, if $X \in \mathbb{C}^{n\times m}$ has the property that, for all b, the norm $\|Ax - b\|$ is smallest when $x = Xb$, then $X \in A\{1,3\}$.*

Since $A^{(1,3)}$-inverse of a matrix is not unique, a system of linear equations can have many least-squares solutions. However, it is shown that among all least-squares solutions of a given system of linear equations, there exists only one such solution of minimum norm.

Definition 1.7. *Let $A \in \mathbb{C}^{m\times n}$ and $b \in \mathbb{C}^m$. A vector $\hat{x}$, which satisfies the equality*

$$\|\hat{x}\|^2 = \min_{x\in\mathbb{C}^n} \|x\|^2. \tag{1.7}$$

is called a minimum-norm solution of the system (1.2).

The next proposition establishes a relation between $\{1,4\}$-inverses and the minimum-norm solutions of the system (1.2).

Proposition 1.8. [11] *Let $A \in \mathbb{C}^{m\times n}$, $b \in \mathbb{C}^m$. If $Ax = b$ has a solution for x, the unique solution for which $\|x\|$ is smallest is given by $x = A^{(1,4)}b$, where $A^{(1,4)} \in A\{1,4\}$. Conversely, if $X \in \mathbb{C}^{n\times m}$ is such that, whenever $Ax = b$ has a solution, $x = Xb$ is the solution of minimum-norm, then $X \in A\{1,4\}$.*

Joining the results from Proposition 1.7 and Proposition 1.8 we are coming to the most important property of the Moore-Penrose inverse.

Corollary 1.1. [12] *Let $A \in \mathbb{C}^{m\times n}$, $b \in \mathbb{C}^m$. Then, among the least-squares solutions of $Ax = b$, $A^\dagger b$ is the one of minimum-norm. Conversely, if $X \in \mathbb{C}^{n\times m}$ has the property that, for all b, the vector Xb is the minimum-norm least-squares solution of $Ax = b$, then $X = A^\dagger$.*

The next proposition, characterizes the set of all least-squares solutions of a given system of linear equations.

Proposition 1.9. [13, 14] If $\mathcal{R}(A)$ is closed then the set S of all least-squares solutions of the system $Ax = b$ is given by

$$S = A^\dagger b \oplus \mathcal{N}(A) = \{A^\dagger b + (I - A^\dagger A)y : \; y \in \mathcal{H}\},$$

where $\mathcal{N}(A)$ denotes the null space of A.

The following two propositions can be found in Groetsch [15] and hold for operators and matrices:

Proposition 1.10. *Let $A \in \mathcal{B}(\mathcal{H})$ and $b \in \mathcal{H}$. Then, for $u \in \mathcal{H}$, the following are equivalent:*

(i) $Au = P_{R(A)}b$
(ii) $\|Au - b\| \leq \| Ax - b\|, \forall x \in \mathcal{H}$
(iii) $A^*Au = A^*b$.

Let $\mathbb{B} = \{u \in \mathcal{H} | T^*Tu = T^*b\}$. This set of solutions is closed and convex, therefore, it has a unique vector with minimal norm. In the literature, Groetsch [15], $\mathbb{B}$ is known as the set of the generalized solutions.

Proposition 1.11. *Let $A \in \mathcal{B}(\mathcal{H}), b \in \mathcal{H}$, and the equation $Ax = b$. Then, if $A^\dagger$ is the generalized inverse of A, we have that $A^\dagger b = u$, where u is the minimal norm solution.*

This property has an application in the problem of minimizing a symmetric positive definite quadratic form $\langle x, Tx\rangle$ subject to linear constraints, assumed consistent.

Another approach to the same problem is the use of a $T^{(1,3)}$ inverse. In this case $T^{(1,3)}b$ is a least squares solution for every $b \in \mathcal{H}$. The following Proposition can be found in [11], Chapter 3.

Proposition 1.12. [11] *Let $T \in B(\mathcal{H})$ with closed range and $b \in \mathcal{H}$. A vector x is a least squares solution of the equation $Tx = b$ iff $Tx = P_{R(T)}b = TT^{(1,3)}b$.*
Then, the general least squares solution is

$$x = T^{(1,3)}b + (I - T^{(1,3)}T)y,$$

where y is an arbitrary vector in $\mathcal{H}$.

A vector x is a least-squares solution of $Tx = b$ if and only if x is a solution of the normal equation $T^*Tx = T^*b$. Therefore, the least squares solutions set, defined in Proposition 1.10, is identical with the set defined in Proposition 1.12. In addition, we will also make use of a $T^{(1,4)}$ inverse. In this case $T^{(1,4)}b$ is the minimal norm solution of the equation $Tx = b$ for every $b \in \mathcal{R}(T)$. The following Proposition can also be found in [11], Chapter 3.

Proposition 1.13. [11] *Let $T \in B(\mathcal{H})$ with closed range and $b \in \mathcal{H}$. If the equation $Tx = b$ has a solution for x, the unique solution for which $\|x\|$ is smallest is given by $x = T^{(1,4)}b$.*

In the rest of this section we will also need to present the notion as well as the basic properties of the weighted Moore-Penrose inverse of a matrix $A \in \mathbb{C}^{m\times n}$ with respect to two Hermitian positive definite matrices $M \in \mathbb{C}^{m\times m}$ *and* $N \in \mathbb{C}^{n\times n}$ denoted by $X = A^{\dagger}_{M,N}$ satisfying the following four equations (See [16] page 118, exercise 30, or [17] section 3. For computational methods see e.g. [18], and for more on this subject, see [19, 20]):

$$AXA = A, \quad XAX = X, \quad (MAX)^* = MAX, \quad (NXA)^* = NXA. \tag{1.8}$$

It is also known (see e.g., [11]) that

$$A^{\dagger}_{M,N} = N^{-\frac{1}{2}}\left(M^{\frac{1}{2}}AN^{-\frac{1}{2}}\right)^{\dagger}M^{\frac{1}{2}}.$$

In this case, $A^{\dagger}_{M,N}b$ is the M-least squares solution of $Ax = b$ which has minimal N-norm.

This notion can be extended in the case when M and N are positive semidefinite matrices: in this case, X is a matrix such that Xb is a minimal N semi-norm, M-least squares solution of $Ax = b$. Subsequently, X must satisfy four conditions from (1.8) (See [16] page 118, exercises 31-34). When N is positive definite, then there exists a unique solution for X.

Another result used in our work is that, wherever a square root of a positive operator A is used, and since EP operators have index equal to 1, we have $\mathcal{R}(A) = \mathcal{R}(A^2)$ (see Ben Israel [11], pages 156-157).

As mentioned above, a necessary condition for the existence of a bounded generalized inverse is that the operator has closed range. Nevertheless, the range of the product of two operators with closed range is not always closed. In Bouldin [21] an equivalent condition is given. This condition is restated in Proposition 1.14.

Proposition 1.14. [21] *Let A and B be operators with closed range, and let*

$$H_i = \mathcal{N}(A) \cap (\mathcal{N}(A) \cap \mathcal{R}(B))^{\perp} = \mathcal{N}(A) \ominus \mathcal{R}(B).$$

The angle between H_i and $\mathcal{R}(B)$ is positive if and only if AB has closed range.

A similar result can be found in Izumino [22], this time using orthogonal projections.

Proposition 1.15. [22] *Let A and B be operators with closed range. Then, AB has closed range if and only if $A^{\dagger}ABB^{\dagger}$ has closed range.*

We will use the above two results to prove the existence of the Moore- Penrose inverse of appropriate operators which will be used in our work.

Another tool, used in this work, is the reverse order law for the Moore-Penrose inverses. In general, the reverse order law does not hold. Conditions which enable the reverse order law are described in Proposition 1.16. This proposition is a restatement of a part of Bouldin's theorem [23] that holds for both operators and matrices.

Proposition 1.16. *Let A, B be bounded operators on $\mathcal{H}$ with closed range. Then the reverse order low $(AB)^\dagger = B^\dagger A^\dagger$ holds if and only if the following three conditions hold:*

(i) *The range of AB is closed,*

(ii) *$A^\dagger A$ commutes with BB^*,*

(iii) *$BB^\dagger$ commutes with A^*A.*

A corollary of Proposition 1.14 is the following proposition that can be found in Karanasios-Pappas [24] and we will use it in our case. We will denote by $\mathrm{Lat}\, T$ the set of all closed subspaces of the underlying Hilbert space $\mathcal{H}$ invariant under T.

Proposition 1.17. *Let $A, T \in \mathcal{B}(\mathcal{H})$ be two operators such that A is invertible and T has closed range. Then*

$$(TA)^\dagger = A^{-1}T^\dagger \quad \text{if and only if} \quad \mathcal{R}(T) \in \mathit{Lat}\,(AA^*).$$

1.2.2. The Drazin Inverse

Apart from the Moore-Penrose inverse and $A^{(i,j,k)}$ inverses, a very useful kind of inverse, with properties analogous to the usual inverse, is the Drazin inverse. Let $A \in \mathbb{R}^{n\times n}$ and $k = \mathrm{ind}(A)$. The matrix $X \in \mathbb{R}^{n\times n}$ satisfying the conditions

$$(1^k)\quad A^kXA = A^k \qquad (2)\quad XAX = X \qquad (5)\quad AX = XA$$

is called the Drazin inverse of the matrix A and it is denoted by A^D.

Proposition 1.18. *If A is a matrix of index k, then the vector A^Db is a solution of the equation*

$$A^{k+1}x = A^kb, \tag{1.9}$$

for all b, in which case the equation (1.9) and the vector A^Db are respectively called the general normal equation and the Drazin-inverse solution of the system $Ax = b$.

In the next lemma we give the main properties of the Drazin inverse.

Lemma 1.4. *Let $A \in \mathbb{C}^{n\times n}$ and $p = \mathrm{ind}(A)$. The following statements are valid:*

1) *$A^lXA = A^l$ for all $l \geq p$*

2) *$\mathcal{R}(A^l) = \mathcal{R}(A^{l+1})$, $\mathcal{N}(A^l) = \mathcal{N}(A^{l+1})$ and $\mathrm{rank}(A^l) = \mathrm{rank}(A^{l+1})$, for all $l \geq p$. Moreover, p is the smallest integer for which the equalities hold.*

3) *The matrix A can be written in the following way:*

$$A \sim \begin{bmatrix} A_1 & 0 \\ 0 & N \end{bmatrix} : \begin{bmatrix} \mathcal{R}(A^p) \\ \mathcal{N}(A^p) \end{bmatrix} \to \begin{bmatrix} \mathcal{R}(A^p) \\ \mathcal{N}(A^p) \end{bmatrix}, \tag{1.10}$$

where A_1 is invertible, and N is nilpotent matrix. The Drazin inverse of A is defined by

$$A^D \sim \begin{bmatrix} A_1^{-1} & 0 \\ 0 & 0 \end{bmatrix} : \begin{bmatrix} \mathcal{R}(A^p) \\ \mathcal{N}(A^p) \end{bmatrix} \to \begin{bmatrix} \mathcal{R}(A^p) \\ \mathcal{N}(A^p) \end{bmatrix};$$

4) *for all* $\lambda \neq 0$, *a vector* x *is a* λ^{-1}*-vector of* A^D *of grade* s *if and only if it is a* λ*-vector of* A *of grade* s, *and* x *is a* 0*-vector of* A^D *if and only if it is a* 0*-vector of* A *(without regard to grade).*

It is important to note that the block form (1.10) of the matrix A can be easily obtained by the Jordan decomposition of A.

Despite the spectral properties, the Drazin inverse, in some cases, it also provides a solution of a given system of linear equations. Namely for $A \in \mathbb{C}^{n\times n}$ and $b \in \mathbb{C}^n$, as it was shown in [16], $A^D b$ is a solution of the following system

$$Ax = b, \quad \text{where } b \in \mathcal{R}(A^p)\,, p = \operatorname{ind}(A). \tag{1.11}$$

and we call it the *Drazin-inverse solution* of the system (1.11). Also, since this is the only case, when the Drazin-inverse provides a solution to the given system, we call the system (1.11), a *Drazin-consistent system.*

The Drazin inverse has many applications in the theory of finite Markov chains as well as in the study of differential equations and singular linear difference equations [16], cryptography [25] etc.

An application of the Drazin inverse in solving a given system of linear equations naturally arises from the minimal properties of the Drazin inverse. For this purpose, we present main results from the paper [26], where corresponding results for the Drazin-inverse solution, to the ones presented for the Moore-Penrose inverse solution, are established.

Theorem 1.1. *Let* $A \in \mathbb{C}^{n\times n}$ *with* $p = \operatorname{ind}(A)$. *Then* $A^D b$ *is the unique solution in* $\mathcal{R}(A^p)$ *of the system*

$$A^{p+1}x = A^p b. \tag{1.12}$$

Theorem 1.2. *Let* $A \in \mathbb{C}^{n\times n}$, $b \in \mathbb{C}^n$ *and* $p = \operatorname{ind}(A)$. *The set of all solutions of the equation* (1.12) *is given by*

$$x = A^D b + \mathcal{N}(A^p). \tag{1.13}$$

Since (1.12) is analogous to (1.9), we shall call the solution (1.12) as the *generalized normal equations* of (1.11).

Let $A = PJP^{-1}$ be the Jordan decomposition of the matrix A. We denote $\|x\|_P = \|P^{-1}x\|$.

Theorem 1.3. [26] *Let* $A \in \mathbb{R}^{n\times n}$ *with* $p = \operatorname{ind}(A)$. *Then* $\hat{x}$ *satisfies*

$$\|b - A\hat{x}\|_P = \min_{u\in\mathcal{N}(A)+\mathcal{R}(A^{p-1})} \|b - Ax\|_P$$

if and only if $\hat{x}$ *is the solution of the equation*

$$A^{p+1}x = A^p b, \quad x \in \mathcal{N}(A) + \mathcal{R}(A^{p-1}).$$

Moreover, the Drazin-inverse solution $x = A^D b$ *is the unique minimal* P*-norm solution of the generalized normal equations* (1.12).

Corollary 1.2. [26] *Let $A \in \mathbb{C}^{n\times n}$, $p = \operatorname{ind}(A)$ and $b \in \mathcal{R}(A)$. Then the inequality $\|x\|_P \geq \|A^D b\|_P$ holds for all solutions x of the system* (1.12), *i.e., $A^D b$ is the unique minimum P-norm solution of the system of equations* (1.12).

Lemma 1.5. [27] *Let $A \in \mathbb{R}^{n\times n}$ with* $\operatorname{ind}(A) = k$. *Then the general solution of*

$$Ax = b, \;\; b \in \mathcal{R}(A^k), \; k = \operatorname{ind}(A) \tag{1.14}$$

is given by

$$x = A^D b + A^{k-1}\left(I - A^D A\right) z,$$

where z is an arbitrary vector. In particular, the minimal P-norm solution of (1.14) *is presented by $x_{\text{opt}} = A^D b$.*

A unified representation theorem for the Drazin inverse was derived in [28]. This general representation of the Drazin inverses leads to a number of specific expressions and computational procedures for computing the Drazin inverse.

1.3. The $A^{(2)}_{T,S}$-Inverse

Recall that, for an arbitrary matrix $A \in \mathbb{C}^{m\times n}$, the set of all outer inverses (or also called $\{2\}$-inverses) is defined by the following

$$A\{2\} = \{X \in \mathbb{C}^{n\times m} : \; XAX = X\}. \tag{1.15}$$

With $A\{2\}_s$ we denote the set of all outer inverses of rank s and the symbol $A^{(2)}$ stands for an arbitrary outer inverse of A.

Proposition 1.19. [11] *Let $A \in \mathbb{C}^{m\times n}_r$, U is a subspace of $\mathbb{C}^n$ of dimension $t \leq r$ and V is a subspace of $\mathbb{C}^m$ of dimension $m - t$, then A has a $\{2\}$-inverse X such that $\mathcal{R}(X) = U$ and $\mathcal{N}(X) = V$ if and only if $AU \oplus V = \mathbb{C}^m$, in which case X is unique and it is denoted by $A^{(2)}_{U,V}$.*

Lemma 1.6. *Let $A \in \mathbb{C}^{m\times n}$ be an arbitrary matrix, U is a subspace of $\mathbb{C}^n$ and V is a subspace of $\mathbb{C}^m$ such that $AU \oplus V = \mathbb{C}^m$. Then the matrix A can be written in the following way:*

$$A \sim \begin{bmatrix} A_1 & 0 \\ 0 & A_2 \end{bmatrix} : \begin{bmatrix} U \\ \mathcal{N}(A^{(2)}_{U,V}A) \end{bmatrix} \to \begin{bmatrix} AU \\ V \end{bmatrix}, \tag{1.16}$$

where A_1 is invertible. Moreover,

$$A^{(2)}_{U,V} \sim \begin{bmatrix} A_1^{-1} & 0 \\ 0 & 0 \end{bmatrix} : \begin{bmatrix} AU \\ V \end{bmatrix} \to \begin{bmatrix} U \\ \mathcal{N}(A^{(2)}_{U,V}A) \end{bmatrix}.$$

The outer generalized inverse with prescribed range U and null-space V is a generalized inverse of special interest in matrix theory. The reason of the importance of this inverse is the fact that: the Moore-Penrose inverse $A^{\dagger}$, the weighted Moore-Penrose inverse $A^{\dagger}_{M,N}$, the Drazin inverse A^D, the group inverse $A^{\#}$, the Bott-Duffin inverse $A^{(-1)}_{(L)}$ and the generalized Bott-Duffin inverse $A^{(+)}_{(L)}$; are all $\{2\}$-generalized inverses of A with prescribed range and null space.

Lemma 1.7. *Let $A \in \mathbb{C}_r^{m\times n}$ and $p = \operatorname{ind}(A)$. Then the following representations are valid:*

1) $A^{\dagger} = A^{(2)}_{\mathcal{R}(A^*),\mathcal{N}(A^*)}$,

2) $A^{\dagger}_{M,N} = A^{(2)}_{\mathcal{R}(N^{-1}A^*M),\mathcal{N}(N^{-1}A^*M)}$,

3) $A^{D} = A^{(2)}_{\mathcal{R}(A^k),\mathcal{N}(A^k)}$,

4) $A^{\#} = A^{(2)}_{\mathcal{R}(A),\mathcal{N}(A)}$ *if and only if* $p = 1$.

1.4. Semidefinite Programming

The source of main topic of this work is included in the so called semidefinite programming. In order to clarify restated results, it is necessary to restate basic facts and notions.

The scalar (inner) product of two matrices $A, B \in \mathbb{C}^{m\times n}$ is defined by

$$\langle A, B\rangle = \sum_{i=1}^{m}\sum_{j=1}^{n} a_{ij}b_{ij} = \operatorname{Tr}(A^T B).$$

Frobenius norm of a matrix A is defined as $\|A\|_F = \sqrt{\langle A, A\rangle}$.

The semidefinite programming (SDP) problem is the problem of optimizing a linear function of a symmetric matrix subject to linear constraints. Also, it is assumed that the matrix of variables is symmetric positive semidefinite. The unconstrained semidefinite optimization problem can be stated in the general form

$$\begin{aligned} &\text{minimize} \quad f(X) \\ &\text{subject to} \quad X \succeq 0, \end{aligned} \tag{1.17}$$

where $f(X) : \mathbb{R}^{n\times n} \to \mathbb{R}$ is a convex and differentiable function over the cone of positive semidefinite matrices. A constrained version of the problem (1.17) is called a semidefinite program (SDP) if both the function f as well as the constraints are linear and possesses the form

$$\begin{aligned} &\text{minimize} \quad f(X) \\ &\text{subject to} \quad h_i(X) = 0,\ i \in I \\ &\qquad\qquad\quad\ X \succeq 0. \end{aligned} \tag{1.18}$$

Here, X belongs to the space of symmetric $n \times n$ matrices, denoted by $\mathbb{S}^{n\times n}$, each of the functions h_i is real-valued affine function on $\mathbb{S}^{n\times n}$ and I denotes the set of indices.

The typical form of a semidefinite program is a minimization problem of the form

$$\begin{aligned} &\text{minimize} \quad \langle C, X\rangle \\ &\text{subject to} \quad \langle A_j, X\rangle = b_j,\ j = 1, \ldots, m \\ &\qquad\qquad\quad\ X \succeq 0. \end{aligned} \tag{1.19}$$

Here $A_1, \ldots, A_m \in \mathbb{S}^n$ are given $n\times n$ symmetric matrices and $b = \left[b_1, \ldots, b_m\right]^T \in \mathbb{R}^m$ is a given vector.

The SDP problem has been studied extensively due to its practical applicability in various fields [29–31]. Various algorithms have been proposed for solving SDP, where the interior point method is one of the efficient methods for SDP, and it possesses polynomial complexity (see, for example, [29–32]). An algorithm for solving general large–scale unconstrained semidefinite optimization problems efficiently is proposed in [33]. The algorithm is based on a hybrid approach and combines Sparse approximate solutions to semidefinite programs, proposed by Hazan (2008), with a standard quasi-Newton algorithm.

Semidefinite programming theoretically subsumes other convex techniques such as linear, quadratic, and second-order cone programming.

In the present monograph, we are interested to solve some SDP problems whose solutions are based on the usage of various classes of generalized inverses.

1.5. Organization of the Paper

The organization of the remainder of this work is the following. In the second section we give an overview of notation and definitions and some known results, which are related to our analysis. The third section is devoted to the T-restricted weighted Moore-Penrose Inverse, which is introduced and investigated in [34,35]. The fourth section is presenting the T-restricted weighted Drazin inverse of a matrix. Some possible generalizations of results surveyed in sections 3 and 4 as well as opportunities for future research are presented in Section 5. Finally we will end this work with several conclusions.

2. Overview of Known Results of Quadratic Optimization and Preliminary Results

According to the minimality of the Frobenius norm of the pseudo-inverse [36], the generalized inverse can be computed as a solution of a certain matrix-valued quadratic convex programming problem with equality constraints. If $m \geq n$ then $A^\dagger$ is a solution of the following optimization problem (2.1) with respect to $X \in \mathbb{C}^{n\times m}$

$$\begin{aligned} &\text{minimize} \quad \|X\|_F^2 \\ &\text{subject to} \quad A^*AX = A^*, \end{aligned} \tag{2.1}$$

where $\|X\|_F$ denotes the Frobenius norm of X. In the case $m < n$, if Y is a solution of the following optimization problem (2.2) with respect to $X \in \mathbb{C}^{m\times n}$

$$\begin{aligned} &\text{minimize} \quad \|X\|_F^2 \\ &\text{subject to} \quad AA^*X = A, \end{aligned} \tag{2.2}$$

then $A^\dagger = Y^*$.

An interesting usage of inner generalized inverses in constrained quadratic optimization problems restricted by some linear constraints is investigated in [38]. For this purpose, D.J. Evans introduced the restricted inverse defined in [38] as in the following proposition.

Proposition 2.1. ([38, Theorem 1]) *Let A be positive semidefinite and C be singular. Then the following constrained quadratic programming problem*

$$\begin{aligned} &\min\ q(x) = \frac{1}{2}\langle x, Tx\rangle + \langle p, x\rangle = \frac{1}{2}x^T Tx + p^T x \\ &\text{s.t. } C^T x = d \end{aligned} \tag{2.3}$$

has the solution given by

$$\hat{x} = (C^T)^{(1)}d - P_{N(C^T)}\left(P_{N(C^T)}TP_{N(C^T)}\right)^{(1)} P_{N(C^T)}\left[p + T(C^T)^{(1)}d\right].$$

If the constraint set $C^T x = d$ from (2.3) is considered in the particular form $Ax = b$, required in (1.9), we obtain $C^T = A$ and $d = b$. In addition, we have that $p = 0$, so that corresponding solution of the minimization problem, given by Proposition 2.1, is equal to

$$\hat{x} = A^{(1)}b - P_{N(A)}\left(P_{N(A)}TP_{N(A)}\right)^{(1)} P_{N(A)}T(A)^{(1)}b. \tag{2.4}$$

If the constraint set $C^T x = d$ from (2.3) is considered in the particular form $A^{k+1}x = A^k b$, required in (1.9), we obtain $C^T = A^{k+1}$ and $d = A^k b$. In addition, we have that $p = 0$, so that the solution of the minimization problem, given by Proposition 2.1, is equal to

$$\hat{x} = (A^{k+1})^{(1)}A^k b - P_{N(A^{k+1})}(P_{N(A^{k+1})}TP_{N(A^{k+1})})^{(1)}P_{N(A^{k+1})}A(A^{k+1})^{(1)}A^k b. \tag{2.5}$$

Nevertheless, since the minimizing vector given by (2.4),(2.5) is derived using a $\{1\}$-inverse, solutions (2.4),(2.5) are not of minimal norm, neither a least squares solution. The solutions given by $\{1\}$-inverses are general solutions of a system of linear equations. So, we do not expect that the minimizing value of $\Phi(x)$ using this kind of inverse will give lower values than the ones given by the generalized inverse introduced in [39,40], which is based on the usage of outer inverses.

The authors of the paper [41] were considered the minimization problem

$$\text{Minimize}\quad F^*RF,\ \ F \in \mathbb{C}^{M\times m} : F^*Q = C. \tag{2.6}$$

Here, R is an $M\times M$ Hermitian matrix, Q and C are $M\times n$ and $m\times n$ matrices respectively satisfying $m < M$ and $n < M$. In other words, main goal of the paper [41] is to find an $M \times m$ minimizer of the quadratic form F^*RF subject to the set of n linear constraints included in the matrix constraint $F^*Q = C$.

In applications where R may be assumed to be strictly positive definite and well-conditioned, problem (2.6) has a unique solution derived in [42]:

$$\begin{aligned} F_0 &= R^{-1}Q\left(Q^*R^{-1}Q\right)^{-1}C^*, \\ F_0^*RF_0 &= \left(Q^*R^{-1}Q\right)^{-1}. \end{aligned}$$

In the general case, the matrix R is ill-conditioned or even exactly singular, a particular minimizer of (2.6) is given by

$$\begin{aligned} F_0 &= (R + QPQ^*)^\dagger Q\left(Q^*(R + QPQ^*)^\dagger Q\right)^{-1}C^*, \\ F_0^*RF_0 &= C\left(\left(Q^*(R + QPQ^*)^\dagger Q\right)^{-1} - I_n\right)C^*. \end{aligned}$$

where P is an arbitrary $n \times n$ positive definite matrix.

The constrained quadratic problem of the general form

$$\text{minimize } \Phi(x) = \langle x, Tx \rangle + \langle p, x \rangle + a, \tag{2.7}$$

subject to the constraint

$$x \in S, \;\; S = \{x : \; x \in \mathbb{R}^n, Ax = b\} \tag{2.8}$$

has been frequently investigated in the literature. It is assumed that $a \in \mathbb{R}$, p is a given vector and T is a positive definite or positive semi-definite real matrix or an operator acting on a separable real Hilbert space.

One approach to solve the problem

$$\text{minimize } \Phi(x), \; x \in S \tag{2.9}$$

are the, so called, Penalty methods [8], which are actually based on the idea of approximating the original problem, to the problem of unconstrained optimization, and then using respective methods in order to solve it.

Quadratic minimization under linear constraints has various applications in electrical circuits, signal processing and linear estimation (see e.g. [5,43,44])

The fact that the unique solution of the problem (2.7) (without constraints) is also the unique solution to the linear equation

$$Tx = p, \tag{2.10}$$

and hence the quadratic minimization problem is equivalent to a linear equation problem, motivates the idea of using generalized inverses as a possible methodology for finding the solution of the problem. An application of the Moore-Penrose inverse for finding a solution of the problem (2.7), where T is positive definite matrix, was presented by Manherz and Hakimi [5]. The special case when $p = 0$ and $a = 0$, was investigated in [9]. The authors in [35], generalized these results to positive operators acting on separable complex Hilbert spaces, and then proposed a new approach for positive semi-definite operators, where the minimization is considered for all vectors belonging to $\mathcal{N}(T)^{\perp}$. Dependence of the stationary points of Φ on perturbations of the operator T is studied in [45].

When the set S is nonempty, the solution of the problem (2.9) is given in [5]. Moreover, a more general approach of the same problem when T is singular is examined in [46] making use of $\{1, 3\}$ and $\{1, 4\}$ inverses.

Now, let us suppose that the set S is empty, then the problem (2.9) does not have a solution. However, the practical problems that appear can result with a model given by (2.9), such that the system $Ax = b$ is not consistent. In this case, the constraint set S does not have a solution, and consequently our problem does not have a solution. For that purpose, in the present article we analyze different sets of constraints, which give approximation to the original problem. This approach have led us to dependency between the solution of the problem given by (2.9), and the Drazin inverse solution $A^D b$ of the system (2.8).

The special case when it is actually a solution of the original problem is also analyzed. The main idea consists of finding a solution to the problem (2.7) such that minimizes the vector $Ax - b$ with respect to the P-norm, where P is the Jordan basis of the matrix A. Consequently, instead of analyzing the constraint set S, it is possible to analyze the normal Drazin equation of the system $Ax = b$:

$$S_D = \{x : \ x \in \mathbb{R}^n, \ A^{k+1}x = A^k b, \ k \geq \text{ind}(A)\}. \tag{2.11}$$

In order to find an approximate solution of (2.9), in the present paper we solve the problem by considering the following minimization problem

$$\text{minimize} \ \ \Phi(x), \ x \in S_D. \tag{2.12}$$

Obviously the set S_D is nonempty.

Several results on the problem that we will examine in this work are listed in the rest of this section.

Let T be a symmetric positive definite matrix. Then, T can be written as $T = UDU^*$, where U is unitary and D is diagonal. Let $D^{\frac{1}{2}}$ denote the positive solution of $X^2 = D$, and let $D^{-\frac{1}{2}}$ denote $\left(D^{\frac{1}{2}}\right)^{-1}$, which exists since T is positive definite.

In order to have a more general idea of this problem we will at first examine it for the infinite dimensional case and then we will consider matrices in the place of operators.

We will consider the case when the positive operator T is singular, that is, T is positive semidefinite. In this case, since $\mathcal{N}(T) \neq \emptyset$, we have that $\langle x, Tx \rangle = 0$, for all $x \in \mathcal{N}(T)$ and so, the problem

$$\text{minimize} \ \ \Phi_0(x) = \langle x, Tx \rangle, \ x \in S$$

has many solutions when $\mathcal{N}(T) \cap S \neq \emptyset$.

An approach to this problem in both the finite and infinite dimensional case would be to look among the vectors $x \in \mathcal{N}(T)^{\perp} = \mathcal{R}(T^*) = \mathcal{R}(T)$ for a minimizing vector for $\langle x, Tx \rangle$. In other words, we will look for the minimum of $\langle x, Tx \rangle$ under the constraints $Ax = b, x \in \mathcal{R}(T)$.

Using the fact that T is an EP operator, we will make use of the first two conditions in the following proposition that can be found in Drivaliaris et al [47]:

Proposition 2.2. [47] *Let $T \in \mathcal{B}(\mathcal{H})$ with closed range. Then the following conditions are equivalent:*

(i) *T is EP.*

(ii) *There exist Hilbert spaces $\mathcal{K}_1$ and $\mathcal{L}_1$, unitary $U_1 \in \mathcal{B}(\mathcal{K}_1 \oplus \mathcal{L}_1, \mathcal{H})$ and isomorphism $A_1 \in \mathcal{B}(\mathcal{K}_1)$ such that*

$$T = U_1(A_1 \oplus 0)U_1^*. \tag{2.13}$$

(iii) *There exist Hilbert spaces $\mathcal{K}_2$ and $\mathcal{L}_2$, $U_2 \in \mathcal{B}(\mathcal{K}_2 \oplus \mathcal{L}_2, \mathcal{H})$ isomorphism and $A_2 \in \mathcal{B}(\mathcal{K}_2)$ isomorphism such that*

$$T = U_2(A_2 \oplus 0)U_2^*.$$

(iv) *There exist Hilbert spaces* $\mathcal{K}_3$ *and* $\mathcal{L}_3$, $U_3 \in \mathcal{B}(\mathcal{K}_3 \oplus \mathcal{L}_3, \mathcal{H})$ *injective and* $A_3 \in \mathcal{B}(\mathcal{K}_3)$ *isomorphism such that*

$$T = U_3(A_3 \oplus 0)U_3^*.$$

Proof. We present a sketch of the proof for (i)⇒(ii):

Let $\mathcal{K}_1 = \mathcal{R}(T)$, $\mathcal{L}_1 = \mathcal{N}(T)$, $U_1 : \mathcal{K}_1 \oplus \mathcal{L}_1 \to \mathcal{H}$ with

$$U_1(x_1, x_2) = x_1 + x_2,$$

for all $x_1 \in \mathcal{R}(T)$ and $x_2 \in \mathcal{N}(T)$, and $A_1 = T|_{\mathcal{R}(T)} : \mathcal{R}(T) \to \mathcal{R}(T)$. Since T is EP, $\mathcal{R}(T) \oplus^{\perp} \mathcal{N}(T) = \mathcal{H}$ and thus U_1 is unitary. Moreover it is easy to see that $U_1^* x = (P_T x, P_{\mathcal{N}(T)} x)$, for all $x \in \mathcal{H}$. It is obvious that A_1 is an isomorphism. A simple calculation leads to (2.13). ■

It is easy to see that when $T = U_1(A_1 \oplus 0)U_1^*$ and T is positive, so is A_1, since $\langle x, Tx \rangle = \langle x_1, A_1 x_1 \rangle, x_1 \in \mathcal{R}(T)$.

In what follows, T will denote a singular positive operator with a canonical form $T = U_1(A_1 \oplus 0)U_1^*$, R is the unique solution of the equation $R^2 = A_1$ and we can define

$$V = \begin{bmatrix} R & 0 \\ 0 & 0 \end{bmatrix}, \tag{2.14}$$

which implies

$$V^{\dagger} = \begin{bmatrix} R^{-1} & 0 \\ 0 & 0 \end{bmatrix}$$

As in the previous cases, since the two operators A and R are arbitrary, one does not expect that the range of their product will always be closed.

Using Proposition 2.2, in Theorem 2.1 from [35] is derived the following result concerning constrained optimization of quadratic forms.

Theorem 2.1. [35] *Let* $T = U_1(A_1 \oplus 0)U_1^* \in \mathcal{B}(\mathcal{H})$ *be a singular positive operator, and the equation* $Ax = b$, *with* $A \in \mathcal{B}(\mathcal{H})$ *singular with closed range and* $b \in \mathcal{H}$. *If the set*

$$S = \{x \in \mathcal{N}(T)^{\perp} : Ax = b\}$$

is not empty, then the problem

$$\text{minimize } \langle x, Tx \rangle, \ x \in S$$

has the unique solution

$$\hat{x} = U_1 V^{\dagger} \left(A U_1 V^{\dagger} \right)^{\dagger} b,$$

where V *is defined in* (2.14), *under the assumption that* $P_{A^*} P_T$ *has closed range.*

Proof. We have that

$$\begin{aligned} \langle x, Tx \rangle &= \langle x, U_1(A_1 \oplus 0)U_1^* x \rangle \\ &= \langle U_1^* x, (A_1 \oplus 0)U_1^* x \rangle \\ &= \langle U_1^* x, (R^2 \oplus 0)U_1^* x \rangle. \end{aligned}$$

Further $U_1^* x = (x_1, x_2)$ and

$$\langle U_1^* x, (A_1 \oplus 0) U_1^* x \rangle = \langle x_1, A_1 x_1 \rangle, \; x_1 \in \mathcal{R}(T).$$

Therefore,

$$\begin{aligned} \langle x, Tx \rangle &= \langle (R \oplus 0) U_1^* x, (R \oplus 0) U_1^* x \rangle \\ &= \langle R x_1, R x_1 \rangle \\ &= \langle y, y \rangle, \end{aligned}$$

where $y = Rx_1$, with $x_1 \in \mathcal{N}(T)^{\perp}$.

The problem of minimizing $\langle x, Tx \rangle$ is equivalent of minimizing $\| \, y \, \|^2$ where

$$y = Rx_1 = (R \oplus 0) U_1^* x \Longleftrightarrow x = U_1 (R^{-1} \oplus 0) y = U_1 V^{\dagger} y.$$

As before, the minimal norm solution $\hat{y}$ is equal to $\hat{y} = \left(A U_1 V^{\dagger} \right)^{\dagger} b$. Therefore,

$$\hat{x_1} = U_1 V^{\dagger} \left(A U_1 V^{\dagger} \right)^{\dagger} b,$$

with $\hat{x_1} \in S$.

We still have to prove that $AU_1 V^{\dagger}$ has closed range. Using Proposition 1.14, the range of $U_1 V^{\dagger}$ is closed since

$$\begin{aligned} H_i &= \mathcal{N}(U_1^*) \cap \left(\mathcal{N}(U_1^*) \cap \mathcal{R}(V^{\dagger}) \right)^{\perp} \\ &= 0 \end{aligned}$$

and so, the angle between U_1^* and $V^{\dagger}$ is equal to $\frac{\pi}{2}$.
From Proposition 1.15 the operator $P_{A^*} P_T$ must have closed range because

$$\begin{aligned} A^{\dagger} A U_1 V^{\dagger} (U_1 V^{\dagger})^{\dagger} &= P_{A^*} U_1 P_R U_1^* \\ &= P_{A^*} U_1 P_{A_1} U_1^* \\ &= P_{A^*} P_T, \end{aligned}$$

making use of Proposition 1.17 and the fact $\mathcal{R}(R) = \mathcal{R}(A_1) = \mathcal{R}(T)$. ∎

Corollary 2.1. [35] *Under all the assumptions of Theorem 2.1 we have that the minimum value of* $f(x) = \langle x, Tx \rangle, x \in S$ *is equal to* $\| \, \left(A U_1 V^{\dagger} \right)^{\dagger} b \, \|^2$.

Proof. One can verify that

$$f_{min}(x) = \langle \hat{x}, T\hat{x} \rangle = \langle U_1 V^{\dagger} (A U_1 V^{\dagger})^{\dagger} b, T U_1 V^{\dagger} (A U_1 V^{\dagger})^{\dagger} b \rangle.$$

Since $T = U_1 (R^2 \oplus 0) U_1^*$ it can be further derived

$$\begin{aligned} f_{min}(x) &= \left\langle U_1 V^{\dagger} \left(A U_1 V^{\dagger} \right)^{\dagger} b, U_1 (R \oplus 0) (A U_1 V^{\dagger})^{\dagger} b \right\rangle \\ &= \left\langle P_T (A U_1 V^{\dagger})^{\dagger} b, \left(A U_1 V^{\dagger} \right)^{\dagger} b \right\rangle \\ &= \left\| \left(A U_1 V^{\dagger} \right)^{\dagger} b \right\|^2. \end{aligned}$$

Since

$$V^\dagger(R \oplus 0) = (I \oplus 0) = P_T$$

and

$$\mathcal{R}(AU_1V^\dagger)^\dagger = \mathcal{R}(AU_1V^\dagger)^* = \mathcal{R}(RU_1A^*) \subseteq \mathcal{R}(R) = \mathcal{R}(T),$$

the following holds:

$$P_T\left(AU_1V^\dagger\right)^\dagger b = (AU_1V^\dagger)^\dagger b.$$

The proof is complete. ■

The following basic result, presented in Proposition 2.3, can be found in [9]. This result is a starting point in the investigation of quadratic forms.

Proposition 2.3. [9] *Consider the equation $Ax = b$. If the set*

$$S = \{x : Ax = b\}$$

is not empty, then the problem

$$\text{minimize } \langle x, Tx\rangle,\ x \in S \tag{2.15}$$

has the unique solution

$$x = UD^{-\frac{1}{2}}(AUD^{-\frac{1}{2}})^\dagger b,$$

where $T = UDU^*$ is the decomposition of T derived in Proposition 1.3.

Proof. Using $T = UDU^*$, one can verify

$$\langle x, Tx\rangle = \langle v, v\rangle,$$

so that the minimization (2.15) can be presented in the equivalent form

$$\begin{aligned} &\text{minimize} \quad \langle v, v\rangle,\\ &\text{such that} \quad AU^*D^{-1/2}v = a, \end{aligned}$$

where

$$v = D^{1/2}Ux \Longleftrightarrow x = U^*D^{-1/2}v.$$

Using the Moore-Penrose solution of the last equation it can be derived

$$v = \left(AU^*D^{-1/2}\right)^\dagger a,$$

which can be used to complete the proof. ■

The following Proposition 2.4 can be found in Manherz and Hakimi [5], and it also represents the starting point in the constrained minimization of quadratic forms.

Proposition 2.4. [5] *Let $T \in \mathbb{R}^{m\times m}$ be positive definite, $A \in \mathbb{R}^{n\times m}$ and consider the equation $Ax = b$ with $b \in \mathbb{R}^{n\times 1}$.*

If the set

$$S = \{x : Ax = b\}$$

is nonempty, then the optimization problem

$$\text{minimize} \quad \Phi(x) = \langle x, Tx\rangle + \langle p, x\rangle + a,\ x \in S,$$

with $p \in \mathbb{R}^{m\times 1}$ and $a \in \mathbb{R}$ has the unique solution

$$x = T^{-\frac{1}{2}}\left(AT^{-\frac{1}{2}}\right)^{\dagger}\left(\frac{1}{2}AT^{-1}p + b\right) - \frac{1}{2}T^{-1}p.$$

Proof. A minimizer of the function $\Phi(x)$ is also minimizer of

$$x^T Tx + p^T x + \frac{1}{4}p^T T^{-1}p = \left\| T^{1/2}x + \frac{1}{2}T^{-1/2}p \right\|.$$

Let us use the substitution

$$y = T^{1/2}x + \frac{1}{2}T^{-1/2}p.$$

This implies

$$x = T^{-1/2}y - \frac{1}{2}T^{-1}p$$

and produces the optimization problem with respect to y:
Minimize $\|y\|^2$ subject to constraint

$$AT^{-1/2}y = \frac{1}{2}AT^{-1}p + b.$$

Using the Moore-Penrose solution of the last equation it follows

$$y_0 = \left(AT^{-1/2}\right)^{\dagger}\left(\frac{1}{2}AT^{-1}p + b\right),$$

which leads to our original attention. ∎

A study of a minimization problem for a matrix-valued function under linear constraints, in the case of a singular matrix, was presented in [43]. More precisely, the authors of [43] considered the problem of minimizing the matrix valued function $W^T RW$, $W \in \mathbb{R}^{M\times m}$, where $R \in \mathbb{R}^{M\times M}$ is a positive semidefinite symmetric matrix and W belongs to a set of linear constraints

$$S = \left\{W \in \mathbb{R}^{M\times m} : C^T W = F\right\},\ C \in \mathbb{R}^{M\times n}, F \in \mathbb{R}^{n\times m}.$$

The main result from [43] is presented in the next statement.

Theorem 2.2. [43] *Let $C \in \mathbb{R}^{M\times M}$ be a positive semidefinite symmetric matrix and the matrices $W \in \mathbb{R}^{M\times m}$, $C \in \mathbb{R}^{M\times n}$ satisfy $m < M, n < M$. If the set*

$$S = \{W : \ C^T W = F, \ \mathcal{R}(W) \subseteq \mathcal{R}(R)\}$$

is not empty, then the problem:

$$\text{minimize} \quad W^T RW, \ W \in S$$

has the unique solution

$$\hat{W} = R^\dagger C \left(C^T R^\dagger C\right)^\dagger F.$$

In the case when S is empty, the constraint must be replaced by the equation

$$C^T W = F_1 = P_{\mathcal{R}(C^T R^\dagger C)} F.$$

Two applications of the proposed minimization method, in Linear Regression and B-spline smoothing, were presented in [43].

Let T be a symmetric positive definite matrix. Let U be the unitary matrix and D the diagonal matrix such that $T = U^* DU$. Let $D^{\frac{1}{2}}$ be the positive solution of the equation $R^2 = D$, and let $D^{-\frac{1}{2}}$ denotes the matrix $(D^{\frac{1}{2}})^{-1} = R^{-1}$. Since T is a positive definite matrix the existence of the matrices T^{-1} and R^{-1} is ensured. Similarly, by $X^{-\frac{1}{2}}$ we denote the matrix $(T^{\frac{1}{2}})^{-1} = X^{-1}$.

Lemma 2.1. *The minimization of the functional $\langle x, Tx\rangle$ is equivalent to the problem of finding a value x of minimum $U^* D^{-\frac{1}{2}}$-norm.*

Proof. Using

$$\begin{aligned}\langle x, Tx\rangle &= \langle x, U^* DUx\rangle \\ &= \left\langle D^{\frac{1}{2}}Ux, D^{\frac{1}{2}}Ux\right\rangle,\end{aligned}$$

we conclude that the minimization of $\langle x, Tx\rangle$ is equivalent with minimizing $\left\|D^{\frac{1}{2}}Ux\right\|^2$. Now, using

$$\| D^{\frac{1}{2}}Ux \|^2 = \|x\|^2_{U^* D^{-\frac{1}{2}}},$$

we conclude that the original problem have led us to the problem of finding a minimizer of $\|x\|^2_{U^* D^{-\frac{1}{2}}}$, which completes the proof. ■

Now, the problem (2.9) could be rewritten as the following multicriteria optimization problem:

Stage 1: minimize $\|Ax - b\|_2$;

Stage 2: minimize $\{\|x\|_{U^* D^{-\frac{1}{2}}}$ among all solutions in Stage 1$\}$.

In the case when the system $Ax = b$ is not consistent in Stage 1 and instead $\|\cdot\|_{U^* D^{-\frac{1}{2}}}$ we use the 2-norm, the stated multicriteria problem reduces to well–known multicriteria problem corresponding to the Moore-Penrose inverse. Therefore, the Moore–Penrose inverse is a solution of the optimization problem

$$\text{minimize } \langle x, x\rangle, \ x \in S,$$

where the 2-norm is assumed.

3. The T-Restricted Weighted Moore-Penrose Inverse

The results surveyed in this section are based on the papers [34, 35].

From all the above discussion, we can at this point translate the results presented to the finite dimensional case, that is, making use of matrices instead of operators. Theorem 3.1 translates Theorem 2.1 for the case of matrices. A useful result is presented before the theorem.

Proposition 3.1. *Let T be an EP matrix. Then, it holds that* $(T^\dagger)^{\frac{1}{2}} = \left(T^{\frac{1}{2}}\right)^\dagger$.

Theorem 3.1. [35] *Let $T \in \mathbb{R}^{n\times n}$ be a singular positive matrix, and the linear system $Ax = b$ is defined by a singular matrix $A \in \mathbb{R}^{m\times n}$ and a vector $b \in \mathbb{R}^n$. If the set*

$$S = \{x \in \mathcal{N}(T)^\perp : Ax = b\}$$

is not empty, then the problem

$$\text{minimize } \langle x, Tx\rangle, \quad x \in S$$

has the unique solution

$$\begin{aligned}\hat{u} &= \left(T^\dagger\right)^{\frac{1}{2}} \left(A\left(T^\dagger\right)^{\frac{1}{2}}\right)^\dagger b\\ &= \left(T^{\frac{1}{2}}\right)^\dagger \left(A\left(T^{\frac{1}{2}}\right)^\dagger\right)^\dagger b.\end{aligned}$$

Definition 3.1. [34] *Let $T \in \mathbb{R}^{n\times n}$ be a positive semidefinite symmetric matrix and $A \in \mathbb{R}^{m\times n}$. Then the $n \times m$ matrix*

$$\begin{aligned}\hat{A}^\dagger_{I_m,T} &:= \left(T^\dagger\right)^{\frac{1}{2}} \left(A\left(T^\dagger\right)^{\frac{1}{2}}\right)^\dagger\\ &= \left(T^{\frac{1}{2}}\right)^\dagger \left(A\left(T^{\frac{1}{2}}\right)^\dagger\right)^\dagger\end{aligned} \tag{3.1}$$

is called the T-restricted weighted Moore–Penrose inverse of A.

Remark 3.1. *The generalized inverse $\hat{A}^\dagger_{I_m,T} b$ is a minimal T semi- norm least squares solution of $Ax = b$, restricted on the range of T.*

Based on Theorem 3.1, similarly as the weighted Moore-Penrose inverse, we can extend this notion to the *N-restricted weighted inverse* with M positive definite and N positive semidefinite:

$$\hat{A}^\dagger_{M,N} = \left(N^\dagger\right)^{\frac{1}{2}} \left(M^{\frac{1}{2}} A (N^\dagger)^{\frac{1}{2}}\right)^\dagger M^{\frac{1}{2}}. \tag{3.2}$$

The generalized inverse defined in (3.2) initiates a minimal N semi-norm, M-least squares solution $\hat{A}^\dagger_{M,N} b$ of $Ax - b$, but restricted on the range of N.

In [34] it is verified that the solution $\hat{u}$, defined in Theorem 3.1, satisfies the constraint $Ax = b$. Indeed,

$$A\hat{u} - A(T^\dagger)^{\frac{1}{2}}(A(T^\dagger)^{\frac{1}{2}})^\dagger b = P_{AT} b$$

and since the set $S = \{x \in \mathcal{R}(T) : Ax = b\}$ is not empty, we have that b must be equal to ATw, for some w and therefore $P_{AT}b = b$.

The matrix $\hat{A}^{\dagger}_{I_m,T}$ does not satisfy all four conditions of equation (1.8) as it is an inverse restricted to the range of T. More precisely, the following is satisfied.

Proposition 3.2. [34] *Let $T \in \mathbb{R}^{m\times m}$ be positive semidefinite, $A \in \mathbb{R}^{n\times m}$ and the equation $Ax = b$. The T-restricted weighted inverse $\hat{A}^{\dagger}_{I,T}$ satisfies the following basic properties:*

(i) $A\hat{A}^{\dagger}_{I,T}A = P_{AT}A$.

(ii) $T\hat{A}^{\dagger}_{I,T}A\hat{A}^{\dagger}_{I,T} = T\hat{A}^{\dagger}_{I,T}$.

(iii) $(A\hat{A}^{\dagger}_{I,T})^* = (A\hat{A}^{\dagger}_{I,T})$.

(iv) $\hat{A}^{\dagger}_{I,T}A\hat{A}^{\dagger}_{I,T} = \hat{A}^{\dagger}_{I,T}P_{AT}$.

Proof.

(i) $A\hat{A}^{\dagger}_{I,T}A = A(T^{\dagger})^{\frac{1}{2}}(A(T^{\dagger})^{\frac{1}{2}})^{\dagger}A = P_{AT^{\dagger}}A = P_{AT}A.$

(ii)
$$\begin{aligned} T\hat{A}^{\dagger}_{I,T}A\hat{A}^{\dagger}_{I,T} &= T(T^{\dagger})^{\frac{1}{2}}(A(T^{\dagger})^{\frac{1}{2}})^{\dagger}A(T^{\dagger})^{\frac{1}{2}}(A(T^{\dagger})^{\frac{1}{2}})^{\dagger} \\ &= T(T^{\dagger})^{\frac{1}{2}}(A(T^{\dagger})^{\frac{1}{2}})^{\dagger} \\ &= T\hat{A}^{\dagger}_{I,T}. \end{aligned}$$

(iii) $(A\hat{A}^{\dagger}_{I,T})^* = (A(T^{\dagger})^{\frac{1}{2}}(A(T^{\dagger})^{\frac{1}{2}})^{\dagger})^* = (P_{AT^{\dagger}})^* = P_{AT^{\dagger}} = A\hat{A}^{\dagger}_{I,T}.$

(iv)
$$\begin{aligned} \hat{A}^{\dagger}_{I,T}A\hat{A}^{\dagger}_{I,T} &= (T^{\dagger})^{\frac{1}{2}}(A(T^{\dagger})^{\frac{1}{2}})^{\dagger}A(T^{\dagger})^{\frac{1}{2}}(A(T^{\dagger})^{\frac{1}{2}})^{\dagger} \\ &= (T^{\dagger})^{\frac{1}{2}}(A(T^{\dagger})^{\frac{1}{2}})^{\dagger}P_{AT^{\frac{1}{2}}} \\ &= \hat{A}^{\dagger}_{I,T}P_{AT}. \end{aligned}$$

■

From the properties of $\hat{A}^{\dagger}_{I,T}$ presented in Proposition 3.2, it is clear that $\hat{A}^{\dagger}_{I,T}$ is not an $\{i, j, k\}$ inverse of A. Nevertheless, many of the already known properties of the generalized inverses also hold for the T-restricted weighted inverse, with slight modifications, as we can see in the following proposition.

Proposition 3.3. [34] *Let $T \in \mathbb{R}^{m\times m}$ be positive semidefinite and $A \in \mathbb{R}^{n\times m}$. The T-restricted weighted inverse $\hat{A}^{\dagger}_{I,T}$ has the following properties:*

(i) *If $\hat{A}^{\dagger}_{I,T} = \hat{A}^{\dagger}_{I,S}$ holds for two positive semidefinite matrices S, T then*

$$\mathcal{R}(AT) = \mathcal{R}(AS).$$

(ii) *Similarly to the well-known formula $TT^{\dagger} = P_T$ the property*

$$A\hat{A}^{\dagger}_{I,T} = P_{AT}$$

is satisfied.

(iii) *If A is an $\mathbb{R}^{m\times m}$ matrix and $\hat{A}^{\dagger}_{I,T}A = A\hat{A}^{\dagger}_{I,T}$, then*

$$P_{AT}(T^{\dagger})^{\frac{1}{2}} = (T^{\dagger})^{\frac{1}{2}}P_{TA^*}.$$

Proof.

(i) Let the two positive semidefinite matrices S, T such that $\hat{A}^{\dagger}_{I,T} = \hat{A}^{\dagger}_{I,S}$. Then, $A\hat{A}^{\dagger}_{I,T} = A\hat{A}^{\dagger}_{I,S} \Rightarrow P_{AT} = P_{AS}$.

(iii) If $\hat{A}^{\dagger}_{I,T}A = A\hat{A}^{\dagger}_{I,T}$ then

$$(T^{\dagger})^{\frac{1}{2}}\left(A\left(T^{\dagger}\right)^{\frac{1}{2}}\right)^{\dagger}A = A(T^{\dagger})^{\frac{1}{2}}\left(A\left(T^{\dagger}\right)^{\frac{1}{2}}\right)^{\dagger},$$

which implies

$$(T^{\dagger})^{\frac{1}{2}}\left(A\left(T^{\dagger}\right)^{\frac{1}{2}}\right)^{\dagger}A(T^{\dagger})^{\frac{1}{2}} = A(T^{\dagger})^{\frac{1}{2}}\left(A(T^{\dagger})^{\frac{1}{2}}\right)^{\dagger}(T^{\dagger})^{\frac{1}{2}}$$

and so,

$$(T^{\dagger})^{\frac{1}{2}}P_{(A(T^{\dagger})^{\frac{1}{2}})^*} = P_{AT}(T^{\dagger})^{\frac{1}{2}}.$$

But, since $\mathcal{R}((A(T^{\dagger})^{\frac{1}{2}})^*) = \mathcal{R}(TA^*)$ we have that $P_{AT}(T^{\dagger})^{\frac{1}{2}} = (T^{\dagger})^{\frac{1}{2}}P_{TA^*}$.

■

Example 3.1. Let $\mathcal{H} = \mathbb{R}^3$, the matrix A is equal to

$$A = \begin{bmatrix} -2 & 3 & 5 \\ 1 & 0 & -1 \end{bmatrix}$$

and the positive semidefinite matrix T is chosen as

$$T = \begin{bmatrix} 7 & -3 & -7 \\ -3 & 36 & 3 \\ -7 & 3 & 7 \end{bmatrix}$$

Here, the equation is $Ax = b$ where

$$b = \begin{bmatrix} 19 \\ -2 \end{bmatrix},$$

is consistent.

We will compute the T-restricted weighted inverse $\hat{A}^{\dagger}_{I_2,T}$. Consequently,

$$\hat{A}^{\dagger}_{I_2,T} = \begin{bmatrix} 0 & \frac{1}{2} \\ \frac{1}{3} & \frac{7}{6} \\ 0 & -\frac{1}{2} \end{bmatrix}.$$

Then, $\hat{u} = \hat{A}^{\dagger}_{I_2,T}b = (-1,4,1)^T$ is the minimal T semi-norm least squares solution of $Ax = b$, restricted on the range of T.

It is easy to see that all vectors $u \in \mathcal{R}(T)$ are of the general form $u = (x, y, -x)^T$, $x, y \in \mathbb{R}$, so the solution $\hat{u}$ has the expected form.

In Figure 1 it is observable in blue the quadratic form $\Phi(x) = \langle x, Tx \rangle$ for all the vectors $x \in \mathcal{N}(T)^{\perp}$ and in red the set of all vectors u satisfying the constraint $Au = b$. As we can see the line is tangent to the surface therefore there is only one solution which is the vector found, $\hat{u} = \hat{A}^{\dagger}_{I_2,T}b$. In this case, $\|\hat{u}\|^2_T = 652$.

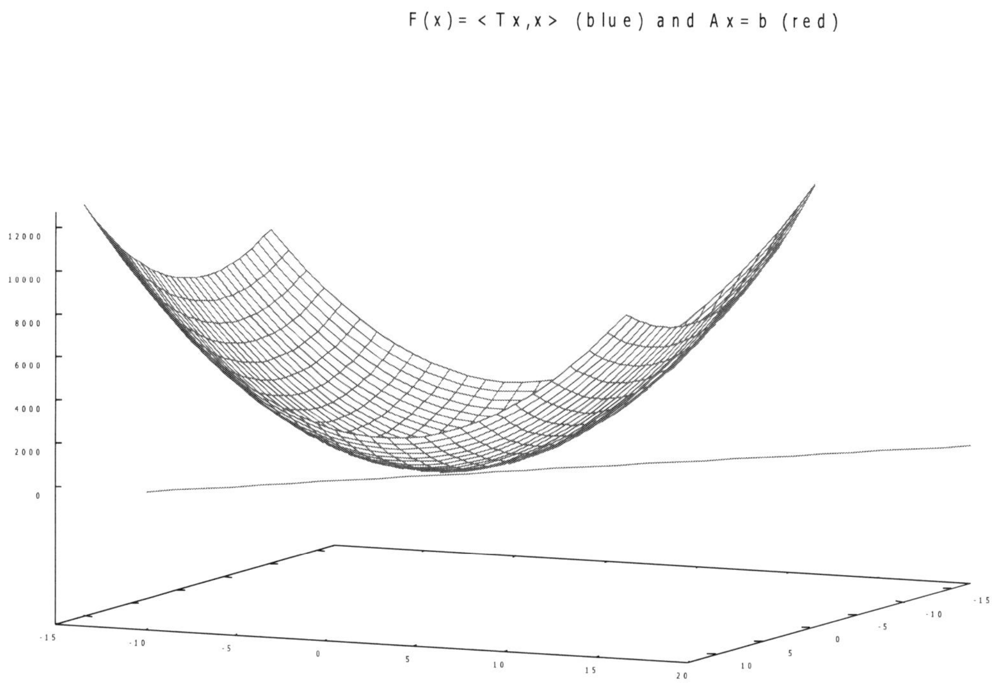

Figure 1. Constrained minimization of $\|.\|_T$, $u \in \mathcal{N}(T)^{\perp}$ under $Ax = b$.

3.1. Relations of $\hat{A}^{\dagger}_{I,T}$ with the V-Orthogonal Projector

For every matrix $X \in \mathbb{R}^{n \times p}$ and a positive semidefinite matrix $V \in \mathbb{R}^{n \times n}$, the matrix

$$P_{X:V} = X(X^*VX)^{\dagger}X^*V \tag{3.3}$$

is called the V-orthogonal projector with respect to the semi-norm $\|.\|_V$ (see e.g., [48], or [17] section 3). Let us mention that the V-orthogonal projector is unique in the case $\text{rank}(VX) = \text{rank}(X)$.

Relations between $\hat{A}^{\dagger}_{I,T}$ and $P_{A:T}$ are investigated in Lemma 3.1. We will make use of the following proposition.

Proposition 3.4. ([48, Theorem 7]) *Let $P_{X:V}$ be as given, and suppose that* $\operatorname{rank}(VX) = \operatorname{rank}(X)$. *Then*

$$P_{X:V} = X\left(V^{\frac{1}{2}}X\right)^{\dagger} V^{\frac{1}{2}}.$$

Using the above notation, we can see that similarly to the Moore-Penrose inverse property $T^{\dagger}T = P_{T^*}$ we have the following statement.

Lemma 3.1. [34] *Let $T \in \mathbb{R}^{m\times m}$ be a positive semidefinite matrix and $A \in \mathbb{R}^{n\times m}$. If*

$$\operatorname{rank}(T^{\dagger}A^*) = \operatorname{rank}(A^*),$$

then the T-restricted weighted inverse $\hat{A}^{\dagger}_{I,T}$ has the property

$$\hat{A}^{\dagger}_{I,T}A = P^*_{A^*:T^{\dagger}}.$$

Proof. Under the stated assumptions, it is observable that $X \equiv A^*$ and $V \equiv T^{\dagger}$. Therefore,

$$P_{A^*:T^{\dagger}} = A^*((T^{\dagger})^{\frac{1}{2}}A^*)^{\dagger}(T^{\dagger})^{\frac{1}{2}}.$$

On the other hand, one can verify

$$P^*_{A^*:T^{\dagger}} = (T^{\dagger})^{\frac{1}{2}}(A(T^{\dagger})^{\frac{1}{2}})^{\dagger}A = \hat{A}^{\dagger}_{I,T}A,$$

which completes the proof. ■

Remark 3.2. *The relation* $\operatorname{rank}(T^{\dagger}A^*) = \operatorname{rank}(A^*)$ *can be replaced by*

$$\mathcal{N}(T) \cap \mathcal{N}(A)^{\perp} = \mathcal{N}(T) \cap \mathcal{R}(A^*) = \{0\}.$$

Proof. *Since*

$$\operatorname{rank}(T^{\dagger}A^*) = \operatorname{rank}(A^*) - \dim(\mathcal{N}(T^{\dagger}) \cap \mathcal{R}(A^*)),$$

it is possible to verify $\mathcal{N}(T^{\dagger}) \cap \mathcal{R}(A^*) = \{0\}$ but since T is positive, $\mathcal{N}(T^{\dagger}) = \mathcal{N}(T)$. So, $\operatorname{rank}(T^{\dagger}A^*) = \operatorname{rank}(A^*)$ *is equivalent to $\mathcal{N}(T) \cap \mathcal{R}(A^*) = \{0\}$.* ■

By the above remark, we can have many results related to the V-orthogonal projector, using Theorems 7 and 8 from [48].

Proposition 3.5. [34] *Let $T \in \mathbb{R}^{m\times m}$ be a positive semidefinite matrix and $A \in \mathbb{R}^{n\times m}$, such that $\mathcal{N}(T) \cap \mathcal{R}(A^*) = \{0\}$. Then, the following statements hold:*

(i)] $AP_{A^*:T^{\dagger}} = P_{AT}A$.

(ii) *In the case when $A\hat{A}^{\dagger}_{I,T} = \hat{A}^{\dagger}_{I,T}A$ we have that $P^*_{A^*:T^{\dagger}} = P_{AT}$.*

(iii) $\hat{A}^{\dagger}_{I,T}P_{AT}A = \hat{A}^{\dagger}_{I,T}A$.

(iv) *The matrix $\hat{A}^{\dagger}_{I,T}A$ is Hermitian.*

(v) $\hat{A}^{\dagger}_{I,T}A = P_T$.

Proof.

(i) from Proposition 3.1 it can be verified $AP_{A^*:T^\dagger} = A\hat{A}^\dagger_{I,T}A$, which is equal to $P_{AT}A$ from Proposition 3.2.

(ii) If $A\hat{A}^\dagger_{I,T} = \hat{A}^\dagger_{I,T}A$ then from propositions 3.1 and 3.3 we conclude $P_{AT} = P^*_{A^*:T^\dagger}$.

(iii) From [48], Theorem 7, we have that $P^2_{A^*:T^\dagger} = P_{A^*:T^\dagger}$ and so

$$(\hat{A}^\dagger_{I,T}A)^2 = (\hat{A}^\dagger_{I,T}A) \Rightarrow \hat{A}^\dagger_{I,T}A\hat{A}^\dagger_{I,T}A = \hat{A}^\dagger_{I,T}A.$$

Therefore $\hat{A}^\dagger_{I,T}P_{AT}A = \hat{A}^\dagger_{I,T}A$.

(iv) From [48], Theorem 8, we have that $P_{A^*:T^\dagger} = P'_{A^*:T^\dagger}$ and so $\hat{A}^\dagger_{I,T}A$ is Hermitian.

(v) From [48], Theorem 8, one can verify $P_{A^*:T^\dagger} = P_{T^\dagger} = P_T$ and so $\hat{A}^\dagger_{I,T}A = P_T$.

■

An important paper for the interested reader relating seminorms and generalized inverses is [49].

3.2. Minimization of Quadratic Functionals

All the results presented in this section come from [46].

The next step for generalizing the so far presented results is to study the problem

$$\text{minimize} \quad \Phi(x) = \langle x, Tx\rangle + \langle p, x\rangle + a, \; x \in S \cap \mathcal{N}(T)^\perp. \tag{3.4}$$

This approach can be seen as a more general case of Theorem 3.1. Let T be a positive definite symmetric matrix. An interesting case to examine is when T is singular and positive semidefinite with a nonempty kernel, $\mathcal{N}(T) \neq \{0\}$. In this case we have that $\langle x, Tx\rangle = 0$ for all $x \in \mathcal{N}(T)$ and so, a first approach in both the finite and infinite dimensional case would be to look among the vectors $x \in \mathcal{N}(T)^\perp = \mathcal{R}(T^*) = \mathcal{R}(T)$ for a minimizing vector for $\Phi(x)$. In this case, generalized inverses will be used.

A first approach to the problem stated is to use an $\{1,3\}$ and an $\{1,4\}$-inverse in order to find the set of least squares solutions and the minimal norm solution among them, without using the constraint $x \in \mathcal{N}(T)^\perp$.

Theorem 3.2. [46] *Let $T \in \mathcal{B}(\mathcal{H})$ be singular positive semidefinite with a non empty kernel, $\mathcal{N}(T) \neq \{0\}$, with $X^2 = T$. Let also $A \in \mathcal{B}(\mathcal{H})$, where A is singular and consider the equation $Ax = b$.*

If the set $S = \{x : Ax = b\}$ is nonempty, then the minimization problem (3.4) *with $p \in \mathcal{H}$ and $a \in \mathbb{R}$ has a least squares solution*

$$\hat{x} = X^{(1,3)}\left(AX^{(1,3)}\right)^{(1,4)}\left(\frac{1}{2}AX^{(1,3)}X^\dagger p + b\right) - \frac{1}{2}X^{(1,3)}X^\dagger p.$$

Proof. Let $x, p \in \mathcal{H}$ and an arbitrary constant $a \in \mathbb{R}$. Hence, if there exists a vector $\hat{x}$ that minimizes $\Phi(x)$, it would also minimize

$$\Psi(x) = \langle Tx, x\rangle + \langle x, p\rangle + \frac{1}{4}\langle T^{\dagger}p, p\rangle.$$

We can easily see that

$$\left\| Xx + \frac{1}{2}X^{\dagger}p \right\|^2 = \Psi(x).$$

The substitution $y = Xx + \frac{1}{2}X^{\dagger}p$ implies that $y \in \mathcal{R}(X) = \mathcal{R}(T)$.
We have that

$$Xx = y - \frac{1}{2}X^{\dagger}p.$$

Hence,

$$x = X^{(1,3)}y - \frac{1}{2}X^{(1,3)}X^{\dagger}p.$$

Since $Ax = b$, we have that $AX^{(1,3)}y - \frac{1}{2}AX^{(1,3)}X^{\dagger}p = b$, and therefore we can find the minimal norm solution for y using a $\{1, 4\}$ inverse. So, $\hat{y}$ is equal to

$$\hat{y} = \left(AX^{(1,3)}\right)^{(1,4)}\left(b + \frac{1}{2}AX^{(1,3)}X^{\dagger}p\right),$$

and therefore, $\Psi(x)$ is minimized.

By substitution, we have that

$$\hat{x} = X^{(1,3)}\left(AX^{(1,3)}\right)^{(1,4)}\left(\frac{1}{2}AX^{(1,3)}X^{\dagger}p + b\right) - \frac{1}{2}X^{(1,3)}X^{\dagger}p.$$

The proof is complete. ∎

We can also find a more general set of solutions, according to Proposition 1.12, since the set of least squares solutions possesses more general form.

Corollary 3.1. [46] *The set of least squares solutions of* (3.4) *is defined by*

$$\tilde{x} = X^{(1,3)}\left(AX^{(1,3)}\right)^{(1,4)}\left(\frac{1}{2}AX^{(1,3)}X^{\dagger}pT + b\right) - \left(AX^{(1,3)}\right)^{(1,4)}ARw - \frac{1}{2}X^{(1,3)}X^{\dagger}p,$$

with an arbitrary vector w *and* $R = (I - X^{(1,3)}X)$.

Proof. From Proposition 1.12 it is not difficult to verify

$$Xx = y - \frac{1}{2}X^{\dagger}p \Longrightarrow x = X^{(1,3)}y - \frac{1}{2}X^{(1,3)}X^{\dagger}p + Rw,$$

where $R = (I - X^{(1,3)}X)$ and w is an arbitrary vector from $\mathcal{H}$.
Therefore, $Ax = b$ implies

$$AX^{(1,3)}y = \frac{1}{2}AX^{(1,3)}X^{\dagger}p + b - ARw.$$

So, a minimal norm solution among all vectors y is

$$\hat{y} = \left(AX^{(1,3)}\right)^{(1,4)} \left(b + \frac{1}{2}AX^{(1,3)}X^{\dagger}p_1\right) - \left(AX^{(1,3)}\right)^{(1,4)} ARw.$$

By substitution we have the general form of the least squares solutions set which minimizes $\Psi(x)$:

$$\tilde{x} = X^{(1,3)} \left(AX^{(1,3)}\right)^{(1,4)} \left(\frac{1}{2}AX^{(1,3)}X^{\dagger}p + b\right) - \left(AX^{(1,3)}\right)^{(1,4)} ARw - \frac{1}{2}X^{(1,3)}X^{\dagger}p,$$

which completes the proof. ■

As a special case of the above set, we can have a unique least squares minimal norm solution, using the Moore-Penrose inverse. Moreover, since $\mathcal{N}(T) = \mathcal{N}(T^{\dagger})$ the vectors examined in this case satisfy also the additional property $x \in \mathcal{N}(T)^{\perp}$.

Theorem 3.3. [46] *Let $T \in \mathcal{B}(\mathcal{H})$ be singular positive semidefinite with a non empty kernel, $\mathcal{N}(T) \neq \{0\}$, with $X^2 = T$. Let also $A \in \mathcal{B}(\mathcal{H})$, where A is singular and consider the equation $Ax = b$.*

If the set $S = \{x : Ax = b\}$ is not empty, then the optimization problem (3.4) *with $p \in \mathcal{H}$ and $a \in \mathbb{R}$ has the unique solution*

$$\hat{x} = X^{\dagger}(AX^{\dagger})^{\dagger} \left(\frac{1}{2}AT^{\dagger}p + b\right) - \frac{1}{2}T^{\dagger}p,$$

assuming that the operator $P_{\mathcal{R}(A^*)}P_{\mathcal{R}(T)}$ has closed range.

Proof. Since we will restrict the minimization for all vectors $x \in \mathcal{N}(T)^{\perp}$ we have that $\langle x, p\rangle = \langle x, p_1\rangle$, where $p_1 = P_{\mathcal{R}(T)}p$ for all vectors $p \in \mathcal{H}$, according to the decomposition $p = p_1 + p_2 \in \mathcal{R}(T) \oplus \mathcal{N}(T)$.

Let $x, p \in \mathcal{H}$ and an arbitrary constant $a \in \mathbb{R}$. Hence if there exists a vector $\hat{x}$ that minimizes $\Phi(x)$, it would also minimize

$$\begin{aligned}\Psi(x) &= \langle Tx, x\rangle + \langle x, p\rangle + \frac{1}{4}\langle T^{\dagger}p, p\rangle \\ &= \langle Tx, x\rangle + \langle x, p_1\rangle + \frac{1}{4}\langle T^{\dagger}p_1, p_1\rangle.\end{aligned}$$

We can easily see that

$$\left\| Xx + \frac{1}{2}X^{\dagger}p_1 \right\|^2 = \Psi(x),$$

and

$$X = U(R \oplus O)U^*, \qquad X^{\dagger} = U(R^{-1} \oplus O)U^*.$$

The substitution

$$y = Xx + \frac{1}{2}X^{\dagger}p_1 \tag{3.5}$$

implies $y \in \mathcal{R}(X) = \mathcal{R}(T)$.

Further,

$$Xx = y - \frac{1}{2}X^\dagger p_1 \Leftrightarrow U(R \oplus O)U^* x = y - \frac{1}{2}U(R^{-1} \oplus O)U^* p_1.$$

Hence,

$$x = U(R^{-1} \oplus O)U^* y - \frac{1}{2}U(R^{-1} \oplus O)(R^{-1} \oplus O)U^* p_1$$

and so, since $y, p_1 \in \mathcal{R}(T)$

$$x = X^\dagger y - \frac{1}{2}T^\dagger p_1, \qquad with \qquad x \in \mathcal{R}(T^\dagger) = \mathcal{R}(T) = \mathcal{N}(T)^\perp.$$

Since $Ax = b$, we have that $AX^\dagger y = b + \frac{1}{2}AT^\dagger p_1$ and therefore by Proposition 1.11 the minimal norm solution exists and is equal to

$$\hat{y} = (AX^\dagger)^\dagger \left(b + \frac{1}{2}AT^\dagger p_1 \right),$$

and therefore, $\Psi(x)$ is minimized.

Now, the substitution (3.5) leads to

$$\hat{x} = X^\dagger (AX^\dagger)^\dagger \left(\frac{1}{2}AT^\dagger p_1 + b \right) - \frac{1}{2}T^\dagger p_1$$

and since $T^\dagger p = T^\dagger p_1$ for all

$$p \in \mathcal{H} \Longrightarrow \hat{x} = X^\dagger (AX^\dagger)^\dagger \left(\frac{1}{2}AT^\dagger p + b \right) - \frac{1}{2}T^\dagger p.$$

The only thing that needs to be proved is the fact that the operator $AX^\dagger$ has closed range and so its Moore-Penrose Inverse is bounded. Since X is positive, therefore EP, $\mathcal{R}(X) = \mathcal{R}(X^2) = \mathcal{R}(T)$ and so X has closed range.

Moreover, since the two operators A and $X^\dagger$ are arbitrary, one does not expect that the range of their product will always be closed. From Proposition 1.15, this is equivalent to the fact that the operator $P_{\mathcal{R}(A^*)}P_{\mathcal{R}(T)}$ has closed range because

$$A^\dagger AX^\dagger (X^\dagger)^\dagger = A^\dagger AX^\dagger X = A^\dagger AU(R^{-1} \oplus 0)(R \oplus 0)U^* = P_{\mathcal{R}(A^*)}P_{\mathcal{R}(T)}$$

and the proof is completed. ■

Corollary 3.2. *The matrix* $\hat{A}^\dagger_{I_m,T}$ *defined by* (3.1) *is a minimizer of the optimization problem*

$$\min \Psi(Z) = \langle Z, TZ \rangle,$$

where T *is a positive* $n \times n$ *matrix and* $X \in \mathbb{R}^{n \times m}$, *subject to the constraint set*

$$Z \in \Omega_I, \ \Omega_I = \{Z : \ Z \in \mathbb{R}^{n \times m}, \ AZ = I, \quad A \in \mathbb{R}^{m \times n}_r\}.$$

Theorem 3.4. *If $A \in \mathbb{R}^{m\times n}$ is given matrix, then*

$$\hat{A}^{\dagger}_{I_m,T} = A^{(2)}_{U,V}, \quad U = \mathcal{R}(T^{\dagger}A^*), V = \mathcal{N}(TA^*) = \mathcal{N}(T^{\dagger}A^*). \tag{3.6}$$

Proof. Definition of $\hat{A}^{\dagger}_{I_m,T}$ gives us $\mathcal{R}(\hat{A}^{\dagger}_{I_m,T}) = \mathcal{R}(T^{\dagger}A^*)$ and

$$\mathcal{N}\left(\left(A\left(T^{\frac{1}{2}}\right)^{\dagger}\right)^{\dagger}\right) \subseteq \mathcal{N}(\hat{A}^{\dagger}_{I_m,T}) \subseteq \mathcal{N}\left(A\left(T^{\frac{1}{2}}\right)^{\dagger}\left(A\left(T^{\frac{1}{2}}\right)^{\dagger}\right)^{\dagger}\right)$$
$$= \mathcal{N}\left(\left(A\left(T^{\frac{1}{2}}\right)^{\dagger}\right)^{\dagger}\right).$$

Hence,

$$\mathcal{N}\left(\hat{A}^{\dagger}_{I_m,T}\right) = \mathcal{N}\left(\left(T^{\frac{1}{2}}\right)^{\dagger}A^*\right) = N(TA^*).$$

However, $A^{(2)}$ inverse with pre described range and null space is uniquely determined and it allows us to conclude that $\hat{A}^{\dagger}_{I_m,T} = A^{(2)}_{U,V}$. ■

Corollary 3.3 gives a full-rank representation of the T-minimal G-constrained inverse and can be useful in its numerical computations.

Corollary 3.3. *Let A, G and T satisfy the conditions of Definition 3.1. If $T^{\dagger}A^* = PQ$ is an arbitrary full-rank decomposition of $T^{\dagger}A^*$ then:*

(1) *QAP is an invertible complex matrix;*

(2) $\hat{A}^{(2)}_{G,T} = P(QAP)^{-1}Q$.

Proof. The proof follows from Theorem 3.4 and [50, Theorem 3.1]. ■

Full-rank representation $\hat{A}^{\dagger}_{I_m,T}$ defined in Corollary 3.3 allows us to define Algorithm 3.1. for its computation. Full-rank factorization of the matrix $T^{\dagger}A^*$ is based on its QR decomposition. Construction of the full-rank factorization of $T^{\dagger}A^*$ using its QR decomposition is reused from [51]

Numerical results on test matrices from [52, 53] are presented in [39].

4. The T-Restricted Weighted Drazin Inverse

All the results presented in this section are based on [39, 40].

A further step will be now to approach the problem of finding an approximate solution of the quadratic problem with the objective function $\Phi(x)$ when the set S is empty, by analyzing the generalized normal equation of the system $Ax = b$. Then, according to results presented in Subsection 1.2.2, it is reasonable to choose a new the constraint set S_D which is defined in (2.11). The minimization problem with the goal function (2.7) subject to the constraint set S_D is considered in [39]. The generalized inverse arising from the results derived in [39] is defined and investigated in [40].

Also, an additional reason for replacing the constraint set S by the constraint set S_D is evident. Namely, if $Ax = b$ is a consistent or inconsistent singular linear system, then the linear system of equations $A^{k+1}x = A^k b$, where $k \geq \operatorname{ind}(A)$ is consistent [54].

Algorithm 3.1. Computing the generalized inverse $\hat{A}^{\dagger}_{I_m,T}$.

Require: The matrix A of dimensions $m \times n$ and of rank r and positive semidefinite $n \times n$ matrix T.

1: Choose arbitrary, but fixed, $n \times m$ matrix G of rank $0 < s \leq r$.

2: Compute $W = T^{\dagger} A^*$.

3: Compute the QR decomposition of the matrix W in the form

$$WP = QR, \tag{3.7}$$

where P is an $m \times m$ permutation matrix, $Q \in \mathbb{C}^{n \times n}$, $Q^*Q = I_n$ and $R \in \mathbb{C}^{n \times m}$ is an upper trapezoidal matrix.

4: Compute $t = \text{rank}(T^{\dagger} A^*)$.

5: Assume that Q and R are partitioned as

$$Q = \begin{bmatrix} Q_1 & Q_2 \end{bmatrix}, \quad R = \begin{bmatrix} R_{11} & R_{12} \\ O & O \end{bmatrix} = \begin{bmatrix} R_1 \\ O \end{bmatrix}, \tag{3.8}$$

where Q_1 consists of the first t columns of the matrix Q and $R_{11} \in \mathbb{C}^{t \times t}$ is nonsingular. Generate the full–rank decomposition of the matrix W by means of

$$W = Q_1(R_1 P^*). \tag{3.9}$$

6: Solve the matrix equation $R_1 P^* A Q_1 X = R_1 P^*$.

7: Compute the output $\hat{A}^{\dagger}_{I_m,T} = Q_1 X$.

Reasons become more apparent in the case when some additional assumptions on the matrix A are imposed. Campbell in [16] showed that $A^D b$ is a solution of linear system $Ax = b$ if and only if $b \in \mathcal{R}(A^k)$, $k = \text{ind}(A)$ and proved that $A^D b$ is the unique solution of $Ax = b$ provided that $x \in \mathcal{R}(A^k)$.

Proposition 4.1. [16, Pages 123, 203] *A vector x is a Drazin-inverse solution $A^D b$ of $Ax = b$ if x is a solution of the normal Drazin equation* (1.9) *of the linear system $Ax = b$. If $b \in \mathcal{R}(A^k)$ then $A^D b$ is the unique solution of $Ax = b$ which belongs to $\mathcal{R}(A^k)$.*

It is assumed that $T \in \mathbb{R}^{n \times n}$ is a positive semidefinite matrix. Let $A \in \mathbb{R}^{n \times n}$ be such that $\text{ind}(A) = k$ and $x, b \in \mathbb{R}^n$. As in the previous case, we consider the minimization of the functional $\Phi(x)$. Since $\mathcal{N}(T) \neq \emptyset$, there exists vectors belonging to $\mathcal{N}(T)$. Clearly, $\langle x, Tx \rangle = 0$ is satisfied for each $x \in \mathcal{N}(T)$ and so, we will investigate the minimization problem

$$\text{minimize } \Phi(x), \; x \in S_D \cap \mathcal{N}(T)^{\perp}. \tag{4.1}$$

Of course, the additional assumption $S_D \cap \mathcal{N}(T)^{\perp} \neq \emptyset$ must be assumed in this case.

For the sake of completeness, we restate necessary facts required in the rest of this section. Let us mention that for an EP matrix T we have $T^{\dagger} = T^{\#}$. The matrix X satisfying $X^2 = T$ is also EP since X is symmetric. Operators T, X and $X^{\dagger} = X^{\#}$ can be considered in the form

$$T = U(A_1 \oplus O)U^T,$$
$$X = U(R \oplus O)U^T, \quad X^\dagger = U(R^{-1} \oplus O)U^T, \; R^2 = A_1.$$

Before the review of the optimization results and the properties of the corresponding generalized inverse, it will be useful to restate an interesting and useful property of the constraint set S_D. This property makes possible several equivalent definitions of the set S_D.

Lemma 4.1. [39] *For a given matrix A, let $k = \operatorname{ind}(A)$. The constraint set S_D is equivalent to the following two sets:*

$$S_D = S_{D_1} = \{x \mid x \in \mathbb{R}^n, A^k x = A^D A^k b\}$$
$$= S_{D_2} = \{x \mid x \in \mathbb{R}^n, A^D A^k x = A^D A^{k-1} b\}, \; k \geq 1.$$

Proof. In the first step of the proof it is necessary to verify $S_D = S_{D_1}$. For this purpose, let us consider arbitrary $x \in S_D$. Multiplying $A^{k+1}x = A^k b$ by A^D from the left and using the property (1^k) of the Drazin inverse, we immediately obtain $x \in S_{D_1}$. On the other hand, let us choose arbitrary $x \in S_{D_1}$. Multiplying from the left both sides of the equation $A^k x = A^D A^k b$ by A, it is easy to conclude that x satisfies

$$A^{k+1}x = AA^D A^k b = A^k b,$$

which, by definition, means $x \in S_D$.

In the second part of the proof it is necessary to verify $S_{D_1} = S_{D_2}$, for each $k \geq 1$. Multiplying $A^k x = A^D A^k b$ from the left by A^D we obtain

$$A^D A^k x = A^D A^D A^k b = A^D A A^D A^{k-1} b = A^D A^{k-1} b.$$

On the other hand, multiplying both sides of $A^D A^k x = A^D A^{k-1} b$ by A we get

$$AA^D A^k x = AA^D A^{k-1} b \Longrightarrow A^k x = AA^D A^{k-1} b = A^D A^k b,$$

which completes the proof. ■

4.1. The Symmetric Positive Semidefinite Case

Theorem 4.1 is the main result from [39]. This statement gives two approximate solutions which minimize $\Phi(x)$ under the constraint set S_D. The first approximation is presented in three equivalent ways. The results of Lemma 4.1 are very important in deriving these equivalent representations.

The next auxiliary result gives decompositions of positive semidefinite matrices, i.e., positive matrices.

Proposition 4.2. *The following statements hold:*

a) If $T \in \mathbb{R}^{n\times n}$ is a positive semidefinite matrix, then there exists an orthogonal matrix U and invertible diagonal matrix T_1 such that

$$T = U^T(T_1 \oplus O)U = U^T \begin{bmatrix} T_1 & 0 \\ 0 & 0 \end{bmatrix} U.$$

Also, there exists a unique matrix X such that $X^2 = T$ which is also an EP matrix, and which satisfies the following

$$X = U^T(R \oplus O)U, \quad X^\dagger = U^T(R^{-1} \oplus O)U, \text{ where } R^2 = T_1.$$

b) If $T \in \mathbb{R}^{n\times n}$ is a positive definite matrix, then there exist an orthogonal matrix U and a diagonal matrix D such that

$$T = U^T DU. \tag{4.2}$$

Also, there exists a unique matrix X such that $X^2 = T$ which is also an EP matrix, and which satisfies the following

$$X = U^T RU \text{ where } R^2 = D.$$

Theorem 4.1. [39] *Let $A \in \mathbb{C}^{n\times n}$ be a given matrix of index $k = \operatorname{ind}(A)$. The following two vectors, denoted by $\hat{x}_1$ and $\hat{x}_2$, are approximate solutions to the problem* (4.1)*:*

$$\begin{aligned}
\hat{x}_1 &= X^\dagger \left(A^{k+1}X^\dagger\right)^\dagger A^k \left(b + \frac{1}{2}AT^\dagger p_1\right) - \frac{1}{2}T^\dagger p_1 && (4.3)\\
&= X^\dagger \left(A^{k}X^\dagger\right)^\dagger A^D A^k \left(b + \frac{1}{2}AT^\dagger p_1\right) - \frac{1}{2}AT^\dagger p_1 && (4.4)\\
&= X^\dagger \left(A^D A^{k}X^\dagger\right)^\dagger A^D A^{k-1} \left(b + \frac{1}{2}AT^\dagger p_1\right) - \frac{1}{2}T^\dagger p_1, \ k \geq 1 && (4.5)\\
\hat{x}_2 &= X^\dagger \left(A^{k}X^\dagger\right)^D A^D A^k \left(b + \frac{1}{2}AT^\dagger p_1\right) - \frac{1}{2}AT^\dagger p_1, && (4.6)
\end{aligned}$$

where $X^2 = T$ and $p_1 = P_{\mathcal{R}(T)}(p)$.

Proof. In this case, $\mathcal{R}(T) = \mathcal{R}(T^*)$ and $\mathcal{N}(T)$ are complementary spaces. Therefore, an arbitrary vector $p \in \mathbb{C}^n$ can be decomposed as $p = p_1 + p_2 \in \mathcal{R}(T) \oplus \mathcal{N}(T)$ and $x \in \mathcal{N}(T)^\perp$. This implies $\langle x, p\rangle = \langle x, p_1\rangle$.

The rest of the proof is based on the principles from [5]. The vector $\hat{x}$ which minimizes $\Phi(x)$ also minimizes

$$\begin{aligned}
\Psi(x) &= \langle x, Tx\rangle + \langle x, p_1\rangle + \frac{1}{4}\langle T^\dagger p_1, p_1\rangle \\
&= \left\| Xx + \frac{1}{2}X^\dagger p_1 \right\|^2, \ p_1 \in \mathcal{R}(T).
\end{aligned}$$

Let us denote by O the zero block of an appropriate size. It is convenient to denote by

$$y = Xx + \frac{1}{2}X^\dagger p_1 = Xx + \frac{1}{2}U\left(R^{-1} \oplus O\right)U^T p_1.$$

Then it can be obtained $y, p_1 \in \mathcal{R}(X) = \mathcal{R}(T)$ and

$$\begin{aligned} x &= U\left(R^{-1} \oplus O\right) U^T y - \frac{1}{2} U\left(R^{-1} \oplus O\right) U^T \cdot U\left(R^{-1} \oplus O\right) U^T p_1 \\ &= X^\dagger y - \frac{1}{2} T^\dagger p_1, \\ & \quad x \in \mathcal{R}(T^\dagger) = \mathcal{R}(T^T) = \mathcal{R}(T) = \mathcal{N}(T)^\perp. \end{aligned} \tag{4.7}$$

Now, after the substitution of x defined by (4.7) into $A^{k+1}x = A^k b$, one can verify

$$A^{k+1} X^\dagger y = A^k b + \frac{1}{2} A^{k+1} T^\dagger p_1. \tag{4.8}$$

In the rest of the proof it is possible to use equation (4.8) and later apply (4.7) in order to derive the solutions (4.3) - (4.6).

First, it is well known that the the vector $\hat{x} = A^\dagger b$ represents the least-squares solution of minimal norm of the equation $Ax = b$. Therefore, by exploiting the best-approximate solution of (4.8), with respect to the variable y, and applying equation (4.7), we immediately obtain the solution $\hat{x}_1$ given by (4.3).

In order to verify (4.4), let us consider (4.8) in the form

$$A\left(A^k X^\dagger y\right) = A^k \left(b + \frac{1}{2} A T^\dagger p_1\right).$$

Since

$$A^k \left(b + \frac{1}{2} A T^\dagger p_1\right) \in \mathcal{R}(A^k)$$

and

$$A^k X^\dagger y \in \mathcal{R}(A^k)$$

the Drazin inverse solution of (4.8) is the unique that belongs to $\mathcal{R}(A^k)$. Therefore, the unique Drazin inverse solution of (4.8) is given by

$$A^k X^\dagger y = A^D A^k \left(b + \frac{1}{2} A T^\dagger p_1\right). \tag{4.9}$$

Finally, if we look at the best-approximate solution of the equation (4.9), we obtain

$$\hat{y} = \left(A^k X^\dagger\right)^\dagger A^D A^k \left(b + \frac{1}{2} A T^\dagger p_1\right)$$

and

$$\hat{x}_1 = X^\dagger \left(A^k X^\dagger\right)^D A^D A^k \left(b + \frac{1}{2} A T^\dagger p_1\right) - \frac{1}{2} A T^\dagger p_1,$$

which is just the desired representation (4.4).

The final part of the proof is a verification that the solutions (4.3) and (4.4) are identical. Let us remind that the solution (4.3) is derived by finding the minimizer x given by (4.7), under the constraint $A^{k+1}x = A^k b$. It is not difficult to verify that the solution given by (4.4), also, can be derived by finding the minimizer x given by (4.7), under the constraint

$A^k x = A^D A^k b$. Indeed, in this case, by finding the best-approximate solution $\hat{y}$ of the system

$$A^k X^\dagger y = A^D A^k b + \frac{1}{2} A^k T^\dagger p_1$$

we obtain

$$\hat{x}_1 = X^\dagger \left(A^k X^\dagger\right)^\dagger \left(A^D A^k b + \frac{1}{2} A^k T^\dagger p_1\right) - \frac{1}{2} T^\dagger p_1,$$

which coincides with (4.4). The equality of the representations (4.3) and (4.4) follows from the equivalence of the constraint sets S_D and S_{D_1}.

Going further, one can verify that (4.5) is the best-approximate solution of the system $A^D A^k y = A^D A^{k-1} b$. From $S_D = S_{D_2}$ follows the equality of (4.3) and (4.5) for $k = 1$. If we continue on this way, we will get the whole set of representations (4.5).

In order to derive the representation (4.6), we again consider (4.9). However, in this case instead of finding the best-approximate solution of the previous equation we look at its Drazin-inverse solution, which is given by

$$\hat{y} = \left(A^k X^\dagger\right)^D A^D A^k \left(b + \frac{1}{2} A T^\dagger p_1\right).$$

This is justified by the fact that both sides of the equation (4.9) belong to $\mathcal{R}(A^k)$. By substituting $\hat{y}$ from the last identity in (4.7), we obtain $\hat{x}_2$ in the form (4.6). ■

Corollary 4.1 gives representations of Theorem 4.1 in the particular case $p = 0$, which frequently appears in practical situations.

Corollary 4.1. [39] *Let $A \in \mathbb{C}^{n\times n}$ be a given matrix and let $k = \mathrm{ind}(A)$. The following two vectors, denoted by $\hat{x}_1$ and $\hat{x}_2$, are approximate solutions to the problem* (4.1), *in the case $p = 0$*

$$\hat{x}_1 = X^\dagger \left(A^{k+1} X^\dagger\right)^\dagger A^k b \quad (4.10)$$

$$= X^\dagger \left(A^k X^\dagger\right)^\dagger A^D A^k b \quad (4.11)$$

$$= X^\dagger \left(A^D A^k X^\dagger\right)^\dagger A^D A^{k-1} b,\ k \geq 1; \quad (4.12)$$

$$\hat{x}_2 = X^\dagger \left(A^k X^\dagger\right)^D A^D A^k b, \quad (4.13)$$

where $X^2 = T$.

Corollary 4.2. [39] *For a given matrix A, let $k = \mathrm{ind}(A)$. The solution $\hat{x}_1$ defined in Theorem 4.1 can be given with the following representation*

$$\hat{x}_1 = U^T V^\dagger \left(A^{k+1} U^T V^\dagger\right)^\dagger A^k \left(b + \frac{1}{2} A U^T (V^\dagger)^2 U p_1\right) - \frac{1}{2} U^T (V^\dagger)^2 U p_1 \quad (4.14)$$

$$= U^T V^\dagger \left(A^k U^T V^\dagger\right)^\dagger A^D A^k \left(b + \frac{1}{2} A U^T (V^\dagger)^2 U p_1\right)$$

$$- \frac{1}{2} A U^T (V^\dagger)^2 U p_1$$

Table 1. Minimal values for indicative matrix sizes

Matrix size	$\min_{\{1\}-inv} \Phi(\breve{x})$	$\min_D \Phi(\breve{x})$
10	$1.5123e+06$	13.6404
50	$1.4548e+15$	160.8436
100	$3.9246e+09$	$1.524.4161$
200	$9.3852e+20$	$2.1914e+04$

$$
\begin{aligned}
&= U^T V^\dagger \left(A^D A^k U^T V^\dagger\right)^\dagger A^D A^{k-1}\left(b+\frac{1}{2}AU^T(V^\dagger)^2 U p_1\right)\\
&-\frac{1}{2}U^T(V^\dagger)^2 U p_1, \\
&k \geq 1
\end{aligned}
\tag{4.15}
$$

where $X^2 = T$, $V = R \oplus O$, $S = T_1 \oplus O$ and $p_1 = P_{\mathcal{R}(T)}(p)$.

Proof. Using the notation above, it is obvious that $X^\dagger = U^T V^\dagger U$ and $T^\dagger = U^T(V^\dagger)^2 U$. Let us denote $U_1 = A^{k+1}U^T V^\dagger$ and $U_2 = U$. Noting that U is an orthogonal matrix, from which $U^\dagger = U^{-1} = U^T$ holds, we conclude that $U_2 U_2^T = U_2^T U_2 = I$. From here we immediately see that the Bouldin's theorem for the reverse-order law of the matrices U_1 and U_2 holds, which is sufficient to complete the proof, by using Theorem 4.1. ∎

Example 4.1. *In Table 1, we present the minimizing values $\Phi(\breve{x})$ for random tested singular matrices A and positive semidefinite matrices T. For comparison reasons we have calculated the minimizing values with the restricted $\{1\}$-inverse from* [38] *and with the proposed T-restricted weighted Drazin inverse. The notation $\min_{\{1\}-inv} \Phi(\breve{x})$ refers to the restricted $\{1\}$-inverse proposed in* [38] *and the notation $\min_D \Phi(\breve{x})$ means the minimum defined by the T-restricted weighted Drazin inverse.*

4.2. The Symmetric Positive Definite Case

In this subsection it is assumed that $T \in \mathbb{R}^{n\times n}$ is a positive matrix. Let $A \in \mathbb{R}^{n\times n}$ be such that $\operatorname{ind}(A) = k$ and $x, b \in \mathbb{R}^n$. Since $\mathcal{N}(T) = \{0\}$, we have that $\Phi(x) = a$ if and only if $x = 0$. So, in what follows we suppose that $x \neq 0$ and we will investigate the quadratic minimization problem

$$
\text{minimize } \Phi(x),\ x \in S_D \cap \{0\}^C, \tag{4.16}
$$

In this section, it is assumed that U, D, X and R represent the notions from Proposition 4.2 part b).

Corollary 4.3. [39] *The following two vectors, denoted by $\hat{x}_1$ and $\hat{x}_2$, are solutions to the* *problem* (4.16)*:*

$$\hat{x}_1 = T^{-\frac{1}{2}}\left(A^{k+1}T^{-\frac{1}{2}}\right)^{\dagger} A^k\left(b+\frac{1}{2}AT^{-1}p\right)-\frac{1}{2}T^{-1}p \tag{4.17}$$

$$= T^{-\frac{1}{2}}\left(A^{k}T^{-\frac{1}{2}}\right)^{\dagger} A^k\left(A^Db+\frac{1}{2}T^{-1}p\right)-\frac{1}{2}T^{-1}p \tag{4.18}$$

$$= T^{-\frac{1}{2}}\left(A^DA^{k}T^{-\frac{1}{2}}\right)^{\dagger} A^DA^{k-1}\left(b+\frac{1}{2}AT^{-1}p\right)-\frac{1}{2}T^{-1}p,\ k\geq 1 \tag{4.19}$$

$$\hat{x}_2 = T^{-\frac{1}{2}}\left(A^{k}T^{-\frac{1}{2}}\right)^{D} A^DA^k\left(b+\frac{1}{2}AT^{-1}p\right)-\frac{1}{2}AT^{-1}p. \tag{4.20}$$

Proof. The matrix X defined by $X^2 = T$ is invertible. It is equal to $X = U^TD^{\frac{1}{2}}U$ and its inverse is

$$X^{-1} = U^TD^{-\frac{1}{2}}U = T^{-\frac{1}{2}}.$$

The rest of the proof follows from Theorem 4.1. ■

Corollary 4.4. [39] *The following two vectors, denoted by $\hat{x}_1$ and $\hat{x}_2$, are solutions to the problem* (4.16)*, in the case $p = 0$:*

$$\hat{x}_1 = T^{-\frac{1}{2}}\left(A^{k+1}T^{-\frac{1}{2}}\right)^{\dagger} A^kb \tag{4.21}$$

$$= T^{-\frac{1}{2}}\left(A^{k}T^{-\frac{1}{2}}\right)^{\dagger} A^DA^kb \tag{4.22}$$

$$= T^{-\frac{1}{2}}\left(A^DA^{k}T^{-\frac{1}{2}}\right)^{\dagger} A^DA^{k-1}b,\ k\geq 1 \tag{4.23}$$

$$\hat{x}_2 = T^{-\frac{1}{2}}\left(A^{k}T^{-\frac{1}{2}}\right)^{D} A^DA^kb. \tag{4.24}$$

Corollary 4.5. [39] *The solution $\hat{x}_1$ defined in Corollary 4.3 can be given with the following representation*

$$\hat{x}_1 = U^TD^{-\frac{1}{2}}\left(A^{k+1}U^TD^{-\frac{1}{2}}\right)^{\dagger}\left(A^kb+\frac{1}{2}A^{k+1}U^TD^{-1}Up\right)-\frac{1}{2}U^TD^{-1}Up \tag{4.25}$$

$$= U^TD^{-\frac{1}{2}}\left(A^{k}U^TD^{-\frac{1}{2}}\right)^{\dagger}\left(A^DA^kb+\frac{1}{2}A^{k}U^TD^{-1}Up\right)-\frac{1}{2}U^TD^{-1}Up \tag{4.26}$$

$$= U^TD^{-\frac{1}{2}}\left(A^DA^{k}U^TD^{-\frac{1}{2}}\right)^{\dagger}\left(A^DA^{k-1}b+\frac{1}{2}A^DA^{k}U^TD^{-1}Up\right)-\frac{1}{2}U^TD^{-1}Up,\quad k\geq 1, \tag{4.27}$$

where $T^{-1} = U^TD^{-1}U$.

Corollary 4.6. [39] *The solution $\hat{x}_1$ defined in Corollary 4.4 can be given with the following representation*

$$\hat{x}_1 = U^TD^{-\frac{1}{2}}\left(A^{k+1}U^TD^{-\frac{1}{2}}\right)^{\dagger} A^kb \tag{4.28}$$

$$= U^TD^{-\frac{1}{2}}\left(A^{k}U^TD^{-\frac{1}{2}}\right)^{\dagger} A^DA^kb \tag{4.29}$$

$$= U^TD^{-\frac{1}{2}}\left(A^DA^{k}U^TD^{-\frac{1}{2}}\right)^{\dagger} A^DA^{k-1}b,\ k\geq 1. \tag{4.30}$$

Remark 4.1. *The solution represented by* (4.17) *and* (4.15) *could be derived from the corresponding approximations corresponding to the restrictions $x \in S$ (Proposition 2.3 and Proposition 2.4), by simply replacing A with A^{k+1} and b with $A^k b$. However, with our approach we introduce several additional representations. Moreover, the solutions* (4.17) *and* (4.15) *are derived from more general results derived for the positive semidefinite case.*

In the following theorem we show that the solution of the constrained quadratic minimization can be represented by an outer inverses with prescribed range and null space, i.e. as a certain $A^{(2)}_{T,S}$-inverse solution.

Theorem 4.2. [39] *The solution of the problem* (4.16) *given by* (4.28) *can be represented as $A^{(2)}_{T,S}$-inverse solution, i.e.,*

$$\hat{x}_1 = A^{(2)}_{\bar{T}\cdot\mathcal{R}((A^k)^T),\mathcal{N}(A^k)} b,$$

where

$$\bar{T} = U^T D^{-\frac{1}{2}} (D^{-\frac{1}{2}})^T U = T^{-1} = U^T D^{-1} U.$$

Proof. Let us denote

$$X = U^T D^{-\frac{1}{2}} \left(A^k\, A\, U^T D^{-\frac{1}{2}} \right)^{\dagger} A^k.$$

Consequently, the solution (4.15) can be represented in the form Xb. Simple algebra reveals that matrix X satisfies the equation $XAX = X$, i.e., it is $\{2\}$-inverse of A.

Next we will determine the range and the null-space of X. Let us notice that

$$\mathcal{R}(A^{k+1} U^T D^{-\frac{1}{2}}) = \mathcal{R}(A^{k+1}) = \mathcal{R}(A^k). \tag{4.31}$$

By using the singular-value decomposition of the matrix $A^{k+1} U^T D^{-\frac{1}{2}}$ we see that

$$A^{k+1} U^T D^{-\frac{1}{2}} \sim \begin{bmatrix} A_1 & 0 \\ 0 & 0 \end{bmatrix} : \begin{bmatrix} \mathcal{R}\left(\left(A^{k+1} U^T D^{-\frac{1}{2}}\right)^T\right) \\ \mathcal{N}(A^{k+1} U^T D^{-\frac{1}{2}}) \end{bmatrix} \to \begin{bmatrix} \mathcal{R}(A^{k+1} U^T D^{-\frac{1}{2}}) \\ \mathcal{N}\left(\left(A^{k+1} U^T D^{-\frac{1}{2}}\right)^T\right) \end{bmatrix}, \tag{4.32}$$

where A_1 is an invertible matrix, from which, by using (4.31), follows that

$$(A^{k+1} U^T D^{-\frac{1}{2}})^{\dagger} \sim \begin{bmatrix} (A_1)^{-1} & 0 \\ 0 & 0 \end{bmatrix} : \begin{bmatrix} \mathcal{R}(A^k) \\ \mathcal{N}\left(\left(A^{k+1} U^T D^{-\frac{1}{2}}\right)^T\right) \end{bmatrix} \to \begin{bmatrix} \mathcal{R}\left(\left(A^{k+1} U^T D^{-\frac{1}{2}}\right)^T\right) \\ \mathcal{N}\left(A^{k+1} U^T D^{-\frac{1}{2}}\right) \end{bmatrix}. \tag{4.33}$$

From here, it is clear that

$$\mathcal{R}((A^{k+1} U^T D^{-\frac{1}{2}})^{\dagger} A^k) = \mathcal{R}((A^{k+1} U^T D^{-\frac{1}{2}})^T) = (D^{-\frac{1}{2}})^T U \cdot \mathcal{R}((A^k)^T).$$

Finally,

$$\begin{aligned} \mathcal{R}(X) &= \mathcal{R}\left(U^T D^{-\frac{1}{2}} \left(A^{k+1} U^T D^{-\frac{1}{2}} \right)^{\dagger} A^k \right) = U^T D^{-\frac{1}{2}} \cdot \mathcal{R}\left(\left(A^{k+1} U^T D^{-\frac{1}{2}} \right)^{\dagger} A^k \right) \\ &= U^T D^{-\frac{1}{2}} (D^{-\frac{1}{2}})^T U \cdot \mathcal{R}((A^k)^T) \end{aligned}$$

If we denote $\bar{D} = D^{-\frac{1}{2}}(D^{-\frac{1}{2}})^T$ and $\bar{T} = U^T\bar{D}U$, we obtain $\mathcal{R}(X) = \bar{T}\cdot\mathcal{R}((A^k)^T)$. Clearly if $T \in \mathbb{R}^{n\times n}$, then $(D^{-\frac{1}{2}})^T = D^{-\frac{1}{2}}$ and $\bar{D} = D^{-1}$.

In order to find the null-space of X we proceed as follows.

Clearly, $\mathcal{N}(A^k) \subset \mathcal{N}((A^{k+1}\,U^T D^{-\frac{1}{2}})^\dagger A^k)$. On the other hand, let us suppose that $x \in \mathcal{N}((A^{k+1}\,U^T D^{-\frac{1}{2}})^\dagger A^k)$. If we assume that $x \notin \mathcal{N}(A^k)$ then $A^k x \neq 0$ and $A^k x \in \mathcal{R}(A^k)$, since $\mathcal{R}(A^k) \oplus \mathcal{N}((A^k)^T) = \mathbb{R}^{n\times n}$ it follows that $A^k x \notin \mathcal{N}((A^k)^T) = \mathcal{N}((A^k\,A\,U^T D^{-\frac{1}{2}})^\dagger)$ which is contradiction. So, in this case $x \in \mathcal{N}(A^k)$ must hold, which completes the proof. ■

4.3. Definition of the Restricted Weighted Drazin Inverse

From the results presented in Subsection 4.1, we can see that a new kind of inverse, similar to the restricted weighted generalized inverse, can be defined:

Definition 4.1. [40] *Let $T \in \mathbb{R}^{n\times n}$ be a positive semidefinite Hermitian matrix and let $A \in \mathbb{R}^{n\times n}$ be singular. Then the $n \times n$ matrix*

$$\begin{aligned}\hat{A}^D_{A^k,T} &:= \left(T^{\frac{1}{2}}\right)^\dagger \left(A^{k+1}\left(T^{\frac{1}{2}}\right)^\dagger\right)^\dagger A^k \\ &= \left(T^\dagger\right)^{\frac{1}{2}} \left(A^{k+1}\left(T^\dagger\right)^{\frac{1}{2}}\right)^\dagger A^k\end{aligned} \tag{4.34}$$

is called the T-restricted weighted Drazin inverse of A.

The vector $\hat{u}^D = \hat{A}^D_{A^k,T}b$ is a minimal T semi-norm solution of the consistent equation $A^{k+1}x = A^k b$, restricted on $\mathcal{N}(T)^\perp = \mathcal{R}(T) = \mathcal{R}(T^*)$.

Example 4.2. *Let $\mathcal{H} = \mathbb{R}^3$, the matrix*

$$A = \begin{bmatrix} 1 & 0 & 1 \\ 0 & 2 & -1 \\ 3 & -4 & 5 \end{bmatrix}$$

and the positive semidefinite matrix

$$T = \begin{bmatrix} 2 & 3 & -6 \\ 3 & 45 & -9 \\ -6 & -9 & 18 \end{bmatrix}$$

In this case one can verify $\operatorname{ind}(A) = 1$.

The equation is $Ax = b$ where $b = \begin{bmatrix} 4 \\ 1 \\ 9 \end{bmatrix}$, which is inconsistent.

Therefore, we will use as a constraint the equation $A^2x = Ab$ which is the Drazin normal equation:

$$\begin{bmatrix} 4 & -4 & 6 \\ -3 & 8 & -7 \\ 18 & -28 & 32 \end{bmatrix} \vec{x} = \begin{bmatrix} 13 \\ -7 \\ 53 \end{bmatrix}.$$

We will compute the T-restricted weighted Drazin inverse of A, $\hat{A}^D_{A^k,T}$ using Definition 4.1. We have that

$$\hat{A}^D_{A^k,T} = \begin{bmatrix} -0.2 & -0.2 & -0.1 \\ 0.45 & 0.7 & 0.1 \\ 0.6 & 0.6 & 0.3 \end{bmatrix}.$$

The vector

$$\hat{u} = \hat{A}^D_{A^k,T} b = \begin{bmatrix} -1.9 \\ 3.4 \\ 5.7 \end{bmatrix}$$

is a minimal T semi-norm solution of the consistent equation $A^2x = Ab$, restricted on $\mathcal{N}(T)^\perp = \mathcal{R}(T) = \mathcal{R}(T^*)$.

The vectors in $\mathcal{N}(T)$ have the form $v = (x, 0, \frac{x}{3})^T, x \in \mathbb{R}$, therefore the vectors $u \in \mathcal{R}(T)$ have the form $u = (x, y, -3x)^T, x, y \in \mathbb{R}$, so the solution $\hat{u}$ has the expected form.

In addition, we have that the normal Drazin equation $A^2\hat{u} = Ab$ is also satisfied.

In the present survey paper we derive an exact minimal property of the generalized inverse $\hat{A}^D_{A^k,T}$. This result is given in Corollary 4.7. It is well known that the vec-operator applied on a matrix A, denoted by $\mathrm{vec}(A)$, stacks the columns into a vector.

Corollary 4.7. *Let $A \in \mathbb{C}^{n\times n}$ is a given matrix satisfying* $\mathrm{ind}(A) = k$. *The matrix $\hat{A}^D_{A^k,T}$, defined by* (4.34), *is a minimizer of the minimization problem*

$$\min \Psi(Z) = \langle Z, TZ\rangle = \mathrm{Tr}(Z^T TZ) = \mathrm{vec}(Z^T)\mathrm{vec}(TZ),$$

where T is a positive semidefinite $n \times n$ matrix and $Z \in \mathbb{R}^{n\times n}$, subject to the constraint set

$$Z \in \Omega_{A^k}, \quad \Omega_{A^k} = \{Z : \ Z \in \mathbb{R}^{n\times n}, \ A^{k+1}Z = A^k, \ A \in \mathbb{R}^{n\times n}_r\}.$$

4.3.1. Properties of the T-Restricted Weighted Drazin Inverse

A natural question to ask is: what kind of generalized inverses are $\hat{A}^D_{A^k,T}$ and $\hat{A}^\dagger_{I_m,T}$? It is easy to verify that both of these generalized inverses satisfy the Penrose equation (2).

Proposition 4.3. [40] *The generalized inverses $\hat{A}^D_{A^k,T}$ and $\hat{A}^\dagger_{I_m,T}$ are outer inverses of A, for arbitrarily chosen matrix T.*

Proof. The proof can be verified by replacing definitions (4.34) and (3.1) in the matrix equation $XAX = X$. ■

Therefore, a new application of outer inverses is presented in this work: application in the minimization of quadratic forms under linear constraints.

We examine further properties of the $\hat{A}^D_{A^k,T}$ inverse. In what follows P_A denotes the orthogonal projector onto the range of A.

Proposition 4.4. [40] *The matrix $\hat{A}^D_{A^k,T}$ has the following properties:*

(i) $\hat{A}^D_{A^k,T} A(T^\dagger)^{\frac{1}{2}} = (T^\dagger)^{\frac{1}{2}} P_{(A^{k+1}T)^*}$.

(ii) $A^{k+1}\hat{A}^D_{A^k,T} = A^k$.

(iii) $A^{k+1}\hat{A}^D_{A^k,T}AT = A^{k+1}T$.

(iv) $A^k\hat{A}^D_{A^k,T}A = A^k$.

Proof.

(i) $\hat{A}^D_{A^k,T}A(T^\dagger)^{\frac{1}{2}} = (T^\dagger)^{\frac{1}{2}}\left(A^{k+1}(T^\dagger)^{\frac{1}{2}}\right)^\dagger A^kA(T^\dagger)^{\frac{1}{2}} = (T^\dagger)^{\frac{1}{2}}P_{(A^{k+1}T)^*}$.

(ii) We have that

$$A^{k+1}\hat{A}^D_{A^k,T} = A^{k+1}(T^\dagger)^{\frac{1}{2}}\left(A^{k+1}(T^\dagger)^{\frac{1}{2}}\right)^\dagger A^k = P_{A^{k+1}T}A^k.$$

Moreover, since the minimization takes place for the vectors $x \in \mathcal{N}(T)^\perp = \mathcal{R}(T)$, we have that $x = Tu$ for some $u \in \mathcal{H}$ and therefore $P_{A^{k+1}T}x = P_{A^{k+1}T}Tu$. So, the orthogonal projector $P_{A^{k+1}T}$ which projects all vectors on the range of the matrix $A^{k+1}T$ is equal to the orthogonal projector onto the range of A^{k+1} for all the vectors $x = Tu$. (It holds that $P_T(Tx) = P_Tx = Tx$).

It is easy to see that every vector $w \in \mathcal{H}$ can be written in the form $w = x + v$ with $x \in \mathcal{N}(T)^\perp$ and $v \in \mathcal{N}(T)$. Therefore, $Tw = Tx$.

Consequently, $P_{A^{k+1}T}w = P_{A^{k+1}T}x = P_{A^{k+1}T}Tu = P_{A^{k+1}}Tu$. In other words $P_{A^{k+1}T}x = P_{A^{k+1}}x$ for all $x \in \mathcal{N}(T)^\perp$ and for every $w \in \mathcal{H}$ of the form $w = x + v$.

So, using the relation $A^{k+1}\hat{A}^D_{A^k,T} = P_{A^{k+1}T}A^k$, we have that

$$A^{k+1}\hat{A}^D_{A^k,T}w = P_{A^{k+1}T}A^kw = P_{A^{k+1}T}A^kx = P_{A^{k+1}}A^kx.$$

Since $\mathcal{R}(A^{k+1}) = \mathcal{R}(A^k)$ we get $P_{A^{k+1}}A^k = A^k$.

(iii) $A^{k+1}\hat{A}^D_{A^k,T}AT = P_{A^{k+1}T}A^{k+1}T = A^{k+1}T$.

(iv) We have from (ii) that $A^{k+1}\hat{A}^D_{A^k,T}A = P_{A^{k+1}T}A^{k+1}$ and so, using the same idea again, we have $A^{k+1}\hat{A}^D_{A^k,T}A = A^{k+1}$. By multiplying this equality from the left with A^D we get the result.

■

We also investigate relations between $\hat{A}^D_{A^k,T}$ and the V-orthogonal projector $P_{X:V}$, defined in (3.3). The V-orthogonal projector becomes unique under the assumption $\text{rank}(VX) = \text{rank}(X)$, as it is stated in Proposition 3.4.

Theorem 4.3. [40] *Let $T \in \mathbb{R}^{n\times n}$ be a positive semidefinite matrix and $A \in \mathbb{R}^{n\times n}$. If*

$$\mathcal{N}(T) \cap \mathcal{R}(A^k) = \{0\}$$

then

$$\hat{A}^D_{A^k,T}A = P^*_{(A^{k+1})^*:T^\dagger}.$$

Proof. When $\mathcal{N}(T) \cap \mathcal{R}(A^k) = \{0\}$ then it holds that

$$\text{rank}(T^\dagger A^{k+1}) = \text{rank}(A^{k+1}),$$

so following Proposition 3.4 it is not difficult to verify that

$$P_{(A^{k+1})^*:T^\dagger} = \left(A^{k+1}\right)^* \left((T^\dagger)^{\frac{1}{2}} \left(A^{k+1}\right)^*\right)^\dagger (T^\dagger)^{\frac{1}{2}}.$$

Therefore,

$$P^*_{(A^{k+1})^*:T^\dagger} = (T^\dagger)^{\frac{1}{2}} \left(A^{k+1} (T^\dagger)^{\frac{1}{2}}\right)^\dagger A^k A,$$

which completes the proof. ■

How is it possible to compute the generalized inverse defined in (4.34)? Straight computation by means of (4.34) requires two sequential computations of the Moore-Penrose inverse. This is a time consuming job which is sensitive to the effects of rounding off errors. To avoid this difficulty, we define an alternative representation of the matrix expression (4.34) and define corresponding algorithm for its computation.

Theorem 4.4. *If* $A \in \mathbb{R}^{n\times n}$ *and* $\text{ind}(A) = k$*, then*

$$\hat{A}^D_{A^k,T} = A^{(2)}_{U,V},\ U = \mathcal{R}\left(T^\dagger (A^*)^{k+1} A^k\right), V = \mathcal{N}\left(T^\dagger (A^*)^{k+1} A^k\right).$$

Proof. Since $\text{ind}(A) = k = \text{ind}(A^*)$, it follows

$$\mathcal{N}\left((A^*)^{k+1}\right) = \mathcal{N}\left((A^*)^k\right),\ \mathcal{R}\left((A^*)^{k+1}\right) = \mathcal{R}\left((A^*)^k\right).$$

Also, it is possible to verify $\mathcal{R}\left(T^\dagger (A^*)^{k+1}\right) = \mathcal{R}\left(T^\dagger (A^*)^k\right)$.
But, $\mathcal{R}\left(\hat{A}^D_{A^k,T}\right) = \mathcal{R}(T^\dagger (A^k)^*) = U$ and

$$\begin{aligned}\mathcal{N}\left(\left(A^{k+1}\left(T^{\frac{1}{2}}\right)^\dagger\right)^\dagger A^k\right) &\subseteq \mathcal{N}(\hat{A}^D_{A^k,T}) \subseteq \left(A^{k+1}\left(T^{\frac{1}{2}}\right)^\dagger \left(A^{k+1}\left(T^{\frac{1}{2}}\right)^\dagger\right)^\dagger A^k\right)\\ &= \mathcal{N}\left(\left(A^{k+1}\left(T^{\frac{1}{2}}\right)^\dagger\right)^\dagger A^k\right).\end{aligned}$$

Hence,

$$\mathcal{N}(\hat{A}^D_{A^k,T}) = \mathcal{N}\left(\left(A^{k+1}\left(T^{\frac{1}{2}}\right)^\dagger\right)^\dagger A^k\right) = \mathcal{N}\left(\left(T^{\frac{1}{2}}\right)^\dagger (A^*)^{k+1} A^k\right) = V.$$

Evidently, $\hat{A}^D_{A^k,T}$ is $A^{(2)}_{U,V}$ with previously described U and V. ■

Corollary 3.3 gives a full-rank representation of the T-minimal G-constrained inverse and can be useful in its numerical computations.

Corollary 4.8. *Let* A *and* T *satisfy the conditions of Definition 4.1. If* $T^\dagger A^* G^* G = PQ$ *is an arbitrary full-rank decomposition of* $T^\dagger (A^*)^{k+1} A^k$ *then:*

(1) QAP *is an invertible complex matrix;*

(2) $\hat{A}^{(2)}_{G,T} = P(QAP)^{-1}Q$.

Full-rank representation of the outer inverse $\hat{A}^{(2)}_{G,T}$ defined in Corollary 4.8 allows us to define Algorithm 4.1. for its computation. Full-rank factorization of $T^{\dagger}(A^*)^{k+1}A^k$ is defined using its QR decomposition.

Algorithm 4.1. Computing the $\hat{A}^D_{A^k,T}$ outer inverse.

Require: The matrix A of dimensions $m \times n$ and of rank r and positive semidefinite $n \times n$ matrix T.

1: Choose arbitrary, but fixed, $n \times m$ matrix G of rank $0 < s \leq r$.
2: Compute $W = T^{\dagger}(A^*)^{k+1}A^k$.
3: Compute the QR decomposition of the matrix W in the form (3.7), where P is an $m \times m$ permutation matrix, $Q \in \mathbb{C}^{n\times n}$, $Q^*Q = I_n$ and $R \in \mathbb{C}^{n\times m}$ is an upper trapezoidal matrix.
4: Compute $t = \text{rank}(T^{\dagger}(A^*)^{k+1}A^k)$.
5: Assume that Q and R are partitioned as in . Generate the full–rank decomposition of the matrix W by means of

$$W = Q_1(R_1P^*). \tag{4.1}$$

6: Solve the matrix equation $R_1P^*AQ_1X = R_1P^*$.
7: Compute the output $\hat{A}^D_{A^k,T} = Q_1X$.

Numerical results on test matrices from [52, 53] are presented in [40].

M. Drazin in [55], introduced the notion of a (b, c)−inverse of an element a in a semigroup:

Definition 4.2. *Let S be any semigroup and let $a, b, c, y \in S$. Then we shall call y a (b, c)−inverse of a if both of the following two relations hold:*

1. $y \in (bSy) \cap (ySc)$.
2. $yab = b$ *and* $cay = c$.

In what follows we will show that $\hat{A}^D_{A^k,T}$ is a particular (b, c)−inverse of A.

Proposition 4.5. [40] *Let S be the semigroup of all real square matrices and let also $b = \hat{A}^D_{A^k,T}$ and $c = A^k$. Then, $\hat{A}^D_{A^k,T}$ is a (b, c)−inverse of A.*

Proof. The proof comes easily from propositions 4.3 and 4.4. ■

We will now make use of Theorem 3.2, in order to generalize its results. In what follows, T will denote a positive semidefinite matrix having a canonical form $T = U(A_1 \oplus 0)U^*$, R is the unique solution of the equation $R^2 = A_1$ and we can define

$$V = \begin{bmatrix} R & 0 \\ 0 & 0 \end{bmatrix} = R \oplus 0,$$

so that

$$V^{\dagger} = \begin{bmatrix} R^{-1} & 0 \\ 0 & 0 \end{bmatrix} = R^{-1} \oplus 0.$$

Note that in the following theorem, Theorem 4.5, we avoid the constraint $x \in \mathcal{R}(T)$, and hence we can find a more general set of minimizing vectors belonging only to the set S_D.

Theorem 4.5. [40] *Let $T \in \mathbb{R}^{n\times n}$ be a positive semidefinite matrix, $A \in \mathbb{R}^{n\times n}$ be singular and consider the non consistent equation $Ax = b$. By considering the set of the Drazin normal equation* (2.11) *we have that the problem of minimizing $\Phi(x) = \langle x, Tx\rangle$, $x \in \mathcal{S}_D$ has a set of solutions defined by*

$$\begin{aligned}\hat{x} = UV^{(1,3)} &\left(\left(A^{k+1}UV^{(1,3)}\right)^{(1,4)} A^k \left(b - AUw + AV^{(1,3)}w\right)\right)\\ &+ U(I - V^{(1,3)}V)w,\end{aligned} \tag{4.2}$$

where $w \in \mathbb{R}^n$ is an arbitrary vector.

Proof. It is not difficult to verify

$$\langle x, Tx\rangle = \langle x, U(A_1 \oplus 0)U^*x\rangle = \langle U^*x, (A_1 \oplus 0)U^*x\rangle = \left\langle U^*x, (R^2 \oplus 0)U^*x\right\rangle.$$

We have that $U^*x = (x_1, x_2)$, $x_1 \in \mathcal{R}(T)$, $x_2 \in \mathcal{N}(T)$ and

$$\langle U^*x, (A_1 \oplus 0)U^*x\rangle = \langle x_1, A_1x_1\rangle.$$

Therefore

$$\langle x, Tx\rangle = \langle (R \oplus 0)U^*x, (R \oplus 0)U^*x\rangle = \langle Rx_1, Rx_1\rangle = \langle y, y\rangle,$$

where $y = Rx_1$, with $x_1 \in \mathcal{N}(T)^{\perp}$. Therefore, the problem of minimizing $\langle x, Tx\rangle$ is equivalent to minimizing $\| y \|^2$ where $y = Rx_1 = VU^*x$.

So, we can have the set of least squares solutions using a $\{1,3\}$-inverse:

$$x = UV^{(1,3)}y + U(I - V^{(1,3)}V)w, \tag{4.3}$$

where w is an arbitrary vector.

Since $A^{k+1}x = A^k b$ we have $A^{k+1}(UV^{(1,3)}y + U(I - V^{(1,3)}V)w) = A^k b$. Therefore

$$A^{k+1}UV^{(1,3)}y + A^{k+1}U(I - V^{(1,3)}V)w = A^k b$$

and the minimal norm solution $\hat{y}$ can be found using a $\{1,4\}$-inverse this time:

$$\hat{y} = (A^{k+1}UV^{(1,3)})^{(1,4)}A^k(b - AUw + AV^{(1,3)}Vw).$$

Therefore, applying (4.3) it is not difficult to verify that $\hat{x}$ is defined by (4.2). ■

Corollary 4.9. *If $A \in \mathbb{R}^{n\times n}$ is an arbitrary matrix such that* $\mathrm{ind}(A) = k$ *and $T \in \mathbb{R}^{n\times n}$ is a positive semidefinite matrix, then $\hat{A}^D_{A^k,T}$ is a $\left(T^\dagger(A^*)^{k+1}A^k, T^\dagger(A^*)^{k+1}A^k\right)$-inverse of A.*

4.3.2. Properties in the Symmetric Positive Definite Case

Let T be a symmetric positive definite matrix. Then, there exist an orthogonal matrix U and a diagonal matrix D such that the decomposition $T = U^T DU$ holds. Let $D^{\frac{1}{2}}$ be the positive solution of the equation $R^2 = D$, and let $D^{-\frac{1}{2}}$ denotes the matrix $(D^{\frac{1}{2}})^{-1} = R^{-1}$.

Since T is positive definite matrix the existence of the matrices T^{-1} and R^{-1} is ensured. Similarly, by $X^{-\frac{1}{2}}$ we denote the matrix $(T^{\frac{1}{2}})^{-1} = X^{-1}$.

One solution of the matrix equation $X^2 = T$ is equal to $X = U^T D^{\frac{1}{2}} U$, where $D^{\frac{1}{2}}$ is a diagonal matrix in which each element along the diagonal is a square root of the corresponding element of D.

Let us consider the function $\Phi(x)$. Since $\mathcal{N}(T) = \{0\}$, we have that $\Phi(x) = a$ if and only if $x = 0$. So, further in the paper we suppose that $x \neq 0$.

Corollary 4.10. [40] *Let T be a positive definite matrix which satisfies decomposition of the form* (4.2). *The following two vectors, denoted by $\hat{x}_1$ and $\hat{x}_2$, are solutions to the problem* (2.12) *with T being a symmetric positive definite matrix*

$$\begin{aligned}\hat{x}_1 &= T^{-\frac{1}{2}} \left(A^{k+1} T^{-\frac{1}{2}}\right)^{\dagger} A^k \left(b + \frac{1}{2} A T^{-1} p\right) - \frac{1}{2} T^{-1} p && (4.4)\\ &= T^{-\frac{1}{2}} \left(A^{k} T^{-\frac{1}{2}}\right)^{\dagger} A^k \left(A^D b + \frac{1}{2} A T^{-1} p\right) - \frac{1}{2} T^{-1} p && (4.5)\\ &= T^{-\frac{1}{2}} \left(A^D A^{k} T^{-\frac{1}{2}}\right)^{\dagger} A^{k-1} \left(A^D b + \frac{1}{2} A T^{-1} p\right) - \frac{1}{2} T^{-1} p,\ k \geq 1 && (4.6)\\ \hat{x}_2 &= T^{-\frac{1}{2}} \left(A^{k} T^{-\frac{1}{2}}\right)^{D} A^D A^k \left(b + \frac{1}{2} A T^{-1} p\right) - \frac{1}{2} A T^{-1} p. && (4.7)\end{aligned}$$

Proof. The matrix X defined by $X^2 = T$ is invertible. It is equal to $X = U^T D^{\frac{1}{2}} U$ and its inverse is $X^{-1} = U^T D^{-\frac{1}{2}} U = T^{-\frac{1}{2}}$. The rest of the proof follows from Theorem 4.1. ■

Corollary 4.11. [40] *The following two vectors, denoted by $\hat{x}_1$ and $\hat{x}_2$, are solutions to the problem* (2.12) *with T being a positive definite matrix for which the decomposition of the form* (4.2) *holds, in the case $p = 0, a = 0$.*

$$\begin{aligned}\hat{x}_1 &= T^{-\frac{1}{2}} (A^{k+1} T^{-\frac{1}{2}})^{\dagger} A^k b && (4.8)\\ &= T^{-\frac{1}{2}} \left(A^{k} T^{-\frac{1}{2}}\right)^{\dagger} A^D A^k b && (4.9)\\ &= T^{-\frac{1}{2}} \left(A^D A^{k} T^{-\frac{1}{2}}\right)^{\dagger} A^D A^{k-1} b,\ k \geq 1 && (4.10)\\ \hat{x}_2 &= T^{-\frac{1}{2}} \left(A^{k} T^{-\frac{1}{2}}\right)^{D} A^D A^k b. && (4.11)\end{aligned}$$

Theorem 4.6. [40] *Under the assumptions of Corollary 4.10 the solution $\hat{x}_1$ of the problem* (2.12) *with T being a positive definite matrix for which decomposition of the form* (4.2) *holds, has the following equivalent representations:*

$$\begin{aligned}\hat{x}_1 &= U^T D^{-\frac{1}{2}} \left(A^{k+1} U^T D^{-\frac{1}{2}}\right)^{\dagger} \left(A^k b + \frac{1}{2} A^{k+1} U^T D^{-1} U p\right) - \frac{1}{2} U^T D^{-1} U p && (4.12)\\ &= U^T D^{-\frac{1}{2}} \left(A^{k} U^T D^{-\frac{1}{2}}\right)^{\dagger} \left(A^D A^k b + \frac{1}{2} A^{k} U^T D^{-1} U p\right) - \frac{1}{2} U^T D^{-1} U p && (4.13)\\ &= U^T D^{-\frac{1}{2}} \left(A^D A^{k} U^T D^{-\frac{1}{2}}\right)^{\dagger} \left(A^D A^{k-1} b + \frac{1}{2} A^{k} U^T D^{-1} U p\right) - \frac{1}{2} U^T D^{-1} U p,\\ &\quad k \geq 1, && (4.14)\end{aligned}$$

where $T^{-1} = U^T D^{-1} U$.

Proof. Since $T^{-\frac{1}{2}} = U^T D^{-\frac{1}{2}} U$, let us denote $U_1 = A^{k+1} U^T D^{-\frac{1}{2}}$ and $U_2 = U$. Noting that

$$\mathcal{R}({U_1}^T U_1 U_2) \subset \mathcal{R}(U_2) = \mathbb{R}^{n \times n}, \ \ \mathcal{R}(U_2 {U_2}^T {U_1}^T) = \mathcal{R}({U_1}^T),$$

we conclude that the reverse-order law for the matrices U_1 and U_2 in (4.4), (4.5) and (4.6) holds. Applying the reverse-order law on the representations (4.4), (4.5) and (4.6) we immediately obtain desired representations. ■

Corollary 4.12. [40] *Under the assumptions of Theorem 2.3 the solution $\hat{x}_1$ of the problem* (2.12) *with T being a positive definite matrix for which the decomposition of the form* (4.2) *holds, has the following equivalent representations:*

$$\hat{x}_1 = U^T D^{-\frac{1}{2}} \left(A^{k+1} U^T D^{-\frac{1}{2}} \right)^{\dagger} A^k b \tag{4.15}$$

$$= U^T D^{-\frac{1}{2}} \left(A^k U^T D^{-\frac{1}{2}} \right)^{\dagger} A^D A^k b \tag{4.16}$$

$$= U^T D^{-\frac{1}{2}} \left(A^D A^k U^T D^{-\frac{1}{2}} \right)^{\dagger} A^D A^{k-1} b, \ k \geq 1. \tag{4.17}$$

5. Some Possible Perspectives

In this section we describe several possibilities for further investigations on the considered topic.

(A) In the general case, it is possible to consider the minimization of the functional $\Phi(x)$ under a more general constraint set

$$\begin{aligned} x \in \mathcal{S}_G, \ \mathcal{S}_G = \big\{ x : \ x \in \mathbb{R}^n, \ GAx = Gb, \ A \in \mathbb{R}_r^{m \times n}, \ G \in \mathbb{R}_s^{n \times m}, \ 0 < s \leq r, \\ \operatorname{rank}(GA) = \operatorname{rank}(G) \big\} \end{aligned} \tag{5.1}$$

which includes all previously considered constraints, namely $x \in S$ and $x \in S_D$. Under the assumption $\operatorname{rank}(GA) = \operatorname{rank}(G)$, taking into account

$$GAx \in \mathcal{R}(GA), \quad Gb \in \mathcal{R}(G) = \mathcal{R}(GA),$$

we conclude that the system (5.1) is consistent. Since $\langle x, Tx \rangle = 0$ for all $x \in \mathcal{N}(T)$, we will investigate the constrained quadratic minimization problem

$$\text{minimize } \ \Phi_0(x), \ x \in \mathcal{S}_G \cap \mathcal{N}(T)^{\perp}. \tag{5.2}$$

(B) Moreover, instead of analyzing the set S, it is possible to analyze the following set, which presents the sets of all solutions of the normal equation of the system $Ax = b$:

$$S_M = \{ x : \ A^* A x = A^* b \}. \tag{5.3}$$

(C) In the case $A \in \mathbb{C}^{n \times n}$ it seems interesting to generalize the constraint set Ω_{A^k} and consider the constraint set

$$\Omega_Y = \Big\{ Y : \ AY = D, \mathcal{R}(Y) \subseteq \mathcal{R}(A^k), \ \mathcal{R}(D) \subseteq \mathcal{R}(A^k). \Big\} \tag{5.4}$$

Any solution of this problem should lead to a generalization of the following result from [56].

Theorem 5.1. [56] *Let* $A \in \mathbb{C}^{n\times n}$ *with* $\operatorname{ind}(A) = k$, $D \in \mathbb{C}^{n\times m}$ *and* $\mathcal{R}(D) \subseteq \mathcal{R}(A^k)$. *Then the restricted matrix equation*

$$AY = D, \ \mathcal{R}(Y) \subseteq \mathcal{R}(A^k) \tag{5.5}$$

has a unique solution

$$Y = A^D D. \tag{5.6}$$

An analogous generalization of the dual result and a general result from [56] can be derived.

Theorem 5.2. [56] *Let* $B \in \mathbb{C}^{m\times m}$ *with* $\operatorname{ind}(B) = k$, $D \in \mathbb{C}^{n\times m}$ *and* $\mathcal{N}(D) \supseteq \mathcal{N}(B^k)$. *Then the restricted matrix equation*

$$XB = D, \ \mathcal{N}(X) \supseteq \mathcal{N}(B^k) \tag{5.7}$$

has a unique solution

$$Y = DB^D. \tag{5.8}$$

Theorem 5.3. [56] *Let* $A \in \mathbb{C}^{n\times n}$ *with* $\operatorname{ind}(B) = k_1$, $B \in \mathbb{C}^{m\times m}$ *with* $\operatorname{ind}(B) = k_1$, $k = \max\{k_1, k_2\}$ *and* $D \in \mathbb{C}^{n\times m}$. *Suppose* $\mathcal{R}(D) \subseteq \mathcal{R}(A^k)$ *and* $\mathcal{N}(D) \supseteq \mathcal{N}(B^k)$. *Then the restricted matrix equation*

$$AXB = D, \ \mathcal{R}(X) \supseteq \mathcal{R}(A^k), \mathcal{N}(X) \supseteq \mathcal{N}(B^k) \tag{5.9}$$

has a unique solution

$$Y = A^D DB^D. \tag{5.10}$$

5.1. Definition of the Minimal (T, G) Inverse

In this subsections we investigate several preliminary results concerning proposition **(A)**.

Lemma 5.1. *The vector*

$$\hat{x} = X^\dagger \left(GAX^\dagger\right)^\dagger Gb, \ \ X^2 = T \tag{5.11}$$

is a solution of the problem (5.2).

Proof. Let $X^2 = T$ and $x \in \mathcal{N}(T)^\perp = \mathcal{R}(T)$. Then

$$\Phi_0(x) = \langle x, Tx\rangle = \langle x, X^2x\rangle = \langle Xx, Xx\rangle.$$

If $y = Xx$, then the problem of minimizing $\Phi_0(x)$ is equivalent to the problem of minimizing $\|y\|$. Since $y = Xx$ and $x \in \mathcal{R}(X)$, we have that $x = X^\dagger y$. In addition, because $x \in S_G \cap \mathcal{N}(T)^\perp$, it follows that $GAx = Gb$ and so, $GAX^\dagger y = Gb$.

In this case, the minimal norm solution for y will be $y = (GAX^\dagger)^\dagger Gb$, and so $Xx = (GAX^\dagger)^\dagger Gb$ and therefore the minimal norm solution of (5.2) is the vector $\hat{x}$ defined by (4.10). ■

Definition 5.1. *Let $T \in \mathbb{R}^{n\times n}$ be a positive semidefinite Hermitian matrix, $G \in \mathbb{R}^{n\times m}$ be a singular matrix and let $A \in \mathbb{R}^{m\times n}$ be also singular, such that* $\mathrm{rank}(GA) = \mathrm{rank}(G)$. *Then, the $n \times m$ matrix*

$$\begin{aligned} \hat{A}^{(2)}_{G,T} &:= \left(T^{\dagger}\right)^{\frac{1}{2}} \left(GA\left(T^{\dagger}\right)^{\frac{1}{2}}\right)^{\dagger} G \\ &= \left(T^{\frac{1}{2}}\right)^{\dagger} \left(GA\left(T^{\frac{1}{2}}\right)^{\dagger}\right)^{\dagger} G \end{aligned} \tag{5.12}$$

is called the T-minimal G-constrained inverse of A (or the minimal (T, G) inverse of A).

Corollary 5.1. *The matrix defined by* (5.12) *is a minimizer of the minimization problem* $\min \Psi(Z) = \langle Z, TZ\rangle$, *where T is a positive $n \times n$ matrix and $X \in \mathbb{R}^{n\times m}$, subject to the constraint set*

$$Z \in \Omega_G, \quad \Omega_G = \{Z : X \in \mathbb{R}^{n\times m},\ GAZ = G,\ A \in \mathbb{R}^{m\times n}_r,\ G \in \mathbb{R}^{n\times m}_s,\ 0 < s \le r\}.$$

Remark 5.1. *The vector $\hat{u}^{(2)} = \hat{A}^{(2)}_{G,T} b$ is a minimal T semi-norm solution of the consistent equation $GAx = Gb$, restricted on $\mathcal{N}(T)^{\perp} = \mathcal{R}(T) = \mathcal{R}(T^*)$.*

In the general case, it is possible to generalize the results derived in [57] concerning with the general solution of a general restricted matrix equation

$$AXB = D, \mathcal{R}(X) \subseteq T, \mathcal{N}(X) \supseteq S, \tag{5.13}$$

where $A \in \mathbb{C}^{m\times n}$, $B \in \mathbb{C}^{p\times q}$, $D \in \mathbb{C}^{m\times q}$ and T, S are subspaces of $\mathbb{C}^n$ and $\mathbb{C}^p$, respectively.

5.2. Particular Cases of the Minimal (T, G) Inverse

In accordance with previously presented results, it is clear that such a constrained quadratic programming problem implies a definition of a new kind of generalized inverse. In the present article we define corresponding generalized inverse, called the minimal (T, G) inverse of A, and investigate its basic properties.

The particular case $G^*G = I_m$ gives the results corresponding to constrained optimization from [35, 46] and produces the restricted generalized inverse $\hat{A}^{\dagger}_{I_m,T}$ from [34].

In the case $m = n$, for the particular choice $G = A^k$ we obtain the results of constrained optimization form [39] and the generalized inverse $\hat{A}^{D}_{A^k,T}$, which is introduced and investigated in [40].

Surprising minimality property of a particular class of $\{2, 3\}$ inverses can be obtained from $\hat{A}^{(2)}_{G,I}$ (where the choice $T = I$ is assumed). Corollary 5.2 shows that $A^{(2,3)}_{\mathcal{N}(GA)^{\perp},\mathcal{N}(G)} b$ is a minimizer of $\langle x, x\rangle$ subject to constraint $GAx = Gb$.

Corollary 5.2. *Let $A \in \mathbb{C}^{m\times n}$ be given and $G \in \mathbb{C}^{n\times m}$ is chosen matrix satisfying* $\mathrm{rank}(GA) = \mathrm{rank}(G)$. *Then the following statement is valid in the case $T = I$:*

$$\hat{A}^{(2)}_{G,I} = A^{(2,3)}_{\mathcal{N}(GA)^{\perp},\mathcal{N}(G)}. \tag{5.14}$$

Proof. It is clear that

$$\hat{A}^{(2)}_{G,I} = (GA)^{\dagger}G.$$

The rest of the proof can be completed using known representations of $\{2,3\}$-inverses from [58]. ∎

Corollary 5.3. *Let U be appropriate positive semidefinite $n \times n$ matrix. For a given $A \in \mathbb{C}^{m\times n}$ in the case $G = U^{\frac{1}{2}}$ the outer inverse $\hat{A}^{(2)}_{G,I}$ becomes the weighted Moore-Penrose inverse:*

$$\hat{A}^{(2)}_{G,T} = A^{\dagger}_{U,T}. \tag{5.15}$$

Proof. Follows from known representation of the weighted Moore-Penrose inverse with singular weights from [59]. ∎

Corollary 5.4. *Let $A \in \mathbb{C}^{m\times n}$ be given and $G \in \mathbb{C}^{n\times m}$ satisfies* $\operatorname{rank}(GA) = \operatorname{rank}(G)$ *and $G^*G = I_m$. Then*

$$\hat{A}^{(2)}_{G,I} = A^{\dagger}. \tag{5.16}$$

Proof. Follows from the representation of outer inverses and uniqueness of $A^{\dagger} = A^{(2)}_{\mathcal{R}(A^*),\mathcal{N}(A^*)}$. ∎

6. Conclusion

We investigate the quadratic minimization problem under linear constraints. Our global objective is to define and investigate the restricted generalized inverses corresponding to minimization of constrained quadratic form.

Results derived in [34, 35] are surveyed in the third section. The constraint set defined by the linear system $Ax = b$ is considered in this section. As a consequence, the T-restricted weighted generalized inverse of a singular matrix A with respect to a positive semidefinite matrix T is considered. It is interesting to mention that the matrix T defines a seminorm for the space. The generalized inverse is denoted by $\hat{A}^{\dagger}_{I_m,T}b$ and it represents a minimal T semi-norm least squares solution of $Ax = b$, restricted on the range of T. It is assumed that T is positive semidefinite, so the minimal seminorm solution is considered for all vectors belonging to $\mathcal{N}(T)^{\perp}$.

The originality of the results derived in [39, 40] is the assumption that the constraint equation $Ax = b$ is not consistent. In this case, the constraints set S is empty. This situation may appear in many practical applications. Therefore, the minimization is an approximate solution of the original problem, and it is obtained using the Drazin inverse solution of the set S. This intention leads to the usage of the constraint set S_D, defined by the linear system $A^{k+1}x = A^k b$.

Several possibilities for further investigations on the quadratic minimization under linear constraint sets are presented and preliminary considered.

The proposed solutions of considered optimization problems can find applications also in many financial problems, apart from the usual matrix optimization areas such as statistical modeling, linear regression, electrical networks, filter design, etc.

References

[1] D. Watkins, *Fundamentals of Matrix Computations*, Wiley-Interscience New York 2002.

[2] J. Nocedal, S.J. Wright, *Numerical Optimization*, Second Edition, Soringer Series in Operations Research, New York, Berlin, Heidelberg, Barcelona, Hong Kong, London, Milan, Paris, Singapore, Tokyo Springer-Verlag, 1999.

[3] M. Broom, On the number of local maxima of a constrained quadratic form, *Proc. Roy. Soc. London Ser. A* 443 no.1919 (1993) 573–584.

[4] W. La Cruz, Extension of the Barzilai- Borwein method for quadratic forms in finite Euclidean spaces, *Numer. Funct. Anal. Optim.* 30 (2009) 306–321.

[5] R. Manherz, S. Hakimi, The Generalized Inverse in Network Analysis and Quadratic-Error Minimization Problems, *IEEE Trans. Circuit Theory* 16 (1969) 559–562.

[6] N. Minamide, K. Nakamura, Minimum error control problem in Banach space, *Technical Report* 16, Nagoya University, Nagoya, Japan, 1969.

[7] N. Minamide, K. Nakamura, A restricted pseudoinverse and its applications to constrained minima, *SIAM J. Appl. Math.* 19 (1970) 167–177.

[8] D. Luenberger, *Optimization by Vector Space Methods*, Wiley Publ. New York, 1969.

[9] A. Ben-Israel, A. Charnes, Generalized Inverses and the Bott-Duffin Network Analysis, *J. Math. Analysis and Appl.* 7 (1963) 428–435.

[10] H. Markowitz, Portfolio Selection, *Cowles Foundation Monograph* N.16. New York John Wiley and Sons Inc., 1959.

[11] A. Ben-Israel, T.N.E. Greville, *Generalized Inverses: Theory and Applications*, Springer-Verlag, Berlin, 2002.

[12] R. Penrose, A generalized inverse for matrices, *Proc. Cambridge Philos. Soc.* 51 (1955) 406–413.

[13] M. Z. Nashed, Steepest descent for singular linear operators equations, *SIAM J. Numer. Anal.* 7 (1970) 358–362.

[14] M.Z. Nashed, Generalized Inverses and Applications, Edited by M. Zuhair Nashed, *Proceedings of an Advanced Seminar*, Academic Press, New York, San Francisko, London, 1976.

[15] C.W. Groetsch, *Generalized inverses of linear operators*, Marcel Dekker Inc. New York, 1977.

[16] S.L. Campbell, C. D. Meyer, *Generalized Inverses of Linear Transformations*, Dover Publications Inc., New York, 1991.

[17] Y. Tian, H. Wang, Characterizations of EP matrices and weighted-EP matrices, *Linear Algebra Appl.* 434 (2011) 1295-1318.

[18] M. Tasić, P. S. Stanimirović, M. Petković, Symbolic computation of weighted Moore-Penrose inverse using partitioning method, *Applied Math. Comput.* 189 (2007) 615-640.

[19] L. Elden, Perturbation theory for the least squares problem with equality constraints, *SIAM J. Numer. Anal.* 17 (1980) 338–350.

[20] L. Elden, A weighted pseudoinverse, generalized singular values, and constrained least squares problems *BIT* 22 (1982) 487–502.

[21] R. Bouldin, The product of operators with closed range, *Tohoku Math. J.* 25(4) (1973) 359-363.

[22] S. Izumino, The product of operators with closed range and an extension of the reverse order law, *Tohoku Math. J.* 34 (1982) 43-52.

[23] R. Bouldin, The pseudo-inverse of a product, *SIAM J. Appl. Math.* 24(4) (1973) 489–495.

[24] S. Karanasios, D. Pappas, Generalized inverses and special type operator Algebras, *Facta Univ. Ser. Math. Inform.* 21 (2006) 41–48.

[25] J. Levine, R.E. Hartwig, Applications of Drazin inverse to the Hill cryptographic systems, *Cryptologia* 4 (1980) 71–85.

[26] Y. Wei, H. Wu, Additional results on index splitting for Drazin inverse of singular linear system, *Electron. J. Linear Algebra* 95 (1998) 115–124.

[27] Y. Wei, Index splitting for the Drazin inverse and the singular linear system, *Appl. Math. Comput.* 95 (1998) 115–124.

[28] Y. Wei, H. Wu, The representation and approximation for Drazin inverse, *J. Comput. Appl. Math.* 126 (2000) 417–432.

[29] F. Alizad, Interior point methods in semidefinite programming with applications to combinatorial optimization, *SIAM J. Optim.* 5 (1995) 13–51.

[30] A. Ben-Tal, A. Nemirovski, Lectures on Modern Convex Optimization: Analysis, Algorithms, and Engineering Applications, *SIAM*, Philadelphia, PA, 2001.

[31] M.J. Todd, Semidefinite optimization, *Acta Numer.*, 10 (2001) 515-560.

[32] M.J. Todd, Dual versus prima-dual interior-point methods for linear and conic programming, *Math. Program.* 111 (2008) 301–313.

[33] S. Laue, A Hybrid Algorithm for Convex Semidefinite Optimization, *29 th International Conference on Machine Learning*, Edinburgh, Scotland, UK, 2012.

[34] V.N. Katsikis, D. Pappas, The restricted weighted generalized inverse of a matrix, *Electron. J. Linear Algebra* 22 (2011) 1156–1167.

[35] D. Pappas, Minimization of Constrained Quadratic forms in Hilbert Spaces, *Ann. Funct. Anal.* 2(1) (2011) 1-12.

[36] D.R. Jensen, Minimal properties of Moore-Penrose inverses, *Linear Algebra Appl.* 196 (1994), 175–182.

[37] S. Boyd, L. Vandenberghe, *Convex Optimization*, Cambridge University Press, Cambridge, 2006.

[38] D.J. Evans, Wenyu Sun, R.J.B. De Sampaio, J. Y. Yuan, The restricted generalized inverses corresponding to constrained quadratic system, *Int. J. Comput. Math.* 62 (1996) 285–296.

[39] P. S. Stanimirović, D. Pappas, S. Miljković, Minimization of quadratic forms by using the Drazin inverse, *Linear Multilinear Algebra* 62 (2014) 252–266.

[40] P.S. Stanimirović, D. Pappas, V. Katsikis, Generalized inverse restricted by the normal Drazin equation, *Linear Multilinear Algebra*, (2014) Accepted, (ID: 908873 DOI:10.1080/03081087.2014.908873).

[41] A. Gorokhov, P. Stoica, Generalized quadratic minimization and a signal processing application, EUSIPCO '98, 9th European Sign. *Proc. Conference*, Rhodes, Greece (1998).

[42] O. L. Frost III, An algorithm for linear constrained adaptive array processing, *Proc. IEEE* 60 (1972) 926–935.

[43] D. Pappas, A. Perperoglou, Constrained matrix optimization with applications, *J. Appl. Math. Comput.* 40 (2013) 357–369.

[44] P. Stoika, A. Jakobsson, J. Li, Matrix Optimization Result with DSP Applications, *Dig. Signal Proc., A review Journal* 6 (1996) 195–201.

[45] R. Rostamian, Continuity properties of stationary points of quadratic functionals in Hilbert space, *Numerical Functional Analysis and Optimization* 3(2) (1981) 147–167.

[46] D. Pappas, Restricted Linear Constrained Minimization of quadratic functionals, *Linear Multilinear Algebra* 62 (2013) 1394–1407.

[47] D. Drivaliaris, S. Karanasios, D. Pappas, Factorizations of EP operators, *Linear Algebra Appl.* 429 (2008) 1555-1567.

[48] Y. Tian, Y. Takane, On V-orthogonal projectors associated with a semi-norm, *Ann. Inst. Stat. Math.* 61 (2009) 517–530.

[49] S.K. Mitra, C.R. Rao, Projections under seminorms and generalized Moore-Penrose inverses, *Linear Algebra Appl.* 9 (1974) 155–167.

[50] X. Sheng, G. Chen, Full-rank representation of generalized inverse $A^{(2)}_{T,S}$ and its applications, *Comput. Math. Appl.* 54 (2007) 1422–1430.

[51] P.S. Stanimirović, D. Pappas, V.N. Katsikis, I.P. Stanimirović, Full–rank representations of outer inverses based on the QR decomposition, *Appl. Math. Comput.* 218 (2012) 10321–10333.

[52] N.J. Higham, *The Matrix Computation Toolbox*, Available from: http://www.maths.manchester.ac.uk/higham/mctoolbox/

[53] *Matrix Market*, National Institute of Standards and Technology, Gaithersburg, MD.

[54] M. Nikuie, M.K. Mirnia, Y. Mahmoudi, Some results about the index of matrix and Drazin inverse, *Mathematical Sciences* 4 (2010) 283–294.

[55] M. Drazin, A class of outer generalized inverses, *Linear Algebra Appl.* 436 (2012) 1909–1923.

[56] G. Wang, Z. Xu, Solving a kind of restricted matrix equations and Cramer rule, *Appl. Math. Comput.* 162 (2005) 329–338.

[57] Y.-Lin Chen, Representations and cramer rules for the solution of a restricted Matrix Equation, *Linear Multilinear Algebra* 35 (1993) 339–354.

[58] P.S. Stanimirović, D.S. Cvetković-Ilić, S. Miljković, M. Miladinović, Full-rank representations of {2,3}, {2,4}-generalized inverses and successive matrix squaring algorithm, *App. Math. Comp.* 217 (2011) 9358–9367.

[59] J.F. Ward, T.L. Boullion, T.O. Lewis, Weighted Pseudoinverses with Singular Weights, *SIAM Journal on Applied Mathematics* 21 (1971) 480–482.

In: Advances in Linear Algebra Research
Editor: Ivan Kyrchei, pp. 57-78
ISBN: 978-1-63463-565-3

Chapter 2

THE STUDY OF THE INVARIANTS OF HOMOGENEOUS MATRIX POLYNOMIALS USING THE EXTENDED HERMITE EQUIVALENCE $\mathcal{E}_{rh}$

***Grigoris I. Kalogeropoulos*[1], *Athanasios D. Karageorgos*[1] *and Athanasios A. Pantelous*[2,3,*]**
[1]Department of Mathematics, University of Athens, Greece
[2]Department of Mathematical Sciences,
University of Liverpool, United Kingdom
[3]Institute for Risk and Uncertainty, University of Liverpool, United Kingdom

Abstract

In systems and control theory, Linear Time Invariant (LTI) descriptor (Differential-Algebraic) systems are intimately related to the matrix pencil theory. Actually, a large number of systems are reduced to the study of differential (difference) systems $S(F,G)$ of the form:

$$S(F,G) : F\dot{x}(t) - Gx(t) \ \text{(or the dual } Fx = G\dot{x}(t)\text{)},$$

and

$$S(F,G) : Fx_{k+1} = Gx_k \ \text{(or the dual } Fx_k = Gx_{k+1}\text{)}, \ F,G \in \mathbb{C}^{m \times n}$$

and their properties can be characterized by the homogeneous pencil $sF - \hat{s}G$. An essential problem in matrix pencil theory is the study of invariants of $sF - \hat{s}G$ under the *bilinear strict equivalence*. This problem is equivalent to the study of complete *Projective Equivalence* (PE), $\mathcal{E}_{\mathcal{P}}$, defined on the set $\mathbb{C}_r$ of complex homogeneous binary polynomials of fixed homogeneous degree r. For a $f(s,\hat{s}) \in \mathbb{C}_r$, the study of invariants of the PE class $\mathcal{E}_{\mathcal{P}}$ is reduced to a study of invariants of matrices of the set $\mathbb{C}^{k \times 2}$ (for $k \geqslant 3$ with all 2×2-minors non-zero) under the *Extended Hermite Equivalence* (EHE), $\mathcal{E}_{rh}$. In this chapter, we present a review of the most interesting properties of the PE and the EHE classes. Moreover, the appropriate projective transformation $d \in RGL(1, \mathbb{C}/\mathbb{R})$ is provided analytically [1].

*E-mail address: A.Pantelous@liverpool.ac.uk

Keywords: extended Hermite equivalence; matrix pencil theory; bilinear-strict equivalence; elementary divisors; linear systems

AMS Subject Classification: 14N05; 51N15; 15A22; 34A30

1. Introduction

The matrix pencils $sF-G$ and $F-\hat{s}G$ can be defined by the homogeneous matrix pencil $sF-\hat{s}G$ (or equivalently by an ordered pair of matrices $(F,-G)$), since the existence of the important notion of duality -the so-called *elementary divisor* type duality or *integrator-differentiator* type duality- has already been demonstrated in the literature; see [2]. Thus, the dual matrix pencils $sF-G$ and $F-\hat{s}G$ are related by the special type of bilinear transformation: $s \to \frac{1}{\hat{s}}$, which clearly transforms relatively, the points $0,\ \infty,\ a \neq 0$ of the compactified complex plain $(\mathbb{C} \cup \{\infty\})$ (or of the Riemann sphere) to the points $\infty,\ 0,\ \frac{1}{a}$; see for further details [3].

In the existing literature, the study of bilinear-strict equivalence of matrix pencils has been initiated by the work of Turnbull and Aitken ([4]). In this early research work, the co-variance property of invariant polynomials and the invariant of minimal indices are established. Kalogeropoulos (see [3]) defines a complete set of invariants for the bilinear-strict equivalence class of a matrix pencil (or equivalently of an ordered pair (F,G)).

Let us define $(F,G) \in \mathbb{F}^{m\times n} \times \mathbb{F}^{m\times n}$ (where $\mathbb{F}$ is a field, i.e., $\mathbb{F}=\mathbb{R}$ or $\mathbb{C}$) and $(s,\hat{s})$ be a pair of indeterminates. Thus the polynomial matrix $sF-\hat{s}G \in \mathbb{F}^{m\times n}[s,\hat{s}]$ is defined as the *homogeneous matrix pencil* of the pair (F,G). Clearly, $sF-\hat{s}G$ is a matrix over the ring $\mathbb{F}[s,\hat{s}]$, i.e., polynomials in $(s,\hat{s})$ with coefficients from $\mathbb{F}$, which might also be viewed as a matrix over the (algebraic) rings $\mathbb{F}(s)[\hat{s}]$ or $\mathbb{F}(\hat{s})[s]$. Now, the following series of definitions are significant for what they follow.

Definition 1. Denote

$$\mathcal{L} \triangleq \left\{L : L = (F,-G)\,;\ F,G \in \mathbb{F}^{m\times n}\right\},$$

to be the set of ordered pairs of $m\times n$-matrices and

$$\Theta \triangleq \{\theta : \theta = (s,\hat{s})\},$$

to be the set of ordered pairs of indeterminates.

Now, for every $L=(F,-G)\in\mathcal{L}$ and $\theta=(s,\hat{s})\in\Theta$, $[L]=\begin{bmatrix} F & -G \end{bmatrix} \in \mathbb{F}^{m\times 2n}$ is called the matrix representation of L. The homogeneous polynomial matrix

$$L_\theta \triangleq L(s,\hat{s}) = \begin{bmatrix} F & -G \end{bmatrix} \begin{bmatrix} sI_n \\ \hat{s}I_n \end{bmatrix} = sF-\hat{s}G,$$

is referred as the θ-matrix pencil of L.

Definition 2. Define the following sets of matrix pencils

$$\mathcal{L}_\theta \triangleq \{L_\theta : \text{for a } \mathit{fixed}\ \theta\in\Theta \text{ and for } \mathit{every}\ L\in\mathcal{L}\},$$

$$\mathcal{L}(\Theta) \triangleq \{L_\theta : \text{for } \mathit{every}\ \theta\in\Theta \text{ and for } \mathit{every}\ L\in\mathcal{L}\}.$$

In the following part of the introduction, *three* types of equivalence are presented on $\mathcal{L}$ (or equivalently on $\mathcal{L}(\Theta)$). These equivalence relations are generated by the action of appropriate transformation groups acting on $\mathcal{L}$ (or equivalently on $\mathcal{L}(\Theta)$). Consider first the set

$$\mathcal{K} \triangleq \left\{k : k = (M, N) ;\ M \in \mathbb{F}^{m \times m}, N \in \mathbb{F}^{2n \times 2n};\ \det M, \det N \neq 0\right\}$$

and a composition rule $(*)$ defined on $\mathcal{K}$ as follows:

$$* : \mathcal{K} \times \mathcal{K} \rightarrow \mathcal{K} : \text{for every } k_1 = (M_1, N_1) \in \mathcal{K} \text{ and } k_2 = (M_2, N_2) \in \mathcal{K},$$

$$\text{then } k_1 * k_2 \triangleq (M_1, N_1) * (M_2, N_2) = (M_1 M_2, N_2 N_1).$$

We can easily verify that $(\mathcal{K}, *)$ is a group with identity element (I_m, I_{2n}). The action of $\mathcal{K}$ on $\mathcal{L}$ is defined by

$$\circ : \mathcal{K} \times \mathcal{L} \rightarrow \mathcal{L} : \text{for every } k \in \mathcal{K}, L \in \mathcal{L}, \text{then } k \circ L \triangleq k \circ (F, -G) = L' \in \mathcal{L} : [L'] = M [L] N.$$

Here, the action "$\circ$" defines an equivalence relation $\mathcal{E}_{\mathcal{K}}$ on $\mathcal{L}$, and $\mathcal{E}_{\mathcal{K}}(L)$ denotes the equivalence class or orbit of $L \in \mathcal{L}$ under $\mathcal{K}$.

(i) (**Strict Equivalence**) The subgroup $(\mathcal{H}, *)$ of $(\mathcal{K}, *)$, where

$$\mathcal{H} \triangleq \left\{h : h = (R, P) ;\ R \in \mathbb{F}^{n \times n},\ P = \begin{bmatrix} Q & \mathbb{O} \\ \mathbb{O} & Q \end{bmatrix}, Q \in \mathbb{F}^{n \times n};\ \det R,\ \det P \neq 0\right\},$$

is called the ***S****trict*-***E****quivalence* ***G****roup* (**SEG**). The action of $\mathcal{H}$ on $\mathcal{L}$ is defined by

$$\circ : \mathcal{H} \times \mathcal{L} \rightarrow \mathcal{L} : \text{for every } h \in \mathcal{H} \text{ and for a } L \in \mathcal{L}, \textit{then}$$

$$h \circ L \triangleq (R, P) \circ (F, -G) = L' \in \mathcal{L} : \left[L'\right] = R [L] P.$$

The equivalence relation $\mathcal{E}_{\mathcal{H}}$, which is defined on $\mathcal{L}$ as above, is called ***S****trict*-***E****quivalence* (**SE**), see also [5] and [3]. Two pencils $L_\theta^1 = sF_1 - \hat{s}G_1 \in \mathcal{L}_\theta$, and $L_\theta^2 = sF_2 - \hat{s}G_2 \in \mathcal{L}_\theta$ are strictly equivalent, i.e., $L_\theta^1 \mathcal{E}_{\mathcal{H}} L_\theta^2$, if and only if there exists $h \in \mathcal{H} : (F_2, -G_2) = h \circ (F_1, -G_1)$. By $\mathcal{E}_{\mathcal{H}}(F, G)$ is denoted the SE class or orbit of $L_\theta = sF - \hat{s}G$.

(ii) (**Bilinear Equivalence**) The subgroup $(\mathcal{B}, *)$ of $(\mathcal{K}, *)$, where

$$\mathcal{B} \triangleq$$

$$\left\{b : b = \left(I_m, \begin{bmatrix} \alpha I_n & \beta I_n \\ \gamma I_n & \delta I_n \end{bmatrix}\right) = (I_m, T_d), d = \begin{bmatrix} \alpha & \beta \\ \gamma & \delta \end{bmatrix} \in \mathbb{F}^{2 \times 2};\ \det d = \alpha\delta - \beta\gamma \neq 0\right\}$$

is called the **B**ilinear-**E**quivalence **G**roup (BEG), see [6]. Every $b \in \mathcal{B}$ is generated by a projective transformation $d = \begin{bmatrix} \alpha & \beta \\ \gamma & \delta \end{bmatrix} \in \mathbb{F}^{2 \times 2}$. The action of $\mathcal{B}$ on $\mathcal{L}$ is defined by

$$b \circ L \triangleq b \circ (F, -G) = L' = \left(F', -G'\right) \in \mathcal{L} : \left[L'\right] = I_m [L] T_d.$$

The equivalence relation $\mathcal{E}_{\mathcal{B}}$, which is defined on $\mathcal{L}$, is called **B**ilinear-**E**quivalence (**BE**). Two pencils $L_\theta^1 = sF_1 - \hat{s}G_1 \in \mathcal{L}_\theta$, and $L_{\theta'}^2 = \lambda F_2 - \hat{\lambda}G_2 \in \mathcal{L}_{\theta'}$, where $\theta = (s, \hat{s}) \in \Theta$

and $\theta' = \left(\lambda, \hat{\lambda}\right) \in \Theta$ are *bilinearly* equivalent, i.e., $L_\theta^1 \mathcal{E}_{\mathcal{B}} L_{\theta'}^2$, if and only if there exists a transformation $d : (s, \hat{s}) \to \left(\lambda, \hat{\lambda}\right)$ and thus a $b \in \mathcal{B}$ generated by d, such that $(F_2, -G_2) = b \circ (F_1, -G_1)$. By $\mathcal{E}_{\mathcal{B}}(F, G)$ is denoted the BE class or orbit of $L_\theta = sF - \hat{s}G$, or equivalently $L = (F, -G)$. Note that the composition rule $(*)$ is not *commutative* on $\mathcal{K}$; However, it can be easily shown that for every $b \in \mathcal{B}$ and for every $h \in \mathcal{H}$, the equality, $b * h = h * b$ holds, see also the following lemma; see [3].

Lemma 1. *For every $b \in \mathcal{B}$ and for every $h \in \mathcal{H}$, then $b * h = h * b$.*

Proof. Start with

$$
\begin{aligned}
h*b &= \left(R, \begin{bmatrix} Q & \mathbb{O} \\ \mathbb{O} & Q \end{bmatrix}\right) * \left(I_m, \begin{bmatrix} \alpha I_n & \beta I_n \\ \gamma I_n & \delta I_n \end{bmatrix}\right) = \left(R \cdot I_m, \begin{bmatrix} \alpha I_n & \beta I_n \\ \gamma I_n & \delta I_n \end{bmatrix} \cdot \begin{bmatrix} Q & \mathbb{O} \\ \mathbb{O} & Q \end{bmatrix}\right) \\
&= \left(R, \begin{bmatrix} \alpha Q & \beta Q \\ \gamma Q & \delta Q \end{bmatrix}\right) = \left(I_m \cdot R, \begin{bmatrix} Q & \mathbb{O} \\ \mathbb{O} & Q \end{bmatrix} \cdot \begin{bmatrix} \alpha I_n & \beta I_n \\ \gamma I_n & \delta I_n \end{bmatrix}\right) \\
&= \left(I_m, \begin{bmatrix} \alpha I_n & \beta I_n \\ \gamma I_n & \delta I_n \end{bmatrix}\right) * \left(R, \begin{bmatrix} Q & \mathbb{O} \\ \mathbb{O} & Q \end{bmatrix}\right) = b * h.
\end{aligned}
$$

□

(iii) (Bilinear-Strict Equivalence) The subgroup $(\mathcal{H} - \mathcal{B}, *)$ of $(\mathcal{K}, *)$, where

$$\mathcal{H} - \mathcal{B} \triangleq \{r : r = h * b; \text{ for every } h \in \mathcal{H} \text{ and for every } b \in \mathcal{B}\}$$

is called the **B**ilinear-**S**trict **E**quivalence Group (**BSEG**). The action of $\mathcal{H} - \mathcal{B}$ on $\mathcal{L}$ is defined by

$$\circ : \mathcal{H} - \mathcal{B} \times \mathcal{L} \to \mathcal{L} : \text{for every } r \in \mathcal{H} - \mathcal{B},$$

and for a $L = (F, -G) \in \mathcal{L}$, then

$$
\begin{aligned}
& r \circ L \triangleq (h * b) \circ (F, -G) = h \circ \{b \circ (F, -G)\} = b \circ \{h \circ (F, -G)\} \\
&= L' = (F', -G') \in \mathcal{L} : [L'] = I_m R [L] F T_d,
\end{aligned}
$$

or equivalently

$$
\begin{aligned}
\begin{bmatrix} F' & -G' \end{bmatrix} &\triangleq I_m R \begin{bmatrix} F & -G \end{bmatrix} Q \begin{bmatrix} \alpha I_n & \beta I_n \\ \gamma I_n & \delta I_n \end{bmatrix} \\
&= \begin{bmatrix} \alpha RFQ - \gamma RGQ & \beta RFQ - \delta RG\, Q \end{bmatrix},
\end{aligned}
$$

where $b = \left(I_m, \begin{bmatrix} \alpha I_n & \beta I_n \\ \gamma I_n & \delta I_n \end{bmatrix}\right)$ and $h = \left(R, \begin{bmatrix} Q & \mathbb{O} \\ \mathbb{O} & Q \end{bmatrix}\right)$.

The equivalence relation $\mathcal{E}_{\mathcal{H}-\mathcal{B}}$, which is defined on $\mathcal{L}$, is called **B**ilinear-**S**trict **E**quivalence (BSE), see [6]. Two pencils $L_\theta^1 = sF_1 - \hat{s}G_1 \in \mathcal{L}_\theta$, and $L_{\theta'}^2 = \lambda F_2 - \hat{\lambda} G_2 \in \mathcal{L}_{\theta'}$, where $\theta = (s, \hat{s}) \in \Theta$ and $\theta' = \left(\lambda, \hat{\lambda}\right) \in \Theta$ are *bilinearly-strict* equivalent, i.e., $L_\theta^1 \mathcal{E}_{\mathcal{H}-\mathcal{B}} L_\theta^2$, if and only if there exists a transformation $d : (s, \hat{s}) \to \left(\lambda, \hat{\lambda}\right)$ and thus a $b \in \mathcal{B}$ generated by d, and an $h \in \mathcal{H}$ such that

$$(F_2, -G_2) = (h * b) \circ (F_1, -G_1).$$

The matrix form of the above condition is given by $\left[L^2_{\theta'}\right] = \left[(h * b) \circ L^1_{\theta}\right]$, or equivalently

$$\begin{bmatrix} F_2 & -G_2 \end{bmatrix} \triangleq R \begin{bmatrix} F_1 & -G_1 \end{bmatrix} \begin{bmatrix} \alpha Q & \beta Q \\ \gamma Q & \delta Q \end{bmatrix} =$$
$$\begin{bmatrix} \alpha R F_1 Q - \gamma R G_1 Q & \beta R F_1 Q - \delta R G_1 \ Q \end{bmatrix}.$$

By $\mathcal{E}_{\mathcal{H}-\mathcal{B}}(F, G)$ is denoted the BSE class or orbit of $L_\theta = sF - \hat{s}G$, or equivalently of $L = (F, -G)$.

Proposition 1. *[3] Let $L \in \mathcal{L}$, $b \in \mathcal{B}$, and $h \in \mathcal{H}$*
(i) If $L \xrightarrow{b} b \circ L \triangleq L^b$, then $\mathcal{E}_{\mathcal{H}}(L) \xrightarrow{b} \mathcal{E}_{\mathcal{H}}\left(L^b\right)$ is a bijection.
(ii) If $L \xrightarrow{h} h \circ L \triangleq L^h$, then $\mathcal{E}_{\mathcal{B}}(L) \xrightarrow{h} \mathcal{E}_{\mathcal{B}}\left(L^h\right)$ is a bijection.

Proof. (i) Let $L' \in \mathcal{E}_{\mathcal{H}}\left(L^b\right)$, then there exists $h \in \mathcal{H} : L' = h \circ \left(L^b\right)$ and since $L^b = b \circ L$, it derives that $L' = h \circ (b \circ L) = b \circ (h \circ L) = b \circ \left(L^h\right)$, where $L^h = h \circ L \in \mathcal{E}_{\mathcal{H}}(L)$, so $b \circ \left(L^h\right) = L'$. Now, if we assume that there exists $L^{h'} \in \mathcal{E}_{\mathcal{H}}(L)$, such that $b \circ \left(L^{h'}\right) = L'$, then $b \circ (h' \circ L) = L' = h \circ (b \circ L)$ or $h' \circ \left(L^b\right) = h \circ \left(L^b\right)$.

Now, we further assume that $h = (R, P)$, $h' = (R', P')$, then $R\left[L^b\right]P = R'\left[L^b\right]P'$ implies that $\left(R'^{-1}R, PP'^{-1}\right) = (I_m, I_{2n})$ and thus $R' = R$, $P' = P$ or $h = h'$. The proof of part (ii) follows along similar steps. □

Note that the action of $h = (R, P) \in \mathcal{H}$ on the pencil $L_\theta = sF - \hat{s}G$ may be interpreted as

$$b \circ L_\theta = R \begin{bmatrix} F & -G \end{bmatrix} \begin{bmatrix} Q & \mathbb{O} \\ \mathbb{O} & Q \end{bmatrix} \begin{bmatrix} sI_n \\ \hat{s}I_n \end{bmatrix} = R\left(sF - \hat{s}G\right)Q = sF' - \hat{s}G' = L'_\theta,$$

and thus SE implies a coordinate transformation in the domain and co-domain of the ordered pair $(F, -G)$, but not a change in the indeterminates $(s, \hat{s})$. However, the action of $b = (I_m, T_d)$ on $L_\theta = sF - \hat{s}G$ may be interpreted as

$$\begin{aligned} b \circ L(s, \hat{s}) &= I_m \begin{bmatrix} F & -G \end{bmatrix} \begin{bmatrix} \alpha I_n & \beta I_n \\ \gamma I_n & \delta I_n \end{bmatrix} \begin{bmatrix} \lambda I_n \\ \hat{\lambda} I_n \end{bmatrix} \\ &= \lambda\left(\alpha F - \gamma G\right) - \hat{\lambda}\left(-\beta F + \delta G\right) \\ &= \lambda F' - \hat{\lambda} G' = {L'}_{\theta'}, \end{aligned}$$

which clearly expresses a change in the indeterminates from $(s, \hat{s}) \xrightarrow{d} \left(\lambda, \hat{\lambda}\right)$ by the bi-linear transformation:

$$d : \begin{bmatrix} s \\ \hat{s} \end{bmatrix} = \begin{bmatrix} \alpha & \beta \\ \gamma & \delta \end{bmatrix} \begin{bmatrix} \lambda \\ \hat{\lambda} \end{bmatrix} \Leftrightarrow \begin{cases} s = \alpha\lambda + \beta\hat{\lambda} \\ \hat{s} = \gamma\lambda + \delta\hat{\lambda} \end{cases} \tag{1}$$

where $\alpha, \beta, \gamma, \delta \in \mathbb{F}$ and $\alpha\delta - \beta\gamma \neq 0$.

Thus, as a conclusion of the introductive section, our bilinear transformation expresses a coordinate type of transformation in the indeterminates but not in the domain and co-domain of the pair $(F, -G)$. The action of $r = h * b$ on the matrix pencil $L_\theta = sF - \hat{s}G$

has the features of both $\mathcal{H}$ and $\mathcal{B}$ groups and thus it implies a coordinate transformation in the domain and co-domain of $(F, -G)$ and a change of indeterminates from $(s, \hat{s})$ to $\left(\lambda, \hat{\lambda}\right)$ according to the expression (1). The nature of the transformation d, that generates the transformation $b \in \mathcal{B}$, is discussed next.

It is known, see for instance [7], that a n-dimensional projective domain over a field $\mathbb{F}$ or a projective space $\mathbb{P}_n(\mathbb{F})$ is a set of entities (usually called points of the space) that admits a certain class $\{\mathcal{R}\}$ of representation by homogeneous $(x_o, x_1, \ldots, x_n)$ in $\mathbb{F}$.

This class is such that, if $\mathcal{R}_0$ is any representation, the whole class $\{\mathcal{R}\}$ consists of all those representations that can be obtained from $\mathcal{R}_0$ by using a non-singular linear transformation, i.e.,

$$x_i' = \sum_{j=0}^{n} a_{ij} x_j, \text{for } i = 0, 1, \ldots, n.$$

Thus, the representation $\mathcal{R}$ of $\mathbb{P}_n(\mathbb{F})$ are connected by a group of non-singular linear transformations. This group is referred to as the *general projective group* and it is denoted by $PGL(n; \mathbb{F})$. In our case, $n = 1$ and $\mathbb{F}$ will be replaced by $\mathbb{C}/\mathbb{R}$, i.e., the projective domain $\mathbb{P}_1(\mathbb{C})$ is the *projective straight* line on the compactified complex plain $(\mathbb{C} \cup \{\infty\})$; the $\{\mathcal{R}\}$ is the class of all bilinear transformations $d : (s, \hat{s}) \to \left(\lambda, \hat{\lambda}\right)$ defined by

$$d : s = \alpha\lambda + \beta\hat{\lambda}, \hat{s} = \gamma\lambda + \delta\hat{\lambda};\ \alpha, \beta, \gamma, \delta \in \mathbb{R},\ \alpha\delta - \beta\gamma \neq 0.$$

Consequently, we are interested for subgroups of $\{\mathcal{R}\}$, $\{\mathcal{R}_{\mathbb{R}}\}$ for which $\alpha, \beta, \gamma, \delta \in \mathbb{R}$. The nature of homogeneous coordinates of points in a line and the geometric meaning of d has been discussed analytically in [8].

2. Significant Properties of the Set of SE Invariants under BE

In this section, the basic properties of the set of SE invariants are revisited. Now, let us remind that $L = (F, -G) \in \mathcal{J}, \theta = (s, \hat{s}) \in \Theta$ and $L_\theta \triangleq L(s, \hat{s}) = sF - \hat{s}G$ be the associated pencil for the relative pair of intermediates $\theta = (s, \hat{s})$, then we will assume that $p = rank_{\mathbb{F}(s,\hat{s})} L_\theta \leqslant \min\{m, n\}$.

The Smith form of $L(s,\ \hat{s})$ over $\mathbb{F}[s,\ \hat{s}]$ is defined by

$$S(s,\ \hat{s}) = \begin{bmatrix} S^*{}_p(s,\ \hat{s}) & \mathbb{O}_{p,\ n-p} \\ \mathbb{O}_{m-p,\ p} & \mathbb{O}_{m-p,\ n-p} \end{bmatrix},$$

where

$$S^*{}_p(s,\ \hat{s}) = diag\left\{\widetilde{f}_1(s,\ \hat{s}),\ \widetilde{f}_2(s,\ \hat{s}),\ \ldots,\ \widetilde{f}_p(s,\ \hat{s})\right\} \quad \text{and} \quad \widetilde{f}_i(s,\ \hat{s}) \in \mathbb{F}[s,\ \hat{s}],$$

are the *invariant* polynomial of $L(s,\ \hat{s})$ over $\mathbb{F}[s,\ \hat{s}]$ with the property that $\widetilde{f}_i(s,\ \hat{s})$ divides $\widetilde{f}_{i+1}(s,\ \hat{s})$ for every $i = 1, 2, ..., p$ and $\widetilde{f}_i(s,\ \hat{s}) = 0$, for $i > p$.

The set of $\left\{\widetilde{f}_i(s,\ \hat{s}), i = 1, 2, ..., p\right\}$ is defined by the standard Smith algorithm,

$$\widetilde{f}_i(s,\ \hat{s}) \triangleq \frac{d_i(s,\ \hat{s})}{d_{i-1}(s,\ \hat{s})}, \text{ for } i = 1, 2, ..., p \text{ and } d_0(s,\ \hat{s}) = 1,$$

where $d_i(s,\ \hat{s})$ is the ith-determinantal divisor of L_θ, which is the greatest common divisor (GCD) of all $i \times i$- minors of L_θ; see for instance for more details [9] and the references therein.

Let k is non-zero trivial elements in $\left\{\widetilde{f}_i(s,\ \hat{s}),\ i = 1, 2, ..., p\right\}$, i.e.,

$$S^*{}_p(s,\ \hat{s}) = diag\{\underbrace{1,\ 1,\ \ldots,\ 1}_{k}, f_1(s,\ \hat{s}),\ \ldots,\ f_{p-k}(s,\ \hat{s})\},$$

k is the power of L_θ and the ordered set $\mathcal{F}(F, G) = \{f_i(s,\ \hat{s}); i = 1, 2, ..., p-k\}$ is called as the *homogeneous invariant polynomial* set of L_θ. Note that the order is defined by the divisibility property, i.e., $f_i(s,\ \hat{s}) \mid f_{i+1}(s,\ \hat{s})$ for every $i = 1, 2, ..., p-k-1$.

Here, it is important to denote the set of column and row minimal indices, respectively $\mathcal{I}_c(F, G)$ (set of c.m.i) and $\mathcal{I}_r(F, G)$ (set of r.m.i) of L_θ. It is a well-known result, see [5] and [4], that the set $\mathcal{F}(F, G)$, $\mathcal{I}_c(F, G)$, and $\mathcal{I}_r(F, G)$ form a complete and independent set of invariants for the SE class $\mathcal{E}_\mathcal{H}(F, G)\mathcal{E}_\mathcal{H}(F, G)$. In the next lines, the action of $b \in \mathcal{B}$ into the set of SE invariants is investigated. At the end of this section, we will transfer our study into the determination of a projective transformation, $d \in PGL(1, \mathbb{C}/\mathbb{R})$, onto homogeneous polynomials, see also [5] and [6].

Let us define $\mathbb{R}_d\{\theta\}$ be the set of homogeneous polynomials of degree d with coefficients on $\mathbb{R}$ for all the possible indeterminates $\theta = (s, \hat{s}) \in \Theta$. Now, the action of d on the polynomial $f(s,\ \hat{s}) \in \mathbb{R}_d\{\theta\}$ is defined by (2):

$$d \circ f(s,\ \hat{s}) = \tilde{f}(\lambda,\ \hat{\lambda}) = f(\alpha\lambda + \beta\hat{\lambda},\ \gamma\lambda + \delta\hat{\lambda}). \tag{2}$$

Two polynomials, $f(s,\ \hat{s})$ and $g(\lambda,\ \hat{\lambda}) \in \mathbb{R}_d\{\theta\}$, are projectively equivalent (PE), $f(s,\ \hat{s})\mathcal{E}_\mathcal{P} g(\lambda,\ \hat{\lambda})$, if there is $d \in PGL(1, \mathbb{C}/\mathbb{R})$ and a $c \in \mathbb{R}\backslash\{0\}$ such that

$$d \circ f(s,\ \hat{s}) = c \cdot g(\lambda,\ \hat{\lambda}). \tag{3}$$

Clearly, (3) plays a significant role in this paper, since (3) defines the Projective Equivalence (PE) on $\mathbb{R}_d\{\theta\}$. Let

$$\mathcal{F}_1 = \{f_i(s,\ \hat{s}) \in \mathbb{R}_{d_i}\{\theta\}; i = 1, 2, ..., p\},\ \mathcal{F}_2 = \left\{\tilde{f}_i(\lambda,\ \hat{\lambda}) \in \mathbb{R}_{d_i}\{\theta\}; i = 1, 2, ..., p\right\}$$

be two ordered sets of homogeneous polynomials.

$\mathcal{F}_1$ and $\mathcal{F}_2$ are *projectively* equivalent, $\mathcal{F}_1\mathcal{E}_\mathcal{P}\mathcal{F}_2$, if and only if $f_i(s,\ \hat{s})\mathcal{E}_\mathcal{P}\tilde{f}_i(\lambda,\ \hat{\lambda})$ for every $i = 1, 2, ..., p$ and for the same transformation $d \in PGL(1, \mathbb{C}/\mathbb{R})$. The projective equivalence class of $f(s,\ \hat{s})$ and the set $\mathcal{F}$ are denoted by $\mathcal{E}_\mathcal{P}(f)$ and $\mathcal{E}_\mathcal{P}(\mathcal{F})$, respectively.

Lemma 2. *[3] Let $f_1(s,\ \hat{s}) \in \mathbb{R}_{d_1}(\theta)$, and $f_2(s,\ \hat{s}) \in \mathbb{R}_{d_2}(\theta)$ and $g(s,\ \hat{s}) \in \mathbb{R}_p(\theta)$ be the GCD. Let $d \in PGL\left(1, \mathbb{C}/\mathbb{R}\right) : (s, \hat{s}) \xrightarrow{d} \left(\lambda, \hat{\lambda}\right)$, and*

$$\tilde{f}_1(\lambda,\ \hat{\lambda}) = d \circ f_1(s,\ \hat{s}), \quad \tilde{f}_2(\lambda,\ \hat{\lambda}) = d \circ f_2(s,\ \hat{s}) \quad and \quad \tilde{g}(\lambda,\ \hat{\lambda}) = d \circ g(s,\ \hat{s}).$$

Then $\tilde{g}(\lambda,\ \hat{\lambda})$ is a GCD of $\widetilde{f}_1(\lambda,\ \hat{\lambda}),\ \ \widetilde{f}_2(\lambda,\ \hat{\lambda})$.

Proof. Since $g(s,\ \hat{s})$ is a GCD of $f_1(s,\ \hat{s})$ and $f_2(s,\ \hat{s})$, then

$$f_1(s,\ \hat{s}) = h_1(s,\ \hat{s}) \cdot g(s,\ \hat{s}) \text{ and } f_2(s,\ \hat{s}) = h_2(s,\ \hat{s}) \cdot g(s,\ \hat{s})$$

where $h_1(s,\ \hat{s})$ and $h_2(s,\ \hat{s})$ are homogeneous polynomials.

Clearly, $d \circ f_1(s,\ \hat{s}) = (d \circ h_1(s,\ \hat{s}))\,(d \circ g(s,\ \hat{s}))$. Now, we assume that

$$\widetilde{h}_1(\lambda,\ \hat{\lambda}) = d \circ h_1(s,\ \hat{s}) \text{ and } \widetilde{g}(\lambda,\ \hat{\lambda}) = d \circ g(s,\ \hat{s}).$$

Similarly, we have $d \circ f_2(s,\ \hat{s}) = (d \circ h_2(s,\ \hat{s}))\,(d \circ g(s,\ \hat{s}))$ and $d \circ h_2(s,\ \hat{s}) = \widetilde{h}_2(\lambda,\ \hat{\lambda})$.

Then, we obtain

$$\widetilde{f}_1(\lambda,\ \hat{\lambda}) = \widetilde{h}_1(\lambda,\ \hat{\lambda}) \cdot \widetilde{g}(\lambda,\ \hat{\lambda}) \text{ and } \widetilde{f}_2(\lambda,\ \hat{\lambda}) = \widetilde{h}_2(\lambda,\ \hat{\lambda}) \cdot \widetilde{g}(\lambda,\ \hat{\lambda}).$$

Consequently, it is derived that

$$\widetilde{g}(\lambda,\ \hat{\lambda}) \,\Big|\, \widetilde{f}_1(\lambda,\ \hat{\lambda}), \text{ and } \widetilde{g}(\lambda,\ \hat{\lambda}) \,\Big|\, \widetilde{f}_2(\lambda,\ \hat{\lambda}).$$

Let us consider that the GCD of $\widetilde{f}_1(\lambda,\ \hat{\lambda})$ and $\widetilde{f}_2(\lambda,\ \hat{\lambda})$ is $\tilde{g}'(\lambda,\ \hat{\lambda})$.

Since $\tilde{g}(\lambda,\ \hat{\lambda})$ is the GCD of $\widetilde{f}_1(\lambda,\ \hat{\lambda})$ and $\widetilde{f}_2(\lambda,\ \hat{\lambda})$, it holds that

$$\tilde{g}(\lambda,\ \hat{\lambda}) \,\Big|\, \tilde{g}'(\lambda,\ \hat{\lambda}).$$

So $\tilde{g}'(\lambda,\ \hat{\lambda}) = \tilde{g}(\lambda,\ \hat{\lambda}) \cdot \widetilde{\phi}(\lambda,\ \hat{\lambda})$, where $\widetilde{\phi}(\lambda,\ \hat{\lambda})$ is a homogeneous polynomial.

By applying the inverse transformation $d^{-1} : (\lambda,\ \hat{\lambda}) \ \rightarrow \ (s,\ \hat{s})$, we obtain

$$d^{-1} \circ \widetilde{g}(\lambda,\ \hat{\lambda}) = \left(d^{-1} \circ \widetilde{g}(\lambda,\ \hat{\lambda})\right)\left(d^{-1} \circ \widetilde{\phi}(\lambda,\ \hat{\lambda})\right) \ \Leftrightarrow \ \bar{g}(s,\ \hat{s}) = g(s,\ \hat{s}) \cdot \phi(s,\ \hat{s})$$

(note that $\bar{g}(s,\ \hat{s}) = d^{-1} \circ \widetilde{g}(\lambda,\ \hat{\lambda})$ and $\phi(s,\ \hat{s}) = d^{-1} \circ \widetilde{\phi}(\lambda,\ \hat{\lambda})$). Thus, we take

$$g(s,\ \hat{s}) \mid \bar{g}(s,\ \hat{s}). \tag{4}$$

However, since we have

$$\widetilde{g}(\lambda,\ \hat{\lambda}) \,\Big|\, \widetilde{f}_1(\lambda,\ \hat{\lambda}) \ \Rightarrow \ d^{-1} \circ \widetilde{g}(\lambda,\ \hat{\lambda}) \,\Big|\, d^{-1} \circ \widetilde{f}_1(\lambda,\ \hat{\lambda}), \text{ then } \bar{g}(s,\ \hat{s}) \mid \overline{f}_1(s,\ \hat{s}).$$

Moreover, we have also

$$\widetilde{g}(\lambda,\ \hat{\lambda}) \,\Big|\, \widetilde{f}_2(\lambda,\ \hat{\lambda}) \ \Rightarrow \ d^{-1} \circ \widetilde{g}(\lambda,\ \hat{\lambda}) \,\Big|\, d^{-1} \circ \widetilde{f}_2(\lambda,\ \hat{\lambda}) \ \Leftrightarrow \ \bar{g}(s,\ \hat{s}) \mid \overline{f}_2(s,\ \hat{s}).$$

From the statement, $g(s,\ \hat{s})$ is the GCD of $f_1(s,\ \hat{s})$ and $f_2(s,\ \hat{s})$, so

$$\bar{g}(s,\ \hat{s}) \mid g(s,\ \hat{s}) \tag{5}$$

From (4) and (5), we obtain

$$\bar{g}(s,\ \hat{s}) = g(s,\ \hat{s}) \ \Rightarrow\ d^{-1} \circ \bar{g}(s,\ \hat{s}) = d^{-1} \circ g(s,\ \hat{s}) \ \Leftrightarrow\ \tilde{g}(\lambda,\ \hat{\lambda}) = \tilde{g}(\lambda,\ \hat{\lambda}).$$

Consequently, $\widetilde{g}(\lambda,\ \hat{\lambda})$ is the GCD of $\widetilde{f_1}(\lambda,\ \hat{\lambda})$ and $\widetilde{f_2}(\lambda,\ \hat{\lambda})$. □

An immediate consequence of the above lemma is the following interesting and known Proposition, see [4].

Proposition 2. *[4] Let $L_1(s,\ \hat{s}) = sF_1 - \hat{s}G_1$, and $L_2(\lambda,\ \hat{\lambda}) = \lambda F_2 - \hat{\lambda}G_2 \in \mathcal{L}(\Theta)$, and define*

$$\mathcal{F}(F_1,\ G_1) = \left\{\widetilde{f_i}(s,\ \hat{s}); i = 1,\ 2,\ \ldots,\ p_1 - k_1\right\},$$
$$\mathcal{F}(F_2,\ G_2) = \left\{\widetilde{f_i}(\lambda,\ \hat{\lambda});\ i = 1,\ 2,\ \ldots,\ p_2 - k_2\right\}$$

be the corresponding homogeneous invariant polynomial sets of $L_1(s,\ \hat{s})$ and $L_2(\lambda,\ \hat{\lambda})$, where (p_1, k_1), (p_2, k_2) are the ordered pairs with rank (p) and power (k), respectively.

If $L_1(s,\ \hat{s})\mathcal{E}_{\mathcal{H}-\mathcal{B}}L_2(\lambda,\ \hat{\lambda})$ for $h \in \mathcal{H}$ and $b \in \mathcal{B}$ is generated by $d \in PGL\,(1, \mathbb{C}/\mathbb{R})$ such that $(s, \hat{s}) \xrightarrow{d} \left(\lambda, \hat{\lambda}\right)$, then

(i) $p_1 = p_2 = p$ and $k_1 = k_2 = k$.

(ii) $\mathcal{F}(F_1,\ G_1)\mathcal{E}_{\mathcal{P}}\mathcal{F}(F_2,\ G_2)$.

Another, quite interesting result of the effect of SBE ($\mathcal{E}_{\mathcal{H}-\mathcal{B}}$) on the sets $\mathcal{I}_c\,(F, G)$ and $\mathcal{I}_r\,(F, G)$ of L_θ is given by the next Proposition, see again [4]. Here, a slightly different proof is presented.

Proposition 3. *Let $L_1(s,\ \hat{s}) = sF_1 - \hat{s}G_1$, and $L_2(\lambda,\ \hat{\lambda}) = \lambda F_2 - \hat{\lambda}G_2 \in \mathcal{L}(\Theta)$, and*

$$\mathcal{I}_c\,(F_1, G_1)\,, \mathcal{I}_c\,(F_2, G_2)\ and\ \mathcal{I}_r\,(F_1, G_1)\,, \mathcal{I}_r\,(F_2, G_2)$$

be the corresponding sets of column minimal indices (cmi) and row minimal indices (rmi) of $L_1(s,\ \hat{s})$ and $L_2(\lambda,\ \hat{\lambda})$, respectively. If $L_1(s,\ \hat{s})\mathcal{E}_{\mathcal{H}-\mathcal{B}}L_2(\lambda,\ \hat{\lambda})$ then

$$\mathcal{I}_c\,(F_1, G_1) = \mathcal{I}_c\,(F_2, G_2)\ and \mathcal{I}_r\,(F_1, G_1) = \mathcal{I}_r\,(F_2, G_2)\,.$$

Proof. Generally speaking, let assume that $U(s,\ \hat{s})$ is a homogeneous minimal basis for $\mathcal{N}_r\,\{sF_1 - \hat{s}G_1\}$ and let $\underline{u}(s,\ \hat{s})$ be a minimal degree vector of $U(s,\ \hat{s})$.

Here, we shall mention that the linear transformation $(s, \hat{s}) \xrightarrow{d} \left(\lambda, \hat{\lambda}\right)$ can not increase the degree of the vector $\tilde{\underline{u}}(\lambda,\ \hat{\lambda}) = d \circ \underline{u}(s,\ \hat{s})$, which lays into the set of $\mathcal{N}_r \left\{\lambda F_2 - \hat{\lambda}G_2\right\}$ (where $\lambda F_2 - \hat{\lambda}G_2$ is the transformed pencil). On contrary, our linear transformation is possible to decrease the degree of $\tilde{\underline{u}}(\lambda,\ \hat{\lambda})$ by canceling some common factors with the vector $\underline{u}(s,\ \hat{s})$. However, considering the last case, if we apply the inverse transformation $d^{-1}: \left(\lambda, \hat{\lambda}\right) \to (s, \hat{s})$, a new vector, $\underline{u}'(s,\ \hat{s})$ in $\mathcal{N}_r\left\{sF_1 - \hat{s}G_1\right\}$ is obtained. This fact contradicts our initial hypothesis, since $\underline{u}(s,\ \hat{s})$ is a minimal degree vector of $U(s,\ \hat{s})$. Consequently, we have proved $\mathcal{I}_c\left(F_1, G_1\right) = \mathcal{I}_c\left(F_2, G_2\right)$.

Using exactly similar approach to the transposed pencil $\left(sF_1 - \hat{s}G_1\right)^t$, the second equality can be straightforwardly proved. □

Indeed, Proposition 2 and 3 express respectively the *covariance* property of the homogeneous invariant polynomials and the *invariance* property of the sets of cmi and rmi of L_θ under $\mathcal{E}_{\mathcal{H}-\mathcal{B}}$ equivalence. Now, by simply combining the two above-mentioned results, the following significant Theorem (criterion) for the $\mathcal{E}_{\mathcal{H}-\mathcal{B}}$ equivalence of the matrix pencils is finally obtained.

Theorem 1. *[3] Let* $L_1(s,\ \hat{s}) = sF_1 - \hat{s}G_1$*, and* $L_2(\lambda,\ \hat{\lambda}) = \lambda F_2 - \hat{\lambda}G_2 \in \mathcal{L}\left(\Theta\right)$.

$L_1(s,\ \hat{s})\mathcal{E}_{\mathcal{H}-\mathcal{B}}L_2(\lambda,\ \hat{\lambda})$ *if and only if the following conditions hold true:*

(i) $\mathcal{I}_c\left(F_1, G_1\right) = \mathcal{I}_c\left(F_2, G_2\right)$ *and* $\mathcal{I}_r\left(F_1, G_1\right) = \mathcal{I}_r\left(F_2, G_2\right)$.

(ii) $\mathcal{F}(F_1,\ G_1)\mathcal{E}_{\mathcal{P}}\mathcal{F}(F_2,\ G_2)$.

Proof. The necessity ("⇒") derives immediately by simply combining Proposition 2 and 3. Now, in order to prove the sufficiency ("⇐"), we will assume that (i) and (ii) hold.

Since $\mathcal{F}(F_1,\ G_1)\mathcal{E}_{\mathcal{P}}\mathcal{F}(F_2,\ G_2)$, then $L_1(s,\ \hat{s})$ and $L_2(\lambda,\ \hat{\lambda})$ have the same rank ($p_1 = p_2 = p$) and the same power ($k_1 = k_2 = k$), and there exists a transformation $d \in PGL\left(1, \mathbb{C}/\mathbb{R}\right)$ such that $(s, \hat{s}) \xrightarrow{d} \left(\lambda, \hat{\lambda}\right)$, and

$$d \circ f_{2,\,i}(\lambda,\ \hat{\lambda}) = C_{2,\,i}\widetilde{f}_{2,\,i}(s,\ \hat{s}) \quad \forall\ i = 1,\ 2,\ \ldots,\ p - k.$$

The transformation $d \in PGL\left(1, \mathbb{C}/\mathbb{R}\right)$ generates a $b \in \mathcal{B}$ such that

$$b \circ L_2(\lambda,\ \hat{\lambda}) = s\tilde{F}_2 - \hat{s}\tilde{G}_2 = \tilde{L}_2(s,\ \hat{s}).$$

Since $\tilde{L}_2(s,\ \hat{s})\mathcal{E}_{\mathcal{B}}L_2(\lambda,\ \hat{\lambda})$, then

$$\mathcal{I}_c\left(\tilde{F}_2, \tilde{G}_2\right) = \mathcal{I}_c\left(F_2, G_2\right) = \mathcal{I}_c\left(F_1, G_1\right),$$

and

$$\mathcal{I}_r\left(\tilde{F}_2, \tilde{G}_2\right) = \mathcal{I}_r\left(F_2, G_2\right) = \mathcal{I}_r\left(F_1, G_1\right).$$

Furthermore, the sets of homogeneous invariant polynomials of $\widetilde{L}_2(s,\ \hat{s})$ and $\widetilde{L}_1(s,\ \hat{s})$ differ only by a scalar value c (see the definition of $\mathcal{E}_{\mathcal{P}}$-equivalence).

Thus, $\widetilde{L}_2(s,\ \hat{s})$ and $\widetilde{L}_1(s,\ \hat{s})$ have the same Smith form over $\mathbb{R}\left[s, \hat{s}\right]$ (or equivalently the same sets of e.d.), and the same sets of cmi and rmi. Consequently, $\tilde{L}_2(s,\ \hat{s})\mathcal{E}_{\mathcal{H}}L_1(s,\ \hat{s})$. Therefore, there exists a $h \in \mathcal{H}$ such that $L_1(s,\ \hat{s}) = h \circ \tilde{L}_2(s,\ \hat{s})$.

Given now that $\tilde{L}_2(s,\ \hat{s}) = b \circ L_2(\lambda,\ \hat{\lambda})$, it follows $L_1(s,\ \hat{s}) = (h * b) \circ L_2(\lambda,\ \hat{\lambda})$, and thus $L_1(s,\ \hat{s})\mathcal{E}_{\mathcal{H}-\mathcal{B}}L_2(\lambda,\ \hat{\lambda})$. □

As a strong result of the above-mentioned theorem, the key factor of the characterization of $\mathcal{E}_{\mathcal{H}-\mathcal{B}}$-equivalence is the $\mathcal{E}_{\mathcal{P}}$-equivalence, which is defined on the set of homogeneous invariant polynomial $\mathcal{F}(F,\ G)$ of the pencil $sF - \hat{s}G$. Consequently, in order to investigate if two pencils are $\mathcal{E}_{\mathcal{H}-\mathcal{B}}$ equivalent, it is equivalent to investigate the conditions under which two polynomials $f(s,\ \hat{s}),\ \tilde{f}(\lambda,\ \hat{\lambda}) \in \mathbb{R}_d(\Theta)$ are $\mathcal{E}_{\mathcal{P}}$ -equivalent. Now, the problem is being transposed to determine the conditions under which $f(s,\ \hat{s})\mathcal{E}_{\mathcal{P}}\tilde{f}(\lambda,\ \hat{\lambda})$. Equivalently, the complete and independent set of invariants for the orbit $\mathcal{E}_{\mathcal{P}}\left(f\left(s, \hat{s}\right)\right)$ should be determined. This new status is being considered next.

3. Projective Equivalence of Homogeneous Binary Polynomials

The aim of this section is to give a number of results for the $\mathcal{E}_{\mathcal{P}}$ - equivalence defined on the set $\mathbb{R}_d(\Theta)$. The origin of $\mathcal{E}_{\mathcal{P}}$ - equivalence stems back to the classical book by Turnbull, see [10], for the algebraic theory of invariants. Initially, we will try to provide all the necessary conditions and definitions in order to become clearer that the determination of the compete set of invariants for $\mathcal{E}_{\mathcal{P}}$-equivalence is the same with the determination of the conditions under which two symmetric sets of points on $\mathbb{C} \cup \{\infty\}$ can be connected under a projective transformation $d \in PGL\left(1, \mathbb{C}/\mathbb{R}\right)$; see [3] and [7].

As we will see in the next sections, the latter problem can be expressed equivalently by the study of invariants of matrices under the notion of *extended Hermite equivalence* (EHE). Let us start with some interesting lemmas; some slightly modifications to proofs of some known results have also been proposed.

Lemma 3. *[9] Let*

$$f(s,\ \hat{s}) = r_1 s^2 + p_1 s\hat{s} + q_1 \hat{s}^2 \text{and } \tilde{f}(\lambda,\ \hat{\lambda}) = r_2\lambda^2 + p_2\lambda\hat{\lambda} + q_2\lambda_2^2 \in \mathbb{R}_2(\Theta)$$

and

$$\Delta_1 = p_1^2 - 4r_1q_1 \text{and } \Delta_2 = p_2^2 - 4r_2q_2$$

be their corresponding determinants, respectively.

If $f(s,\ \hat{s})\mathcal{E}_{\mathcal{P}}\tilde{f}(\lambda,\ \hat{\lambda})$, i.e., $d \circ f(s,\ \hat{s}) = c \cdot \tilde{f}(\lambda,\ \hat{\lambda})$, then

$$\Delta_2 = \frac{(\alpha\delta - \beta\gamma)^2}{c^2}\Delta_1. \tag{6}$$

Proof. We can write the polynomial

$$f(s,\ \hat{s}) = [s,\ \hat{s}] \cdot \begin{bmatrix} r_1 & p_1/2 \\ p_1/2 & q_1 \end{bmatrix} \cdot \begin{bmatrix} s \\ \hat{s} \end{bmatrix} = \underline{s}^t \cdot D_1 \cdot \underline{s},$$

where

$$\underline{s} = \begin{bmatrix} s \\ \hat{s} \end{bmatrix} \text{ and } D_1 = \begin{bmatrix} r_1 & p_1/2 \\ p_1/2 & q_1 \end{bmatrix}.$$

Similarly, we have

$$\tilde{f}\lambda,\ \hat{\lambda}) = [\lambda,\ \hat{\lambda}] \cdot \begin{bmatrix} r_2 & p_2/2 \\ p_2/2 & q_2 \end{bmatrix} \cdot \begin{bmatrix} \lambda \\ \hat{\lambda} \end{bmatrix} = \underline{\lambda}^t \cdot D_2 \cdot \underline{\lambda},$$

where

$$\underline{\lambda} = \begin{bmatrix} \lambda \\ \hat{\lambda} \end{bmatrix} \text{ and } D_2 = \begin{bmatrix} r_2 & p_2/2 \\ p_2/2 & q_2 \end{bmatrix}.$$

Obviously, we take

$$\det D_1 = -\frac{1}{4}\Delta_1 \text{and } \det D_2 = -\frac{1}{4}\Delta_2.$$

The transformation $d \in PGL\left(1, \mathbb{C}/\mathbb{R}\right)$ gives

$$\begin{bmatrix} s \\ \hat{s} \end{bmatrix} = \begin{bmatrix} \alpha & \beta \\ \gamma & \delta \end{bmatrix} \cdot \begin{bmatrix} \lambda \\ \hat{\lambda} \end{bmatrix} \Leftrightarrow \begin{bmatrix} s \\ \hat{s} \end{bmatrix} = [d] \cdot \begin{bmatrix} \lambda \\ \hat{\lambda} \end{bmatrix} \Leftrightarrow \underline{s} = [d] \cdot \underline{\lambda},$$

and

$$|[d]| = \alpha\delta - \beta\gamma \neq 0.$$

Then, we derive

$$d \circ f(s,\ \hat{s}) = ([d] \cdot \underline{\lambda})^t D_1 \left([d] \cdot \underline{\lambda}\right) = \underline{\lambda}^t [d]^t D_1 \, [d] \, \underline{\lambda}.$$

Moreover, we take

$$c\tilde{f}(\lambda,\ \hat{\lambda}) = c \cdot \underline{\lambda}^t \cdot D_2 \cdot \underline{\lambda}.$$

Since, the following equality holds

$$d \circ f(s,\ \hat{s}) = c\tilde{f}(\lambda,\ \hat{\lambda}) \ \Leftrightarrow\ \underline{\lambda}^t [d]^t \cdot D_1 \cdot [d] \cdot \underline{\lambda} = \lambda^t \cdot c \cdot D_2 \cdot \underline{\lambda} \ \ \forall\ \underline{\lambda} \in \Theta.$$

Thus,

$$[d]^t \cdot D_1 [d] = c \cdot D_2 \ \Rightarrow\ \det\left([d]^t \cdot D_1 \cdot [d]\right) = \det(c \cdot D_2) \ \Leftrightarrow$$

$$\det\,[d]^t \cdot \det D_1 \cdot \det[d] = c^2 \cdot \det D_2 \ \Leftrightarrow\ (\alpha\delta - \beta\gamma)^2 \cdot \left(-\frac{1}{4}\right) \cdot \Delta_1 =$$

$$c^2 \left(-\frac{1}{4}\right) \Delta_2 \ \Leftrightarrow\ (\alpha\delta - \beta\gamma)^2 \cdot \Delta_1 = c^2 \cdot \Delta_2 \ \Leftrightarrow\ \Delta_2 = \frac{(\alpha\delta - \beta\gamma)^2}{c^2} \Delta_1.$$

□

Remark 1. The condition (6) implies that the determinant Δ_1 is an invariant of the polynomial $f(s,\ \hat{s}) = rs^2 + ps\hat{s} + q\hat{s}^2$ of weight 2 under the $\mathcal{E}_{\mathcal{P}}$-equivalence, see also [10]. Furthermore, the equality $sign\left\{\Delta_1\right\} = sign\left\{\Delta_2\right\}$ and the *reducibility property* over $\mathbb{R}$ of $f(s,\ \hat{s}) \in \mathbb{R}_2\left\{\Theta\right\}$ are invariant under $\mathcal{E}_{\mathcal{P}}$-equivalence.

A straightforward consequence of Lemma 4 and Remark 1 is the following proposition. Its proof follows immediately by the way the projective transformation $d \in PGL\left(1, \mathbb{C}/\mathbb{R}\right)$ is applied on the unique factorization of $f(s,\ \hat{s})$.

Proposition 4. *[3] Let* $p_i(s,\ \hat{s}) = (\gamma_i s - \delta_i \hat{s})$ *with* $\gamma_i,\ \delta_i \in \mathbb{C}$ *and* $(\gamma_i, \delta_i) \neq (0, 0)$ *be the primes over* $\mathbb{C}$ *of* $f(s,\ \hat{s}) \in \mathbb{R}_d(\theta)$*, i.e.,*

$$f(s,\ \hat{s}) = \prod_{i=1}^{d} p_i(s,\ \hat{s}).$$

If we take

$$\widetilde{f}(\lambda,\ \hat{\lambda}) \in \mathbb{R}_d(\Theta) \text{ and } f(s,\ \hat{s})\mathcal{E}_{\mathcal{P}}\widetilde{f}(\lambda,\ \hat{\lambda}),$$

for a $d \in PGL\,(1, \mathbb{C}/\mathbb{R})$*, then*

(i) *any pair* $p_i(s,\ \hat{s}),\ \ p_j(s,\ \hat{s})$ *(with* $(\gamma_i, \delta_i) \neq \zeta(\gamma_j, \delta_j)\ \forall \zeta \in \mathbb{C}\backslash\{0\}$ *of is being mapped under* d *to a pair of complex conjugate primes of* $\widetilde{f}(\lambda,\ \hat{\lambda})$ *and vice versa.*

(ii) *any pair of complex conjugate primes of* $f(s,\ \hat{s})$*, i.e.,*

$$p(s,\ \hat{s}) = (\gamma s - \delta\hat{s}) \text{ and } \overline{p}(s,\ \hat{s}) = (\overline{\gamma}s - \overline{\delta}\hat{s}),$$

is being mapped under d *to a pair of complex conjugate primes of* $\widetilde{f}(\lambda,\ \hat{\lambda})$ *and vice versa.*

(iii) *any pair* $p(s,\ \hat{s}),\ \ p(s,\ \hat{s})$ *of repeated primes* $(\gamma_i, \delta_i) \neq \zeta(\gamma_j', \delta_j'), \zeta \in \mathbb{C}\backslash\{0\}$ *of* $f(s,\ \hat{s})$ *is being mapped under* d *to a pair of complex conjugate primes of* $\widetilde{f}(\lambda,\ \hat{\lambda})$ *and vice versa.*

Now, consider a polynomial $f(s,\ \hat{s}) \in \mathbb{R}_d(\Theta)$ and we denote with

$$\mathcal{D}_{\mathbb{R}}(f) = \{(\alpha_i s - \beta_i\hat{s})^{\tau_i}\alpha_i,\ \beta_i \in \mathbb{R},\ \ (\alpha_i,\ \beta_i) \neq (0,\ 0) \text{ and } \ i = 1,\ 2,\ \ldots, \mu\}, \quad (7)$$

and

$$\mathcal{D}_{\mathbb{C}}(f) = \left\{(\gamma_i s - \delta_i\hat{s})^{p_i},\ (\bar{\gamma}_i s - \bar{\delta}_i\hat{s})^{p_i},\ \ \gamma_i,\ \delta_i \in \mathbb{C},\ \ (\gamma_i,\ \delta_i) \neq (0,\ 0),\ \ i = 1,\ 2,\ \ldots, \nu\right\}, \quad (8)$$

the set of real and complex e.d. of $f(s,\ \hat{s})$ over $\mathbb{C}$, respectively. Note that $\mathcal{D}_{\mathbb{C}}(f)$ is symmetric, i.e., if $(\gamma_i s - \delta_i\hat{s})^{p_i} \in \mathcal{D}_{\mathbb{C}}(f) \Leftrightarrow (\bar{\gamma}_i s - \bar{\delta}_i\hat{s})^{p_i} \in \mathcal{D}_{\mathbb{C}}(f)$. Moreover, it is true that

$$\sum_{i=1}^{\mu}\tau_i + 2\sum_{i=1}^{\nu}p_i = d. \quad (9)$$

Reminding Proposition 4, the $\mathcal{E}_{\mathcal{P}}$-equivalence can be expressed by defining the sets (7) and (8), see Proposition 5. The proof is a very straightforward result, so it is omitted.

Proposition 5. *Let* $f(s,\ \hat{s}),\ \ \widetilde{f}(\lambda,\ \hat{\lambda}) \in \mathbb{R}_d(\Theta)$ *and*

$$\left(\mathcal{D}_{\mathbb{R}}(f),\ \mathcal{D}_{\mathbb{C}}(f)\right),\quad \left(\mathcal{D}_{\mathbb{R}}(\widetilde{f}),\ \mathcal{D}_{\mathbb{C}}(\widetilde{f})\right)$$

be the unique factorization sets, then $f(s,\ \ \hat{s})\mathcal{E}_{\mathcal{P}}\widetilde{f}(\lambda,\ \ \hat{\lambda})$ *if and only if there exists* $d \in PGL\,(1, \mathbb{C}/\mathbb{R})$ *such that the following conditions hold*

(i) *For every* $e_i(s,\ \hat{s}) \in \mathcal{D}_{\mathbb{R}}(f)$*, there exists a* $\tilde{e}_i(\lambda,\ \hat{\lambda}) \in \mathcal{D}_{\mathbb{R}}(\widetilde{f})$ *such that*

$$e_i(s,\ \hat{s})\mathcal{E}_{\mathcal{P}}\tilde{e}_i(\lambda,\ \hat{\lambda}),$$

or equivalently $d \circ e_i(s,\ \hat{s}) = c_i \cdot \widetilde{e}_i(\lambda,\ \hat{\lambda}),\ \ c_i \in \mathbb{R}\backslash\{0\}$ *and vice versa.*

***(ii)** For every $e_i(s,\ \hat{s}) \in \mathcal{D}_{\mathbb{C}}(f)$, there exists a $\tilde{e}_i'(\lambda,\ \hat{\lambda}) \in \mathcal{D}_{\mathbb{C}}(\tilde{f})$ such that*

$$e_i'(s,\ \hat{s})\mathcal{E}_{\mathcal{P}}\tilde{e}_i'(\lambda,\ \hat{\lambda}),$$

or equivalently $d \circ e_i'(s,\ \hat{s}) = c_i \cdot \tilde{e}_i'(\lambda,\ \hat{\lambda}),\ \ c_i \in \mathbb{C}\backslash\{0\}$ *and vice versa.*

Proposition 5 expresses the *covariance* property of the sets (7) and (8) of $f(s,\ \hat{s})$ under the $\mathcal{E}_{\mathcal{P}}$-equivalence. With the above result, it becomes clear that the study of the invariants of $\mathcal{E}_{\mathcal{P}}(f)$ is equivalent with the study of properties of the elementary divisor sets under a $d \in PGL(1, \mathbb{C}/\mathbb{R})$ transformation.

Before, we go any further, it is important to underline that the real e.d. $(\alpha s - \beta\hat{s})^{\tau}$ can be represented by an ordered triple $(\alpha, \beta; \tau)$ where $\alpha, \beta \in \mathbb{R}$ and $\tau\mathbb{Z}$. Similarly, the pair of complex conjugate e.d. $(\gamma s - \delta\hat{s})^p$, $(\overline{\gamma}s - \overline{\delta}\hat{s})^p$ can be also represented by an ordered triple $(\gamma, \delta; p)$ where $\gamma, \delta \in \mathbb{C}$ and $p \in \mathbb{Z}$. Using now the two representations, we define the following sets.

Definition 3. **(i)** We define the set

$$\mathcal{B}_i{}' \triangleq \left\{(\gamma_j^i,\ \delta_j^i\ ;\ p_i)\ \ with\ \ \gamma_j^i,\ \delta_j^i \in \mathbb{C},\ \ p_i \in \mathbb{Z},\ \ j = 1,\ 2,\ \ldots,\ \nu_i\right.$$

and

$$\left.(\gamma_j^i,\ \delta_j^i) \neq \xi(\gamma_k^i,\ \delta_k^i)\ \ \forall\, j \neq k,\ \ \xi \in \mathbb{C}\backslash\{0\}\right\},$$

as the set of all ordered triples corresponding the elements

$$(\gamma_j^i s - \delta_j^i\hat{s})^{p_i},\quad (\bar{\gamma}_j^i s - \bar{\delta}_j^i\hat{s})^{p_i},$$

in $\mathcal{D}_{\mathbb{C}}(f)$ with the same degree p_i.

An ordering of the set $\mathcal{B}_i{}'$ is defined by any permutation of its elements which is defined by $\pi\,(\mathcal{B}_i{}')$. The set of all such permutations is denoted by $< \mathcal{B}_i{}' >$.

(ii) We define

$$\mathcal{B}_{\mathbb{C}}(f) \triangleq \{\mathcal{B}_1{}',\ \mathcal{B}_2{}',\ \ldots,\ \mathcal{B}_\sigma{}' : p_1 < p_2 < \ldots < p_\sigma\},$$

which corresponds to the set of all pair of complex conjugate e.d. and it is referred as the *unique complex factorization* set ($\mathbb{C}$-UFS) of $f(s,\ \hat{s})$.

The set $\mathcal{J}_{\mathbb{C}}(f) \triangleq (p_1, \nu_1), (p_2, \nu_2), (p_\sigma, \nu_\sigma)$, where ν_i is the number of elements for $\mathcal{B}_i{}'$, is known as the complex list of polynomials. Every permutation of $\mathcal{B}_{\mathbb{C}}(f)$, i.e.,

$$\pi\,(\mathcal{B}_{\mathbb{C}}(f)) = \left\{\pi(\mathcal{B}_1'),\ \ldots,\ \pi(\mathcal{B}_\sigma{}') : \pi(\mathcal{B}_i{}') \in < \mathcal{B}_i{}' >\right\},$$

defines a normal ordering of $\mathcal{B}_{\mathbb{C}}(f)$. Note that the set of all such permutations is denoted by $< \mathcal{B}_{\mathbb{C}}(f) >$.

(iii) Let $\pi\,(\mathcal{B}_{\mathbb{C}}(f))$, where

$$\pi(\mathcal{B}_i{}') = \left\{(\gamma_1^i,\ \delta_1^i;\ p_i),\ \ldots,\ (\gamma_{\nu_i}^i,\ \delta_{\nu_i}^i;\ p_i)\right\} \in < \mathcal{B}_i{}' > .$$

A matrix representation of $\pi(\mathcal{B}_{\mathbb{C}}(f))$ is given by

$$[\mathcal{B}_{\mathbb{C}}^{\pi}(f)] = \begin{bmatrix} \begin{bmatrix} \mathcal{B}_1'^{\pi} \end{bmatrix} \\ \begin{bmatrix} \mathcal{B}_2'^{\pi} \end{bmatrix} \\ \vdots \\ \begin{bmatrix} \mathcal{B}_{\sigma}'^{\pi} \end{bmatrix} \end{bmatrix}, \text{ where } \begin{bmatrix} \mathcal{B}_i'^{\pi} \end{bmatrix} = \begin{bmatrix} \gamma_1^i & \delta_1^i \\ \gamma_2^i & \delta_2^i \\ \vdots & \vdots \\ \gamma_{\nu_i}^i & \delta_{\nu_i}^i \end{bmatrix} \in \mathbb{C}^{\nu_i \times 2}. \tag{10}$$

The matrix (10) is referred as a $(\mathbb{C} - \pi)$ basis matrix of $f(s, \hat{s})$.

(iv) For the set $\mathcal{D}_{\mathbb{R}}(f)$, we define in a similar manner the sets

$$\mathcal{B}_j \triangleq \{(\alpha_j^i, \beta_j^i; \tau_i) : \alpha_j^i, \beta_j^i \in \mathbb{R}, \ \tau_i \in \mathbb{Z} \text{ and } j = 1, 2, \ldots, \mu_j$$

$$\text{with } (\alpha_j^i, \beta_j^i) \neq \zeta(\alpha_k^i, \beta_k^i), \ \forall j \neq k \text{ with } \zeta \in \mathbb{R}\backslash\{0\}\},$$

$$\mathcal{B}_{\mathbb{R}}(f) \triangleq= \{B_1, B_2, \ldots, B_{\rho}; \tau_1 < \tau_2 \ldots < \tau_{\rho}\}$$

and

$$\mathcal{J}_{\mathbb{R}}(f) \triangleq \{(\tau_1, \mu_1), (\tau_2, \mu_2), \ldots, (\tau_{\rho}, \mu_{\rho})\},$$

as well as the notions of normal ordering and of the matrix representation. The sets $\mathcal{B}_{\mathbb{R}}(f)$, and $\mathcal{J}_{\mathbb{R}}(f)$ are referred as the *unique real factorization set* ($\mathbb{C}$-UFS) and the *real list* of $f(s, \hat{s})$, respectively. The matrix $[\mathcal{B}_{\mathbb{R}}^{\pi}(f)]$ is defined similarly as in (10) and it is referred as a $(\mathbb{R} - \pi)$ basis matrix of $f(s, \hat{s})$.

(v) Now, the sets $\mathcal{B}(f) \triangleq \{\mathcal{B}_{\mathbb{R}}(f), \mathcal{B}_{\mathbb{C}}(f)\}$ and $\mathcal{J}(f) = \{\mathcal{J}_{\mathbb{R}}(f), \mathcal{J}_{\mathbb{C}}(f)\}$ are called the *unique factorization* set (UFS) and the list of $f(s, \hat{s})$, respectively.

For every $\pi \in < \mathcal{B}_{\mathbb{R}}(f) >$ and $\pi' \in < \mathcal{B}_{\mathbb{C}}(f) >$, a matrix representation of $\mathcal{B}(f)$ for the pair (π, π') is defined by

$$\left[\mathcal{B}^{\pi, \pi'}(f)\right] = \begin{bmatrix} [\mathcal{B}_{\mathbb{R}}^{\pi}(f)] \\ \left[\mathcal{B}_{\mathbb{C}}^{\pi'}(f)\right] \end{bmatrix}.$$

The matrix $\left[\mathcal{B}^{\pi, \pi'}(f)\right]$ is referred as a (π, π') basis matrix of

Remark 2. Following the above definitions, it is clear that since

$$\#\mathcal{B}_i' = v_i \text{and } \#\mathcal{B}_i = \mu_i, \text{then } \# < \mathcal{B}_i' >= v_i!, \# < \mathcal{B}_i >= \mu_i!,$$

we obtain

$$\# < \mathcal{B}_{\mathbb{C}}(f) >= v_1! v_2! \cdots v_{\sigma}!, \# < \mathcal{B}_{\mathbb{R}}(f) >= \mu_1! \mu_2! \cdots \mu_{\rho}!,$$

and

$$\# < \mathcal{B}(f) >= v_1! v_2! \cdots v_{\sigma}! \mu_1! \mu_2! \cdots \mu_{\rho}!.$$

With the following numerical example, the matrix representation of $\mathcal{B}(f)$ for a pair of (π, π'), where $\pi \in < \mathcal{B}_{\mathbb{R}}(f) >$ and $\pi' \in < \mathcal{B}_{\mathbb{C}}(f) >$ for a particular $f(s, \hat{s})$, is calculated analytically.

Example 2. Assume that we have the following homogeneous polynomial

$$f(s, \hat{s}) =$$
$$(s-2\hat{s})(s+3\hat{s})^2(s-\hat{s})^2(s+j\hat{s})(s-j\hat{s})\,(s+(2+j)\hat{s})\,(s+(2-j)\hat{s})\,(js+\hat{s})^2\cdot$$
$$\cdot(-js+\hat{s})^2((1+j)s-2\hat{s})^2((1-j)s-2\hat{s})^2(3\hat{s}+(1+j)\hat{s})^2(3\hat{s}+(1-j)\hat{s})^2(s+\hat{s})^2,$$

where $j^2 = -1$. Then, the following sets derive

$$\mathcal{B}_1{}' = \{(1,\ j;\ 1)(1,\ -2+j;\ 1\},$$
$$\mathcal{B}_2{}' = \{(j,\ -1;\ 2)(1+j,\ 2;\ 2)(3,\ -1+j;\ 2)\},$$
$$\mathcal{B}_1 = (1, 2; 1),$$
$$\mathcal{B}_2 = (1, 3; 2)(1, 1; 2)(1, 1; 2).$$

Then, the set

$$\mathcal{B}_{\mathbb{C}}(f) = \{\mathcal{B}_1,\ \mathcal{B}_2\}, \text{and } \mathcal{J}_{\mathbb{C}}(f) = \{(1,2)\,,\ (2,3)\}.$$

Moreover,

$$\mathcal{B}_{\mathbb{R}}(f) = \{\mathcal{B}_1,\ \mathcal{B}_2\}, \text{and } \mathcal{J}_{\mathbb{R}}(f) = \{(1,1)\,,\ (2,3)\}.$$

So, we can have a permutation π' such that

$$\left[\mathcal{B}_1'^{\pi'}\right] = \begin{bmatrix} 1 & j \\ 1 & -2+j \end{bmatrix} \in \mathbb{C}^{2\times 2},\ \left[\mathcal{B}_2'^{\pi'}\right] = \begin{bmatrix} j & -1 \\ 1+j & 2 \\ 3 & -1+j \end{bmatrix} \in \mathbb{C}^{3\times 2}.$$

And a permutation π such that

$$[\mathcal{B}_1^{\pi}] = \begin{bmatrix} 1 & 2 \end{bmatrix} \quad and \quad [\mathcal{B}_2^{\pi}] = \begin{bmatrix} 1 & -3 \\ 1 & 1 \\ 1 & -1 \end{bmatrix}.$$

Then, the set

$$\left[\mathcal{B}_{\mathbb{C}}^{\pi'}(f)\right] = \left[\begin{bmatrix} \mathcal{B}_1'^{\pi'} \\ \mathcal{B}_2'^{\pi'} \end{bmatrix}\right] = \left[\begin{array}{cc} 1 & j \\ 1 & -2+j \\ \hline j & 1 \\ 1+j & 2 \\ 3 & -1+j \end{array}\right],$$

and

$$[\mathcal{B}_{\mathbb{R}}^{\pi}(f)] = \left[\begin{array}{cc} 1 & 2 \\ \hline 1 & -3 \\ 1 & 1 \\ 1 & -1 \end{array}\right].$$

Finally, we obtain the matrix representation,

$$\left[\mathcal{B}^{\pi,\,\pi'}(f)\right] = \begin{bmatrix} [\mathcal{B}^{\pi}_{\mathbb{R}}(f)] \\ \left[\mathcal{B}^{\pi'}_{\mathbb{C}}(f)\right] \end{bmatrix} = \left[\begin{array}{cc} 1 & 2 \\ \hline 1 & -3 \\ 1 & 1 \\ 1 & -1 \\ \hline 1 & j \\ 1 & -2+j \\ \hline j & 1 \\ 1+j & 2 \\ 3 & -1+j \end{array}\right].$$

Now, having in our mind the proposed notations and definitions, the main result of this section is presented. The proof of Theorem 3 derives straightforwardly by Propositions 4 and 5, and the fact that the pair (γ, δ) (see Proposition 4) which characterizes an elementary divisor is different from any other pair of the same homogeneous polynomial.

Theorem 3. *Let* $f(s,\ \hat{s}),\ \widetilde{f}(\lambda,\ \hat{\lambda}) \in \mathbb{R}_d(\Theta)$. *Now, let*

$$\{\mathcal{B}_{\mathbb{R}}(f),\ \mathcal{J}_{\mathbb{R}}(f),\ \mathcal{B}_{\mathbb{C}}(f),\ \mathcal{J}_{\mathbb{C}}(f)\}\ \ and\ \ \{\mathcal{B}_{\mathbb{R}}(\widetilde{f}),\ \mathcal{J}_{\mathbb{R}}(\widetilde{f}),\ \mathcal{B}_{\mathbb{C}}(\widetilde{f}),\ \mathcal{J}_{\mathbb{C}}(\widetilde{f})\},$$

be the corresponding sets associated with $f(s,\ \hat{s})$ *and* $\widetilde{f}(\lambda,\ \hat{\lambda})$, *respectively.*

Then $f(s,\ \hat{s})\mathcal{E}_{\mathcal{P}}f(\lambda,\ \hat{\lambda})$ *if and only if the following conditions hold*

(i) $\mathcal{J}_{\mathbb{R}}(f) = \mathcal{J}_{\mathbb{R}}(\widetilde{f})$ *and* $\mathcal{J}_{\mathbb{C}}(f) = \mathcal{J}_{\mathbb{C}}(\widetilde{f})$.

(ii) *There exist the permutations*

$$\pi\left(\mathcal{B}_{\mathbb{R}}(f)\right) \in < \mathcal{B}_{\mathbb{R}}(f) >,\ \widetilde{\pi}\left(\mathcal{B}_{\mathbb{R}}(\widetilde{f})\right) \in < \mathcal{B}_{\mathbb{R}}(\widetilde{f}) >,$$

and

$$\pi'\left(\mathcal{B}_{\mathbb{C}}(f)\right) \in < \mathcal{B}_{\mathbb{C}}(f) >,\ \widetilde{\pi}'\left(\mathcal{B}_{\mathbb{C}}(\widetilde{f})\right) \in < \mathcal{B}_{\mathbb{C}}(\widetilde{f}) >,$$

the transformation $d \in PGL\left(1, \mathbb{C}/\mathbb{R}\right)$ *and* $\zeta_i \in \mathbb{R}\setminus\{0\}$, $\xi_i \in \mathbb{C}\setminus\{0\}$ *such that*

$$\left[\mathcal{B}^{\widetilde{\pi}}_{\mathbb{R}}(\widetilde{f})\right] = diag(\zeta_i)\left[\mathcal{B}^{\pi}_{\mathbb{R}}(f)\right][d], \tag{11}$$

$$\left[\mathcal{B}^{\widetilde{\pi}'}_{\mathbb{C}}(\widetilde{f})\right] = diag(\xi_i)\left[\mathcal{B}^{\pi'}_{\mathbb{C}}(f)\right][d]. \tag{12}$$

Another straightforward result is described in the next Corollary. The proof is omitted.

Corollary 1. *The real and the complex list* $\mathcal{J}_{\mathbb{R}}(f)$, $\mathcal{J}_{\mathbb{C}}(f)$ *of* $f(s,\ \hat{s})$ *are invariants over the* $\mathcal{E}_{\mathcal{P}}\left(f\right)$*-equivalence class.*

Now, let us take two pairs of sets

$$\mathcal{B}_{\mathbb{R}}(f) = \{\mathcal{B}_{\mathbb{R}}(f), \mathcal{B}_{\mathbb{C}}(f)\}, \text{and } \mathcal{B}(\widetilde{f}) = \{\mathcal{B}_{\mathbb{R}}(\widetilde{f}),\ \mathcal{B}_{\mathbb{C}}(\widetilde{f})\}$$

for which

$$\mathcal{J}_{\mathbb{R}}(f)=\mathcal{J}_{\mathbb{R}}(\widetilde{f}) \quad and \quad \mathcal{J}_{\mathbb{C}}(f)=\mathcal{J}_{\mathbb{C}}(\widetilde{f}),$$

and the conditions (ii) of Theorem 3 hold for a transformation $d\in PGL(1,\mathbb{C}/\mathbb{R})$ and nonzero constants ζ_i, ξ_i, then the $\mathcal{B}(f)\mathcal{E}_{\bar{\mathcal{P}}}\mathcal{B}(\widetilde{f})$ is called *normally projective equivalence* (NPE).

Profoundly, the notion of NPE on the set $\mathcal{B}(f)$ is equivalent to the $\mathcal{E}_{\mathcal{P}}$-equivalence notion defined on $\mathbb{R}_d\{\Theta\}$. In other words, $\mathcal{E}_{\bar{\mathcal{P}}}$-equivalence is nothing else but $\mathcal{E}_{\mathcal{P}}$-equivalence defined on the UFS of the polynomials in $\mathbb{R}_d\{\Theta\}$. The reason why we want to use this equivalence has to do with the fact that the NPE reduces the study of invariants of $\mathcal{E}_{\mathcal{P}}(f)$ into a standard matrix algebra problem, which has to do only with the calculation of expressions (11) and (12).

Remark 3. It is worth to point out that the study of $\mathcal{E}_{\mathcal{P}}$-equivalence, as it has been expressed by the significant Theorem 3, is equivalent to a classical algebraic projective geometric problem, i.e., given two symmetric sets of points ζ, $\tilde{\zeta}$ of $\mathbb{C}\cup\{\infty\}$, find the necessary and sufficient conditions for the existence of the projective transformation $d\in PGL(1,\mathbb{C}/\mathbb{R})$: $\zeta\overset{d}{\to}\tilde{\zeta}$.

In this part of the section, in order to be able to calculate analytically the complete set of invariants for $\mathcal{E}_{\mathcal{P}}(f)$, we should remind the extended Hermite equivalence of matrices, which is a straightforward result of $\mathcal{E}_{\bar{\mathcal{P}}}$-equivalence.

Let $T\in\mathbb{C}^{k\times 2}$. The matrix T is called *entirely nonsingular* if none of its 2×2 minors is zero. The set of the entirely non-singular matrices of $k\times 2$ dimension is denoted by $\mathbb{C}_n^{k\times 2}$. The subset of $\mathbb{C}_n^{k\times 2}$, which is constructed by real T matrices is denoted by $\mathbb{R}_n^{k\times 2}$. Clearly, the matrices $\left[\mathcal{B}_{\mathbb{C}}^{\pi'}(f)\right]$, $[\mathcal{B}_{\mathbb{R}}^{\pi}(f)]$ are entirely nonsingular complex and real matrices, respectively.

Definition 4. Let $T_1,\ T_2\in\mathbb{C}^{k\times 2}\left(or\ \mathbb{R}^{k\times 2}\right)$. T_1 and T_2 will be complex (or real) *extended Hermite equivalent*, and they will be denoted by $T_1\mathcal{E}_{eh}^{\mathbb{C}}T_2\ \left(or\ T_1\mathcal{E}_{eh}^{\mathbb{R}}T_2\right)$ if and only if there exist $\xi_i\in\mathbb{C}\backslash\{0\}$, $i=1,2,...,k$ and a matrix $Q\in\mathbb{C}^{2\times 2}$ with $\det(Q)\in\mathbb{R}\backslash\{0\}$ such that

$$T_2=diag\{\xi_i\}\,T_1Q. \tag{13}$$

Remark 4. In the case of $\mathcal{E}_{eh}^{\mathbb{C}}$-equivalence, we can always assume that

$$det(Q)=1.$$

However, in the case of $\mathcal{E}_{eh}^{\mathbb{R}}$-equivalence, we can always assume that

$$det(Q)=1 \text{ or } 1.$$

This is due to the fact that if $det(Q)\neq\pm 1$, then $|\det(Q)|$ can be absorbed by the ξ_i parameters, see next Lemma.

Clearly, the study of $\mathcal{E}_{\bar{\mathcal{P}}}$-equivalence of the set $\mathcal{B}(f)$ is reduced to the study of $\mathcal{E}_{eh}^{\mathbb{R}}$ equivalence over $\left[\mathcal{B}^{\pi,\,\pi'}(f)\right]$.

Lemma 4. *[1] Let $T_1,\ T_2 \in \mathbb{C}^{k\times 2}$ and $T_1\mathcal{E}_{eh}^{\mathbb{R}}T_2$, then*

$$T_2 = diag(\xi_i)T_1Q \quad with \quad \det(Q) = 1 \ \ or \ \ -1 \tag{14}$$

and $Q \in \mathbb{R}^{2\times 2}$.

According to Lemma 4, the definition of the real extended Hermite equivalence can be slightly reformulated.

Definition 5. Two matrices $T_1,\ T_2 \in \mathbb{C}^{k\times 2}$ are called real extended Hermite equivalent, i.e., $T_1\mathcal{E}_{eh}^{\mathbb{R}}T_2$ if and only if there exist $\xi_i \in \mathbb{C}\backslash\{0\}$, $i = 1, 2, ..., k$ and a matrix $Q \in \mathbb{R}^{2\times 2}$ with $\det(Q) = \pm 1$ such that (13) is valid.

Now ending this section, we would like to remind that two matrix pencils,

$$L_1(s,\ \hat{s}) = sF_1 - \hat{s}G_1 \quad and \quad L_2(\lambda,\ \hat{\lambda}) = \lambda F_2 - \hat{\lambda}G_2,$$

are SBE if and only if we have

$$\mathcal{I}_c(F_1, G_1) = \mathcal{I}_c(F_2, G_2),\ \ \mathcal{I}_r(F_1, G_1) = \mathcal{I}_r(F_2, G_2),$$

$$\mathcal{J}_{\mathbb{R}}(f) = \mathcal{J}_{\mathbb{R}}(\widetilde{f}),\ \ \mathcal{J}_{\mathbb{C}}(f) = \mathcal{J}_{\mathbb{C}}(\widetilde{f}),$$

and the permutations $\pi,\ \pi',\ \widetilde{\pi}, \widetilde{\pi}'$ such that $\left[\mathcal{B}^{\widetilde{\pi},\widetilde{\pi}'}(\widetilde{f})\right]\mathcal{E}_{eh}^{\mathbb{R}}\left[\mathcal{B}^{\pi,\pi'}(f)\right]$ for every homogeneous polynomials $f,\ \tilde{f}$ of $L_1(s,\ \hat{s})$, $L_2(\lambda,\ \hat{\lambda})$, respectively.

Following now the above discussion, $\left[\mathcal{B}^{\widetilde{\pi},\widetilde{\pi}'}(\widetilde{f})\right]\mathcal{E}_{eh}^{\mathbb{R}}\left[\mathcal{B}^{\pi,\pi'}(f)\right]$ for every homogeneous polynomials $f,\ \tilde{f}$ if and only if there exist $\xi_i \in \mathbb{C}\backslash\{0\}$, $i = 1, 2, ..., k$ and a matrix $Q \in \mathbb{R}^{2\times 2}$ with $\det(Q) = \pm 1$ such that

$$\left[\mathcal{B}^{\widetilde{\pi},\ \widetilde{\pi}'}(\tilde{f})\right] = diag(\xi_i)\left[\mathcal{B}^{\pi,\ \pi'}(f)\right]Q \tag{15}$$

for every $f,\ \tilde{f}$ and the same $Q \in \mathbb{R}^{2\times 2}$.

In the next section, we will determine analytically the matrix $Q \in \mathbb{R}^{2\times 2}$ and ξ_i when the matrix $\left[\mathcal{B}^{\pi,\pi'}(f)\right]$ exists for a homogeneous polynomial of $L_1(s,\ \hat{s})$, with dimension $k \times 2$ and $k \geq 3$, i.e., with rectangular coefficients. Note that the matrix Q is the projective transformation $d \in PGL(1, \mathbb{C}/\mathbb{R})$.

4. $d \in PGL(1, \mathbb{C}/\mathbb{R})$ When Two Pencils are SBE

In order to calculate the matrix Q, the first step is to provide the following significant Theorem 4 which has been proven in [1].

Theorem 4. *[1] Let $T_1,\ T_2 \in \mathbb{C}_n^{k\times 2}$ with $k \geq 3$ and*

$$T_2 = diag(\xi_i)T_1Q_1 \ and \ T_2 = diag(\zeta_i)T_1Q_2,$$

where $\xi_i,\ \zeta_i \in \mathbb{C}\backslash\{0\},\ i = 1,2,...,k$ *and matrices* $Q_1,\ Q_2 \in \mathbb{R}^{2\times 2}$ *with* $\det(Q_1), \det(Q_2) = \pm 1$. *Then,*

a) $\zeta_i = -\xi_i$ *for every* $i = 1,2,...,k$, *and* $Q_2 = -Q_1$.

b) $\zeta_i = \xi_i$ *for every* $i = 1,2,...,k$, *and* $Q_2 = Q_1$.

Based on the above Theorem, the following corollary is derived, the proof also is given in [1].

Corollary 2. *[1] Let* $T_1,\ T_2 \in \mathbb{C}_n^{k\times 2}$ *with* $k \geq 3$. *If* $T_1\mathcal{E}_{eh}^{\mathbb{R}}T_2$, *then there exist unique* ξ_i, $i = 1,2,\ldots k$ *such as* $\xi_i \in \mathbb{C}\backslash\{0\}$ *and a unique* $Q \in \mathbb{R}^{2\times 2}$ *with* $det(Q) = 1$ *or* 1, *then*

$$T_2 = diag\xi_i T_1 Q.$$

Remark 5. Assume that

$$L_1(s,\ \hat{s}) = sF_1 - \hat{s}G_1 \quad and \quad L_2(\lambda,\ \hat{\lambda}) = \lambda F_2 - \hat{\lambda}G_2,$$

which are BSE, $L_1(s,\ \hat{s})\mathcal{E}_{\mathcal{H}-\mathcal{B}}L_2(\lambda,\ \hat{\lambda})$.

Now, let $\mathcal{F}(F_1,\ G_1) = \{f_1(s,\ \hat{s}),\ \ldots,\ f_{p-k_1}(s,\ \hat{s})\}$ and $\mathcal{F}(F_2,\ G_2) = \left\{\tilde{f}_1(\lambda,\ \hat{\lambda}),\ \ldots,\ \tilde{f}_{p-k_1}(\lambda,\ \hat{\lambda})\right\}$, be the set of homogeneous invariant polynomials of $L_1(s,\ \hat{s})$ and $L_2(\lambda,\ \hat{\lambda})$, respectively. Assume that there exists a polynomial $f_j(s,\ \hat{s})$ for the pencil $L_1(s,\ \hat{s})$ and the permutations π, π' such that $(\pi - \pi')$ be a $k \times 2$ base matrix of $f_j(s,\ \hat{s})$ with $k \geq 3$ for $j = 1,2,\ldots,p-k_1$. Then, since $L_1(s,\ \hat{s})\mathcal{E}_{\mathcal{H}-\mathcal{B}}L_2(\lambda,\ \hat{\lambda})$, there exist permutations $\widetilde{\pi},\ \widetilde{\pi}'$:

$$\left[\mathcal{B}^{\pi,\ \pi'}(f_j)\right]\mathcal{E}_{eh}^{r}\left[\mathcal{B}^{\widetilde{\pi},\ \widetilde{\pi}'}(\widetilde{f}_j)\right] \;\Leftrightarrow\; \left[\mathcal{B}^{\widetilde{\pi},\ \widetilde{\pi}'}(\widetilde{f}_j)\right] = diag\{\xi_i\}\left[\mathcal{B}^{\pi,\ \pi'}(f_j)\right]Q.$$

Note that the matrix $Q \in \mathbb{R}^{2\times 2}$ with $det(Q) = 1$ or -1 and the $\xi_i, i = 1,2,\ldots,k$ are unique (see Corollary 2). Also the matrix Q creates the projective transformation $d \in PGL(1,\mathbb{C}/\mathbb{R})$.

With the next significant Proposition, we provide the analytical formulas for the unique determination of ξ_i and Q. Let $T \in \mathbb{C}_n^{k\times 2}$, then we denote with a_{ij} the determinant of the sub-matrix of T which is constructed by the ith and j^{th} row

$$i < j, i = 1,2,\ldots,k-1 \text{and } j = 2,3,\ldots,k.$$

Proposition 6. *[1] Let* $T_1,\ T_2 \in \mathbb{C}_n^{k\times 2}$ *with* $k \geq 3$.

If $T_1\mathcal{E}_{eh}^{\mathbb{R}}T_2$ *and* $det(Q) = \pm 1$, *and* ξ_i *the parameters of the real extended Hermite equivalence, then*
(a)

$$\xi_1^2 = \frac{1}{\det(Q)}\frac{\alpha_{1,\,i_1}^{(2)}}{\alpha_{1,\,i_1}^{(1)}}\cdot\frac{\alpha_{1,\,i_2}^{(2)}}{\alpha_{1,\,i_2}^{(1)}}\cdot\frac{\alpha_{i,\,i_2}^{(1)}}{\alpha_{i,\,i_2}^{(2)}} \tag{16}$$

for every i_1, i_2 *with* $1 < i_1 < i_2$ *and* $i_1 = 2, 3, \ldots, (k-1)$, $i_2 = 3, 4, 5, \ldots, k$.

(b)

$$\xi_i = \frac{\alpha_{1,\,i}^{(2)}}{\alpha_{1,\,i}^{(1)}} \cdot \frac{1}{\det(Q)\,\xi_1} \tag{17}$$

for every $i = 2, \ldots, k$ *and* $\alpha_{i_1,\,i_2}^{(1)}$ *which is the determinant of the sub-matrix of* T_1 *which is defined by the* i_1^{th} *and* i_2^{th} *row, and* $\alpha_{i_1,\,i_2}^{(2)}$ *respectively the determinant of the matrix* T_2 *with* $i_1 < i_2$.

So far, we have determined analytically the extended Hermite equivalence in $\mathbb{R}$ for two matrices. The parameters ξ_i are calculated from equation (16); see also Proposition 6. The last part of this section deals with the calculation of matrix Q, such as the projective transformation $d \in PGL(1, \mathbb{C}/\mathbb{R})$ is fully determined. Since,

$$\left[\mathcal{B}^{\pi,\,\pi'}(f_j)\right] \mathcal{E}_{eh}^{r} \left[\mathcal{B}^{\widetilde{\pi},\,\widetilde{\pi}'}(f_j)\right] \;\Leftrightarrow\; \left[\mathcal{B}^{\widetilde{\pi},\,\widetilde{\pi}'}(\widetilde{f}_j)\right] = diag(\xi_i) \left[\mathcal{B}^{\pi,\,\pi'}(f_j)\right] \cdot Q,$$

for $j = 1, 2, \ldots, (p - k_1)$, where the base matrices have dimension $k \times 2$ with $k \geq 3$.

Now, if we choose the first two rows of $\left[\mathcal{B}^{\widetilde{\pi},\,\widetilde{\pi}'}(\widetilde{f}_j)\right]$ and $\left[\mathcal{B}^{\pi,\,\pi'}(f_j)\right]$, then (18) holds

$$\Pi_{12}(\widetilde{f}_j) = \begin{bmatrix} \xi_1 & 0 \\ 0 & \xi_2 \end{bmatrix} \cdot \Pi_{12}(f_j) \cdot Q, \tag{18}$$

where $\Pi_{12}(\widetilde{f}_j)$, $\Pi_{12}(f_j)$ are sub-matrices of $\left[\mathcal{B}^{\widetilde{\pi},\,\widetilde{\pi}'}(\widetilde{f}_j)\right]$ and $\left[\mathcal{B}^{\pi,\,\pi'}(f_j)\right]$, respectively. Note that $\Pi_{12}(\widetilde{f}_j)$, $\Pi_{12}(f_j)$ are non-singular 2×2 matrices, and

$$\begin{bmatrix} \xi_1 & 0 \\ 0 & \xi_2 \end{bmatrix} \quad with \quad \begin{bmatrix} \xi_1 & 0 \\ 0 & \xi_2 \end{bmatrix}^{-1} = \begin{bmatrix} \frac{1}{\xi_1} & 0 \\ 0 & \frac{1}{\xi_2} \end{bmatrix}.$$

Finally, the matrix Q is determined by (19)

$$Q = \Pi_{12}^{-1}(f_j) \begin{bmatrix} \frac{1}{\xi_1} & 0 \\ 0 & \frac{1}{\xi_2} \end{bmatrix} \Pi_{12}\left(\tilde{f}_j\right). \tag{19}$$

References

[1] A.D. Karageorgos, A.A. Pantelous, G.I. Kalogeropoulos, *The study of invariants of matrices under the notion of extended Hermite equivalence (EHE)*, Report - University of Athens, 2014.

[2] N. Karcanias, G.E. Hayton, Generalized autonomous differential systems, algebraic duality and geometric theory, *Proceedings of IFAC VIII*, Triennial World Congress, Kyoto, Japan, 1981.

[3] G.I. Kalogeropoulos, *Matrix pencils and linear systems*, Ph.D Thesis, City University, London UK, 1985.

[4] H.W. Turnbull, A.C. Aitken, *An introduction to the theory of canonical matrices*, Dover, 1961.

[5] F.R. Gantmacher ,*The theory of matrices* II, Chelsea, New York, 1959.

[6] N. Karcanias, G. Kalogeropoulos, Projective equivalence of homogeneous binary polynomials, *Linear Algebra and its Applications* 121 (1989) 433–453.

[7] G.J. Semple, T.G. Kneebone, *Algebraic projective geometry*, Oxford University Press UK, 1952.

[8] G.I. Kalogeropoulos, A.D. Karageorgos, A.A. Pantelous, A stabilization criterion for matrix pencils under bilinear transformation, *Linear Algebra and its Applications* 428 (11-12) (2008) 2852–2862.

[9] M. Mitrouli, G.I. Kalogeropoulos,, A compound matrix algorithm for the computation of the Smith form of a polynomial matrix, *Numerical Algorithms* 7 (2) (1994) 145–159.

[10] H.W. Turnbull, *The theory of determinants, matrices and invariants*, Blackie and Son Ltd, London, UK, 1928.

In: Advances in Linear Algebra Research
Editor: Ivan Kyrchei, pp. 79-132
ISBN: 978-1-63463-565-3

Chapter 3

CRAMER'S RULE FOR GENERALIZED INVERSE SOLUTIONS

Ivan I. Kyrchei*
Pidstryhach Institute for Applied Problems
of Mechanics and Mathematics, Lviv, Ukraine

Abstract

By a generalized inverse of a given matrix, we mean a matrix that exists for a larger class of matrices than the nonsingular matrices, that has some of the properties of the usual inverse, and that agrees with inverse when given matrix happens to be nonsingular. In theory, there are many different generalized inverses that exist. We shall consider the Moore Penrose, weighted Moore-Penrose, Drazin and weighted Drazin inverses.

New determinantal representations of these generalized inverse based on their limit representations are introduced in this chapter. Application of this new method allows us to obtain analogues classical adjoint matrix. Using the obtained analogues of the adjoint matrix, we get Cramer's rules for the least squares solution with the minimum norm and for the Drazin inverse solution of singular linear systems. Cramer's rules for the minimum norm least squares solutions and the Drazin inverse solutions of the matrix equations $\mathbf{AX} = \mathbf{D}$, $\mathbf{XB} = \mathbf{D}$ and $\mathbf{AXB} = \mathbf{D}$ are also obtained, where $\mathbf{A}$, $\mathbf{B}$ can be singular matrices of appropriate size. Finally, we derive determinantal representations of solutions of the differential matrix equations, $\mathbf{X}' + \mathbf{AX} = \mathbf{B}$ and $\mathbf{X}' + \mathbf{XA} = \mathbf{B}$, where the matrix $\mathbf{A}$ is singular.

Keywords: generalized inverse; Drazin inverse; weighted Drazin inverse; Moore-Penrose inverse; weighted Moore-Penrose inverse; system of linear equations; Cramer's Rule; matrix equation; generalized inverse solution; least squares solution; Drazin inverse solution; differential matrix equation

AMS Subject Classification: 15A09; 15A24

*E-mail address: kyrchei@online.ua

1. Preface

It's well-known in linear algebra, an n-by-n square matrix $\mathbf{A}$ is called invertible (also nonsingular or nondegenerate) if there exists an n-by-n square matrix $\mathbf{X}$ such that

$$\mathbf{AX} = \mathbf{XA} = \mathbf{I}_n.$$

If this is the case, then the matrix $\mathbf{X}$ is uniquely determined by $\mathbf{A}$ and is called the inverse of $\mathbf{A}$, denoted by $\mathbf{A}^{-1}$.

By a generalized inverse of a given matrix, we mean a matrix that exists for a larger class of matrices than the nonsingular matrices, that has some of the properties of the usual inverse, and that agrees with inverse when given matrix happens to be nonsingular.

For any matrix $\mathbf{A} \in \mathbb{C}^{m\times n}$ consider the following equations in $\mathbf{X}$:

$$\mathbf{AXA} = \mathbf{A}; \tag{1.1}$$

$$\mathbf{XAX} = \mathbf{X}; \tag{1.2}$$

$$(\mathbf{AX})^* = \mathbf{AX}; \tag{1.3}$$

$$(\mathbf{XA})^* = \mathbf{XA}. \tag{1.4}$$

and if $m = n$, also

$$\mathbf{AX} = \mathbf{AX}; \tag{1.5}$$

$$\mathbf{A}^{k+1}\mathbf{X} = \mathbf{A}^k. \tag{1.6}$$

For a sequence $\mathcal{G}$ of $\{1, 2, 3, 4, 5\}$ the set of matrices obeying the equations represented in $\mathcal{G}$ is denoted by $\mathbf{A}\{\mathcal{G}\}$. A matrix from $\mathbf{A}\{\mathcal{G}\}$ is called an $\mathcal{G}$-inverse of $\mathbf{A}$ and denoted by $\mathbf{A}^{(\mathcal{G})}$.

Consider some principal cases.

If $\mathbf{X}$ satisfies all the equations (1.1)-(1.4) is said to be **the Moore-Penrose inverse** of $\mathbf{A}$ and denote $\mathbf{A}^+ = \mathbf{A}^{(1,2,3,4)}$. The MoorePenrose inverse was independently described by E. H. Moore [1] in 1920, Arne Bjerhammar [2] in 1951 and Roger Penrose [3] in 1955. R. Penrose introduced the characteristic equations (1.1)-(1.4).

If $\det \mathbf{A} \neq 0$, then $\mathbf{A}^+ = \mathbf{A}^{-1}$.

The group inverse $\mathbf{A}^g$ is the unique $\mathbf{A}^{(1,2,5)}$ inverse of $\mathbf{A}$, and exists if and only if $Ind\,\mathbf{A} = \min\{k : \operatorname{rank}\mathbf{A}^{k+1} = \operatorname{rank}\mathbf{A}^k\} = 1$.

A matrix $\mathbf{X} = \mathbf{A}^D$ is said to be **the Drazin inverse** of $\mathbf{A}$ if (1.6) (for some positive integer k), (1.2) and (1.5) are satisfied, where $k = Ind\,\mathbf{A}$. It is named after Michael P. Drazin [4]. In particular, when $Ind\mathbf{A} = 1$, then the matrix $\mathbf{X}$ is the group inverse, $\mathbf{X} = \mathbf{A}^g$. If $Ind\mathbf{A} = 0$, then $\mathbf{A}$ is nonsingular, and $\mathbf{A}^D \equiv \mathbf{A}^{-1}$.

Let Hermitian positive definite matrices $\mathbf{M}$ and $\mathbf{N}$ of order m and n, respectively, be given. For any matrix $\mathbf{A} \in \mathbb{C}^{m\times n}$, **the weighted Moore-Penrose inverse** of $\mathbf{A}$ is the unique solution $\mathbf{X} = \mathbf{A}^+_{M,N}$ of the matrix equations (1.1) and (1.2) and the following equations in $\mathbf{X}$ [5]:

$$(3M)\ (\mathbf{MAX})^* = \mathbf{MAX};\ (4N)\ (\mathbf{NXA})^* = \mathbf{NXA}.$$

In particular, when $\mathbf{M} = \mathbf{I}_m$ and $\mathbf{N} = \mathbf{I}_n$, the matrix $\mathbf{X}$ satisfying the equations (1.1), (1.2), (3M), (4N) is the Moore-Penrose inverse $\mathbf{A}^+$.

The weighted Drazin inverse is being considered as well.

To determine the inverse and to give its analytic solution, we calculate a matrix of cofactors, known as an adjugate matrix or a classical adjoint matrix. The classical adjoint of $\mathbf{A}$, denote $Adj[\mathbf{A}]$, is the transpose of the cofactor matrix, then $\mathbf{A}^{-1} = \frac{Adj[\mathbf{A}]}{|\mathbf{A}|}$. Representation an inverse matrix by its classical adjoint matrix also plays a key role for Cramer's rule of systems of linear equations or matrices equations.

Obviously, the important question is the following: what are the analogues for the adjoint matrix of generalized inverses and, consequently, for Cramer's rule of generalized inverse solutions of matrix equations?

This is the main goal of the chapter.

In this chapter we shall adopt the following notation. Let $\mathbb{C}^{m\times n}$ be the set of m by n matrices with complex entries, $\mathbb{C}_r^{m\times n}$ be a subset of $\mathbb{C}^{m\times n}$ in which any matrix has rank r, $\mathbf{I}_m$ be the identity matrix of order m, and $\|.\|$ be the Frobenius norm of a matrix.

Denote by $\mathbf{a}_{.j}$ and $\mathbf{a}_{i.}$ the jth column and the ith row of $\mathbf{A} \in \mathbb{C}^{m\times n}$, respectively. Then $\mathbf{a}^*_{.j}$ and $\mathbf{a}^*_{i.}$ denote the jth column and the ith row of a conjugate and transpose matrix $\mathbf{A}^*$ as well. Let $\mathbf{A}_{.j}(\mathbf{b})$ denote the matrix obtained from $\mathbf{A}$ by replacing its jth column with the vector $\mathbf{b}$, and by $\mathbf{A}_{i.}(\mathbf{b})$ denote the matrix obtained from $\mathbf{A}$ by replacing its ith row with $\mathbf{b}$.

Let $\alpha := \{\alpha_1, \ldots, \alpha_k\} \subseteq \{1, \ldots, m\}$ and $\beta := \{\beta_1, \ldots, \beta_k\} \subseteq \{1, \ldots, n\}$ be subsets of the order $1 \le k \le \min\{m, n\}$. Then $\left|\mathbf{A}^\alpha_\beta\right|$ denotes the minor of $\mathbf{A}$ determined by the rows indexed by α and the columns indexed by β. Clearly, $|\mathbf{A}^\alpha_\alpha|$ denotes a principal minor determined by the rows and columns indexed by α. The cofactor of a_{ij} in $\mathbf{A} \in \mathbb{C}^{n\times n}$ is denoted by $\frac{\partial}{\partial a_{ij}}|\mathbf{A}|$.

For $1 \le k \le n$, $L_{k,n} := \{\,\alpha : \alpha = (\alpha_1, \ldots, \alpha_k),\ 1 \le \alpha_1 \le \ldots \le \alpha_k \le n\}$ denotes the collection of strictly increasing sequences of k integers chosen from $\{1, \ldots, n\}$. Let $N_k := L_{k,m} \times L_{k,n}$. For fixed $\alpha \in L_{p,m}$, $\beta \in L_{p,n}$, $1 \le p \le k$, let

$$I_{k,\,m}(\alpha) := \{I : I \in L_{k,\,m}, I \supseteq \alpha\},$$
$$J_{k,\,n}(\beta) := \{J : J \in L_{k,\,n}, J \supseteq \beta\},$$
$$N_k(\alpha, \beta) := I_{k,\,m}(\alpha) \times J_{k,\,n}(\beta)$$

For case $i \in \alpha$ and $j \in \beta$, we denote

$$I_{k,m}\{i\} := \{\alpha : \alpha \in L_{k,m}, i \in \alpha\}, J_{k,n}\{j\} := \{\beta : \beta \in L_{k,n}, j \in \beta\},$$
$$N_k\{i, j\} := I_{k,\,m}\{i\} \times J_{k,\,n}\{j\}.$$

The chapter is organized as follows. In Section 2 determinantal representations by analogues of the classical adjoint matrix for the Moore Penrose, weighted Moore-Penrose, Drazin and weighted Drazin inverses are obtained.

In Section 3 we show that the obtained analogues of the adjoint matrix for the generalized inverse matrices enable us to obtain natural analogues of Cramer's rule for generalized inverse solutions of systems of linear equations and demonstrate it in two examples.

In Section 4, we obtain analogs of the Cramer rule for generalized inverse solutions of the matrix equations, $\mathbf{AX} = \mathbf{B}$, $\mathbf{XA} = \mathbf{B}$ and $\mathbf{AXB} = \mathbf{D}$, namely for the minimum norm least squares solutions and the Drazin inverse solutions. We show numerical examples to illustrate the main results as well.

In Section 5, we use the determinantal representations of the Drazin inverse solution to solutions of the following differential matrix equations, $\mathbf{X}'+\mathbf{AX}=\mathbf{B}$ and $\mathbf{X}'+\mathbf{XA}=\mathbf{B}$, where $\mathbf{A}$ is singular. It is demonstrated in the example.

Facts set forth in Sections 2 and 3 were partly published in [6], in Section 4 were published in [7, 8] and in Sections 5 were published in [8].

Note that we obtained some of the submitted results for matrices over the quaternion skew field within the framework of the theory of the column and row determinants [9, 10, 11, 12, 13, 14].

2. Analogues of the Classical Adjoint Matrix for Generalized Inverse Matrices

For determinantal representations of the generalized inverse matrices as analogues of the classical adjoint matrix, we apply the method, which consists on the limit representation of the generalized inverse matrices, lemmas on rank of some matrices and on characteristic polynomial. We used this method at first in [6] and then in [8]. Liu et al. in [15] deduce the new determinantal representations of the outer inverse $\mathbf{A}^{(2)}_{T,S}$ based on these principles as well. In this chapter we obtain detailed determinantal representations by analogues of the classical adjoint matrix for the Moore Penrose, weighted Moore-Penrose, Drazin and weighted Drazin inverses.

2.1. Analogues of the Classical Adjoint Matrix for the Moore - Penrose Inverse

Determinantal representation of the Moore - Penrose inverse was studied in [1],[16, 17, 18, 19]. The main result consists in the following theorem.

Theorem 2.1. *The Moore - Penrose inverse $\mathbf{A}^{+}=(a^{+}_{ij})\in\mathbb{C}^{n\times m}$ of $\mathbf{A}\in\mathbb{C}^{m\times n}_{r}$ has the following determinantal representation*

$$a^{+}_{ij}=\frac{\sum\limits_{(\alpha,\beta)\in N_r\{j,i\}}\left|(\mathbf{A}^{*})^{\beta}_{\alpha}\right|\frac{\partial}{\partial a_{j\,i}}\left|\mathbf{A}^{\alpha}_{\beta}\right|}{\sum\limits_{(\gamma,\delta)\in N_r}\left|(\mathbf{A}^{*})^{\delta}_{\gamma}\right|\left|\mathbf{A}^{\gamma}_{\delta}\right|},\quad 1\leq i,j\leq n.$$

This determinantal representation of the Moore - Penrose inverse is based on corresponding full-rank representation [16]: if $\mathbf{A}=\mathbf{PQ}$, where $\mathbf{P}\in\mathbb{C}^{m\times r}_{r}$ and $\mathbf{Q}\in\mathbb{C}^{r\times n}_{r}$, then

$$\mathbf{A}^{+}=\mathbf{Q}^{*}(\mathbf{P}^{*}\mathbf{AQ}^{*})^{-1}\mathbf{P}^{*}.$$

For a better understanding of the structure of the Moore - Penrose inverse we consider it by singular value decomposition of $\mathbf{A}$. Let

$$\begin{array}{l}\mathbf{AA}^{*}\mathbf{u}_i=\sigma_i^2\mathbf{u}_i,\quad i=\overline{1,m}\\ \mathbf{A}^{*}\mathbf{A}\mathbf{v}_i=\sigma_i^2\mathbf{v}_i,\quad i=\overline{1,n},\end{array}$$

$$\sigma_1\leq\sigma_2\leq\ldots\sigma_r>0=\sigma_{r+1}=\sigma_{r+2}=\ldots$$

and the singular value decomposition (SVD) of $\mathbf{A}$ is $\mathbf{A} = \mathbf{U\Sigma V}^*$, where

$$\begin{array}{ll} \mathbf{U} = [\mathbf{u}_1\, \mathbf{u}_2 ... \mathbf{u}_m] \in \mathbb{C}^{m\times m}, & \mathbf{U}^*\mathbf{U} = \mathbf{I}_m, \\ \mathbf{V} = [\mathbf{v}_1\, \mathbf{v}_2 ... \mathbf{v}_n] \in \mathbb{C}^{n\times n}, & \mathbf{V}^*\mathbf{V} = \mathbf{I}_n, \end{array}$$

$$\mathbf{\Sigma} = \operatorname{diag}(\sigma_1, \sigma_2, ..., \sigma_r) \in \mathbb{C}^{m\times n}.$$

Then [3], $\mathbf{A}^+ = \mathbf{V\Sigma}^+\mathbf{U}^*$, where $\mathbf{\Sigma}^+ = \operatorname{diag}(\sigma_1^{-1}, \sigma_2^{-1}, ..., \sigma_r^{-1})$.

We need the following limit representation of the Moore-Penrose inverse.

Lemma 2.2. *[20] If* $\mathbf{A} \in \mathbb{C}^{m\times n}$, *then*

$$\mathbf{A}^+ = \lim_{\lambda\to 0} \mathbf{A}^* \left(\mathbf{AA}^* + \lambda\mathbf{I}\right)^{-1} = \lim_{\lambda\to 0} \left(\mathbf{A}^*\mathbf{A} + \lambda\mathbf{I}\right)^{-1} \mathbf{A}^*,$$

where $\lambda \in \mathbb{R}_+$, *and* $\mathbb{R}_+$ *is the set of positive real numbers.*

Corollary 2.3. *[21] If* $\mathbf{A} \in \mathbb{C}^{m\times n}$, *then the following statements are true.*

i) If $\operatorname{rank} \mathbf{A} = \mathrm{n}$, *then* $\mathbf{A}^+ = \left(\mathbf{A}^*\mathbf{A}\right)^{-1} \mathbf{A}^*$.

ii) If $\operatorname{rank} \mathbf{A} = \mathrm{m}$, *then* $\mathbf{A}^+ = \mathbf{A}^* \left(\mathbf{AA}^*\right)^{-1}$.

iii) If $\operatorname{rank} \mathbf{A} = \mathrm{n} = \mathrm{m}$, *then* $\mathbf{A}^+ = \mathbf{A}^{-1}$.

We need the following well-known theorem about the characteristic polynomial and lemmas on rank of some matrices.

Theorem 2.4. *[22] Let* d_r *be the sum of principal minors of order* r *of* $\mathbf{A} \in \mathbb{C}^{n\times n}$. *Then its characteristic polynomial* $p_{\mathbf{A}}(t)$ *can be expressed as* $p_{\mathbf{A}}(t) = \det(t\mathbf{I} - \mathbf{A}) = t^n - d_1 t^{n-1} + d_2 t^{n-2} - \ldots + (-1)^n d_n$.

Lemma 2.5. *If* $\mathbf{A} \in \mathbb{C}_r^{m\times n}$, *then* $\operatorname{rank} \left(\mathbf{A}^*\mathbf{A}\right)_{.\,i}\left(\mathbf{a}_{.j}^*\right) \le r$.

Proof. Let $\mathbf{P}_{i\,k}(-a_{j\,k}) \in \mathbb{C}^{n\times n}$, $(k \neq i)$, be the matrix with $-a_{j\,k}$ in the (i,k) entry, 1 in all diagonal entries, and 0 in others. It is the matrix of an elementary transformation. It follows that

$$\left(\mathbf{A}^*\mathbf{A}\right)_{.\,i}\left(\mathbf{a}_{.j}^*\right) \cdot \prod_{k\neq i} \mathbf{P}_{i\,k}(-a_{j\,k}) = \begin{pmatrix} \sum\limits_{k\neq j} a_{1k}^* a_{k1} & \ldots & a_{1j}^* & \ldots & \sum\limits_{k\neq j} a_{1k}^* a_{kn} \\ \ldots & \ldots & \ldots & \ldots & \ldots \\ \sum\limits_{k\neq j} a_{nk}^* a_{k1} & \ldots & \underset{i-th}{a_{nj}^*} & \ldots & \sum\limits_{k\neq j} a_{nk}^* a_{kn} \end{pmatrix}.$$

The obtained above matrix has the following factorization.

$$\begin{pmatrix} \sum\limits_{k\neq j} a_{1k}^* a_{k1} & \ldots & a_{1j}^* & \ldots & \sum\limits_{k\neq j} a_{1k}^* a_{kn} \\ \ldots & \ldots & \ldots & \ldots & \ldots \\ \sum\limits_{k\neq j} a_{nk}^* a_{k1} & \ldots & \underset{i-th}{a_{nj}^*} & \ldots & \sum\limits_{k\neq j} a_{nk}^* a_{kn} \end{pmatrix} =$$

$$= \begin{pmatrix} a_{11}^* & a_{12}^* & \dots & a_{1m}^* \\ a_{21}^* & a_{22}^* & \dots & a_{2m}^* \\ \dots & \dots & \dots & \dots \\ a_{n1}^* & a_{n2}^* & \dots & a_{nm}^* \end{pmatrix} \begin{pmatrix} a_{11} & \dots & 0 & \dots & a_{n1} \\ \dots & \dots & \dots & \dots & \dots \\ 0 & \dots & 1 & \dots & 0 \\ \dots & \dots & \dots & \dots & \dots \\ a_{m1} & \dots & \underset{i-th}{0} & \dots & a_{mn} \end{pmatrix} j-th.$$

Denote by $\tilde{\mathbf{A}} := \begin{pmatrix} a_{11} & \dots & 0 & \dots & a_{1n} \\ \dots & \dots & \dots & \dots & \dots \\ 0 & \dots & 1 & \dots & 0 \\ \dots & \dots & \dots & \dots & \dots \\ a_{m1} & \dots & \underset{i-th}{0} & \dots & a_{mn} \end{pmatrix} j-th$. The matrix $\tilde{\mathbf{A}}$ is obtained from $\mathbf{A}$ by replacing all entries of the jth row and of the ith column with zeroes except that the (j, i) entry equals 1. Elementary transformations of a matrix do not change its rank. It follows that $\operatorname{rank} (\mathbf{A}^*\mathbf{A})_{.i} \left(\mathbf{a}^*_{.j}\right) \leq \min \left\{\operatorname{rank} \mathbf{A}^*, \operatorname{rank} \tilde{\mathbf{A}}\right\}$. Since $\operatorname{rank} \tilde{\mathbf{A}} \geq \operatorname{rank} \mathbf{A} = \operatorname{rank} \mathbf{A}^*$ and $\operatorname{rank} \mathbf{A}^*\mathbf{A} = \operatorname{rank} \mathbf{A}$ the proof is completed. ■ The following lemma can be proved in the same way.

Lemma 2.6. *If* $\mathbf{A} \in \mathbb{C}^{m\times n}_r$, *then* $\operatorname{rank} (\mathbf{A}\mathbf{A}^*)_{i.} \left(\mathbf{a}^*_{j.}\right) \leq r$.

Analogues of the characteristic polynomial are considered in the following two lemmas.

Lemma 2.7. *If* $\mathbf{A} \in \mathbb{C}^{m\times n}$ *and* $\lambda \in \mathbb{R}$, *then*

$$\det \left((\lambda \mathbf{I}_n + \mathbf{A}^*\mathbf{A})_{.i} \left(\mathbf{a}^*_{.j}\right)\right) = c_1^{(ij)} \lambda^{n-1} + c_2^{(ij)} \lambda^{n-2} + \ldots + c_n^{(ij)}, \tag{2.1}$$

where $c_n^{(ij)} = \left|(\mathbf{A}^*\mathbf{A})_{.i} \left(\mathbf{a}^*_{.j}\right)\right|$ *and* $c_s^{(ij)} = \sum\limits_{\beta \in J_{s,n}\{i\}} \left|\left((\mathbf{A}^*\mathbf{A})_{.i} \left(\mathbf{a}^*_{.j}\right)\right)^\beta_\beta\right|$ *for all* $s = \overline{1, n-1}$, $i = \overline{1, n}$, *and* $j = \overline{1, m}$.

Proof. Denote $\mathbf{A}^*\mathbf{A} = \mathbf{V} = (v_{ij}) \in \mathbb{C}^{n\times n}$. Consider $(\lambda \mathbf{I}_n + \mathbf{V})_{.i} (\mathbf{v}_{.i}) \in \mathbb{C}^{n\times n}$. Taking into account Theorem 2.4 we obtain

$$|(\lambda \mathbf{I}_n + \mathbf{V})_{.i} (\mathbf{v}_{.i})| = d_1 \lambda^{n-1} + d_2 \lambda^{n-2} + \ldots + d_n, \tag{2.2}$$

where $d_s = \sum\limits_{\beta \in J_{s,n}\{i\}} |(\mathbf{V})^\beta_\beta|$ is the sum of all principal minors of order s that contain the i-th column for all $s = \overline{1, n-1}$ and $d_n = \det \mathbf{V}$. Since $\mathbf{v}_{.i} = \sum\limits_l \mathbf{a}^*_{.l} a_{li}$, where $\mathbf{a}^*_{.l}$ is the lth column-vector of $\mathbf{A}^*$ for all $l = \overline{1, n}$, then we have on the one hand

$$\begin{gathered} |(\lambda \mathbf{I} + \mathbf{V})_{.i} (\mathbf{v}_{.i})| = \sum_l |(\lambda \mathbf{I} + \mathbf{V})_{.l} (\mathbf{a}^*_{.l} a_{li})| = \\ \sum_l |(\lambda \mathbf{I} + \mathbf{V})_{.i} (\mathbf{a}^*_{.l})| \cdot a_{li} \end{gathered} \tag{2.3}$$

Having changed the order of summation, we obtain on the other hand for all $s = \overline{1, n-1}$

$$\begin{gathered} d_s = \sum_{\beta \in J_{s,n}\{i\}} \left|(\mathbf{V})^\beta_\beta\right| = \sum_{\beta \in J_{s,n}\{i\}} \sum_l \left|(\mathbf{V}_{.i} (\mathbf{a}^*_{.l} a_{li}))^\beta_\beta\right| = \\ \sum_l \sum_{\beta \in J_{s,n}\{i\}} \left|(\mathbf{V}_{.i} (\mathbf{a}^*_{.l}))^\beta_\beta\right| \cdot a_{li}. \end{gathered} \tag{2.4}$$

By substituting (2.3) and (2.4) in (2.2), and equating factors at $a_{l\,i}$ when $l = j$, we obtain the equality (2.1). ■

By analogy can be proved the following lemma.

Lemma 2.8. *If* $\mathbf{A} \in \mathbb{C}^{m\times n}$ *and* $\lambda \in \mathbb{R}$, *then*

$$\det\left((\lambda \mathbf{I}_m + \mathbf{A}\mathbf{A}^*)_{j\,.}\left(\mathbf{a}^*_{i.}\right)\right) = r_1^{(ij)}\lambda^{m-1} + r_2^{(ij)}\lambda^{m-2} + \ldots + r_m^{(ij)},$$

where $r_m^{(ij)} = \left|(\mathbf{A}\mathbf{A}^*)_{j\,.}\left(\mathbf{a}^*_{i.}\right)\right|$ *and* $r_s^{(ij)} = \sum\limits_{\alpha \in I_{s,m}\{j\}} \left|\left((\mathbf{A}\mathbf{A}^*)_{j\,.}\left(\mathbf{a}^*_{i.}\right)\right)^{\alpha}_{\alpha}\right|$ *for all* $s = \overline{1, n-1}$, $i = \overline{1,n}$, *and* $j = \overline{1,m}$.

The following theorem and remarks introduce the determinantal representations of the Moore-Penrose by analogs of the classical adjoint matrix.

Theorem 2.9. *If* $\mathbf{A} \in \mathbb{C}^{m\times n}_r$ *and* $r < min\{m, n\}$, *then the Moore-Penrose inverse* $\mathbf{A}^+ = \left(a^+_{ij}\right) \in \mathbb{C}^{n\times m}$ *possess the following determinantal representations:*

$$a^+_{ij} = \frac{\sum\limits_{\beta \in J_{r,\,n}\{i\}} \left|\left((\mathbf{A}^*\mathbf{A})_{\,.\,i}\left(\mathbf{a}^*_{.j}\right)\right)^{\beta}_{\beta}\right|}{\sum\limits_{\beta \in J_{r,\,n}} \left|(\mathbf{A}^*\mathbf{A})^{\beta}_{\beta}\right|}, \tag{2.5}$$

or

$$a^+_{ij} = \frac{\sum\limits_{\alpha \in I_{r,m}\{j\}} \left|\left((\mathbf{A}\mathbf{A}^*)_{j\,.}\left(\mathbf{a}^*_{i.}\right)\right)^{\alpha}_{\alpha}\right|}{\sum\limits_{\alpha \in I_{r,\,m}} \left|(\mathbf{A}\mathbf{A}^*)^{\alpha}_{\alpha}\right|}. \tag{2.6}$$

for all $i = \overline{1,n}$, $j = \overline{1,m}$.

Proof. At first we shall obtain the representation (2.5). If $\lambda \in \mathbb{R}_+$, then the matrix $(\lambda\mathbf{I} + \mathbf{A}^*\mathbf{A}) \in \mathbb{C}^{n\times n}$ is Hermitian and $\operatorname{rank}(\lambda\mathbf{I} + \mathbf{A}^*\mathbf{A}) = n$. Hence, there exists its inverse

$$(\lambda\mathbf{I} + \mathbf{A}^*\mathbf{A})^{-1} = \frac{1}{\det(\lambda\mathbf{I} + \mathbf{A}^*\mathbf{A})}\begin{pmatrix} L_{11} & L_{21} & \ldots & L_{n\,1} \\ L_{12} & L_{22} & \ldots & L_{n\,2} \\ \ldots & \ldots & \ldots & \ldots \\ L_{1\,n} & L_{2\,n} & \ldots & L_{n\,n} \end{pmatrix},$$

where L_{ij} $(\forall i, j = \overline{1,n})$ is a cofactor in $\lambda\mathbf{I} + \mathbf{A}^*\mathbf{A}$. By Lemma 2.2, $\mathbf{A}^+ = \lim\limits_{\lambda\to 0}(\lambda\mathbf{I} + \mathbf{A}^*\mathbf{A})^{-1}\mathbf{A}^*$, so that

$$\mathbf{A}^+ = \lim_{\lambda\to 0}\begin{pmatrix} \frac{\det(\lambda\mathbf{I}+\mathbf{A}^*\mathbf{A})_{.1}\left(\mathbf{a}^*_{.1}\right)}{\det(\lambda\mathbf{I}+\mathbf{A}^*\mathbf{A})} & \ldots & \frac{\det(\lambda\mathbf{I}+\mathbf{A}^*\mathbf{A})_{.1}\left(\mathbf{a}^*_{.m}\right)}{\det(\lambda\mathbf{I}+\mathbf{A}^*\mathbf{A})} \\ \ldots & \ldots & \ldots \\ \frac{\det(\lambda\mathbf{I}+\mathbf{A}^*\mathbf{A})_{.n}\left(\mathbf{a}^*_{.1}\right)}{\det(\lambda\mathbf{I}+\mathbf{A}^*\mathbf{A})} & \ldots & \frac{\det(\lambda\mathbf{I}+\mathbf{A}^*\mathbf{A})_{.n}\left(\mathbf{a}^*_{.m}\right)}{\det(\lambda\mathbf{I}+\mathbf{A}^*\mathbf{A})} \end{pmatrix}. \tag{2.7}$$

From Theorem 2.4 we get

$$\det(\lambda\mathbf{I} + \mathbf{A}^*\mathbf{A}) = \lambda^n + d_1\lambda^{n-1} + d_2\lambda^{n-2} + \ldots + d_n,$$

where d_r ($\forall r = \overline{1, n-1}$) is a sum of principal minors of $\mathbf{A}^*\mathbf{A}$ of order r and $d_n = \det \mathbf{A}^*\mathbf{A}$. Since $\operatorname{rank} \mathbf{A}^*\mathbf{A} = \operatorname{rank} \mathbf{A} = r$, then $d_n = d_{n-1} = \ldots = d_{r+1} = 0$ and

$$\det(\lambda \mathbf{I} + \mathbf{A}^*\mathbf{A}) = \lambda^n + d_1\lambda^{n-1} + d_2\lambda^{n-2} + \ldots + d_r\lambda^{n-r}. \tag{2.8}$$

In the same way, we have for arbitrary $1 \leq i \leq n$ and $1 \leq j \leq m$ from Lemma 2.7

$$\det(\lambda \mathbf{I} + \mathbf{A}^*\mathbf{A})_{.i}\left(\mathbf{a}^*_{.j}\right) = l_1^{(ij)}\lambda^{n-1} + l_2^{(ij)}\lambda^{n-2} + \ldots + l_n^{(ij)},$$

where for an arbitrary $1 \leq k \leq n-1$, $l_k^{(ij)} = \sum\limits_{\beta \in J_{k,\,n}\{i\}} \left|\left((\mathbf{A}^*\mathbf{A})_{.i}(\mathbf{a}^*_{.j})\right)^{\beta}_{\beta}\right|$, and $l_n^{(ij)} = \det(\mathbf{A}^*\mathbf{A})_{.i}\left(\mathbf{a}^*_{.j}\right)$. By Lemma 2.5, $\operatorname{rank}(\mathbf{A}^*\mathbf{A})_{.i}\left(\mathbf{a}^*_{.j}\right) \leq r$ so that if $k > r$, then $\left|\left((\mathbf{A}^*\mathbf{A})_{.i}(\mathbf{a}^*_{.j})\right)^{\beta}_{\beta}\right| = 0$, $(\forall \beta \in J_{k,\,n}\{i\}, \forall i = \overline{1, n}, \forall j = \overline{1, m})$. Therefore if $r+1 \leq k < n$, then $l_k^{(ij)} = \sum\limits_{\beta \in J_{k,\,n}\{i\}} \left|\left((\mathbf{A}^*\mathbf{A})_{.i}(\mathbf{a}^*_{.j})\right)^{\beta}_{\beta}\right| = 0$ and $l_n^{(ij)} = \det(\mathbf{A}^*\mathbf{A})_{.i}\left(\mathbf{a}^*_{.j}\right) = 0$, $\left(\forall i = \overline{1, n}, \forall j = \overline{1, m}\right)$. Finally we obtain

$$\det(\lambda \mathbf{I} + \mathbf{A}^*\mathbf{A})_{.i}\left(\mathbf{a}^*_{.j}\right) = l_1^{(ij)}\lambda^{n-1} + l_2^{(ij)}\lambda^{n-2} + \ldots + l_r^{(ij)}\lambda^{n-r}. \tag{2.9}$$

By replacing the denominators and the numerators of the fractions in entries of matrix (2.7) with the expressions (2.8) and (2.9) respectively, we get

$$\mathbf{A}^+ = \lim_{\lambda \to 0} \begin{pmatrix} \frac{l_1^{(11)}\lambda^{n-1}+\ldots+l_r^{(11)}\lambda^{n-r}}{\lambda^n+d_1\lambda^{n-1}+\ldots+d_r\lambda^{n-r}} & \ldots & \frac{l_1^{(1m)}\lambda^{n-1}+\ldots+l_r^{(1m)}\lambda^{n-r}}{\lambda^n+d_1\lambda^{n-1}+\ldots+d_r\lambda^{n-r}} \\ \ldots & \ldots & \ldots \\ \frac{l_1^{(n1)}\lambda^{n-1}+\ldots+l_r^{(n1)}\lambda^{n-r}}{\lambda^n+d_1\lambda^{n-1}+\ldots+d_r\lambda^{n-r}} & \ldots & \frac{l_1^{(nm)}\lambda^{n-1}+\ldots+l_r^{(nm)}\lambda^{n-r}}{\lambda^n+d_1\lambda^{n-1}+\ldots+d_r\lambda^{n-r}} \end{pmatrix} =$$

$$= \begin{pmatrix} \frac{l_r^{(11)}}{d_r} & \ldots & \frac{l_r^{(1m)}}{d_r} \\ \ldots & \ldots & \ldots \\ \frac{l_r^{(n1)}}{d_r} & \ldots & \frac{l_r^{(nm)}}{d_r} \end{pmatrix}.$$

From here it follows (2.5).

We can prove (2.6) in the same way. ■

Corollary 2.10. *If* $\mathbf{A} \in \mathbb{C}^{m\times n}_r$ *and* $r < \min\{m, n\}$ *or* $r = m < n$, *then the projection matrix* $\mathbf{P} = \mathbf{A}^+\mathbf{A}$ *can be represented as*

$$\mathbf{P} = \left(\frac{p_{ij}}{d_r(\mathbf{A}^*\mathbf{A})}\right)_{n\times n},$$

where $\mathbf{d}_{.j}$ *denotes the* j*th column of* $(\mathbf{A}^*\mathbf{A})$ *and, for arbitrary* $1 \leq i, j \leq n$, $p_{ij} = \sum\limits_{\beta \in J_{r,n}\{i\}} \left|\left((\mathbf{A}^*\mathbf{A})_{.i}(\mathbf{d}_{.j})\right)^{\beta}_{\beta}\right|$.

Proof. Representing the Moore - Penrose inverse $\mathbf{A}^+$ by (2.5), we obtain

$$\mathbf{P} = \frac{1}{d_r(\mathbf{A}^*\mathbf{A})}\begin{pmatrix} l_{11} & l_{12} & \dots & l_{1m} \\ l_{21} & l_{22} & \dots & l_{2m} \\ \dots & \dots & \dots & \dots \\ l_{n1} & l_{n2} & \dots & l_{nm} \end{pmatrix}\begin{pmatrix} a_{11} & a_{12} & \dots & a_{1n} \\ a_{21} & a_{22} & \dots & a_{2n} \\ \dots & \dots & \dots & \dots \\ a_{m1} & a_{m2} & \dots & a_{mn} \end{pmatrix}.$$

Therefore, for arbitrary $1 \le i, j \le n$ we get

$$p_{ij} = \sum_k \sum_{\beta \in J_{r,n}\{i\}} \left|\left((\mathbf{A}^*\mathbf{A})_{.i}(\mathbf{a}^*_{.k})\right)^\beta_\beta\right| \cdot a_{kj} =$$

$$= \sum_{\beta \in J_{r,n}\{i\}} \sum_k \left|\left((\mathbf{A}^*\mathbf{A})_{.i}(\mathbf{a}^*_{.k} \cdot a_{kj})\right)^\beta_\beta\right| = \sum_{\beta \in J_{r,n}\{i\}} \left|\left((\mathbf{A}^*\mathbf{A})_{.i}(\mathbf{d}^*_{.j})\right)^\beta_\beta\right|.$$

■ Using the representation (2.6) of the Moore - Penrose inverse the following corollary can be proved in the same way.

Corollary 2.11. *If $\mathbf{A} \in \mathbb{C}^{m\times n}_r$, where $r < \min\{m,n\}$ or $r = n < m$, then a projection matrix $\mathbf{Q} = \mathbf{A}\mathbf{A}^+$ can be represented as*

$$\mathbf{Q} = \left(\frac{q_{ij}}{d_r(\mathbf{A}\mathbf{A}^*)}\right)_{m\times m},$$

where $\mathbf{g}_{i.}$ denotes the ith row of $(\mathbf{A}\mathbf{A}^*)$ and, for arbitrary $1 \le i,j \le m$, $q_{ij} = \sum_{\alpha \in I_{r,m}\{j\}} \left|\left((\mathbf{A}\mathbf{A}^*)_{j.}(\mathbf{g}_{i.})\right)^\alpha_\alpha\right|$.

Remark 2.12. *If $\operatorname{rank}\mathbf{A} = n$, then from Corollary 2.3 we get $\mathbf{A}^+ = (\mathbf{A}^*\mathbf{A})^{-1}\mathbf{A}^*$. Representing $(\mathbf{A}^*\mathbf{A})^{-1}$ by the classical adjoint matrix, we have*

$$\mathbf{A}^+ = \frac{1}{\det(\mathbf{A}^*\mathbf{A})}\begin{pmatrix} \det(\mathbf{A}^*\mathbf{A})_{.1}(\mathbf{a}^*_{.1}) & \dots & \det(\mathbf{A}^*\mathbf{A})_{.1}(\mathbf{a}^*_{.m}) \\ \dots & \dots & \dots \\ \det(\mathbf{A}^*\mathbf{A})_{.n}(\mathbf{a}^*_{.1}) & \dots & \det(\mathbf{A}^*\mathbf{A})_{.n}(\mathbf{a}^*_{.m}) \end{pmatrix}. \tag{2.10}$$

If $n < m$, then (2.5) is valid.

Remark 2.13. *As above, if $\operatorname{rank}\mathbf{A} = m$, then*

$$\mathbf{A}^+ = \frac{1}{\det(\mathbf{A}\mathbf{A}^*)}\begin{pmatrix} \det(\mathbf{A}\mathbf{A}^*)_{1.}(\mathbf{a}^*_{1.}) & \dots & \det(\mathbf{A}\mathbf{A}^*)_{m.}(\mathbf{a}^*_{1.}) \\ \dots & \dots & \dots \\ \det(\mathbf{A}\mathbf{A}^*)_{1.}(\mathbf{a}^*_{n.}) & \dots & \det(\mathbf{A}\mathbf{A}^*)_{m.}(\mathbf{a}^*_{n.}) \end{pmatrix}. \tag{2.11}$$

If $n > m$, then (2.6) is valid as well.

Remark 2.14. *By definition of the classical adjoint $Adj(\mathbf{A})$ for an arbitrary invertible matrix $\mathbf{A} \in \mathbb{C}^{n\times n}$ one may put, $Adj(\mathbf{A}) \cdot \mathbf{A} = \det \mathbf{A} \cdot \mathbf{I}_n$.*

If $\mathbf{A} \in \mathbb{C}^{m\times n}$ and $\operatorname{rank}\mathbf{A} = n$, then by Corollary 2.3, $\mathbf{A}^+\mathbf{A} = \mathbf{I}_n$. Representing the matrix $\mathbf{A}^+$ by (2.10) as $\mathbf{A}^+ = \frac{\mathbf{L}}{\det(\mathbf{A}^*\mathbf{A})}$, we obtain $\mathbf{L}\mathbf{A} = \det(\mathbf{A}^*\mathbf{A}) \cdot \mathbf{I}_n$. This means that the matrix $\mathbf{L} = (l_{ij}) \in \mathbb{C}^{n\times m}$ is a left analogue of $Adj(\mathbf{A})$, where $\mathbf{A} \in \mathbb{C}^{m\times n}_n$, and $l_{ij} = \det(\mathbf{A}^*\mathbf{A})_{.i}\left(\mathbf{a}^*_{.j}\right)$ for all $i = \overline{1,n}$, $j = \overline{1,m}$.

If rank $\mathbf{A} = m$, *then by Corollary 2.3,* $\mathbf{A}\mathbf{A}^{+} = \mathbf{I}_m$. *Representing the matrix* $\mathbf{A}^{+}$ *by (2.11) as* $\mathbf{A}^{+} = \frac{\mathbf{R}}{\det(\mathbf{A}\mathbf{A}^{*})}$, *we obtain* $\mathbf{A}\mathbf{R} = \mathbf{I}_m \cdot \det(\mathbf{A}\mathbf{A}^{*})$. *This means that the matrix* $\mathbf{R} = (r_{ij}) \in \mathbb{C}^{m\times n}$ *is a right analogue of* $Adj(\mathbf{A})$, *where* $\mathbf{A} \in \mathbb{C}^{m\times n}_{m}$, *and* $r_{ij} = \det(\mathbf{A}\mathbf{A}^{*})_{j.}(\mathbf{a}^{*}_{i.})$ *for all* $i = \overline{1,n}$, $j = \overline{1,m}$.

If $\mathbf{A} \in \mathbb{C}^{m\times n}_{r}$ *and* $r < \min\{m,n\}$, *then by (2.5) we have* $\mathbf{A}^{+} = \frac{\mathbf{L}}{d_r(\mathbf{A}^{*}\mathbf{A})}$, *where* $\mathbf{L} = (l_{ij}) \in \mathbb{C}^{n\times m}$ *and* $l_{ij} = \sum\limits_{\beta\in J_{r,n}\{i\}} \left|\left((\mathbf{A}^{*}\mathbf{A})_{.i}\left(\mathbf{a}^{*}_{.j}\right)\right)^{\beta}_{\beta}\right|$ *for all* $i = \overline{1,n}$, $j = \overline{1,m}$. *From Corollary 2.10 we get* $\mathbf{L}\mathbf{A} = d_r(\mathbf{A}^{*}\mathbf{A})\cdot\mathbf{P}$. *The matrix* $\mathbf{P}$ *is idempotent. All eigenvalues of an idempotent matrix chose from 1 or 0 only. Thus, there exists an unitary matrix* $\mathbf{U}$ *such that*

$$\mathbf{L}\mathbf{A} = d_r(\mathbf{A}^{*}\mathbf{A})\,\mathbf{U}\mathbf{diag}(1,\ldots,1,0,\ldots,0)\mathbf{U}^{*},$$

where $\mathbf{diag}(1,\ldots,1,0,\ldots,0) \in \mathbb{C}^{n\times n}$ *is a diagonal matrix. Therefore, the matrix* $\mathbf{L}$ *can be considered as a left analogue of* $Adj(\mathbf{A})$, *where* $\mathbf{A} \in \mathbb{C}^{m\times n}_{r}$.

In the same way, if $\mathbf{A} \in \mathbb{C}^{m\times n}_{r}$ *and* $r < \min\{m,n\}$, *then by (2.5) we have* $\mathbf{A}^{+} = \frac{\mathbf{R}}{d_r(\mathbf{A}\mathbf{A}^{*})}$, *where* $\mathbf{R} = (r_{ij}) \in \mathbb{C}^{n\times m}$, $r_{ij} = \sum\limits_{\alpha\in I_{r,m}\{j\}} |((\mathbf{A}\mathbf{A}^{*})_{j.}(\mathbf{a}^{*}_{i.}))^{\alpha}_{\alpha}|$ *for all* $i = \overline{1,n}$, $j = \overline{1,m}$. *From Corollary 2.11 we get* $\mathbf{A}\mathbf{R} = d_r(\mathbf{A}\mathbf{A}^{*})\cdot\mathbf{Q}$. *The matrix* $\mathbf{Q}$ *is idempotent. There exists an unitary matrix* $\mathbf{V}$ *such that*

$$\mathbf{A}\mathbf{R} = d_r(\mathbf{A}\mathbf{A}^{*})\,\mathbf{V}\mathbf{diag}(1,\ldots,1,0,\ldots,0)\mathbf{V}^{*},$$

where $\mathbf{diag}(1,\ldots,1,0,\ldots,0) \in \mathbb{C}^{m\times m}$. *Therefore, the matrix* $\mathbf{R}$ *can be considered as a right analogue of* $Adj(\mathbf{A})$ *in this case.*

Remark 2.15. *To obtain an entry of* $\mathbf{A}^{+}$ *by Theorem 2.1 one calculates* $(C^{r}_{n}C^{r}_{m} + C^{r-1}_{n-1}C^{r-1}_{m-1})$ *determinants of order* r. *Whereas by the equation (2.5) we calculate as much as* $(C^{r}_{n} + C^{r-1}_{n-1})$ *determinants of order* r *or we calculate the total of* $(C^{r}_{m} + C^{r-1}_{m-1})$ *determinants by (2.6). Therefore the calculation of entries of* $\mathbf{A}^{+}$ *by Theorem 2.9 is easier than by Theorem 2.1.*

2.2. Analogues of the Classical Adjoint Matrix for the Weighted Moore-Penrose Inverse

Let Hermitian positive definite matrices $\mathbf{M}$ and $\mathbf{N}$ of order m and n, respectively, be given. The weighted Moore-Penrose inverse $\mathbf{X} = \mathbf{A}^{+}_{M,N}$ can be explicitly expressed from the weighted singular value decomposition due to Van Loan [23].

Lemma 2.16. *Let* $\mathbf{A} \in \mathbb{C}^{m\times n}_{r}$. *There exist* $\mathbf{U} \in \mathbb{C}^{m\times m}$, $\mathbf{V} \in \mathbb{C}^{n\times n}$ *satisfying* $\mathbf{U}^{*}\mathbf{M}\mathbf{U} = \mathbf{I}_m$ *and* $\mathbf{V}^{*}\mathbf{N}^{-1}\mathbf{V} = \mathbf{I}_n$ *such that*

$$\mathbf{A} = \mathbf{U}\begin{pmatrix} \mathbf{D} & \mathbf{0} \\ \mathbf{0} & \mathbf{0} \end{pmatrix}\mathbf{V}^{*}.$$

Then the weighted Moore-Penrose inverse $\mathbf{A}^{+}_{M,N}$ *can be represented*

$$\mathbf{A}^{+}_{M,N} = \mathbf{N}^{-1}\mathbf{V}\begin{pmatrix} \mathbf{D}^{-1} & \mathbf{0} \\ \mathbf{0} & \mathbf{0} \end{pmatrix}\mathbf{U}^{*}\mathbf{M},$$

where $\mathbf{D} = diag(\sigma_1, \sigma_2, ..., \sigma_r)$, $\sigma_1 \geq \sigma_2 \geq ... \geq \sigma_r > 0$ *and* σ_i^2 *is the nonzero eigenvalues of* $\mathbf{N}^{-1}\mathbf{A}^*\mathbf{M}\mathbf{A}$.

For the weighted Moore-Penrose inverse $\mathbf{X} = \mathbf{A}^+_{M,N}$, we have the following limit representation.

Lemma 2.17. *([24], Corollary 3.4.) Let* $\mathbf{A} \in \mathbb{C}^{m\times n}$, $\mathbf{A}^\sharp = \mathbf{N}^{-1}\mathbf{A}^*\mathbf{M}$. *Then*

$$\mathbf{A}^+_{M,N} = \lim_{\lambda\to 0}(\lambda\mathbf{I} + \mathbf{A}^\sharp\mathbf{A})^{-1}\mathbf{A}^\sharp.$$

By analogy to Lemma 2.17 can be proved the following lemma.

Lemma 2.18. *Let* $\mathbf{A} \in \mathbb{C}^{m\times n}$, $\mathbf{A}^\sharp = \mathbf{N}^{-1}\mathbf{A}^*\mathbf{M}$. *Then*

$$\mathbf{A}^+_{M,N} = \lim_{\lambda\to 0}\mathbf{A}^\sharp(\lambda\mathbf{I} + \mathbf{A}\mathbf{A}^\sharp)^{-1}.$$

Denote by $\mathbf{a}^\sharp_{.j}$ and $\mathbf{a}^\sharp_{i.}$ the jth column and the ith row of $\mathbf{A}^\sharp$ respectively. By putting $\mathbf{A}^\sharp$ instead $\mathbf{A}^*$, we obtain the proofs of the following two lemmas and theorem similar to the proofs of Lemmas 2.5, 2.6, 2.7, 2.8 and Theorem 2.9, respectively.

Lemma 2.19. *If* $\mathbf{A} \in \mathbb{C}^{m\times n}_r$ *and* $\mathbf{A}^\sharp$ *is defined as above, then*

$$\operatorname{rank}\left(\mathbf{A}^\sharp\mathbf{A}\right)_{.i}\left(\mathbf{a}^\sharp_{.j}\right) \leq \operatorname{rank}\left(\mathbf{A}^\sharp\mathbf{A}\right),$$

$$\operatorname{rank}\left(\mathbf{A}\mathbf{A}^\sharp\right)_{j.}\left(\mathbf{a}^\sharp_{i.}\right) \leq \operatorname{rank}\left(\mathbf{A}\mathbf{A}^\sharp\right),$$

for all $i = \overline{1, n}$ *and* $j = \overline{1, m}$

Analogues of the characteristic polynomial are considered in the following lemma.

Lemma 2.20. *If* $\mathbf{A} \in \mathbb{C}^{m\times n}$ *and* $\lambda \in \mathbb{R}$, *then*

$$\det\left(\left(\lambda\mathbf{I}_n + \mathbf{A}^\sharp\mathbf{A}\right)_{.i}\left(\mathbf{a}^\sharp_{.j}\right)\right) = c_1^{(ij)}\lambda^{n-1} + c_2^{(ij)}\lambda^{n-2} + \ldots + c_n^{(ij)},$$

$$\det\left((\lambda\mathbf{I}_m + \mathbf{A}\mathbf{A}^\sharp)_{j.}(\mathbf{a}^\sharp_{i.})\right) = r_1^{(ij)}\lambda^{m-1} + r_2^{(ij)}\lambda^{m-2} + \ldots + r_m^{(ij)},$$

where $c_n^{(ij)} = \left|\left(\mathbf{A}^\sharp\mathbf{A}\right)_{.i}\left(\mathbf{a}^\sharp_{.j}\right)\right|$, $r_m^{(ij)} = |(\mathbf{A}\mathbf{A}^*)_{j.}(\mathbf{a}^*_{i.})|$ *and* $c_s^{(ij)} = \sum\limits_{\beta\in J_{s,n}\{i\}}\left|\left(\left(\mathbf{A}^\sharp\mathbf{A}\right)_{.i}\left(\mathbf{a}^\sharp_{.j}\right)\right)^\beta_\beta\right|$, $r_t^{(ij)} = \sum\limits_{\alpha\in I_{t,m}\{j\}}\left|\left((\mathbf{A}\mathbf{A}^\sharp)_{j.}(\mathbf{a}^\sharp_{i.})\right)^\alpha_\alpha\right|$ *for all* $s = \overline{1, n-1}$, $t = \overline{1, m-1}$, $i = \overline{1, n}$, *and* $j = \overline{1, m}$.

The following theorem introduce the determinantal representations of the weighted Moore-Penrose by analogs of the classical adjoint matrix.

Theorem 2.21. *If* $\mathbf{A} \in \mathbb{C}_r^{m\times n}$ *and* $r < min\{m,n\}$, *then the weighted Moore-Penrose inverse* $\mathbf{A}_{M,N}^{+} = \left(\tilde{a}_{ij}^{+}\right) \in \mathbb{C}^{n\times m}$ *possess the following determinantal representation:*

$$\tilde{a}_{ij}^{+} = \frac{\sum\limits_{\beta\in J_{r,\,n}\{i\}} \left|\left(\left(\mathbf{A}^{\sharp}\mathbf{A}\right)_{.\,i}\left(\mathbf{a}_{.j}^{\sharp}\right)\right)_{\beta}^{\beta}\right|}{\sum\limits_{\beta\in J_{r,\,n}} \left|\left(\mathbf{A}^{\sharp}\mathbf{A}\right)_{\beta}^{\beta}\right|}, \tag{2.12}$$

or

$$\tilde{a}_{ij}^{+} = \frac{\sum\limits_{\alpha\in I_{r,m}\{j\}} \left|\left((\mathbf{A}\mathbf{A}^{\sharp})_{j\,.}\,(\mathbf{a}_{i.}^{\sharp})\right)_{\alpha}^{\alpha}\right|}{\sum\limits_{\alpha\in I_{r,\,m}} |(\mathbf{A}\mathbf{A}^{\sharp})_{\alpha}^{\alpha}|}, \tag{2.13}$$

for all $i = \overline{1,n}$, $j = \overline{1,m}$.

2.3. Analogues of the Classical Adjoint Matrix for the Drazin Inverse

The Drazin inverse can be represented explicitly by the Jordan canonical form as follows.

Theorem 2.22. *[25] If* $\mathbf{A} \in \mathbb{C}^{n\times n}$ *with* $Ind\,\mathbf{A} = k$ *and*

$$\mathbf{A} = \mathbf{P}\begin{pmatrix}\mathbf{C} & \mathbf{0}\\ \mathbf{0} & \mathbf{N}\end{pmatrix}\mathbf{P}^{-1}$$

where $\mathbf{C}$ *is nonsingular and* $\operatorname{rank}\mathbf{C} = \operatorname{rank}\mathbf{A}^{k}$, *and* $\mathbf{N}$ *is nilpotent of order* k, *then*

$$\mathbf{A}^{D} = \mathbf{P}\begin{pmatrix}\mathbf{C}^{-1} & \mathbf{0}\\ \mathbf{0} & \mathbf{0}\end{pmatrix}\mathbf{P}^{-1}. \tag{2.14}$$

Stanimirovic' [26] introduced a determinantal representation of the Drazin inverse by the following theorem.

Theorem 2.23. *The Drazin inverse* $\mathbf{A}^{D} = \left(a_{ij}^{D}\right)$ *of an arbitrary matrix* $\mathbf{A} \in \mathbb{C}^{n\times n}$ *with* $Ind\mathbf{A} = k$ *possesses the following determinantal representation*

$$a_{ij}^{D} = \frac{\sum\limits_{(\alpha,\beta)\in N_{r_k}\{j,i\}} \left|(\mathbf{A}^{s})_{\alpha}^{\beta}\right| \frac{\partial}{\partial a_{j\,i}} \left|\mathbf{A}_{\beta}^{\alpha}\right|}{\sum\limits_{(\gamma,\,\delta)\in N_{r_k}} \left|(\mathbf{A}^{s})_{\gamma}^{\delta}\right| \left|\mathbf{A}_{\delta}^{\gamma}\right|}, \quad 1 \le i,j \le n; \tag{2.15}$$

where $s \ge k$ *and* $r_k = rank\mathbf{A}^{s}$.

This determinantal representations of the Drazin inverse is based on a full-rank representation.

We use the following limit representation of the Drazin inverse.

Lemma 2.24. *[27] If $\mathbf{A} \in \mathbb{C}^{n\times n}$, then*

$$\mathbf{A}^{D} = \lim_{\lambda\to 0}\left(\lambda\mathbf{I}_n + \mathbf{A}^{k+1}\right)^{-1}\mathbf{A}^{k},$$

where $k = Ind\,\mathbf{A}$, $\lambda \in \mathbb{R}_{+}$, and $\mathbb{R}_{+}$ is a set of the real positive numbers.

Since the equation (1.6) can be replaced by follows

$$\mathbf{X}\mathbf{A}^{k+1} = \mathbf{A}^{k},$$

the following lemma can be obtained by analogy to Lemma 2.24.

Lemma 2.25. *If $\mathbf{A} \in \mathbb{C}^{n\times n}$, then*

$$\mathbf{A}^{D} = \lim_{\lambda\to 0}\mathbf{A}^{k}\left(\lambda\mathbf{I}_n + \mathbf{A}^{k+1}\right)^{-1},$$

where $k = Ind\,\mathbf{A}$, $\lambda \in \mathbb{R}_{+}$, and $\mathbb{R}_{+}$ is a set of the real positive numbers.

Denote by $\mathbf{a}_{.j}^{(k)}$ and $\mathbf{a}_{i.}^{(k)}$ the jth column and the ith row of $\mathbf{A}^{k}$ respectively. We consider the following auxiliary lemma.

Lemma 2.26. *If $\mathbf{A} \in \mathbb{C}^{n\times n}$ with $Ind\,\mathbf{A} = k$, then for all $i, j = \overline{1,n}$*

$$\operatorname{rank}\mathbf{A}_{i.}^{k+1}\left(\mathbf{a}_{j.}^{(k)}\right) \leq \operatorname{rank}\mathbf{A}^{k+1}.$$

Proof. The matrix $\mathbf{A}_{i.}^{k+1}\left(\mathbf{a}_{j.}^{(k)}\right)$ may by represent as follows

$$\begin{pmatrix} \sum\limits_{s=1}^{n} a_{1s}a_{s1}^{(k)} & \ldots & \sum\limits_{s=1}^{n} a_{1s}a_{sn}^{(k)} \\ \ldots & \ldots & \ldots \\ a_{j1}^{(k)} & \ldots & a_{jn}^{(k)} \\ \ldots & \ldots & \ldots \\ \sum\limits_{s=1}^{n} a_{ns}a_{s1}^{(k)} & \ldots & \sum\limits_{s=1}^{n} a_{ns}a_{sn}^{(k)} \end{pmatrix}$$

Let $\mathbf{P}_{l\,i}\left(-a_{l\,j}\right) \in \mathbb{C}^{n\times n}$, $(l \neq i)$, be a matrix with $-a_{l\,j}$ in the (l, i) entry, 1 in all diagonal entries, and 0 in others. It is a matrix of an elementary transformation. It follows that

$$\mathbf{A}_{i.}^{k+1}\left(\mathbf{a}_{j.}^{(k)}\right) \cdot \prod_{l\neq i}\mathbf{P}_{l\,i}\left(-a_{l\,j}\right) = \begin{pmatrix} \sum\limits_{s\neq j}^{n} a_{1s}a_{s1}^{(k)} & \ldots & \sum\limits_{s\neq j}^{n} a_{1s}a_{sn}^{(k)} \\ \ldots & \ldots & \ldots \\ a_{j1}^{(k)} & \ldots & a_{jn}^{(k)} \\ \ldots & \ldots & \ldots \\ \sum\limits_{s\neq j}^{n} a_{ns}a_{s1}^{(k)} & \ldots & \sum\limits_{s\neq j}^{n} a_{ns}a_{sn}^{(k)} \end{pmatrix} \quad ith$$

The obtained above matrix has the following factorization.

$$\begin{pmatrix} \sum\limits_{s\neq j}^{n} a_{1s}a_{s1}^{(k)} & \dots & \sum\limits_{s\neq j}^{n} a_{1s}a_{sn}^{(k)} \\ \dots & \dots & \dots \\ a_{j1}^{(k)} & \dots & a_{jn}^{(k)} \\ \dots & \dots & \dots \\ \sum\limits_{s\neq j}^{n} a_{ns}a_{s1}^{(k)} & \dots & \sum\limits_{s\neq j}^{n} a_{ns}a_{sn}^{(k)} \end{pmatrix} =$$

$$\begin{pmatrix} a_{11} & \dots & 0 & \dots & a_{1n} \\ \dots & \dots & \dots & \dots & \dots \\ 0 & \dots & 1 & \dots & 0 \\ \dots & \dots & \dots & \dots & \dots \\ a_{n1} & \dots & 0 & \dots & a_{nn} \end{pmatrix} \begin{pmatrix} a_{11}^{(k)} & a_{12}^{(k)} & \dots & a_{1n}^{(k)} \\ a_{21}^{(k)} & a_{22}^{(k)} & \dots & a_{2n}^{(k)} \\ \dots & \dots & \dots & \dots \\ a_{n1}^{(k)} & a_{n2}^{(k)} & \dots & a_{nn}^{(k)} \end{pmatrix}$$

Denote the first matrix by

$$\tilde{\mathbf{A}} := \begin{pmatrix} a_{11} & \dots & 0 & \dots & a_{1n} \\ \dots & \dots & \dots & \dots & \dots \\ 0 & \dots & 1 & \dots & 0 \\ \dots & \dots & \dots & \dots & \dots \\ a_{n1} & \dots & \underset{jth}{0} & \dots & a_{nn} \end{pmatrix} ith.$$

The matrix $\tilde{\mathbf{A}}$ is obtained from $\mathbf{A}$ by replacing all entries of the ith row and the jth column with zeroes except for 1 in the (i,j) entry. Elementary transformations of a matrix do not change its rank. It follows that $\operatorname{rank} \mathbf{A}_{i.}^{k+1}\left(\mathbf{a}_{j.}^{(k)}\right) \leq \min\left\{\operatorname{rank} \mathbf{A}^{k}, \operatorname{rank} \tilde{\mathbf{A}}\right\}$. Since $\operatorname{rank} \tilde{\mathbf{A}} \geq \operatorname{rank} \mathbf{A}^{k}$ the proof is completed. ■

The following lemma is proved similarly.

Lemma 2.27. *If $\mathbf{A} \in \mathbb{C}^{n\times n}$ with $Ind\,\mathbf{A} = k$, then for all $i,j = \overline{1,n}$*

$$\operatorname{rank} \mathbf{A}_{.i}^{k+1}\left(\mathbf{a}_{.j}^{(k)}\right) \leq \operatorname{rank} \mathbf{A}^{k+1}.$$

Lemma 2.28. *If $\mathbf{A} \in \mathbb{C}^{n\times n}$ and $\lambda \in \mathbb{R}$, then*

$$\det\left((\lambda \mathbf{I}_n + \mathbf{A}^{k+1})_{j.}\,(\mathbf{a}_{i.}^{(k)})\right) = r_1^{(ij)}\lambda^{n-1} + r_2^{(ij)}\lambda^{n-2} + \ldots + r_n^{(ij)}, \tag{2.16}$$

where $r_n^{(ij)} = \left|\mathbf{A}_{j.}^{k+1}(\mathbf{a}_{i.}^{(k)})\right|$ and $r_s^{(ij)} = \sum\limits_{\alpha\in I_{s,n}\{j\}} \left|\left(\mathbf{A}_{j.}^{k+1}(\mathbf{a}_{i.}^{(k)})\right)_{\alpha}^{\alpha}\right|$ for all $s = \overline{1,n-1}$ and $i,j = \overline{1,n}$.

Proof. Consider the matrix $\left((\lambda \mathbf{I}_n + \mathbf{A}^{k+1})_{j.}\,(\mathbf{a}_{j.}^{(k)})\right) \in \mathbb{C}^{n\times n}$. Taking into account Theorem 2.4 we obtain

$$\left|\left((\lambda \mathbf{I}_n + \mathbf{A}^{k+1})_{j.}\,(\mathbf{a}_{j.}^{(k)})\right)\right| = d_1\lambda^{n-1} + d_2\lambda^{n-2} + \ldots + d_n, \tag{2.17}$$

where $d_s = \sum\limits_{\alpha \in I_{s,n}\{j\}} \left|\left(\mathbf{A}^{k+1}\right)_{\alpha}^{\alpha}\right|$ is the sum of all principal minors of order s that contain the j-th row for all $s = \overline{1, n-1}$ and $d_n = \det \mathbf{A}^{k+1}$. Since $\mathbf{a}_{j.}^{(k+1)} = \sum\limits_{l} a_{jl}\mathbf{a}_{l.}^{(k)}$, where $\mathbf{a}_{l.}^{(k)}$ is the lth row-vector of $\mathbf{A}^k$ for all $l = \overline{1, n}$, then we have on the one hand

$$\left|\left((\lambda \mathbf{I}_n + \mathbf{A}^{k+1})_{j.}(\mathbf{a}_{j.}^{(k)})\right)\right| = \sum_{l} \left|(\lambda \mathbf{I} + \mathbf{A}^{k+1})_{l.}\left(a_{jl}\mathbf{a}_{l.}^{(k)}\right)\right| = \sum_{l} a_{jl} \cdot \left|(\lambda \mathbf{I} + \mathbf{A}^{k+1})_{l.}\left(\mathbf{a}_{l.}^{(k)}\right)\right| \tag{2.18}$$

Having changed the order of summation, we obtain on the other hand for all $s = \overline{1, n-1}$

$$d_s = \sum_{\alpha \in I_{s,n}\{j\}} \left|\left(\mathbf{A}^{k+1}\right)_{\alpha}^{\alpha}\right| = \sum_{\alpha \in I_{s,n}\{j\}} \sum_{l} \left|\left(\mathbf{A}_{j.}^{k+1}\left(a_{jl}\mathbf{a}_{l.}^{(k)}\right)\right)_{\alpha}^{\alpha}\right| = \sum_{l} a_{jl} \cdot \sum_{\alpha \in I_{s,n}\{j\}} \left|\left(\mathbf{A}_{j.}^{k+1}\left(\mathbf{a}_{l.}^{(k)}\right)\right)_{\alpha}^{\alpha}\right| \tag{2.19}$$

By substituting (2.18) and (2.19) in (2.17), and equating factors at a_{jl} when $l = i$, we obtain the equality (2.16). ■

Theorem 2.29. *If* $Ind\,\mathbf{A} = k$ *and* $\operatorname{rank} \mathbf{A}^{k+1} = \operatorname{rank} \mathbf{A}^k = r \leq n$ *for* $\mathbf{A} \in \mathbb{C}^{n\times n}$*, then the Drazin inverse* $\mathbf{A}^D = \left(a_{ij}^D\right) \in \mathbb{C}^{n\times n}$ *possess the following determinantal representations:*

$$a_{ij}^D = \frac{\sum\limits_{\alpha \in I_{r,n}\{j\}} \left|\left(\mathbf{A}_{j.}^{k+1}\left(\mathbf{a}_{i.}^{(k)}\right)\right)_{\alpha}^{\alpha}\right|}{\sum\limits_{\alpha \in I_{r,n}} \left|\left(\mathbf{A}^{k+1}\right)_{\alpha}^{\alpha}\right|}, [\tag{2.20}$$

and

$$a_{ij}^D = \frac{\sum\limits_{\beta \in J_{r,n}\{i\}} \left|\left(\mathbf{A}_{.i}^{k+1}\left(\mathbf{a}_{.j}^{(k)}\right)\right)_{\beta}^{\beta}\right|}{\sum\limits_{\beta \in J_{r,n}} \left|\left(\mathbf{A}^{k+1}\right)_{\beta}^{\beta}\right|}, \tag{2.21}$$

for all $i, j = \overline{1, n}$.

Proof. At first we shall prove the equation (2.20).

If $\lambda \in \mathbb{R}_+$, then $\operatorname{rank}\left(\lambda \mathbf{I} + \mathbf{A}^{k+1}\right) = n$. Hence, there exists the inverse matrix

$$\left(\lambda \mathbf{I} + \mathbf{A}^{k+1}\right)^{-1} = \frac{1}{\det\left(\lambda \mathbf{I} + \mathbf{A}^{k+1}\right)} \begin{pmatrix} R_{11} & R_{21} & \dots & R_{n1} \\ R_{12} & R_{22} & \dots & R_{n2} \\ \dots & \dots & \dots & \dots \\ R_{1n} & R_{2n} & \dots & R_{nn} \end{pmatrix},$$

where R_{ij} is a cofactor in $\lambda \mathbf{I} + \mathbf{A}^{k+1}$ for all $i, j = \overline{1, n}$. By Theorem 2.25, $\mathbf{A}^D = \lim\limits_{\lambda\to 0} \mathbf{A}^k \left(\lambda \mathbf{I}_n + \mathbf{A}^{k+1}\right)^{-1}$, so that

$$\mathbf{A}^D = \lim_{\lambda\to 0} \frac{1}{\det\left(\lambda \mathbf{I} + \mathbf{A}^{k+1}\right)} \begin{pmatrix} \sum_{s=1}^{n} a_{1s}^{(k)} R_{1s} & \dots & \sum_{s=1}^{n} a_{1s}^{(k)} R_{ns} \\ \dots & \dots & \dots \\ \sum_{s=1}^{n} a_{ns}^{(k)} R_{1s} & \dots & \sum_{s=1}^{n} a_{ns}^{(k)} R_{ns} \end{pmatrix} =$$

$$\lim_{\lambda\to 0}\begin{pmatrix} \frac{\det\left(\lambda\mathbf{I}+\mathbf{A}^{k+1}\right)_{1.}\left(\mathbf{a}_{1.}^{(k)}\right)}{\det\left(\lambda\mathbf{I}+\mathbf{A}^{k+1}\right)} & \dots & \frac{\det\left(\lambda\mathbf{I}+\mathbf{A}^{k+1}\right)_{n.}\left(\mathbf{a}_{n.}^{(k)}\right)}{\det\left(\lambda\mathbf{I}+\mathbf{A}^{k+1}\right)} \\ \dots & \dots & \dots \\ \frac{\det\left(\lambda\mathbf{I}+\mathbf{A}^{k+1}\right)_{1.}\left(\mathbf{a}_{n.}^{(k)}\right)}{\det\left(\lambda\mathbf{I}+\mathbf{A}^{k+1}\right)} & \dots & \frac{\det\left(\lambda\mathbf{I}+\mathbf{A}^{k+1}\right)_{n.}\left(\mathbf{a}_{n.}^{(k)}\right)}{\det\left(\lambda\mathbf{I}+\mathbf{A}^{k+1}\right)} \end{pmatrix} \tag{2.22}$$

Taking into account Theorem 2.4 , we have

$$\det\left(\lambda\mathbf{I}+\mathbf{A}^{k+1}\right)=\lambda^n+d_1\lambda^{n-1}+d_2\lambda^{n-2}+\ldots+d_n,$$

where $d_s=\sum\limits_{\alpha\in I_{s,n}}\left|\left(\mathbf{A}^{k+1}\right)_{\alpha}^{\alpha}\right|$ is a sum of the principal minors of $\mathbf{A}^{k+1}$ of order s, for all $s=\overline{1,n-1}$, and $d_n=\det\mathbf{A}^{k+1}$. Since $\operatorname{rank}\mathbf{A}^{k+1}=r$, then $d_n=d_{n-1}=\ldots=d_{r+1}=0$ and

$$\det\left(\lambda\mathbf{I}+\mathbf{A}^{k+1}\right)=\lambda^n+d_1\lambda^{n-1}+d_2\lambda^{n-2}+\ldots+d_r\lambda^{n-r}. \tag{2.23}$$

By Lemma 2.28 for all $i,j=\overline{1,n}$,

$$\det\left(\lambda\mathbf{I}+\mathbf{A}^{k+1}\right)_{j.}\left(\mathbf{a}_{i.}^{(k)}\right)=l_1^{(ij)}\lambda^{n-1}+l_2^{(ij)}\lambda^{n-2}+\ldots+l_n^{(ij)},$$

where for all $s=\overline{1,n-1}$,

$$l_s^{(ij)}=\sum_{\alpha\in I_{s,n}\{j\}}\left|\left(\mathbf{A}_{j.}^{k+1}\left(\mathbf{a}_{i.}^{(k)}\right)\right)_{\alpha}^{\alpha}\right|,$$

and $l_n^{(ij)}=\det\mathbf{A}_{j.}^{k+1}\left(\mathbf{a}_{i.}^{(k)}\right)$.

By Lemma 2.26, $\operatorname{rank}\mathbf{A}_{j.}^{k+1}\left(\mathbf{a}_{i.}^{(k)}\right)\le r$, so that if $s>r$, then for all $\alpha\in I_{s,n}\{i\}$ and for all $i,j=\overline{1,n}$,

$$\left|\left(\mathbf{A}_{j.}^{k+1}\left(\mathbf{a}_{i.}^{(k)}\right)\right)_{\alpha}^{\alpha}\right|=0.$$

Therefore if $r+1\le s<n$, then for all $i,j=\overline{1,n}$,

$$l_s^{(ij)}=\sum_{\alpha\in I_{s,n}\{j\}}\left|\left(\mathbf{A}_{j.}^{k+1}\left(\mathbf{a}_{i.}^{(k)}\right)\right)_{\alpha}^{\alpha}\right|=0,$$

and $l_n^{(ij)}=\det\mathbf{A}_{j.}^{k+1}\left(\mathbf{a}_{i.}^{(k)}\right)=0$. Finally we obtain

$$\det\left(\lambda\mathbf{I}+\mathbf{A}^{k+1}\right)_{j.}\left(\mathbf{a}_{i.}^{(k)}\right)=l_1^{(ij)}\lambda^{n-1}+l_2^{(ij)}\lambda^{n-2}+\ldots+l_r^{(ij)}\lambda^{n-r}. \tag{2.24}$$

By replacing the denominators and the nominators of the fractions in the entries of the matrix (2.22) with the expressions (2.23) and (2.24) respectively, finally we obtain

$$\mathbf{A}^D=\lim_{\lambda\to 0}\begin{pmatrix} \frac{l_1^{(11)}\lambda^{n-1}+\ldots+l_r^{(11)}\lambda^{n-r}}{\lambda^n+d_1\lambda^{n-1}+\ldots+d_r\lambda^{n-r}} & \dots & \frac{l_1^{(1n)}\lambda^{n-1}+\ldots+l_r^{(1n)}\lambda^{n-r}}{\lambda^n+d_1\lambda^{n-1}+\ldots+d_r\lambda^{n-r}} \\ \dots & \dots & \dots \\ \frac{l_1^{(n1)}\lambda^{n-1}+\ldots+l_r^{(n1)}\lambda^{n-r}}{\lambda^n+d_1\lambda^{n-1}+\ldots+d_r\lambda^{n-r}} & \dots & \frac{l_1^{(nn)}\lambda^{n-1}+\ldots+l_r^{(nn)}\lambda^{n-r}}{\lambda^n+d_1\lambda^{n-1}+\ldots+d_r\lambda^{n-r}} \end{pmatrix}=$$

$$= \begin{pmatrix} \frac{l_r^{(11)}}{d_r} & \dots & \frac{l_r^{(1n)}}{d_r} \\ \dots & \dots & \dots \\ \frac{l_r^{(n1)}}{d_r} & \dots & \frac{l_r^{(nn)}}{d_r} \end{pmatrix},$$

where for all $i, j = \overline{1, n}$,

$$l_r^{(ij)} = \sum_{\alpha \in I_{r,n}\{j\}} \left| \left(\mathbf{A}_{j.}^{k+1} \left(\mathbf{a}_{i.}^{(k)} \right) \right)_{\alpha}^{\alpha} \right|, \; d_r = \sum_{\alpha \in I_{r,n}} \left| \left(\mathbf{A}^{k+1} \right)_{\alpha}^{\alpha} \right|.$$

The equation (2.21) can be proved similarly.

This completes the proof. ■ Using Theorem 2.29 we evidently can obtain determinantal representations of the group inverse and the following determinantal representation of the identities $\mathbf{A}^D\mathbf{A}$ and $\mathbf{A}\mathbf{A}^D$ on $R(\mathbf{A}^k)$

Corollary 2.30. *If* $Ind\,\mathbf{A} = 1$ *and* $\operatorname{rank} \mathbf{A}^2 = \operatorname{rank} \mathbf{A} = r \leq n$ *for* $\mathbf{A} \in \mathbb{C}^{n\times n}$*, then the group inverse* $\mathbf{A}^g = \left(a_{ij}^g\right) \in \mathbb{C}^{n\times n}$ *possess the following determinantal representations:*

$$a_{ij}^g = \frac{\sum\limits_{\alpha \in I_{r,n}\{j\}} \left| \left(\mathbf{A}_{j.}^2 \left(\mathbf{a}_{i.} \right) \right)_{\alpha}^{\alpha} \right|}{\sum\limits_{\alpha \in I_{r,n}} \left| (\mathbf{A}^2)_{\alpha}^{\alpha} \right|}, \tag{2.25}$$

$$a_{ij}^g = \frac{\sum\limits_{\beta \in J_{r,n}\{i\}} \left| \left(\mathbf{A}_{.i}^2 \left(\mathbf{a}_{.j} \right) \right)_{\beta}^{\beta} \right|}{\sum\limits_{\beta \in J_{r,n}} \left| (\mathbf{A}^2)_{\beta}^{\beta} \right|},$$

for all $i, j = \overline{1, n}$.

Corollary 2.31. *If* $Ind\,\mathbf{A} = k$ *and* $\operatorname{rank} \mathbf{A}^{k+1} = \operatorname{rank} \mathbf{A}^k = r \leq n$ *for* $\mathbf{A} \in \mathbb{C}^{n\times n}$*, then the matrix* $\mathbf{A}\mathbf{A}^D = (q_{ij}) \in \mathbb{C}^{n\times n}$ *possess the following determinantal representation*

$$q_{ij} = \frac{\sum\limits_{\alpha \in I_{r,n}\{j\}} \left| \left(\mathbf{A}_{j.}^{k+1} \left(\mathbf{a}_{i.}^{(k+1)} \right) \right)_{\beta}^{\beta} \right|}{\sum\limits_{\alpha \in I_{r,n}} \left| (\mathbf{A}^{k+1})_{\beta}^{\beta} \right|}, \tag{2.26}$$

for all $i, j = \overline{1, n}$.

Corollary 2.32. *If* $Ind\,\mathbf{A} = k$ *and* $\operatorname{rank} \mathbf{A}^{k+1} = \operatorname{rank} \mathbf{A}^k = r \leq n$ *for* $\mathbf{A} \in \mathbb{C}^{n\times n}$*, then the matrix* $\mathbf{A}^D\mathbf{A} = (p_{ij}) \in \mathbb{C}^{n\times n}$ *possess the following determinantal representation*

$$p_{ij} = \frac{\sum\limits_{\beta \in J_{r,n}\{i\}} \left| \left(\mathbf{A}_{.i}^{k+1} \left(\mathbf{a}_{.j}^{(k+1)} \right) \right)_{\beta}^{\beta} \right|}{\sum\limits_{\beta \in J_{r,n}} \left| \left(\mathbf{A}_{.i}^{k+1} \right)_{\beta}^{\beta} \right|}, \tag{2.27}$$

for all $i, j = \overline{1, n}$.

2.4. Analogues of the Classical Adjoint Matrix for the W-Weighted Drazin Inverse

Cline and Greville [28] extended the Drazin inverse of square matrix to rectangular matrix and called it as **the weighted Drazin inverse** (WDI). The W-weighted Drazin inverse of $\mathbf{A} \in \mathbb{C}^{m\times n}$ with respect to $\mathbf{W} \in \mathbb{C}^{n\times m}$ is defined to be the unique solution $\mathbf{X} \in \mathbb{C}^{m\times n}$ of the following three matrix equations:

$$\begin{array}{l} 1)\ (\mathbf{AW})^{k+1}\mathbf{XW} = (\mathbf{AW})^{k}, \\ 2)\ \mathbf{XWAWX} = \mathbf{X}, \\ 3)\ \mathbf{AWX} = \mathbf{XWA}, \end{array} \tag{2.28}$$

where $k = \max\{Ind(\mathbf{AW}), Ind(\mathbf{WA})\}$. It is denoted by $\mathbf{X} = \mathbf{A}_{d,W}$. In particular, when $\mathbf{A} \in \mathbb{C}^{m\times m}$ and $\mathbf{W} = \mathbf{I}_m$, then $\mathbf{A}_{d,W}$ reduce to $\mathbf{A}^D$. If $\mathbf{A} \in \mathbb{C}^{m\times m}$ is non-singular square matrix and $\mathbf{W} = \mathbf{I}_m$, then $Ind(A) = 0$ and $\mathbf{A}_{d,W} = \mathbf{A}^D = \mathbf{A}^{-1}$.

The properties of WDI can be found in (e.g.,[29, 30, 31, 32]). We note the general algebraic structures of the W-weighted Drazin inverse [29]. Let for $\mathbf{A} \in \mathbb{C}^{m\times n}$ and $\mathbf{W} \in \mathbb{C}^{n\times m}$ exist $\mathbf{L} \in \mathbb{C}^{m\times m}$ and $\mathbf{Q} \in \mathbb{C}^{n\times n}$ such that

$$\mathbf{A} = \mathbf{L}\begin{pmatrix} \mathbf{A}_{11} & \mathbf{0} \\ \mathbf{0} & \mathbf{A}_{22} \end{pmatrix}\mathbf{Q}^{-1}, \quad \mathbf{W} = \mathbf{Q}\begin{pmatrix} \mathbf{W}_{11} & \mathbf{0} \\ \mathbf{0} & \mathbf{W}_{22} \end{pmatrix}\mathbf{L}^{-1}.$$

Then

$$\mathbf{A}_{d,W} = \mathbf{L}\begin{pmatrix} (\mathbf{W}_{11}\mathbf{A}_{11}\mathbf{W}_{11})^{-1} & \mathbf{0} \\ \mathbf{0} & \mathbf{0} \end{pmatrix}\mathbf{Q}^{-1},$$

where $\mathbf{L}$, $\mathbf{L}$, $\mathbf{A}_{11}$, $\mathbf{W}_{11}$ are non-singular matrices, and $\mathbf{A}_{22}$, $\mathbf{W}_{22}$ are nilpotent matrices. By [27] we have the following limit representations of the W-weighted Drazin inverse,

$$\mathbf{A}_{d,W} = \lim_{\lambda\to 0}\left(\lambda\mathbf{I}_m + (\mathbf{AW})^{k+2}\right)^{-1}(\mathbf{AW})^k\mathbf{A} \tag{2.29}$$

and

$$\mathbf{A}_{d,W} = \lim_{\lambda\to 0}\mathbf{A}(\mathbf{WA})^k\left(\lambda\mathbf{I}_n + (\mathbf{WA})^{k+2}\right)^{-1} \tag{2.30}$$

where $\lambda \in \mathbb{R}_+$, and $\mathbb{R}_+$ is a set of the real positive numbers.

Denote $\mathbf{WA} =: \mathbf{U}$ and $\mathbf{AW} =: \mathbf{V}$. Denote by $\mathbf{v}_{.j}^{(k)}$ and $\mathbf{v}_{i.}^{(k)}$ the jth column and the ith row of $\mathbf{V}^k$ respectively. Denote by $\bar{\mathbf{V}}^k := (\mathbf{AW})^k\mathbf{A} \in \mathbb{C}^{m\times n}$ and $\bar{\mathbf{W}} = \mathbf{WAW} \in \mathbb{C}^{n\times m}$.

Lemma 2.33. *If* $\mathbf{AW} = \mathbf{V} = (v_{ij}) \in \mathbb{C}^{m\times m}$ *with* $Ind\mathbf{V} = k$, *then*

$$\operatorname{rank}\left(\mathbf{V}^{k+2}\right)_{.i}\left(\bar{\mathbf{v}}_{.j}^{(k)}\right) \le \operatorname{rank}\left(\mathbf{V}^{k+2}\right). \tag{2.31}$$

Proof. We have $\mathbf{V}^{k+2} = \bar{\mathbf{V}}^k\bar{\mathbf{W}}$. Let $\mathbf{P}_{i\,s}\left(-\bar{w}_{j\,s}\right) \in \mathbb{C}^{m\times m}$, $(s \ne i)$, be a matrix with $-\bar{w}_{j\,s}$ in the (i, s) entry, 1 in all diagonal entries, and 0 in others. The matrix $\mathbf{P}_{i\,s}\left(-\bar{w}_{j\,s}\right)$, $(s \ne i)$, is a matrix of an elementary transformation. It follows that

$$\left(\mathbf{V}^{k+2}\right)_{.i}\left(\bar{\mathbf{v}}_{.j}^{(k)}\right)\cdot\prod_{s\ne i}\mathbf{P}_{i\,s}\left(-\bar{w}_{j\,s}\right) = \begin{pmatrix} \sum\limits_{s\ne j}\bar{v}_{1s}^{(k)}\bar{w}_{s1} & \dots & \bar{v}_{1j}^{(k)} & \dots & \sum\limits_{s\ne j}\bar{v}_{1s}^{(k)}\bar{w}_{sm} \\ \dots & \dots & \dots & \dots & \dots \\ \sum\limits_{s\ne j}\bar{v}_{ms}^{(k)}\bar{w}_{s1} & \dots & \underbrace{\bar{v}_{mj}^{(k)}}_{i-th} & \dots & \sum\limits_{s\ne j}\bar{v}_{ms}^{(k)}\bar{w}_{sm} \end{pmatrix}.$$

We have the next factorization of the obtained matrix.

$$
\begin{pmatrix}
\sum\limits_{s\neq j} \bar{v}_{1s}^{(k)}\bar{w}_{s1} & \ldots & \bar{v}_{1j}^{(k)} & \ldots & \sum\limits_{s\neq j} \bar{v}_{1s}^{(k)}\bar{w}_{sm} \\
\ldots & \ldots & \ldots & \ldots & \ldots \\
\sum\limits_{s\neq j} \bar{v}_{ms}^{(k)}\bar{w}_{s1} & \ldots & \bar{v}_{mj}^{(k)} & \ldots & \sum\limits_{s\neq j} \bar{v}_{ms}^{(k)}\bar{w}_{sm}
\end{pmatrix} =
$$
$$
\hspace{4em} i-th
$$

$$
= \begin{pmatrix}
\bar{v}_{11}^{(k)} & \bar{v}_{12}^{(k)} & \ldots & \bar{v}_{1n}^{(k)} \\
\bar{v}_{21}^{(k)} & \bar{v}_{22}^{(k)} & \ldots & \bar{v}_{2n}^{(k)} \\
\ldots & \ldots & \ldots & \ldots \\
\bar{v}_{m1}^{(k)} & \bar{v}_{m2}^{(k)} & \ldots & \bar{v}_{mn}^{(k)}
\end{pmatrix}
\begin{pmatrix}
\bar{w}_{11} & \ldots & 0 & \ldots & \bar{w}_{1m} \\
\ldots & \ldots & \ldots & \ldots & \ldots \\
0 & \ldots & 1 & \ldots & 0 \\
\ldots & \ldots & \ldots & \ldots & \ldots \\
\bar{w}_{n1} & \ldots & 0 & \ldots & \bar{w}_{nm}
\end{pmatrix} j-th.
$$
$$
\hspace{10em} i-th
$$

Denote $\tilde{\mathbf{W}} := \begin{pmatrix}
\bar{w}_{11} & \ldots & 0 & \ldots & \bar{w}_{1m} \\
\ldots & \ldots & \ldots & \ldots & \ldots \\
0 & \ldots & 1 & \ldots & 0 \\
\ldots & \ldots & \ldots & \ldots & \ldots \\
\bar{w}_{n1} & \ldots & 0 & \ldots & \bar{w}_{nm}
\end{pmatrix} j-th$ (with $i-th$ marking the column). The matrix $\tilde{\mathbf{W}}$ is obtained from $\bar{\mathbf{W}} = \mathbf{WAW}$ by replacing all entries of the jth row and the ith column with zeroes except for 1 in the (i, j) entry. Since elementary transformations of a matrix do not change a rank, then $\operatorname{rank} \mathbf{V}^{k+2}_{.\,i}\left(\bar{\mathbf{v}}^{(k)}_{.j}\right) \leq \min\left\{\operatorname{rank} \bar{\mathbf{V}}^{k}, \operatorname{rank} \tilde{\mathbf{W}}\right\}$. It is obvious that

$$
\begin{aligned}
\operatorname{rank} \bar{\mathbf{V}}^{k} &= \operatorname{rank} (\mathbf{AW})^{k}\mathbf{A} \geq \operatorname{rank} (\mathbf{AW})^{k+2}, \\
\operatorname{rank} \tilde{\mathbf{W}} &\geq \operatorname{rank}\, \mathbf{WAW} \geq \operatorname{rank} (\mathbf{AW})^{k+2}.
\end{aligned}
$$

From this the inequality (2.31) follows immediately. ■

The next lemma is proved similarly.

Lemma 2.34. *If* $\mathbf{WA} = \mathbf{U} = (u_{ij}) \in \mathbb{C}^{n\times n}$ *with* $Ind\,\mathbf{U} = k$, *then*

$$
\operatorname{rank}\left(\mathbf{U}^{k+2}\right)_{i\,.}\left(\bar{\mathbf{u}}^{(k)}_{j\,.}\right) \leq \operatorname{rank}\left(\mathbf{U}^{k+2}\right),
$$

where $\bar{\mathbf{U}}^{k} := \mathbf{A}(\mathbf{WA})^{k} \in \mathbb{C}^{m\times n}$

Analogues of the characteristic polynomial are considered in the following two lemmas.

Lemma 2.35. *If* $\mathbf{AW} = \mathbf{V} = (v_{ij}) \in \mathbb{C}^{m\times m}$ *with* $Ind\,\mathbf{V} = k$ *and* $\lambda \in \mathbb{R}$, *then*

$$
\left|\left(\lambda \mathbf{I}_m + \mathbf{V}^{k+2}\right)_{.\,i}\left(\bar{\mathbf{v}}^{(k)}_{.j}\right)\right| = c_1^{(ij)}\lambda^{m-1} + c_2^{(ij)}\lambda^{m-2} + \ldots + c_m^{(ij)}, \tag{2.32}
$$

where $c_m^{(ij)} = \det\left(\mathbf{V}^{k+2}\right)_{.\,i}\left(\bar{\mathbf{v}}^{(k)}_{.j}\right)$ *and* $c_s^{(ij)} = \sum\limits_{\beta\in J_{s,\,m}\{i\}} \det\left(\left(\mathbf{V}^{k+2}\right)_{.\,i}\left(\bar{\mathbf{v}}^{(k)}_{.j}\right)\right)^{\beta}_{\beta}$ *for all* $s = \overline{1, m-1}$, $i = \overline{1, m}$, *and* $j = \overline{1, n}$.

Proof. Consider the matrix $\left(\lambda\mathbf{I}+\mathbf{V}^{k+2}\right)_{.i}(\mathbf{v}_{.i}^{(k+2)}) \in \mathbb{C}^{m\times m}$. Taking into account Theorem 2.4 we obtain

$$\left|\left(\lambda\mathbf{I}+\mathbf{V}^{k+2}\right)_{.i}\left(\mathbf{v}_{.i}^{(k+2)}\right)\right| = d_1\lambda^{m-1}+d_2\lambda^{m-2}+\ldots+d_m, \tag{2.33}$$

where $d_s = \sum\limits_{\beta\in J_{s,m}\{i\}} |\left(\mathbf{V}^{k+2}\right)_\beta^\beta|$ is the sum of all principal minors of order s that contain the i-th column for all $s=\overline{1,m-1}$ and $d_m = \det\left(\mathbf{V}^{k+2}\right)$. Since $\mathbf{v}_{.i}^{(k+2)} = \begin{pmatrix} \sum\limits_l \bar{v}_{1l}^{(k)}\bar{w}_{li} \\ \sum\limits_l \bar{v}_{2l}^{(k)}\bar{w}_{li} \\ \vdots \\ \sum\limits_l \bar{v}_{nl}^{(k)}\bar{w}_{li} \end{pmatrix} = \sum\limits_l \bar{\mathbf{v}}_{.l}^{(k)}\bar{w}_{li}$, where $\bar{\mathbf{v}}_{.l}^{(k)}$ is the lth column-vector of $\bar{\mathbf{V}}^k = (\mathbf{AW})^k\mathbf{A}$ and $\mathbf{WAW}=\bar{\mathbf{W}}=(\bar{w}_{li})$ for all $l=\overline{1,n}$, then we have on the one hand

$$\begin{gathered}\left|\left(\lambda\mathbf{I}+\mathbf{V}^{k+2}\right)_{.i}\left(\mathbf{v}_{.i}^{(k+2)}\right)\right| = \sum_l\left|\left(\lambda\mathbf{I}+\mathbf{V}^{k+2}\right)_{.l}\left(\bar{\mathbf{v}}_{.l}^{(k)}\bar{w}_{li}\right)\right| = \\ \sum_l\left|\left(\lambda\mathbf{I}+\mathbf{V}^{k+2}\right)_{.i}\left(\bar{\mathbf{v}}_{.l}^{(k)}\right)\right|\cdot\bar{w}_{li}\end{gathered} \tag{2.34}$$

Having changed the order of summation, we obtain on the other hand for all $s=\overline{1,m-1}$

$$\begin{gathered}d_s = \sum_{\beta\in J_{s,m}\{i\}}\left|\left(\mathbf{V}^{k+2}\right)_\beta^\beta\right| = \sum_{\beta\in J_{s,m}\{i\}}\sum_l\left|\left(\left(\mathbf{V}^{k+2}\right)_{.i}\left(\bar{\mathbf{v}}_{.l}^{(k)}\bar{w}_{li}\right)\right)_\beta^\beta\right| = \\ \sum_l\sum_{\beta\in J_{s,m}\{i\}}\left|\left(\left(\mathbf{V}^{k+2}\right)_{.i}\left(\bar{\mathbf{v}}_{.l}^{(k)}\right)\right)_\beta^\beta\right|\cdot\bar{w}_{li}.\end{gathered} \tag{2.35}$$

By substituting (2.34) and (2.35) in (2.33), and equating factors at $\bar{w}_{li}$ when $l=j$, we obtain the equality (2.32). ∎ By analogy can be proved the following lemma.

Lemma 2.36. *If* $\mathbf{WA}=\mathbf{U}=(u_{ij})\in\mathbb{C}^{n\times n}$ *with* $Ind\mathbf{U}=k$ *and* $\lambda\in\mathbb{R}$, *then*

$$\left|(\lambda\mathbf{I}+\mathbf{U}^{k+2})_{j.}(\bar{\mathbf{u}}_{i.}^{(k)})\right| = r_1^{(ij)}\lambda^{n-1}+r_2^{(ij)}\lambda^{n-2}+\ldots+r_n^{(ij)},$$

where $r_n^{(ij)} = \left|(\mathbf{U}^{k+2})_{j.}(\bar{\mathbf{u}}_{i.}^{(k)})\right|$ *and* $r_s^{(ij)} = \sum\limits_{\alpha\in I_{s,n}\{j\}}\left|\left((\mathbf{U}^{k+2})_{j.}(\bar{\mathbf{u}}_{i.}^{(k)})\right)_\alpha^\alpha\right|$ *for all* $s=\overline{1,n-1}$, $i=\overline{1,m}$, *and* $j=\overline{1,n}$.

Theorem 2.37. *If* $\mathbf{A}\in\mathbb{C}^{m\times n}$, $\mathbf{W}\in\mathbb{C}^{n\times m}$ *with* $k=\max\{Ind(\mathbf{AW}),Ind(\mathbf{WA})\}$ *and* $\operatorname{rank}(\mathbf{AW})^k=r$, *then the W-weighted Drazin inverse* $\mathbf{A}_{d,W}=\left(a_{ij}^{d,W}\right)\in\mathbb{C}^{m\times n}$ *with respect to* $\mathbf{W}$ *possess the following determinantal representations:*

$$a_{ij}^{d,W} = \frac{\sum\limits_{\beta\in J_{r,m}\{i\}}\left|\left((\mathbf{AW})_{.i}^{k+2}\left(\bar{\mathbf{v}}_{.j}^{(k)}\right)\right)_\beta^\beta\right|}{\sum\limits_{\beta\in J_{r,m}}\left|(\mathbf{AW})^{k+2}{}_\beta^\beta\right|}, \tag{2.36}$$

or

$$a_{ij}^{d,W} = \frac{\sum\limits_{\alpha\in I_{r,n}\{j\}} \left| \left((\mathbf{WA})_{j.}^{k+2} (\bar{\mathbf{u}}_{i.}^{(k)}) \right) {}_{\alpha}^{\alpha} \right|}{\sum\limits_{\alpha\in I_{r,n}} \left| (\mathbf{WA})^{k+2}\, {}_{\alpha}^{\alpha} \right|}. \tag{2.37}$$

where $\bar{\mathbf{v}}_{.j}^{(k)}$ *is the* j*th column of* $\bar{\mathbf{V}}^k = (\mathbf{AW})^k\mathbf{A}$ *for all* $j = 1, ..., m$ *and* $\bar{\mathbf{u}}_{i.}^{(k)}$ *is the* i*th row of* $\bar{\mathbf{U}}^k = \mathbf{A}(\mathbf{WA})^k$ *for all* $i = 1, ..., n$.

Proof. At first we shall prove (2.36). By (2.29),

$$\mathbf{A}_{d,W} = \lim_{\lambda\to 0} \left(\lambda\mathbf{I}_m + (\mathbf{AW})^{k+2} \right)^{-1} (\mathbf{AW})^k \mathbf{A}.$$

Let

$$\left(\lambda\mathbf{I}_m + (\mathbf{AW})^{k+2} \right)^{-1} = \frac{1}{\det\left(\lambda\mathbf{I}_m + (\mathbf{AW})^{k+2}\right)} \begin{pmatrix} L_{11} & L_{21} & \dots & L_{m1} \\ L_{12} & L_{22} & \dots & L_{m2} \\ \dots & \dots & \dots & \dots \\ L_{1m} & L_{2m} & \dots & L_{mm} \end{pmatrix},$$

where L_{ij} is a left ij-th cofactor of a matrix $\lambda\mathbf{I}_m + (\mathbf{AW})^{k+2}$. Then we have

$$\left(\lambda\mathbf{I}_m + (\mathbf{AW})^{k+2}\right)^{-1} (\mathbf{AW})^k\mathbf{A} =$$
$$= \frac{1}{\det\left(\lambda\mathbf{I}_m + (\mathbf{AW})^{k+2}\right)} \begin{pmatrix} \sum\limits_{s=1}^{m} L_{s1}\bar{v}_{s1}^{(k)} & \sum\limits_{s=1}^{m} L_{s1}\bar{v}_{s2}^{(k)} & \dots & \sum\limits_{s=1}^{m} L_{s1}\bar{v}_{sn}^{(k)} \\ \sum\limits_{s=1}^{m} L_{s2}\bar{v}_{s1}^{(k)} & \sum\limits_{s=1}^{m} L_{s2}\bar{v}_{s2}^{(k)} & \dots & \sum\limits_{s=1}^{m} L_{s2}\bar{v}_{sn}^{(k)} \\ \dots & \dots & \dots & \dots \\ \sum\limits_{s=1}^{m} L_{sm}\bar{v}_{s1}^{(k)} & \sum\limits_{s=1}^{m} L_{sm}\bar{v}_{s2}^{(k)} & \dots & \sum\limits_{s=1}^{m} L_{sm}\bar{v}_{sn}^{(k)} \end{pmatrix}.$$

By (2.29), we obtain

$$\mathbf{A}_{d,W} = \lim_{\lambda\to 0} \begin{pmatrix} \frac{\left| \left(\lambda\mathbf{I}_m + (\mathbf{AW})^{k+2}\right)_{.1} \left(\bar{\mathbf{v}}_{.1}^{(k)}\right) \right|}{\left| \left(\lambda\mathbf{I}_m + (\mathbf{AW})^{k+2}\right) \right|} & \dots & \frac{\left| \left(\lambda\mathbf{I}_m + (\mathbf{AW})^{k+2}\right)_{.1} \left(\bar{\mathbf{v}}_{.n}^{(k)}\right) \right|}{\left| \left(\lambda\mathbf{I}_m + (\mathbf{AW})^{k+2}\right) \right|} \\ \dots & \dots & \dots \\ \frac{\left| \left(\lambda\mathbf{I}_m + (\mathbf{AW})^{k+2}\right)_{.n} \left(\bar{\mathbf{v}}_{.1}^{(k)}\right) \right|}{\left| \left(\lambda\mathbf{I}_m + (\mathbf{AW})^{k+2}\right) \right|} & \dots & \frac{\left| \left(\lambda\mathbf{I}_m + (\mathbf{AW})^{k+2}\right)_{.m} \left(\bar{\mathbf{v}}_{.n}^{(k)}\right) \right|}{\left| \left(\lambda\mathbf{I}_m + (\mathbf{AW})^{k+2}\right) \right|} \end{pmatrix}. \tag{2.38}$$

By Theorem 2.4 we have

$$\left| \left(\lambda\mathbf{I}_m + (\mathbf{AW})^{k+2} \right) \right| = \lambda^m + d_1\lambda^{m-1} + d_2\lambda^{m-2} + \ldots + d_m,$$

where $d_s = \sum\limits_{\beta\in J_{s,m}} \left| \left(\lambda\mathbf{I}_m + (\mathbf{AW})^{k+2}\right) {}_{\beta}^{\beta} \right|$ is a sum of principal minors of $(\mathbf{AW})^{k+2}$ of order s for all $s = \overline{1, m-1}$ and $d_m = \left|(\mathbf{AW})^{k+2}\right|$.

Since

$$\operatorname{rank}(\mathbf{AW})^{k+2} = \operatorname{rank}(\mathbf{AW})^{k+1} = \operatorname{rank}(\mathbf{AW})^k = r,$$

then

$$d_m = d_{m-1} = \ldots = d_{r+1} = 0.$$

It follows that $\det\left(\lambda \mathbf{I}_m + (\mathbf{AW})^{k+2}\right) = \lambda^m + d_1\lambda^{m-1} + d_2\lambda^{m-2} + \ldots + d_r\lambda^{m-r}$.

By Lemma 2.35

$$\left|\left(\lambda \mathbf{I}_m + (\mathbf{AW})^{k+2}\right)_{.\,i}\left(\bar{\mathbf{v}}_{.j}^{(k)}\right)\right| = c_1^{(ij)}\lambda^{m-1} + c_2^{(ij)}\lambda^{m-2} + \ldots + c_m^{(ij)}$$

for $i = \overline{1,m}$ and $j = \overline{1,n}$, where $c_s^{(ij)} = \sum\limits_{\beta \in J_{s,\,m}\{i\}} \left|\left((\mathbf{AW})_{.i}^{k+2}\left(\bar{\mathbf{v}}_{.j}^{(k)}\right)\right)_{\beta}^{\beta}\right|$ for all $s = \overline{1, m-1}$ and $c_m^{(ij)} = \left|(\mathbf{AW})_{.i}^{k+2}\left(\bar{\mathbf{v}}_{.j}^{(k)}\right)\right|$.

We shall prove that $c_k^{(ij)} = 0$, when $k \geq r+1$ for $i = \overline{1,m}$ and $j = \overline{1,n}$. By Lemma 2.33 $\left((\mathbf{AW})_{.i}^{k+2}\left(\bar{\mathbf{v}}_{.j}^{(k)}\right)\right) \leq r$, then the matrix $\left((\mathbf{AW})_{.i}^{k+2}\left(\bar{\mathbf{v}}_{.j}^{(k)}\right)\right)$ has no more r linearly independent columns.

Consider $\left((\mathbf{AW})_{.i}^{k+2}\left(\bar{\mathbf{v}}_{.j}^{(k)}\right)\right)_{\beta}^{\beta}$, when $\beta \in J_{s,m}\{i\}$. It is a principal submatrix of $\left((\mathbf{AW})_{.i}^{k+2}\left(\bar{\mathbf{v}}_{.j}^{(k)}\right)\right)$ of order $s \geq r+1$. Deleting both its i-th row and column, we obtain a principal submatrix of order $s-1$ of $(\mathbf{AW})^{k+2}$. We denote it by $\mathbf{M}$. The following cases are possible.

- Let $s = r+1$ and $\det \mathbf{M} \neq 0$. In this case all columns of $\mathbf{M}$ are right-linearly independent. The addition of all of them on one coordinate to columns of $\left((\mathbf{AW})_{.i}^{k+2}\left(\bar{\mathbf{v}}_{.j}^{(k)}\right)\right)_{\beta}^{\beta}$ keeps their right-linear independence. Hence, they are basis in a matrix $\left((\mathbf{AW})_{.i}^{k+2}\left(\bar{\mathbf{v}}_{.j}^{(k)}\right)\right)_{\beta}^{\beta}$, and the i-th column is the right linear combination of its basis columns. From this, $\left|\left((\mathbf{AW})_{.i}^{k+2}\left(\bar{\mathbf{v}}_{.j}^{(k)}\right)\right)_{\beta}^{\beta}\right| = 0$, when $\beta \in J_{s,n}\{i\}$ and $s = r+1$.
- If $s = r+1$ and $\det \mathbf{M} = 0$, than p, $(p \leq r)$, columns are basis in $\mathbf{M}$ and in $\left((\mathbf{AW})_{.i}^{k+2}\left(\bar{\mathbf{v}}_{.j}^{(k)}\right)\right)_{\beta}^{\beta}$. Then $\left|\left((\mathbf{AW})_{.i}^{k+2}\left(\bar{\mathbf{v}}_{.j}^{(k)}\right)\right)_{\beta}^{\beta}\right| = 0$ as well.
- If $s > r+1$, then $\det \mathbf{M} = 0$ and p, $(p < r)$, columns are basis in the both matrices $\mathbf{M}$ and $\left((\mathbf{AW})_{.i}^{k+2}\left(\bar{\mathbf{v}}_{.j}^{(k)}\right)\right)_{\beta}^{\beta}$. Therefore, $\left|\left((\mathbf{AW})_{.i}^{k+2}\left(\bar{\mathbf{v}}_{.j}^{(k)}\right)\right)_{\beta}^{\beta}\right| = 0$.

Thus in all cases we have $\left|\left((\mathbf{AW})_{.i}^{k+2}\left(\bar{\mathbf{v}}_{.j}^{(k)}\right)\right)_{\beta}^{\beta}\right| = 0$, when $\beta \in J_{s,m}\{i\}$ and $r+1 \leq s < m$. From here if $r+1 \leq s < m$, then

$$c_s^{(ij)} = \sum_{\beta \in J_{s,\,m}\{i\}} \left|\left((\mathbf{AW})_{.i}^{k+2}\left(\bar{\mathbf{v}}_{.j}^{(k)}\right)\right)_{\beta}^{\beta}\right| = 0,$$

and $c_m^{(ij)} = \det\left((\mathbf{AW})_{.i}^{k+2}\left(\bar{\mathbf{v}}_{.j}^{(k)}\right)\right) = 0$ for $i = \overline{1,m}$ and $j = \overline{1,n}$.

Hence, $\left|\left(\lambda \mathbf{I} + (\mathbf{AW})^{k+2}\right)_{.\,i}\left(\bar{\mathbf{v}}_{.j}^{(k)}\right)\right| = c_1^{(ij)}\lambda^{m-1} + \ldots + c_r^{(ij)}\lambda^{m-r}$ for $i = \overline{1,m}$ and $j = \overline{1,n}$. By substituting these values in the matrix from (2.38), we obtain

$$\mathbf{A}_{d,W} = \lim_{\lambda\to 0}\begin{pmatrix} \frac{c_1^{(11)}\lambda^{m-1}+...+c_r^{(11)}\lambda^{m-r}}{\lambda^m+d_1\lambda^{m-1}+...+d_r\lambda^{m-r}} & \dots & \frac{c_1^{(1n)}\lambda^{m-1}+...+c_r^{(1n)}\lambda^{m-r}}{\lambda^m+d_1\lambda^{m-1}+...+d_r\lambda^{m-r}} \\ \dots & \dots & \dots \\ \frac{c_1^{(m1)}\lambda^{m-1}+...+c_r^{(m1)}\lambda^{m-r}}{\lambda^m+d_1\lambda^{m-1}+...+d_r\lambda^{m-r}} & \dots & \frac{c_1^{(mn)}\lambda^{m-1}+...+c_r^{(mn)}\lambda^{m-r}}{\lambda^m+d_1\lambda^{m-1}+...+d_r\lambda^{m-r}} \end{pmatrix} = \begin{pmatrix} \frac{c_r^{(11)}}{d_r} & \dots & \frac{c_r^{(1n)}}{d_r} \\ \dots & \dots & \dots \\ \frac{c_r^{(m1)}}{d_r} & \dots & \frac{c_r^{(mn)}}{d_r} \end{pmatrix}.$$

where $c_r^{(ij)} = \sum_{\beta\in J_{r,\,m}\{i\}} \left|\left(\left(\mathbf{A}^{k+1}\right)_{.\,i}\left(\mathbf{a}_{.j}^{(k)}\right)\right)_\beta^\beta\right|$ and $d_r = \sum_{\beta\in J_{r,\,m}} \left|\left(\mathbf{A}^{k+1}\right)_\beta^\beta\right|$. Thus, we have obtained the determinantal representation of $\mathbf{A}_{d,W}$ by (2.36).

By analogy can be proved (2.37). ■

3. Cramer's Rules for Generalized Inverse Solutions of Systems of Linear Equations

An obvious consequence of a determinantal representation of the inverse matrix by the classical adjoint matrix is the Cramer rule. As we know, Cramer's rule gives an explicit expression for the solution of nonsingular linear equations. In [33], Robinson gave an elegant proof of Cramer's rule which aroused great interest in finding determinantal formulas for solutions of some restricted linear equations both consistent and nonconsistent. It has been widely discussed by Robinson [33], Ben-Israel [34], Verghese [35], Werner [36], Chen [37], Ji [38] ,Wang [39], Wei [31].

In this section we demonstrate that the obtained analogues of the adjoint matrix for the generalized inverse matrices enable us to obtain natural analogues of Cramer's rule for generalized inverse solutions of systems of linear equations.

3.1. Cramer's Rule for the Least Squares Solution with the Minimum Norm

Definition 3.1. *Suppose in a complex system of linear equations:*

$$\mathbf{A}\cdot\mathbf{x} = \mathbf{y} \tag{3.1}$$

the coefficient matrix $\mathbf{A}\in\mathbb{C}_r^{m\times n}$ *and a column of constants* $\mathbf{y} = (y_1,\dots,y_m)^T \in \mathbb{C}^m$. *The least squares solution with the minimum norm of (3.1) is the vector* $\mathbf{x}^0\in\mathbb{C}^n$ *satisfying*

$$\left\|\mathbf{x}^0\right\| = \min_{\tilde{\mathbf{x}}\in\mathbb{C}^n}\left\{\|\tilde{\mathbf{x}}\| \;|\; \|\mathbf{A}\cdot\tilde{\mathbf{x}}-\mathbf{y}\| = \min_{\mathbf{x}\in\mathbb{C}^n}\|\mathbf{A}\cdot\mathbf{x}-\mathbf{y}\|\right\},$$

where $\mathbb{C}^n$ *is an* n*-dimension complex vector space.*

If the equation (3.1) has no precision solutions, then $\mathbf{x}^0$ is its optimal approximation. The following important proposition is well-known.

Theorem 3.2. *[21] The vector* $\mathbf{x} = \mathbf{A}^+\mathbf{y}$ *is the least squares solution with the minimum norm of the system (3.1).*

Theorem 3.3. *The following statements are true for the system of linear equations (3.1).*

i) *If* $\operatorname{rank} \mathbf{A} = n$*, then the components of the least squares solution with the minimum norm* $\mathbf{x}^0 = \left(x_1^0, \ldots, x_n^0\right)^T$ *are obtained by the formula*

$$x_j^0 = \frac{\det(\mathbf{A}^*\mathbf{A})_{.j}(\mathbf{f})}{\det \mathbf{A}^*\mathbf{A}}, \quad \left(\forall j = \overline{1,n}\right), \tag{3.2}$$

where $\mathbf{f} = \mathbf{A}^*\mathbf{y}$.

ii) *If* $\operatorname{rank} \mathbf{A} = r \leq m < n$*, then*

$$x_j^0 = \frac{\sum\limits_{\beta \in J_{r,n}\{j\}} \left|\left((\mathbf{A}^*\mathbf{A})_{.j}(\mathbf{f})\right)_\beta^\beta\right|}{d_r(\mathbf{A}^*\mathbf{A})}, \quad \left(\forall j = \overline{1,n}\right). \tag{3.3}$$

Proof. i) If $\operatorname{rank} \mathbf{A} = n$, then we can represent $\mathbf{A}^+$ by (2.10). By multiplying $\mathbf{A}^+$ into $\mathbf{y}$ we get (3.2).

ii) If $\operatorname{rank} \mathbf{A} = k \leq m < n$, then $\mathbf{A}^+$ can be represented by (2.5). By multiplying $\mathbf{A}^+$ into $\mathbf{y}$ the least squares solution with the minimum norm of the linear system (3.1) is given by components as in (3.3). ■ Using (2.7) and (2.11), we can obtain another representation of the Cramer rule for the least squares solution with the minimum norm of a linear system.

Theorem 3.4. *The following statements are true for a system of linear equations written in the form* $\mathbf{x} \cdot \mathbf{A} = \mathbf{y}$.

i) *If* $\operatorname{rank} \mathbf{A} = m$*, then the components of the least squares solution* $\mathbf{x}^0 = \mathbf{y}\mathbf{A}^+$ *are obtained by the formula*

$$x_i^0 = \frac{\det(\mathbf{A}\mathbf{A}^*)_{i.}(\mathbf{g})}{\det \mathbf{A}\mathbf{A}^*}, \quad \left(\forall i = \overline{1,m}\right),$$

where $\mathbf{g} = \mathbf{y}\mathbf{A}^*$.

ii) *If* $\operatorname{rank} \mathbf{A} = r \leq n < m$*, then*

$$x_i^0 = \frac{\sum\limits_{\alpha \in I_{r,m}\{i\}} \left|\left((\mathbf{A}\mathbf{A}^*)_{i.}(\mathbf{g})\right)_\alpha^\alpha\right|}{d_r(\mathbf{A}\mathbf{A}^*)}, \quad \left(\forall i = \overline{1,m}\right).$$

Proof. The proof of this theorem is analogous to that of Theorem 3.3. ■

Remark 3.5. *The obtained formulas of the Cramer rule for the least squares solution differ from similar formulas in [34, 36, 37, 38, 39]. They give a closer analogue to usual Cramer's rule for consistent nonsingular systems of linear equations.*

3.2. Cramer's Rule for the Drazin Inverse Solution

In some situations, however, people pay more attention to the Drazin inverse solution of singular linear systems [40, 41, 42, 43].

Consider a general system of linear equations (3.1), where $\mathbf{A} \in \mathbb{C}^{n\times n}$ and $\mathbf{x}$, $\mathbf{y}$ are vectors in $\mathbb{C}^n$. $R(\mathbf{A})$ denotes the range of $\mathbf{A}$ and $N(\mathbf{A})$ denotes the null space of $\mathbf{A}$.

The characteristic of the Drazin inverse solution $\mathbf{A}^D\mathbf{y}$ is given in [24] by the following theorem.

Theorem 3.6. *Let $\mathbf{A} \in \mathbb{C}^{n\times n}$ with $Ind(A) = k$. Then $\mathbf{A}^D\mathbf{y}$ is both the unique solution in $R(\mathbf{A}^k)$ of*

$$\mathbf{A}^{k+1}\mathbf{x} = \mathbf{A}^{\mathrm{k}}\mathbf{y}, \tag{3.4}$$

and the unique minimal $\mathbf{P}$-norm least squares solution of (3.1).

Remark 3.7. *The $\mathbf{P}$-norm is defined as $\|\mathbf{x}\|_{\mathbf{P}} = \|\mathbf{P}^{-1}\mathbf{x}\|$ for $\mathbf{x} \in \mathbb{C}^n$, where $\mathbf{P}$ is a nonsingular matrix that transforms $\mathbf{A}$ into its Jordan canonical form (2.14).*

In other words, the the Drazin inverse solution $\mathbf{x} = \mathbf{A}^D\mathbf{y}$ is the unique solution of the problem: for a given $\mathbf{A}$ and a given vector $\mathbf{y} \in R(\mathbf{A}^k)$, find a vector $\mathbf{x} \in R(\mathbf{A}^k)$ satisfying $\mathbf{A}\mathbf{x} = \mathbf{y}$ with $Ind\,\mathbf{A} = k$.

In general, unlike $\mathbf{A}^+\mathbf{y}$, the Drazin inverse solution $\mathbf{A}^D\mathbf{y}$ is not a true solution of a singular system (3.1), even if the system is consistent. However, Theorem 3.6 means that $\mathbf{A}^D\mathbf{y}$ is the unique minimal P-norm least squares solution of (3.1).

A determinantal representation of the $\mathbf{P}$-norm least squares solution of a system of linear equations (3.1) by the determinantal representation (2.15) of the Drazin inverse has been obtained in [44].

We give Cramer's rule for the $\mathbf{P}$-norm least squares solution (the Drazin inverse solution) of (3.1) in the following theorem.

Theorem 3.8. *Let $\mathbf{A} \in \mathbb{C}^{n\times n}$ with $Ind(\mathbf{A}) = k$ and $\operatorname{rank}\mathbf{A}^{k+1} = \operatorname{rank}\mathbf{A}^k = r$. Then the unique minimal $\mathbf{P}$-norm least squares solution $\widehat{\mathbf{x}} = (\widehat{x}_1, \dots, \widehat{x}_n)^T$ of the system (3.1) is given by*

$$\widehat{x}_i = \frac{\sum\limits_{\beta\in J_{r,\,n}\{i\}} \left|\left(\mathbf{A}^{k+1}_{.i}(\mathbf{f})\right)^{\beta}_{\beta}\right|}{\sum\limits_{\beta\in J_{r,n}} \left|\left(\mathbf{A}^{k+1}\right)^{\beta}_{\beta}\right|} \quad \forall i = \overline{1,n}, \tag{3.5}$$

where $\mathbf{f} = \mathbf{A}^k\mathbf{y}$.

Proof. Representing the Drazin inverse by (2.21) and by virtue of Theorem 3.6, we have

$$\widehat{\mathbf{x}} = \begin{pmatrix} \widehat{x}_1 \\ \dots \\ \widehat{x}_n \end{pmatrix} = \mathbf{A}^D\mathbf{y} = \frac{1}{d_r\left(\mathbf{A}^{k+1}\right)} \begin{pmatrix} \sum\limits_{s=1}^{n} d_{1s}y_s \\ \dots \\ \sum\limits_{s=1}^{n} d_{ns}y_s \end{pmatrix}.$$

Therefore,

$$\widehat{x}_i = \frac{1}{d_r\left(\mathbf{A}^{k+1}\right)} \sum_{s=1}^{n} \sum_{\beta \in J_{r,n}\{i\}} \left| \left(\mathbf{A}^{k+1}_{.i}\left(\mathbf{a}^{(k)}_{.s}\right)\right)^{\beta}_{\beta} \right| \cdot y_s =$$

$$= \frac{1}{d_r\left(\mathbf{A}^{k+1}\right)} \sum_{\beta \in J_{r,n}\{i\}} \sum_{s=1}^{n} \left| \left(\mathbf{A}^{k+1}_{.i}\left(\mathbf{a}^{(k)}_{.s}\right)\right)^{\beta}_{\beta} \right| \cdot y_s =$$

$$= \frac{1}{d_r\left(\mathbf{A}^{k+1}\right)} \sum_{\beta \in J_{r,n}\{i\}} \sum_{s=1}^{n} \left| \left(\mathbf{A}^{k+1}_{.i}\left(\mathbf{a}^{(k)}_{.s} \cdot y_s\right)\right)^{\beta}_{\beta} \right|.$$

From this (3.5) follows immediately. ■ If we shall present a system of linear equations as,

$$\mathbf{x}\mathbf{A} = \mathbf{y}, \tag{3.6}$$

where $\mathbf{A} \in \mathbb{C}^{n\times n}$ with $Ind(A) = k$ and $\operatorname{rank}\mathbf{A}^{k+1} = \operatorname{rank}\mathbf{A}^{k} = r$, then by using the Drazin inverse determinantal representation (2.20) we have the following analog of Cramer's rule for the Drazin inverse solution of (3.6):

$$\widehat{x}_i = \frac{\sum\limits_{\alpha \in I_{r,n}\{i\}} \left| \left(\mathbf{A}^{k+1}_{i.}\left(\mathbf{g}\right)\right)^{\alpha}_{\alpha} \right|}{\sum\limits_{\alpha \in I_{r,n}} \left| \left(\mathbf{A}^{k+1}\right)^{\alpha}_{\alpha} \right|}, \quad \forall i = \overline{1,n},$$

where $\mathbf{g} = \mathbf{y}\mathbf{A}^{k}$.

3.3. Cramer's Rule for the W-Weighted Drazin Inverse Solution

Consider restricted linear equations

$$\mathbf{WAWx} = \mathbf{y}, \tag{3.7}$$

where $\mathbf{A} \in \mathbb{C}^{m\times n}$, $\mathbf{W} \in \mathbb{C}^{n\times m}$, $k_1 = Ind(\mathbf{AW})$, $k_2 = Ind(\mathbf{WA})$ with $\mathbf{y} \in R((\mathbf{WA})^{k_2})$ and $\operatorname{rank}(\mathbf{WA})^{k_2} = \operatorname{rank}(\mathbf{AW})^{k_1} = r$.

In [31], Wei has showed that there exists an unique solution $\mathbf{A}_{d,W}\mathbf{y}$ of the linear equations (3.7) and given a Cramer rule for the W-weighted Drazin inverse solution of (3.7) by the following theorem.

Theorem 3.9. *Let* A, W *be the same as in (3.7). Suppose that* $\mathbf{U} \in \mathbb{C}^{n\times(n-r)}_{n-r}$ *and* $\mathbf{V}^* \in \mathbb{C}^{m\times(m-r)}_{m-r}$ *be matrices whose columns form bases for* $N((\mathbf{WA})^{k_2})$ *and* $N((\mathbf{AW})^{k_1})$*, respectively. Then the unique W-weighted Drazin inverse solution* $\mathbf{x} = (x_1, ..., x_m$ *of (3.7) satisfies*

$$x_i = \det\begin{pmatrix} \mathbf{WAW}(i \to \mathbf{y}) & \mathbf{U} \\ \mathbf{V}(i \to \mathbf{0}) & \mathbf{0} \end{pmatrix} \Big/ \det\begin{pmatrix} \mathbf{WAW} & \mathbf{U} \\ \mathbf{V} & \mathbf{0} \end{pmatrix}$$

where $i = \overline{1,m}$.

Let $k = \max\{k_1, k_2\}$. Denote $\mathbf{f} = (\mathbf{AW})^k \mathbf{A} \cdot \mathbf{y}$. Then by Theorem 2.37 using the determinantal representation (2.36) of the W-weighted Drazin inverse $\mathbf{A}_{d,W}$, we evidently obtain the following Cramer's rule of the W-weighted Drazin inverse solution of (3.7),

$$x_i = \frac{\sum\limits_{\beta \in J_{r,\,m}\{i\}} \left| \left((\mathbf{AW})^{k+2}_{.\,i} (\mathbf{f}) \right) {}^{\beta}_{\beta} \right|}{\sum\limits_{\beta \in J_{r,\,m}} \left| (\mathbf{AW})^{k+2} {}^{\beta}_{\beta} \right|}, \tag{3.8}$$

where $i = \overline{1, m}$.

Remark 3.10. *Note that for (3.8) unlike Theorem 3.9, we do not need auxiliary matrices* $\mathbf{U}$ *and* $\mathbf{V}$.

3.4. Examples

1. Let us consider the system of linear equations.

$$\begin{cases} 2x_1 - 5x_3 + 4x_4 = 1, \\ 7x_1 - 4x_2 - 9x_3 + 1.5x_4 = 2, \\ 3x_1 - 4x_2 + 7x_3 - 6.5x_4 = 3, \\ x_1 - 4x_2 + 12x_3 - 10.5x_4 = 1. \end{cases} \tag{3.9}$$

The coefficient matrix of the system is $\mathbf{A} = \begin{pmatrix} 2 & 0 & -5 & 4 \\ 7 & -4 & -9 & 1.5 \\ 3 & -4 & 7 & -6.5 \\ 1 & -4 & 12 & -10.5 \end{pmatrix}$. The rank of $\mathbf{A}$ is equal to 3. We have

$$\mathbf{A}^* = \begin{pmatrix} 2 & 7 & 3 & 1 \\ 0 & -4 & -4 & -4 \\ -5 & -9 & 7 & 12 \\ 4 & 1.5 & -6.5 & -10.5 \end{pmatrix}, \mathbf{A}^*\mathbf{A} = \begin{pmatrix} 63 & -44 & -40 & -11.5 \\ -44 & 48 & -40 & 62 \\ -40 & -40 & 299 & -205 \\ -11.5 & 62 & -205 & 170.75 \end{pmatrix}.$$

At first we obtain entries of $\mathbf{A}^+$ by (2.10):

$$d_3(\mathbf{A}^*\mathbf{A}) = \begin{vmatrix} 63 & -44 & -40 \\ -44 & 48 & -40 \\ -40 & -40 & 299 \end{vmatrix} + \begin{vmatrix} 63 & -44 & -11.5 \\ -44 & 48 & 62 \\ -11.5 & 62 & 170.75 \end{vmatrix} +$$
$$+ \begin{vmatrix} 63 & -40 & -11.5 \\ -40 & 299 & -205 \\ -11.5 & -205 & 170.75 \end{vmatrix} + \begin{vmatrix} 48 & -40 & 62 \\ -40 & 299 & -205 \\ 62 & -205 & 170.75 \end{vmatrix} = 102060,$$

$$l_{11} = \begin{vmatrix} 2 & -44 & -40 \\ 0 & 48 & -40 \\ -5 & -40 & 299 \end{vmatrix} + \begin{vmatrix} 2 & -44 & -11.5 \\ 0 & 48 & 62 \\ 4 & 62 & 170.75 \end{vmatrix} + \begin{vmatrix} 2 & -40 & -11.5 \\ -5 & 299 & -205 \\ 4 & -205 & 170.75 \end{vmatrix} =$$
$$= 25779,$$

and so forth. Continuing in the same way, we get

$$\mathbf{A}^{+} = \frac{1}{102060}\begin{pmatrix} 25779 & -4905 & 20742 & -5037 \\ -3840 & -2880 & -4800 & -960 \\ 28350 & -17010 & 22680 & -5670 \\ 39558 & -18810 & 26484 & -13074 \end{pmatrix}.$$

Now we obtain the least squares solution of the system (3.9) by the matrix method.

$$\mathbf{x}^0 = \begin{pmatrix} x_1^0 \\ x_2^0 \\ x_3^0 \\ x_4^0 \end{pmatrix} = \frac{1}{102060}\begin{pmatrix} 25779 & -4905 & 20742 & -5037 \\ -3840 & -2880 & -4800 & -960 \\ 28350 & -17010 & 22680 & -5670 \\ 39558 & -18810 & 26484 & -13074 \end{pmatrix} \cdot \begin{pmatrix} 1 \\ 2 \\ 3 \\ 1 \end{pmatrix} =$$

$$= \frac{1}{102060}\begin{pmatrix} 73158 \\ -24960 \\ 56700 \\ 68316 \end{pmatrix} = \begin{pmatrix} \frac{12193}{17010} \\ -\frac{416}{1071} \\ \frac{5}{9} \\ \frac{5693}{8505} \end{pmatrix}$$

Next we get the least squares solution with minimum norm of the system (3.9) by the Cramer rule (3.3), where

$$\mathbf{f} = \begin{pmatrix} 2 & 7 & 3 & 1 \\ 0 & -4 & -4 & -4 \\ -5 & -9 & 7 & 12 \\ 4 & 1.5 & -6.5 & -10.5 \end{pmatrix} \cdot \begin{pmatrix} 1 \\ 2 \\ 3 \\ 1 \end{pmatrix} = \begin{pmatrix} 26 \\ -24 \\ 10 \\ -23 \end{pmatrix}.$$

Thus we have

$$x_1^0 = \frac{1}{102060}\left(\begin{vmatrix} 26 & -44 & -40 \\ -24 & 48 & -40 \\ 10 & -40 & 299 \end{vmatrix} + \begin{vmatrix} 26 & -44 & -11.5 \\ -24 & 48 & 62 \\ -23 & 62 & 170.75 \end{vmatrix} + \right.$$

$$\left. + \begin{vmatrix} 26 & -40 & -11.5 \\ 10 & 299 & -205 \\ 23 & -205 & 170.75 \end{vmatrix} \right) = \frac{73158}{102060} = \frac{12193}{17010};$$

$$x_2^0 = \frac{1}{102060}\left(\begin{vmatrix} 63 & 26 & -40 \\ -44 & -24 & -40 \\ -40 & 10 & 299 \end{vmatrix} + \begin{vmatrix} 63 & 26 & -11.5 \\ -44 & -24 & 62 \\ -11.5 & -23 & 170.75 \end{vmatrix} + \right.$$

$$\left. + \begin{vmatrix} -24 & -40 & 62 \\ 10 & 299 & -205 \\ -23 & -205 & 170.75 \end{vmatrix} \right) = \frac{-24960}{102060} = -\frac{416}{1071};$$

$$x_3^0 = \frac{1}{102060}\left(\begin{vmatrix} 63 & -44 & 26 \\ -44 & 48 & -24 \\ -40 & -40 & 10 \end{vmatrix} + \begin{vmatrix} 63 & 26 & -11.5 \\ -40 & 10 & -205 \\ -11.5 & -23 & 170.75 \end{vmatrix} + \right.$$

$$+\begin{vmatrix} 48 & -24 & 62 \\ -40 & 10 & -205 \\ 62 & -23 & 170.75 \end{vmatrix}\Bigg) = \frac{56700}{102060} = \frac{5}{9};$$

$$x_4^0 = \frac{1}{102060}\left(\begin{vmatrix} 63 & -44 & 26 \\ -44 & 48 & -24 \\ -11.5 & 62 & -23 \end{vmatrix} + \begin{vmatrix} 63 & -40 & 26 \\ -40 & 299 & 10 \\ -11.5 & -205 & -23 \end{vmatrix} + \right.$$

$$\left. + \begin{vmatrix} 48 & -40 & -24 \\ -40 & 299 & 10 \\ 62 & -205 & -23 \end{vmatrix}\right) = \frac{68316}{102060} = \frac{5693}{8505}.$$

2. Let us consider the following system of linear equations.

$$\begin{cases} x_1 - x_2 + x_3 + x_4 = 1, \\ x_2 - x_3 + x_4 = 2, \\ x_1 - x_2 + x_3 + 2x_4 = 3, \\ x_1 - x_2 + x_3 + x_4 = 1. \end{cases} \tag{3.10}$$

The coefficient matrix of the system is the matrix $\mathbf{A} = \begin{pmatrix} 1 & -1 & 1 & 1 \\ 0 & 1 & -1 & 1 \\ 1 & -1 & 1 & 2 \\ 1 & -1 & 1 & 1 \end{pmatrix}$. It is easy to verify the following:

$$\mathbf{A}^2 = \begin{pmatrix} 3 & -4 & 4 & 3 \\ 0 & 1 & -1 & 0 \\ 4 & -5 & 5 & 4 \\ 3 & -4 & 4 & 3 \end{pmatrix}, \quad \mathbf{A}^3 = \begin{pmatrix} 10 & -14 & 14 & 10 \\ -1 & 2 & -2 & -1 \\ 13 & -18 & 18 & 13 \\ 10 & -14 & 14 & 10 \end{pmatrix},$$

and $\operatorname{rank}\mathbf{A} = 3$, $\operatorname{rank}\mathbf{A}^2 = \operatorname{rank}\mathbf{A}^3 = 2$. This implies $k = Ind(\mathbf{A}) = 2$. We obtain entries of $\mathbf{A}^D$ by (2.21).

$$d_2(\mathbf{A}^3) = \begin{vmatrix} 10 & -14 \\ -1 & 2 \end{vmatrix} + \begin{vmatrix} 10 & 14 \\ 13 & 18 \end{vmatrix} + \begin{vmatrix} 10 & 10 \\ 10 & 10 \end{vmatrix}$$
$$+ \begin{vmatrix} 2 & -2 \\ -18 & 18 \end{vmatrix} + \begin{vmatrix} 2 & -1 \\ -14 & 10 \end{vmatrix} + \begin{vmatrix} 18 & 13 \\ 14 & 10 \end{vmatrix} = 8,$$

$$d_{11} = \begin{vmatrix} 3 & -14 \\ 0 & 2 \end{vmatrix} + \begin{vmatrix} 3 & 14 \\ 4 & 18 \end{vmatrix} + \begin{vmatrix} 3 & 10 \\ 3 & 10 \end{vmatrix} = 4,$$

and so forth.

Continuing in the same way, we get $\mathbf{A}^D = \begin{pmatrix} 0.5 & 0.5 & -0.5 & 0.5 \\ 1.75 & 2.5 & -2.5 & 1.75 \\ 1.25 & 1.5 & -1.5 & 1.25 \\ 0.5 & 0.5 & -0.5 & 0.5 \end{pmatrix}$. Now we obtain the Drazin inverse solution $\widehat{\mathbf{x}}$ of the system (3.10) by the Cramer rule (3.5), where

$$\mathbf{g} = \mathbf{A}^2\mathbf{y} = \begin{pmatrix} 3 & -4 & 4 & 3 \\ 0 & 1 & -1 & 0 \\ 4 & -5 & 5 & 4 \\ 3 & -4 & 4 & 3 \end{pmatrix} \cdot \begin{pmatrix} 1 \\ 2 \\ 3 \\ 1 \end{pmatrix} = \begin{pmatrix} 10 \\ -1 \\ 13 \\ 10 \end{pmatrix}.$$

Thus we have

$$\widehat{x}_1 = \frac{1}{8}\left(\begin{vmatrix} 10 & -14 \\ -1 & 2 \end{vmatrix} + \begin{vmatrix} 10 & 14 \\ 13 & 18 \end{vmatrix} + \begin{vmatrix} 10 & 10 \\ 10 & 10 \end{vmatrix} \right) = \frac{1}{2},$$

$$\widehat{x}_2 = \frac{1}{8}\left(\begin{vmatrix} 10 & 10 \\ -1 & -1 \end{vmatrix} + \begin{vmatrix} -1 & -2 \\ 13 & 18 \end{vmatrix} + \begin{vmatrix} -1 & -1 \\ 10 & 10 \end{vmatrix} \right) = 1,$$

$$\widehat{x}_3 = \frac{1}{8}\left(\begin{vmatrix} 10 & 10 \\ 13 & 13 \end{vmatrix} + \begin{vmatrix} 2 & -1 \\ -18 & 13 \end{vmatrix} + \begin{vmatrix} 13 & 13 \\ 10 & 10 \end{vmatrix} \right) = 1,$$

$$\widehat{x}_4 = \frac{1}{8}\left(\begin{vmatrix} 10 & 10 \\ 10 & 10 \end{vmatrix} + \begin{vmatrix} 2 & -1 \\ -14 & 10 \end{vmatrix} + \begin{vmatrix} 18 & 13 \\ 14 & 10 \end{vmatrix} \right) = \frac{1}{2}.$$

4. Cramer's Rule of the Generalized Inverse Solutions of Some Matrix Equations

Matrix equation is one of the important study fields of linear algebra. Linear matrix equations, such as

$$\mathbf{AX} = \mathbf{C}, \tag{4.1}$$

$$\mathbf{XB} = \mathbf{D}, \tag{4.2}$$

and

$$\mathbf{AXB} = \mathbf{D}, \tag{4.3}$$

play an important role in linear system theory therefore a large number of papers have presented several methods for solving these matrix equations [45, 46, 47, 48, 49]. In [50], Khatri and Mitra studied the Hermitian solutions to the matrix equations (4.1) and (4.3) over the complex field and the system of the equations (4.1) and (4.2). Wang, in [51, 52], and Li and Wu, in [53] studied the bisymmetric, symmetric and skew-antisymmetric least squares solution to this system over the quaternion skew field. Extreme ranks of real matrices in least squares solution of the equation (4.3) was investigated in [54] over the complex field and in [55] over the quaternion skew field.

As we know, the Cramer rule gives an explicit expression for the solution of nonsingular linear equations. Robinson's result ([33]) aroused great interest in finding determinantal representations of a least squares solution as some analogs of Cramer's rule for the matrix equations (for example, [56, 57, 58]). Cramer's rule for solutions of the restricted matrix equations (4.1), (4.2) and (4.3) was established in [59, 60, 61].

In this section, we obtain analogs of the Cramer rule for generalized inverse solutions of the aforementioned equations without any restriction.

We shall show numerical examples to illustrate the main results as well.

4.1. Cramer's Rule for the Minimum Norm Least Squares Solution of Some Matrix Equations

Definition 4.1. *Consider a matrix equation*

$$\mathbf{AX} = \mathbf{B}, \tag{4.4}$$

where $\mathbf{A} \in \mathbb{C}^{m \times n}, \mathbf{B} \in \mathbb{C}^{m \times s}$ *are given,* $\mathbf{X} \in \mathbb{C}^{n \times s}$ *is unknown. Suppose*

$$S_1 = \{\mathbf{X} | \mathbf{X} \in \mathbb{C}^{n \times s}, \|\mathbf{AX} - \mathbf{B}\| = \min\}.$$

Then matrices $\mathbf{X} \in \mathbb{C}^{n \times s}$ *such that* $\mathbf{X} \in S_1$ *are called least squares solutions of the matrix equation (4.4). If* $\mathbf{X}_{LS} = \min_{\mathbf{X} \in S_1} \|\mathbf{X}\|$, *then* $\mathbf{X}_{LS}$ *is called the minimum norm least squares solution of (4.4).*

If the equation (4.4) has no precision solutions, then $\mathbf{X}_{LS}$ is its optimal approximation. The following important proposition is well-known.

Lemma 4.2. *([38]) The least squares solutions of (4.4) are*

$$\mathbf{X} = \mathbf{A}^{+}\mathbf{B} + (\mathbf{I}_n - \mathbf{A}^{+}\mathbf{A})\mathbf{C},$$

where $\mathbf{A} \in \mathbb{C}^{m \times n}, \mathbf{B} \in \mathbb{C}^{m \times s}$ *are given, and* $\mathbf{C} \in \mathbb{C}^{n \times s}$ *is an arbitrary matrix. The least squares minimum norm solution is* $\mathbf{X}_{LS} = \mathbf{A}^{+}\mathbf{B}$.

We denote $\mathbf{A}^*\mathbf{B} =: \hat{\mathbf{B}} = (\hat{b}_{ij}) \in \mathbb{C}^{n \times s}$.

Theorem 4.3. *(i) If* $\operatorname{rank} \mathbf{A} = r \leq m < n$, *then we have for the minimum norm least squares solution* $\mathbf{X}_{LS} = (x_{ij}) \in \mathbb{C}^{n \times s}$ *of (4.4) for all* $i = \overline{1, n}$, $j = \overline{1, s}$

$$x_{ij} = \frac{\sum\limits_{\beta \in J_{r,\,n}\{i\}} \left| \left((\mathbf{A}^*\mathbf{A})_{.\,i} \left(\hat{\mathbf{b}}_{.j} \right) \right) {}^{\beta}_{\beta} \right|}{\sum\limits_{\beta \in J_{r,\,n}} \left| (\mathbf{A}^*\mathbf{A}) {}^{\beta}_{\beta} \right|}. \tag{4.5}$$

(ii) If $\operatorname{rank} \mathbf{A} = n$, *then for all* $i = \overline{1, n}$, $j = \overline{1, s}$ *we have*

$$x_{ij} = \frac{\det(\mathbf{A}^*\mathbf{A})_{.\,i} \left(\hat{\mathbf{b}}_{.j} \right)}{\det(\mathbf{A}^*\mathbf{A})}, \tag{4.6}$$

where $\hat{\mathbf{b}}_{.j}$ *is the* j*th column of* $\hat{\mathbf{B}}$ *for all* $j = \overline{1, s}$.

Proof. i) If $\operatorname{rank} \mathbf{A} = r \leq m < n$, then by Theorem 2.9 we can represent $\mathbf{A}^{+}$ by (2.5). Therefore, we obtain for all $i = \overline{1, n}$, $j = \overline{1, s}$

$$x_{ij} = \sum_{k=1}^{m} a^{+}_{ik} b_{kj} = \sum_{k=1}^{m} \frac{\sum\limits_{\beta \in J_{r,\,n}\{i\}} \left| \left((\mathbf{A}^*\mathbf{A})_{.\,i} (\mathbf{a}^*_{.k}) \right) {}^{\beta}_{\beta} \right|}{\sum\limits_{\beta \in J_{r,\,n}} \left| (\mathbf{A}^*\mathbf{A}) {}^{\beta}_{\beta} \right|} \cdot b_{kj} =$$

$$\frac{\sum\limits_{\beta \in J_{r,\,n}\{i\}} \sum_{k=1}^{m} \left| \left((\mathbf{A}^*\mathbf{A})_{.\,i} (\mathbf{a}^*_{.k}) \right) {}^{\beta}_{\beta} \right| \cdot b_{kj}}{\sum\limits_{\beta \subset J_{r,\,n}} \left| (\mathbf{A}^*\mathbf{A}) {}^{\beta}_{\beta} \right|}.$$

Since $\sum_k \mathbf{a}^*_{.k} b_{kj} = \begin{pmatrix} \sum_k a^*_{1k} b_{kj} \\ \sum_k a^*_{2k} b_{kj} \\ \vdots \\ \sum_k a^*_{nk} b_{kj} \end{pmatrix} = \hat{\mathbf{b}}_{.j}$, then it follows (4.5).

(ii) The proof of this case is similarly to that of (i) by using Corollary 2.3. ■

Definition 4.4. *Consider a matrix equation*

$$\mathbf{XA} = \mathbf{B}, \tag{4.7}$$

where $\mathbf{A} \in \mathbb{C}^{m \times n}, \mathbf{B} \in \mathbb{C}^{s \times n}$ *are given,* $\mathbf{X} \in \mathbb{C}^{s \times m}$ *is unknown. Suppose*

$$S_2 = \{\mathbf{X} | \, \mathbf{X} \in \mathbb{C}^{s \times m}, \|\mathbf{XA} - \mathbf{B}\| = \min\}.$$

Then matrices $\mathbf{X} \in \mathbb{C}^{s \times m}$ *such that* $\mathbf{X} \in S_2$ *are called least squares solutions of the matrix equation (4.7). If* $\mathbf{X}_{LS} = \min_{\mathbf{X} \in S_2} \|\mathbf{X}\|$, *then* $\mathbf{X}_{LS}$ *is called the minimum norm least squares solution of (4.7).*

The following lemma can be obtained by analogy to Lemma 4.2.

Lemma 4.5. *The least squares solutions of (4.7) are*

$$\mathbf{X} = \mathbf{BA}^+ + \mathbf{C}(\mathbf{I}_m - \mathbf{AA}^+),$$

where $\mathbf{A} \in \mathbb{C}^{m \times n}, \mathbf{B} \in \mathbb{C}^{s \times n}$ *are given, and* $\mathbf{C} \in \mathbb{C}^{s \times m}$ *is an arbitrary matrix. The minimum norm least squares solution is* $\mathbf{X}_{LS} = \mathbf{BA}^+$.

We denote $\mathbf{BA}^* =: \check{\mathbf{B}} = (\check{b}_{ij}) \in \mathbb{C}^{s \times m}$.

Theorem 4.6. *(i) If* $\operatorname{rank} \mathbf{A} = r \leq n < m$, *then we have for the minimum norm least squares solution* $\mathbf{X}_{LS} = (x_{ij}) \in \mathbb{C}^{s \times m}$ *of (4.7) for all* $i = \overline{1, s}$, $j = \overline{1, m}$

$$x_{ij} = \frac{\sum\limits_{\alpha \in I_{r,m}\{j\}} \left| \left((\mathbf{AA}^*)_{j\,.} \left(\check{\mathbf{b}}_{i\,.} \right) \right) {\,}^{\alpha}_{\alpha} \right|}{\sum\limits_{\alpha \in I_{r,\,m}} \left| (\mathbf{AA}^*) {\,}^{\alpha}_{\alpha} \right|}. \tag{4.8}$$

(ii) If $\operatorname{rank} \mathbf{A} = m$, *then for all* $i = \overline{1, s}$, $j = \overline{1, m}$ *we have*

$$x_{ij} = \frac{\det(\mathbf{AA}^*)_{j.} \left(\check{\mathbf{b}}_{i\,.} \right)}{\det(\mathbf{AA}^*)}, \tag{4.9}$$

where $\check{\mathbf{b}}_{i.}$ *is the ith row of* $\check{\mathbf{B}}$ *for all* $i = \overline{1, s}$.

Proof. (i) If $\operatorname{rank} \mathbf{A} = r \leq n < m$, then by Theorem 2.9 we can represent $\mathbf{A}^+$ by (2.6). Therefore, for all $i = \overline{1, s}$, $j = \overline{1, m}$ we obtain

$$x_{ij} = \sum_{k=1}^{n} b_{ik} a^+_{kj} = \sum_{k=1}^{n} b_{ik} \cdot \frac{\sum\limits_{\alpha \in I_{r,m}\{j\}} \left| \left((\mathbf{AA}^*)_{j\,.} \left(\mathbf{a}^*_{k\,.} \right) \right) {\,}^{\alpha}_{\alpha} \right|}{\sum\limits_{\alpha \in I_{r,\,m}} \left| (\mathbf{AA}^*) {\,}^{\alpha}_{\alpha} \right|} =$$

$$\frac{\sum_{k=1}^{n} b_{ik} \sum\limits_{\alpha \in I_{r,m}\{j\}} \left| \left((\mathbf{AA}^*)_{j.} (\mathbf{a}^*_{k.}) \right) {}^{\alpha}_{\alpha} \right|}{\sum\limits_{\alpha \in I_{r,m}} |(\mathbf{AA}^*) {}^{\alpha}_{\alpha}|}$$

Since for all $i = \overline{1, s}$

$$\sum_{k} b_{ik}\mathbf{a}^*_{k.} = \left(\sum_{k} b_{ik}a^*_{k1} \quad \sum_{k} b_{ik}a^*_{k2} \quad \cdots \quad \sum_{k} b_{ik}a^*_{km} \right) = \check{\mathbf{b}}_{i.},$$

then it follows (4.8).
(ii) The proof of this case is similarly to that of (i) by using Corollary 2.3. ■

Definition 4.7. *Consider a matrix equation*

$$\mathbf{AXB} = \mathbf{D}, \tag{4.10}$$

where $\mathbf{A} \in \mathbb{C}^{m\times n}_{r_1}, \mathbf{B} \in \mathbb{C}^{p\times q}_{r_2}, \mathbf{D} \in \mathbb{C}^{m\times q}$ *are given,* $\mathbf{X} \in \mathbb{C}^{n\times p}$ *is unknown. Suppose*

$$S_3 = \{\mathbf{X} | \, \mathbf{X} \in \mathbb{C}^{n\times p}, \|\mathbf{AXB} - \mathbf{D}\| = \min\}.$$

Then matrices $\mathbf{X} \in \mathbb{C}^{n\times p}$ *such that* $\mathbf{X} \in S_3$ *are called least squares solutions of the matrix equation (4.10). If* $\mathbf{X}_{LS} = \min_{\mathbf{X}\in S_3}\|\mathbf{X}\|$, *then* $\mathbf{X}_{LS}$ *is called the minimum norm least squares solution of (4.10).*

The following important proposition is well-known.

Lemma 4.8. *([36]) The least squares solutions of (4.10) are*

$$\mathbf{X} = \mathbf{A}^+\mathbf{DB}^+ + (\mathbf{I}_n - \mathbf{A}^+\mathbf{A})\mathbf{V} + \mathbf{W}(\mathbf{I}_p - \mathbf{BB}^+),$$

where $\mathbf{A} \in \mathbb{C}^{m\times n}_{r_1}, \mathbf{B} \in \mathbb{C}^{p\times q}_{r_2}, \mathbf{D} \in \mathbb{C}^{m\times q}$ *are given, and* $\{\mathbf{V}, \mathbf{W}\} \subset \mathbb{C}^{n\times p}$ *are arbitrary quaternion matrices. The minimum norm least squares solution is* $\mathbf{X}_{LS} = \mathbf{A}^+\mathbf{DB}^+$.

We denote $\widetilde{\mathbf{D}} = \mathbf{A}^*\mathbf{DB}^*$.

Theorem 4.9. *(i) If* $\operatorname{rank}\mathbf{A} = r_1 < n$ *and* $\operatorname{rank}\mathbf{B} = r_2 < p$, *then for the minimum norm least squares solution* $\mathbf{X}_{LS} = (x_{ij}) \in \mathbb{C}^{n\times p}$ *of (4.10) we have*

$$x_{ij} = \frac{\sum\limits_{\beta \in J_{r_1,\,n}\{i\}} \left| (\mathbf{A}^*\mathbf{A})_{.\,i} \left(\mathbf{d}^{\mathbf{B}}_{.j} \right) {}^{\beta}_{\beta} \right|}{\sum\limits_{\beta \in J_{r_1,n}} \left| (\mathbf{A}^*\mathbf{A})^{\beta}_{\beta} \right| \sum\limits_{\alpha \in I_{r_2,p}} |(\mathbf{BB}^*)^{\alpha}_{\alpha}|}, \tag{4.11}$$

or

$$x_{ij} = \frac{\sum\limits_{\alpha \in I_{r_2,p}\{j\}} \left| (\mathbf{BB}^*)_{j.} (\mathbf{d}^{\mathbf{A}}_{i.}) {}^{\alpha}_{\alpha} \right|}{\sum\limits_{\beta \in J_{r_1,n}} \left| (\mathbf{A}^*\mathbf{A})^{\beta}_{\beta} \right| \sum\limits_{\alpha \in I_{r_2,p}} |(\mathbf{BB}^*)^{\alpha}_{\alpha}|}, \tag{4.12}$$

where

$$\mathbf{d}^{\mathbf{B}}_{.j} = \left[\sum_{\alpha\in I_{r_2,p}\{j\}} \left|(\mathbf{BB}^*)_{j.}\left(\tilde{\mathbf{d}}_{1.}\right){}^{\alpha}_{\alpha}\right|, \ldots, \sum_{\alpha\in I_{r_2,p}\{j\}} \left|(\mathbf{BB}^*)_{j.}\left(\tilde{\mathbf{d}}_{n.}\right){}^{\alpha}_{\alpha}\right| \right]^T, \quad (4.13)$$

$$\mathbf{d}^{\mathbf{A}}_{i.} = \left[\sum_{\beta\in J_{r_1,n}\{i\}} \left|(\mathbf{A}^*\mathbf{A})_{.i}\left(\tilde{\mathbf{d}}_{.1}\right){}^{\beta}_{\beta}\right|, \ldots, \sum_{\alpha\in I_{r_1,n}\{i\}} \left|(\mathbf{A}^*\mathbf{A})_{.i}\left(\tilde{\mathbf{d}}_{.p}\right){}^{\beta}_{\beta}\right| \right] \quad (4.14)$$

are the column-vector and the row-vector, respectively. $\tilde{\mathbf{d}}_{i.}$ *is the* i*-th row of* $\widetilde{\mathbf{D}}$ *for all* $i = \overline{1,n}$*, and* $\tilde{\mathbf{d}}_{.j}$ *is the* j*-th column of* $\widetilde{\mathbf{D}}$ *for all* $j = \overline{1,p}$*.*

(ii) *If* $\operatorname{rank}\mathbf{A} = n$ *and* $\operatorname{rank}\mathbf{B} = p$*, then for the least squares solution* $\mathbf{X}_{LS} = (x_{ij}) \in \mathbb{C}^{n\times p}$ *of (4.10) we have for all* $i = \overline{1,n}$*,* $j = \overline{1,p}$*,*

$$x_{ij} = \frac{\det\left((\mathbf{A}^*\mathbf{A})_{.i}\left(\mathbf{d}^{\mathbf{B}}_{.j}\right)\right)}{\det(\mathbf{A}^*\mathbf{A})\cdot\det(\mathbf{BB}^*)}, \quad (4.15)$$

or

$$x_{ij} = \frac{\det\left((\mathbf{BB}^*)_{j.}\left(\mathbf{d}^{\mathbf{A}}_{i.}\right)\right)}{\det(\mathbf{A}^*\mathbf{A})\cdot\det(\mathbf{BB}^*)}, \quad (4.16)$$

where

$$\mathbf{d}^{\mathbf{B}}_{.j} := \left[\det\left((\mathbf{BB}^*)_{j.}\left(\tilde{\mathbf{d}}_{1.}\right)\right), \ldots, \det\left((\mathbf{BB}^*)_{j.}\left(\tilde{\mathbf{d}}_{n.}\right)\right)\right]^T, \quad (4.17)$$

$$\mathbf{d}^{\mathbf{A}}_{i.} := \left[\det\left((\mathbf{A}^*\mathbf{A})_{.i}\left(\tilde{\mathbf{d}}_{.1}\right)\right), \ldots, \det\left((\mathbf{A}^*\mathbf{A})_{.i}\left(\tilde{\mathbf{d}}_{.p}\right)\right)\right] \quad (4.18)$$

are respectively the column-vector and the row-vector.

(iii) *If* $\operatorname{rank}\mathbf{A} = n$ *and* $\operatorname{rank}\mathbf{B} = r_2 < p$*, then for the least squares solution* $\mathbf{X}_{LS} = (x_{ij}) \in \mathbb{C}^{n\times p}$ *of (4.10) we have*

$$x_{ij} = \frac{\det\left((\mathbf{A}^*\mathbf{A})_{.i}\left(\mathbf{d}^{\mathbf{B}}_{.j}\right)\right)}{\det(\mathbf{A}^*\mathbf{A}) \sum\limits_{\alpha\in I_{r_2,p}} |(\mathbf{BB}^*)^{\alpha}_{\alpha}|}, \quad (4.19)$$

or

$$x_{ij} = \frac{\sum\limits_{\alpha\in I_{r_2,p}\{j\}} \left|(\mathbf{BB}^*)_{j.}\left(\mathbf{d}^{\mathbf{A}}_{i.}\right){}^{\alpha}_{\alpha}\right|}{\det(\mathbf{A}^*\mathbf{A}) \sum\limits_{\alpha\in I_{r_2,p}} |(\mathbf{BB}^*)^{\alpha}_{\alpha}|}, \quad (4.20)$$

where $\mathbf{d}^{\mathbf{B}}_{.j}$ *is (4.13) and* $\mathbf{d}^{\mathbf{A}}_{i.}$ *is (4.18).*

(iiii) *If* $\operatorname{rank}\mathbf{A} = r_1 < m$ *and* $\operatorname{rank}\mathbf{B} = p$*, then for the least squares solution* $\mathbf{X}_{LS} = (x_{ij}) \in \mathbb{C}^{n\times p}$ *of (4.10) we have*

$$x_{ij} = \frac{\det\left((\mathbf{BB}^*)_{j.}\left(\mathbf{d}^{\mathbf{A}}_{i.}\right)\right)}{\sum\limits_{\beta\in J_{r_1,n}} \left|(\mathbf{A}^*\mathbf{A})^{\beta}_{\beta}\right| \cdot \det(\mathbf{BB}^*)}, \quad (4.21)$$

or

$$x_{ij} = \frac{\sum\limits_{\beta \in J_{r_1,n}\{i\}} \left|(\mathbf{A}^*\mathbf{A})_{.i}\left(\mathbf{d}^{\mathbf{B}}_{.j}\right){}^{\beta}_{\beta}\right|}{\sum\limits_{\beta \in J_{r_1,n}} \left|(\mathbf{A}^*\mathbf{A})^{\beta}_{\beta}\right| \det(\mathbf{B}\mathbf{B}^*)}, \tag{4.22}$$

where $\mathbf{d}^{\mathbf{B}}_{.j}$ *is (4.17) and* $\mathbf{d}^{\mathbf{A}}_{i.}$ *is (4.14).*

Proof. (i) If $\mathbf{A} \in \mathbb{C}^{m\times n}_{r_1}$, $\mathbf{B} \in \mathbb{C}^{p\times q}_{r_2}$ and $r_1 < n$, $r_2 < p$, then by Theorem 2.9 the Moore-Penrose inverses $\mathbf{A}^+ = \left(a^+_{ij}\right) \in \mathbb{C}^{n\times m}$ and $\mathbf{B}^+ = \left(b^+_{ij}\right) \in \mathbb{C}^{q\times p}$ possess the following determinantal representations respectively,

$$a^+_{ij} = \frac{\sum\limits_{\beta \in J_{r_1,n}\{i\}} \left|(\mathbf{A}^*\mathbf{A})_{.i}\left(\mathbf{a}^*_{.j}\right){}^{\beta}_{\beta}\right|}{\sum\limits_{\beta \in J_{r_1,n}} \left|(\mathbf{A}^*\mathbf{A})^{\beta}_{\beta}\right|},$$

$$b^+_{ij} = \frac{\sum\limits_{\alpha \in I_{r_2,p}\{j\}} \left|(\mathbf{B}\mathbf{B}^*)_{j.}(\mathbf{b}^*_{i.})^{\alpha}_{\alpha}\right|}{\sum\limits_{\alpha \in I_{r_2,p}} \left|(\mathbf{B}\mathbf{B}^*)^{\alpha}_{\alpha}\right|}. \tag{4.23}$$

Since by Theorem 4.8 $\mathbf{X}_{LS} = \mathbf{A}^+\mathbf{D}\mathbf{B}^+$, then an entry of $\mathbf{X}_{LS} = (x_{ij})$ is

$$x_{ij} = \sum_{s=1}^{q}\left(\sum_{k=1}^{m} a^+_{ik} d_{ks}\right) b^+_{sj}. \tag{4.24}$$

Denote by $\hat{\mathbf{d}}_{.s}$ the sth column of $\mathbf{A}^*\mathbf{D} =: \hat{\mathbf{D}} = (\hat{d}_{ij}) \in \mathbb{C}^{n\times q}$ for all $s = \overline{1,q}$. It follows from $\sum\limits_{k} \mathbf{a}^*_{.k} d_{ks} = \hat{\mathbf{d}}_{.s}$ that

$$\sum_{k=1}^{m} a^+_{ik} d_{ks} = \sum_{k=1}^{m} \frac{\sum\limits_{\beta \in J_{r_1,n}\{i\}} \left|(\mathbf{A}^*\mathbf{A})_{.i}(\mathbf{a}^*_{.k})^{\beta}_{\beta}\right|}{\sum\limits_{\beta \in J_{r_1,n}} \left|(\mathbf{A}^*\mathbf{A})^{\beta}_{\beta}\right|} \cdot d_{ks} =$$

$$\frac{\sum\limits_{\beta \in J_{r_1,n}\{i\}} \sum\limits_{k=1}^{m} \left|(\mathbf{A}^*\mathbf{A})_{.i}(\mathbf{a}^*_{.k})^{\beta}_{\beta}\right| \cdot d_{ks}}{\sum\limits_{\beta \in J_{r_1,n}} \left|(\mathbf{A}^*\mathbf{A})^{\beta}_{\beta}\right|} = \frac{\sum\limits_{\beta \in J_{r_1,n}\{i\}} \left|(\mathbf{A}^*\mathbf{A})_{.i}\left(\hat{\mathbf{d}}_{.s}\right){}^{\beta}_{\beta}\right|}{\sum\limits_{\beta \in J_{r_1,n}} \left|(\mathbf{A}^*\mathbf{A})^{\beta}_{\beta}\right|} \tag{4.25}$$

Suppose $\mathbf{e}_{s.}$ and $\mathbf{e}_{.s}$ are respectively the unit row-vector and the unit column-vector whose components are 0, except the sth components, which are 1. Substituting (4.25) and (4.23) in (4.24), we obtain

$$x_{ij} = \sum_{s=1}^{q} \frac{\sum\limits_{\beta \in J_{r_1,n}\{i\}} \left|(\mathbf{A}^*\mathbf{A})_{.i}\left(\hat{\mathbf{d}}_{.s}\right){}^{\beta}_{\beta}\right|}{\sum\limits_{\beta \in J_{r_1,n}} \left|(\mathbf{A}^*\mathbf{A})^{\beta}_{\beta}\right|} \frac{\sum\limits_{\alpha \in I_{r_2,p}\{j\}} \left|(\mathbf{B}\mathbf{B}^*)_{j.}(\mathbf{b}^*_{s.})^{\alpha}_{\alpha}\right|}{\sum\limits_{\alpha \in I_{r_2,p}} \left|(\mathbf{B}\mathbf{B}^*)^{\alpha}_{\alpha}\right|}.$$

Since

$$\hat{\mathbf{d}}_{.s} = \sum_{l=1}^{n} \mathbf{e}_{.l}\hat{d}_{ls}, \quad \mathbf{b}^*_{s.} = \sum_{t=1}^{p} b^*_{st}\mathbf{e}_{t.}, \quad \sum_{s=1}^{q} \hat{d}_{ls}b^*_{st} = \widetilde{d}_{lt}, \tag{4.26}$$

then we have

$$x_{ij} =$$

$$\frac{\sum\limits_{s=1}^{q}\sum\limits_{t=1}^{p}\sum\limits_{l=1}^{n}\sum\limits_{\beta\in J_{r_1,\,n}\{i\}} \left|(\mathbf{A}^*\mathbf{A})_{.\,i}(\mathbf{e}_{.l})\,{}_{\beta}^{\beta}\right| \hat{d}_{ls}b^*_{st} \sum\limits_{\alpha\in I_{r_2,p}\{j\}} |(\mathbf{B}\mathbf{B}^*)_{j\,.}(\mathbf{e}_{t.})\,{}_{\alpha}^{\alpha}|}{\sum\limits_{\beta\in J_{r_1,\,n}} \left|(\mathbf{A}^*\mathbf{A})\,{}_{\beta}^{\beta}\right| \sum\limits_{\alpha\in I_{r_2,p}} |(\mathbf{B}\mathbf{B}^*)\,{}_{\alpha}^{\alpha}|} =$$

$$\frac{\sum\limits_{t=1}^{p}\sum\limits_{l=1}^{n}\sum\limits_{\beta\in J_{r_1,\,n}\{i\}} \left|(\mathbf{A}^*\mathbf{A})_{.\,i}(\mathbf{e}_{.l})\,{}_{\beta}^{\beta}\right| \widetilde{d}_{lt} \sum\limits_{\alpha\in I_{r_2,p}\{j\}} |(\mathbf{B}\mathbf{B}^*)_{j\,.}(\mathbf{e}_{t.})\,{}_{\alpha}^{\alpha}|}{\sum\limits_{\beta\in J_{r_1,\,n}} \left|(\mathbf{A}^*\mathbf{A})\,{}_{\beta}^{\beta}\right| \sum\limits_{\alpha\in I_{r_2,p}} |(\mathbf{B}\mathbf{B}^*)\,{}_{\alpha}^{\alpha}|}. \tag{4.27}$$

Denote by

$$d_{it}^{\mathbf{A}} :=$$

$$\sum_{\beta\in J_{r_1,\,n}\{i\}} \left|(\mathbf{A}^*\mathbf{A})_{.\,i}\left(\widetilde{\mathbf{d}}_{.\,t}\right)\,{}_{\beta}^{\beta}\right| = \sum_{l=1}^{n}\sum_{\beta\in J_{r_1,\,n}\{i\}} \left|(\mathbf{A}^*\mathbf{A})_{.\,i}(\mathbf{e}_{.l})\,{}_{\beta}^{\beta}\right| \widetilde{d}_{lt}$$

the t-th component of a row-vector $\mathbf{d}_{i.}^{\mathbf{A}} = (d_{i1}^{\mathbf{A}}, ..., d_{ip}^{\mathbf{A}})$ for all $t = \overline{1,p}$. Substituting it in (4.27), we have

$$x_{ij} = \frac{\sum\limits_{t=1}^{p} d_{it}^{\mathbf{A}} \sum\limits_{\alpha\in I_{r_2,p}\{j\}} |(\mathbf{B}\mathbf{B}^*)_{j\,.}(\mathbf{e}_{t.})\,{}_{\alpha}^{\alpha}|}{\sum\limits_{\beta\in J_{r_1,\,n}} \left|(\mathbf{A}^*\mathbf{A})\,{}_{\beta}^{\beta}\right| \sum\limits_{\alpha\in I_{r_2,p}} |(\mathbf{B}\mathbf{B}^*)\,{}_{\alpha}^{\alpha}|}.$$

Since $\sum\limits_{t=1}^{p} d_{it}^{\mathbf{A}}\mathbf{e}_{t.} = \mathbf{d}_{i.}^{\mathbf{A}}$, then it follows (4.12).

If we denote by

$$d_{lj}^{\mathbf{B}} := \sum_{t=1}^{p} \widetilde{d}_{lt} \sum_{\alpha\in I_{r_2,p}\{j\}} |(\mathbf{B}\mathbf{B}^*)_{j\,.}(\mathbf{e}_{t.})\,{}_{\alpha}^{\alpha}| = \sum_{\alpha\in I_{r_2,p}\{j\}} \left|(\mathbf{B}\mathbf{B}^*)_{j\,.}(\widetilde{\mathbf{d}}_{l.})\,{}_{\alpha}^{\alpha}\right| \tag{4.28}$$

the l-th component of a column-vector $\mathbf{d}_{.j}^{\mathbf{B}} = (d_{1j}^{\mathbf{B}}, ..., d_{jn}^{\mathbf{B}})^T$ for all $l = \overline{1,n}$ and substitute it in (4.27), we obtain

$$x_{ij} = \frac{\sum\limits_{l=1}^{n}\sum\limits_{\beta\in J_{r_1,\,n}\{i\}} \left|(\mathbf{A}^*\mathbf{A})_{.\,i}(\mathbf{e}_{.l})\,{}_{\beta}^{\beta}\right| d_{lj}^{\mathbf{B}}}{\sum\limits_{\beta\in J_{r_1,\,n}} \left|(\mathbf{A}^*\mathbf{A})\,{}_{\beta}^{\beta}\right| \sum\limits_{\alpha\in I_{r_2,p}} |(\mathbf{B}\mathbf{B}^*)\,{}_{\alpha}^{\alpha}|}.$$

Since $\sum\limits_{l=1}^{n} \mathbf{e}_{.l}d_{lj}^{\mathbf{B}} = \mathbf{d}_{.j}^{\mathbf{B}}$, then it follows (4.11).

(ii) If rank $\mathbf{A} = n$ and rank $\mathbf{B} = p$, then by Corollary 2.3 $\mathbf{A}^{+} = (\mathbf{A}^*\mathbf{A})^{-1}\mathbf{A}^*$ and $\mathbf{B}^{+} = \mathbf{B}^*(\mathbf{B}\mathbf{B}^*)^{-1}$. Therefore, we obtain

$$\mathbf{X}_{LS} = (\mathbf{A}^*\mathbf{A})^{-1}\mathbf{A}^*\mathbf{D}\mathbf{B}^*(\mathbf{B}\mathbf{B}^*)^{-1} =$$

$$= \begin{pmatrix} x_{11} & x_{12} & \dots & x_{1p} \\ x_{21} & x_{22} & \dots & x_{2p} \\ \dots & \dots & \dots & \dots \\ x_{n1} & x_{n2} & \dots & x_{np} \end{pmatrix} = \frac{1}{\det(\mathbf{A}^*\mathbf{A})} \begin{pmatrix} L_{11}^{\mathbf{A}} & L_{21}^{\mathbf{A}} & \dots & L_{n1}^{\mathbf{A}} \\ L_{12}^{\mathbf{A}} & L_{22}^{\mathbf{A}} & \dots & L_{n2}^{\mathbf{A}} \\ \dots & \dots & \dots & \dots \\ L_{1n}^{\mathbf{A}} & L_{2n}^{\mathbf{A}} & \dots & L_{nn}^{\mathbf{A}} \end{pmatrix} \times$$

$$\times \begin{pmatrix} \tilde{d}_{11} & \tilde{d}_{12} & \dots & \tilde{d}_{1m} \\ \tilde{d}_{21} & \tilde{d}_{22} & \dots & \tilde{d}_{2m} \\ \dots & \dots & \dots & \dots \\ \tilde{d}_{n1} & \tilde{d}_{n2} & \dots & \tilde{d}_{nm} \end{pmatrix} \frac{1}{\det(\mathbf{B}\mathbf{B}^*)} \begin{pmatrix} R_{11}^{\mathbf{B}} & R_{21}^{\mathbf{B}} & \dots & R_{p1}^{\mathbf{B}} \\ R_{12}^{\mathbf{B}} & R_{22}^{\mathbf{B}} & \dots & R_{p2}^{\mathbf{B}} \\ \dots & \dots & \dots & \dots \\ R_{1p}^{\mathbf{B}} & R_{2p}^{\mathbf{B}} & \dots & R_{pp}^{\mathbf{B}} \end{pmatrix},$$

where $\tilde{d}_{ij}$ is ij-th entry of the matrix $\widetilde{\mathbf{D}}$, $L_{ij}^{\mathbf{A}}$ is the ij-th cofactor of $(\mathbf{A}^*\mathbf{A})$ for all $i, j = \overline{1, n}$ and $R_{ij}^{\mathbf{B}}$ is the ij-th cofactor of $(\mathbf{B}\mathbf{B}^*)$ for all $i, j = \overline{1, p}$. This implies

$$x_{ij} = \frac{\sum\limits_{k=1}^{n} L_{ki}^{\mathbf{A}} \left(\sum\limits_{s=1}^{p} \tilde{d}_{ks} R_{js}^{\mathbf{B}} \right)}{\det(\mathbf{A}^*\mathbf{A}) \cdot \det(\mathbf{B}\mathbf{B}^*)}, \tag{4.29}$$

for all $i = \overline{1, n}$, $j = \overline{1, p}$. We obtain the sum in parentheses and denote it as follows

$$\sum_{s=1}^{p} \tilde{d}_{k\,s} R_{j\,s}^{\mathbf{B}} = \det(\mathbf{B}\mathbf{B}^*)_{j.} \left(\tilde{\mathbf{d}}_{k\,.} \right) := d_{k\,j}^{\mathbf{B}},$$

where $\tilde{\mathbf{d}}_{k\,.}$ is the k-th row-vector of $\tilde{\mathbf{D}}$ for all $k = \overline{1, n}$. Suppose $\mathbf{d}_{.\,j}^{\mathbf{B}} := \left(d_{1\,j}^{\mathbf{B}}, \dots, d_{n\,j}^{\mathbf{B}} \right)^T$ is the column-vector for all $j = \overline{1, p}$. Reducing the sum $\sum\limits_{k=1}^{n} L_{ki}^{\mathbf{A}} d_{k\,j}^{\mathbf{B}}$, we obtain an analog of Cramer's rule for (4.10) by (4.15).

Interchanging the order of summation in (4.29), we have

$$x_{ij} = \frac{\sum\limits_{s=1}^{p} \left(\sum\limits_{k=1}^{n} L_{ki}^{\mathbf{A}} \tilde{d}_{ks} \right) R_{js}^{\mathbf{B}}}{\det(\mathbf{A}^*\mathbf{A}) \cdot \det(\mathbf{B}\mathbf{B}^*)}.$$

We obtain the sum in parentheses and denote it as follows

$$\sum_{k=1}^{n} L_{ki}^{\mathbf{A}} \tilde{d}_{k\,s} = \det(\mathbf{A}^*\mathbf{A})_{.\,i} \left(\tilde{\mathbf{d}}_{.\,s} \right) =: d_{i\,s}^{\mathbf{A}},$$

where $\tilde{\mathbf{d}}_{.\,s}$ is the s-th column-vector of $\tilde{\mathbf{D}}$ for all $s = \overline{1, p}$. Suppose $\mathbf{d}_{i\,.}^{\mathbf{A}} := \left(d_{i\,1}^{\mathbf{A}}, \dots, d_{i\,p}^{\mathbf{A}} \right)$ is the row-vector for all $i = \overline{1, n}$. Reducing the sum $\sum\limits_{s=1}^{n} d_{i\,s}^{\mathbf{A}} R_{js}^{\mathbf{B}}$, we obtain another analog of Cramer's rule for the least squares solutions of (4.10) by (4.16).

(iii) If $\mathbf{A} \in \mathbb{C}_{r_1}^{m\times n}$, $\mathbf{B} \in \mathbb{C}_{r_2}^{p\times q}$ and $r_1 = n$, $r_2 < p$, then by Remark 2.12 and Theorem 2.9 the Moore-Penrose inverses $\mathbf{A}^+ = \left(a_{ij}^+\right) \in \mathbb{C}^{n\times m}$ and $\mathbf{B}^+ = \left(b_{ij}^+\right) \in \mathbb{C}^{q\times p}$ possess the following determinantal representations respectively,

$$a_{ij}^+ = \frac{\det\left(\mathbf{A}^*\mathbf{A}\right)_{.i}\left(\mathbf{a}_{.j}^*\right)}{\det\left(\mathbf{A}^*\mathbf{A}\right)},$$

$$b_{ij}^+ = \frac{\sum\limits_{\alpha\in I_{r_2,p}\{j\}} \left|(\mathbf{B}\mathbf{B}^*)_{j.}\left(\mathbf{b}_{i.}^*\right){}_{\alpha}^{\alpha}\right|}{\sum\limits_{\alpha\in I_{r_2,p}} \left|(\mathbf{B}\mathbf{B}^*){}_{\alpha}^{\alpha}\right|}. \tag{4.30}$$

Since by Theorem 4.8 $\mathbf{X}_{LS} = \mathbf{A}^+\mathbf{D}\mathbf{B}^+$, then an entry of $\mathbf{X}_{LS} = (x_{ij})$ is (4.24). Denote by $\hat{\mathbf{d}}_{.s}$ the s-th column of $\mathbf{A}^*\mathbf{D} =: \hat{\mathbf{D}} = (\hat{d}_{ij}) \in \mathbb{C}^{n\times q}$ for all $s = \overline{1,q}$. It follows from $\sum\limits_k \mathbf{a}_{.k}^* d_{ks} = \hat{\mathbf{d}}_{.s}$ that

$$\sum_{k=1}^m a_{ik}^+ d_{ks} = \sum_{k=1}^m \frac{\det\left(\mathbf{A}^*\mathbf{A}\right)_{.i}\left(\mathbf{a}_{.k}^*\right)}{\det\left(\mathbf{A}^*\mathbf{A}\right)} \cdot d_{ks} = \frac{\det\left(\mathbf{A}^*\mathbf{A}\right)_{.i}\left(\hat{\mathbf{d}}_{.s}\right)}{\det\left(\mathbf{A}^*\mathbf{A}\right)} \tag{4.31}$$

Substituting (4.31) and (4.30) in (4.24), and using (4.26) we have

$$x_{ij} = \sum_{s=1}^q \frac{\det\left(\mathbf{A}^*\mathbf{A}\right)_{.i}\left(\hat{\mathbf{d}}_{.s}\right)}{\det\left(\mathbf{A}^*\mathbf{A}\right)} \frac{\sum\limits_{\alpha\in I_{r_2,p}\{j\}} \left|(\mathbf{B}\mathbf{B}^*)_{j.}\left(\mathbf{b}_{s.}^*\right){}_{\alpha}^{\alpha}\right|}{\sum\limits_{\alpha\in I_{r_2,p}} \left|(\mathbf{B}\mathbf{B}^*){}_{\alpha}^{\alpha}\right|} =$$

$$\frac{\sum\limits_{s=1}^q \sum\limits_{t=1}^p \sum\limits_{l=1}^n \det\left(\mathbf{A}^*\mathbf{A}\right)_{.i}\left(\mathbf{e}_{.l}\right) \hat{d}_{ls} b_{st}^* \sum\limits_{\alpha\in I_{r_2,p}\{j\}} \left|(\mathbf{B}\mathbf{B}^*)_{j.}\left(\mathbf{e}_{t.}\right){}_{\alpha}^{\alpha}\right|}{\det\left(\mathbf{A}^*\mathbf{A}\right) \sum\limits_{\alpha\in I_{r_2,p}} \left|(\mathbf{B}\mathbf{B}^*){}_{\alpha}^{\alpha}\right|} =$$

$$\frac{\sum\limits_{t=1}^p \sum\limits_{l=1}^n \det\left(\mathbf{A}^*\mathbf{A}\right)_{.i}\left(\mathbf{e}_{.l}\right) \widetilde{d}_{lt} \sum\limits_{\alpha\in I_{r_2,p}\{j\}} \left|(\mathbf{B}\mathbf{B}^*)_{j.}\left(\mathbf{e}_{t.}\right){}_{\alpha}^{\alpha}\right|}{\det\left(\mathbf{A}^*\mathbf{A}\right) \sum\limits_{\alpha\in I_{r_2,p}} \left|(\mathbf{B}\mathbf{B}^*){}_{\alpha}^{\alpha}\right|}. \tag{4.32}$$

If we substitute (4.28) in (4.32), then we get

$$x_{ij} = \frac{\sum\limits_{l=1}^n \det\left(\mathbf{A}^*\mathbf{A}\right)_{.i}\left(\mathbf{e}_{.l}\right) d_{lj}^{\mathbf{B}}}{\det\left(\mathbf{A}^*\mathbf{A}\right) \sum\limits_{\alpha\in I_{r_2,p}} \left|(\mathbf{B}\mathbf{B}^*){}_{\alpha}^{\alpha}\right|}.$$

Since again $\sum\limits_{l=1}^n \mathbf{e}_{.l} d_{lj}^{\mathbf{B}} = \mathbf{d}_{.j}^{\mathbf{B}}$, then it follows (4.19), where $\mathbf{d}_{.j}^{\mathbf{B}}$ is (4.13).

If we denote by

$$d_{it}^{\mathbf{A}} :=$$

$$\sum_{l=1}^{n} \det\left(\mathbf{A}^*\mathbf{A}\right)_{.i}\left(\widetilde{\mathbf{d}}_{.t}\right) = \sum_{l=1}^{n} \det\left(\mathbf{A}^*\mathbf{A}\right)_{.i}\left(\mathbf{e}_{.l}\right)\widetilde{d}_{lt}$$

the t-th component of a row-vector $\mathbf{d}_{i.}^{\mathbf{A}} = (d_{i1}^{\mathbf{A}}, ..., d_{ip}^{\mathbf{A}})$ for all $t = \overline{1,p}$ and substitute it in (4.32), we obtain

$$x_{ij} = \frac{\sum\limits_{t=1}^{p} d_{it}^{\mathbf{A}} \sum\limits_{\alpha \in I_{r_2,p}\{j\}} \left|(\mathbf{BB}^*)_{j.}(\mathbf{e}_{t.})\,{}_{\alpha}^{\alpha}\right|}{\det\left(\mathbf{A}^*\mathbf{A}\right) \sum\limits_{\alpha \in I_{r_2,p}} \left|(\mathbf{BB}^*)\,{}_{\alpha}^{\alpha}\right|}.$$

Since again $\sum\limits_{t=1}^{p} d_{it}^{\mathbf{A}}\mathbf{e}_{t.} = \mathbf{d}_{i.}^{\mathbf{A}}$, then it follows (4.20), where $\mathbf{d}_{i.}^{\mathbf{A}}$ is (4.18).

(iiii) The proof is similar to the proof of (iii). ■

4.2. Cramer's Rule of the Drazin Inverse Solutions of Some Matrix Equations

Consider a matrix equation

$$\mathbf{AX} = \mathbf{B}, \tag{4.33}$$

where $\mathbf{A} \in \mathbb{C}^{n\times n}$ with $Ind\,\mathbf{A} = k$, $\mathbf{B} \in \mathbb{C}^{n\times m}$ are given and $\mathbf{X} \in \mathbb{C}^{n\times m}$ is unknown.

Theorem 4.10. *([62], Theorem 1) If the range space $R(\mathbf{B}) \subset R(\mathbf{A}^k)$, then the matrix equation (4.33) with constrain $R(\mathbf{X}) \subset R(\mathbf{A}^k)$ has a unique solution*

$$\mathbf{X} = \mathbf{A}^D\mathbf{B}.$$

We denote $\mathbf{A}^k\mathbf{B} =: \hat{\mathbf{B}} = (\hat{b}_{ij}) \in \mathbb{C}^{n\times m}$.

Theorem 4.11. *If* $\operatorname{rank}\mathbf{A}^{k+1} = \operatorname{rank}\mathbf{A}^k = r \le n$ *for* $\mathbf{A} \in \mathbb{C}^{n\times n}$, *then for the Drazin inverse solution* $\mathbf{X} = \mathbf{A}^D\mathbf{B} = (x_{ij}) \in \mathbb{C}^{n\times m}$ *of (4.33) we have for all* $i = \overline{1,n}$, $j = \overline{1,m}$,

$$x_{ij} = \frac{\sum\limits_{\beta \in J_{r,\,n}\{i\}} \left|\left(\mathbf{A}_{.i}^{k+1}\left(\hat{\mathbf{b}}_{.j}\right)\right)\,{}_{\beta}^{\beta}\right|}{\sum\limits_{\beta \in J_{r,\,n}} \left|\left(\mathbf{A}^{k+1}\right)\,{}_{\beta}^{\beta}\right|}. \tag{4.34}$$

Proof. By Theorem 2.29 we can represent $\mathbf{A}^D$ by (2.21). Therefore, we obtain for all $i = \overline{1,n}$, $j = \overline{1,m}$,

$$x_{ij} = \sum_{s=1}^{n} a_{is}^{D} b_{sj} = \sum_{s=1}^{n} \frac{\sum\limits_{\beta \in J_{r,\,n}\{i\}} \left|\left(\mathbf{A}_{.i}^{k+1}\left(\mathbf{a}_{.s}^{(k)}\right)\right)\,{}_{\beta}^{\beta}\right|}{\sum\limits_{\beta \in J_{r,\,n}} \left|\left(\mathbf{A}^{k+1}\right)\,{}_{\beta}^{\beta}\right|} \cdot b_{sj} =$$

$$\frac{\sum\limits_{\beta \in J_{r,\,n}\{i\}} \sum_{s=1}^{n} \left|\left(\mathbf{A}_{.i}^{k+1}\left(\mathbf{a}_{.s}^{(k)}\right)\right)\,{}_{\beta}^{\beta}\right| \cdot b_{sj}}{\sum\limits_{\beta \in J_{r,\,n}} \left|\left(\mathbf{A}^{k+1}\right)\,{}_{\beta}^{\beta}\right|}.$$

Since $\sum\limits_s \mathbf{a}_{.s}^{(k)} b_{sj} = \begin{pmatrix} \sum\limits_s a_{1s}^{(k)} b_{sj} \\ \sum\limits_s a_{2s}^{(k)} b_{sj} \\ \vdots \\ \sum\limits_s a_{ns}^{(k)} b_{sj} \end{pmatrix} = \hat{\mathbf{b}}_{.j}$, then it follows (4.34). ■

Consider a matrix equation

$$\mathbf{XA} = \mathbf{B}, \tag{4.35}$$

where $\mathbf{A} \in \mathbb{C}^{m\times m}$ with $Ind\,\mathbf{A} = k$, $\mathbf{B} \in \mathbb{C}^{n\times m}$ are given and $\mathbf{X} \in \mathbb{C}^{n\times m}$ is unknown.

Theorem 4.12. *([62], Theorem 2) If the null space $N(\mathbf{B}) \supset N(\mathbf{A}^k)$, then the matrix equation (4.35) with constrain $N(\mathbf{X}) \supset N(\mathbf{A}^k)$ has a unique solution*

$$\mathbf{X} = \mathbf{B}\mathbf{A}^D.$$

We denote $\mathbf{B}\mathbf{A}^k =: \check{\mathbf{B}} = (\check{b}_{ij}) \in \mathbb{C}^{n\times m}$.

Theorem 4.13. *If* $\operatorname{rank} \mathbf{A}^{k+1} = \operatorname{rank} \mathbf{A}^k = r \le m$ *for* $\mathbf{A} \in \mathbb{C}^{m\times m}$*, then for the Drazin inverse solution* $\mathbf{X} = \mathbf{B}\mathbf{A}^D = (x_{ij}) \in \mathbb{C}^{n\times m}$ *of (4.35), we have for all* $i = \overline{1,n}$, $j = \overline{1,m}$,

$$x_{ij} = \frac{\sum\limits_{\alpha \in I_{r,m}\{j\}} \left| \left(\mathbf{A}^{k+1}_{j.} \left(\check{\mathbf{b}}_{i.} \right) \right) {\alpha \atop \alpha} \right|}{\sum\limits_{\alpha \in I_{r,\,m}} \left| \left(\mathbf{A}^{k+1} \right) {\alpha \atop \alpha} \right|}. \tag{4.36}$$

Proof. By Theorem 2.29 we can represent $\mathbf{A}^D$ by (2.20). Therefore, we obtain for all $i = \overline{1,n}$, $j = \overline{1,m}$,

$$x_{ij} = \sum_{s=1}^{m} b_{is} a_{sj}^D = \sum_{s=1}^{m} b_{is} \cdot \frac{\sum\limits_{\alpha \in I_{r,m}\{j\}} \left| \left(\mathbf{A}^{k+1}_{j.} \left(\mathbf{a}_{s.}^{(k)} \right) \right) {\alpha \atop \alpha} \right|}{\sum\limits_{\alpha \in I_{r,\,m}} \left| \left(\mathbf{A}^{k+1} \right) {\alpha \atop \alpha} \right|} =$$

$$\frac{\sum_{s=1}^{m} b_{ik} \sum\limits_{\alpha \in I_{r,m}\{j\}} \left| \left(\mathbf{A}^{k+1}_{j.} \left(\mathbf{a}_{s.}^{(k)} \right) \right) {\alpha \atop \alpha} \right|}{\sum\limits_{\alpha \in I_{r,\,m}} \left| \left(\mathbf{A}^{k+1} \right) {\alpha \atop \alpha} \right|}$$

Since for all $i = \overline{1,n}$

$$\sum_s b_{is} \mathbf{a}_{s.}^{(k)} = \begin{pmatrix} \sum\limits_s b_{is} a_{s1}^{(k)} & \sum\limits_s b_{is} a_{s2}^{(k)} & \cdots & \sum\limits_s b_{is} a_{sm}^{(k)} \end{pmatrix} = \check{\mathbf{b}}_{i.},$$

then it follows (4.36). ■

Consider a matrix equation

$$\mathbf{AXB} = \mathbf{D}, \tag{4.37}$$

where $\mathbf{A} \in \mathbb{C}^{n\times n}$ with $Ind\,\mathbf{A} = k_1$, $\mathbf{B} \in \mathbb{C}^{m\times m}$ with $Ind\mathbf{B} = k_2$ and $\mathbf{D} \in \mathbb{C}^{n\times m}$ are given, and $\mathbf{X} \in \mathbb{C}^{n\times m}$ is unknown.

Theorem 4.14. *([62], Theorem 3) If* $R(\mathbf{D}) \subset R(\mathbf{A}^{k_1})$ *and* $N(\mathbf{D}) \supset N(\mathbf{B}^{k_2})$, $k = max\{k_1, k_2\}$, *then the matrix equation (4.37) with constrain* $R(\mathbf{X}) \subset R(\mathbf{A}^k)$ *and* $N(\mathbf{X}) \supset N(\mathbf{B}^k)$ *has a unique solution*

$$\mathbf{X} = \mathbf{A}^D \mathbf{D} \mathbf{B}^D.$$

We denote $\mathbf{A}^{k_1}\mathbf{D}\mathbf{B}^{k_2} =: \widetilde{\mathbf{D}} = (\widetilde{d}_{ij}) \in \mathbb{C}^{n\times m}$.

Theorem 4.15. *If* $\operatorname{rank}\mathbf{A}^{k_1+1} = \operatorname{rank}\mathbf{A}^{k_1} = r_1 \le n$ *for* $\mathbf{A} \in \mathbb{C}^{n\times n}$, *and* $\operatorname{rank}\mathbf{B}^{k_2+1} = \operatorname{rank}\mathbf{B}^{k_2} = r_2 \le m$ *for* $\mathbf{B} \in \mathbb{C}^{m\times m}$, *then for the Drazin inverse solution* $\mathbf{X} = \mathbf{A}^D\mathbf{D}\mathbf{B}^D =: (x_{ij}) \in \mathbb{C}^{n\times m}$ *of (4.37) we have*

$$x_{ij} = \frac{\sum\limits_{\beta\in J_{r_1,\,n}\{i\}} \left|\mathbf{A}^{k_1+1}_{\,.\,i}\left(\mathbf{d}^{\mathbf{B}}_{.j}\right)\,{}^{\beta}_{\beta}\right|}{\sum\limits_{\beta\in J_{r_1,n}} \left|(\mathbf{A}^{k_1+1})^{\beta}_{\beta}\right| \sum\limits_{\alpha\in I_{r_2,m}} \left|(\mathbf{B}^{k_2+1})^{\alpha}_{\alpha}\right|}, \tag{4.38}$$

or

$$x_{ij} = \frac{\sum\limits_{\alpha\in I_{r_2,m}\{j\}} \left|\mathbf{B}^{k_2+1}_{j\,.}\left(\mathbf{d}^{\mathbf{A}}_{i\,.}\right)\,{}^{\alpha}_{\alpha}\right|}{\sum\limits_{\beta\in J_{r_1,n}} \left|(\mathbf{A}^{k_1+1})^{\beta}_{\beta}\right| \sum\limits_{\alpha\in I_{r_2,m}} \left|(\mathbf{B}^{k_2+1})^{\alpha}_{\alpha}\right|}, \tag{4.39}$$

where

$$\mathbf{d}^{\mathbf{B}}_{.j} = \left[\sum_{\alpha\in I_{r_2,m}\{j\}} \left|\mathbf{B}^{k_2+1}_{j.}\left(\widetilde{\mathbf{d}}_{1.}\right)\,{}^{\alpha}_{\alpha}\right|, \ldots, \sum_{\alpha\in I_{r_2,m}\{j\}} \left|\mathbf{B}^{k_2+1}_{j.}\left(\widetilde{\mathbf{d}}_{n.}\right)\,{}^{\alpha}_{\alpha}\right|\right]^T, \tag{4.40}$$

$$\mathbf{d}^{\mathbf{A}}_{i\,.} = \left[\sum_{\beta\in J_{r_1,n}\{i\}} \left|\mathbf{A}^{k_1+1}_{.i}\left(\widetilde{\mathbf{d}}_{.1}\right)\,{}^{\beta}_{\beta}\right|, \ldots, \sum_{\alpha\in I_{r_1,n}\{i\}} \left|\mathbf{A}^{k_1+1}_{.i}\left(\widetilde{\mathbf{d}}_{.m}\right)\,{}^{\beta}_{\beta}\right|\right]$$

are the column-vector and the row-vector. $\widetilde{\mathbf{d}}_{i.}$ *and* $\widetilde{\mathbf{d}}_{.j}$ *are respectively the* i*-th row and the* j*-th column of* $\widetilde{\mathbf{D}}$ *for all* $i = \overline{1,n}$, $j = \overline{1,m}$.

Proof. By (2.21) and (2.20) the Drazin inverses $\mathbf{A}^D = \left(a^D_{ij}\right) \in \mathbb{C}^{n\times n}$ and $\mathbf{B}^D = \left(b^D_{ij}\right) \in \mathbb{C}^{m\times m}$ possess the following determinantal representations, respectively,

$$a^D_{ij} = \frac{\sum\limits_{\beta\in J_{r_1,\,n}\{i\}} \left|\mathbf{A}^{k_1+1}_{\,.\,i}\left(\mathbf{a}^{(k_1)}_{.j}\right)\,{}^{\beta}_{\beta}\right|}{\sum\limits_{\beta\in J_{r_1,\,n}} \left|(\mathbf{A}^{k_1+1})\,{}^{\beta}_{\beta}\right|},$$

$$b^D_{ij} = \frac{\sum\limits_{\alpha\in I_{r_2,m}\{j\}} \left|\mathbf{B}^{k_2+1}_{j\,.}(\mathbf{b}^{(k_2)}_{i.})\,{}^{\alpha}_{\alpha}\right|}{\sum\limits_{\alpha\in I_{r_2,m}} \left|(\mathbf{B}^{k_2+1})\,{}^{\alpha}_{\alpha}\right|}. \tag{4.41}$$

Then an entry of the Drazin inverse solution $\mathbf{X} = \mathbf{A}^D\mathbf{D}\mathbf{B}^D =: (x_{ij}) \in \mathbb{C}^{n\times m}$ is

$$x_{ij} = \sum_{s=1}^{m} \left(\sum_{t=1}^{n} a_{it}^{D} d_{ts} \right) b_{sj}^{D}. \tag{4.42}$$

Denote by $\hat{\mathbf{d}}_{.s}$ the s-th column of $\mathbf{A}^k \mathbf{D} =: \hat{\mathbf{D}} = (\hat{d}_{ij}) \in \mathbb{C}^{n \times m}$ for all $s = \overline{1, m}$. It follows from $\sum\limits_t \mathbf{a}_{.t}^{D} d_{ts} = \hat{\mathbf{d}}_{.s}$ that

$$\sum_{t=1}^{n} a_{it}^{D} d_{ts} = \sum_{t=1}^{n} \frac{\sum\limits_{\beta \in J_{r_1,n}\{i\}} \left| \mathbf{A}_{.i}^{k_1+1} \left(\mathbf{a}_{.t}^{(k_1)} \right) {}_{\beta}^{\beta} \right|}{\sum\limits_{\beta \in J_{r_1,n}} \left| \left(\mathbf{A}^{k_1+1} \right) {}_{\beta}^{\beta} \right|} \cdot d_{ts} =$$

$$\frac{\sum\limits_{\beta \in J_{r_1,n}\{i\}} \sum\limits_{t=1}^{n} \left| \mathbf{A}_{.i}^{k_1+1} \left(\mathbf{a}_{.t}^{(k_1)} \right) {}_{\beta}^{\beta} \right| \cdot d_{ts}}{\sum\limits_{\beta \in J_{r_1,n}} \left| \left(\mathbf{A}^{k_1+1} \right) {}_{\beta}^{\beta} \right|} = \frac{\sum\limits_{\beta \in J_{r_1,n}\{i\}} \left| \mathbf{A}_{.i}^{k_1+1} \left(\hat{\mathbf{d}}_{.s} \right) {}_{\beta}^{\beta} \right|}{\sum\limits_{\beta \in J_{r_1,n}} \left| \left(\mathbf{A}^{k_1+1} \right) {}_{\beta}^{\beta} \right|} \tag{4.43}$$

Substituting (4.43) and (4.41) in (4.42), we obtain

$$x_{ij} = \sum_{s=1}^{m} \frac{\sum\limits_{\beta \in J_{r_1,n}\{i\}} \left| \mathbf{A}_{.i}^{k_1+1} \left(\hat{\mathbf{d}}_{.s} \right) {}_{\beta}^{\beta} \right|}{\sum\limits_{\beta \in J_{r_1,n}} \left| \left(\mathbf{A}^{k_1+1} \right) {}_{\beta}^{\beta} \right|} \frac{\sum\limits_{\alpha \in I_{r_2,m}\{j\}} \left| \mathbf{B}_{j.}^{k_2+1} (\mathbf{b}_{s.}^{(k_2)}) {}_{\alpha}^{\alpha} \right|}{\sum\limits_{\alpha \in I_{r_2,m}} \left| (\mathbf{B}^{k_2+1}) {}_{\alpha}^{\alpha} \right|}.$$

Suppose $\mathbf{e}_{s.}$ and $\mathbf{e}_{.s}$ are respectively the unit row-vector and the unit column-vector whose components are 0, except the sth components, which are 1. Since

$$\hat{\mathbf{d}}_{.s} = \sum_{l=1}^{n} \mathbf{e}_{.l} \hat{d}_{ls}, \quad \mathbf{b}_{s.}^{(k_2)} = \sum_{t=1}^{m} b_{st}^{(k_2)} \mathbf{e}_{t.}, \quad \sum_{s=1}^{m} \hat{d}_{ls} b_{st}^{(k_2)} = \widetilde{d}_{lt},$$

then we have

$$x_{ij} =$$

$$\frac{\sum\limits_{s=1}^{m} \sum\limits_{t=1}^{m} \sum\limits_{l=1}^{n} \sum\limits_{\beta \in J_{r_1,n}\{i\}} \left| \mathbf{A}_{.i}^{k_1+1} (\mathbf{e}_{.l}) {}_{\beta}^{\beta} \right| \hat{d}_{ls} b_{st}^{(k_2)} \sum\limits_{\alpha \in I_{r_2,m}\{j\}} \left| \mathbf{B}_{j.}^{k_2+1} (\mathbf{e}_{t.}) {}_{\alpha}^{\alpha} \right|}{\sum\limits_{\beta \in J_{r_1,n}} \left| \left(\mathbf{A}^{k_1+1} \right) {}_{\beta}^{\beta} \right| \sum\limits_{\alpha \in I_{r_2,m}} \left| (\mathbf{B}^{k_2+1}) {}_{\alpha}^{\alpha} \right|} =$$

$$\frac{\sum\limits_{t=1}^{m} \sum\limits_{l=1}^{n} \sum\limits_{\beta \in J_{r_1,n}\{i\}} \left| \mathbf{A}_{.i}^{k_1+1} (\mathbf{e}_{.l}) {}_{\beta}^{\beta} \right| \widetilde{d}_{lt} \sum\limits_{\alpha \in I_{r_2,m}\{j\}} \left| \mathbf{B}_{j.}^{k_2+1} (\mathbf{e}_{t.}) {}_{\alpha}^{\alpha} \right|}{\sum\limits_{\beta \in J_{r_1,n}} \left| \left(\mathbf{A}^{k_1+1} \right) {}_{\beta}^{\beta} \right| \sum\limits_{\alpha \in I_{r_2,m}} \left| (\mathbf{B}^{k_2+1}) {}_{\alpha}^{\alpha} \right|}. \tag{4.44}$$

Denote by

$$d_{it}^{\mathbf{A}} :=$$

$$\sum_{\beta \in J_{r_1,n}\{i\}} \left| \mathbf{A}_{.i}^{k_1+1} \left(\widetilde{\mathbf{d}}_{.t} \right) {}_{\beta}^{\beta} \right| = \sum_{l=1}^{n} \sum_{\beta \in J_{r_1,n}\{i\}} \left| \mathbf{A}_{.i}^{k_1+1} (\mathbf{e}_{.l}) {}_{\beta}^{\beta} \right| \widetilde{d}_{lt}$$

the t-th component of a row-vector $\mathbf{d}^{\mathbf{A}}_{i.} = (d^{\mathbf{A}}_{i1}, ..., d^{\mathbf{A}}_{im})$ for all $t = \overline{1, m}$. Substituting it in (4.44), we obtain

$$x_{ij} = \frac{\sum\limits_{t=1}^{m} d^{\mathbf{A}}_{it} \sum\limits_{\alpha \in I_{r_2,m}\{j\}} \left| \mathbf{B}^{k_2+1}_{j.}(\mathbf{e}_{t.})\,{}^{\alpha}_{\alpha} \right|}{\sum\limits_{\beta \in J_{r_1,n}} \left| (\mathbf{A}^{k_1+1})\,{}^{\beta}_{\beta} \right| \sum\limits_{\alpha \in I_{r_2,m}} \left| (\mathbf{B}^{k_2+1})\,{}^{\alpha}_{\alpha} \right|}.$$

Since $\sum\limits_{t=1}^{m} d^{\mathbf{A}}_{it}\mathbf{e}_{t.} = \mathbf{d}^{\mathbf{A}}_{i.}$, then it follows (4.39).

If we denote by

$$d^{\mathbf{B}}_{lj} := \sum_{t=1}^{m} \widetilde{d}_{lt} \sum_{\alpha \in I_{r_2,m}\{j\}} \left| \mathbf{B}^{k_2+1}_{j.}(\mathbf{e}_{t.})\,{}^{\alpha}_{\alpha} \right| = \sum_{\alpha \in I_{r_2,m}\{j\}} \left| \mathbf{B}^{k_2+1}_{j.}(\widetilde{\mathbf{d}}_{l.})\,{}^{\alpha}_{\alpha} \right|$$

the l-th component of a column-vector $\mathbf{d}^{\mathbf{B}}_{.j} = (d^{\mathbf{B}}_{1j}, ..., d^{\mathbf{B}}_{jn})^T$ for all $l = \overline{1, n}$ and substitute it in (4.44), we obtain

$$x_{ij} = \frac{\sum\limits_{l=1}^{n} \sum\limits_{\beta \in J_{r_1,n}\{i\}} \left| \mathbf{A}^{k_1+1}_{.i}(\mathbf{e}_{.l})\,{}^{\beta}_{\beta} \right| d^{\mathbf{B}}_{lj}}{\sum\limits_{\beta \in J_{r_1,n}} \left| (\mathbf{A}^{k_1+1})\,{}^{\beta}_{\beta} \right| \sum\limits_{\alpha \in I_{r_2,m}} \left| (\mathbf{B}^{k_2+1})\,{}^{\alpha}_{\alpha} \right|}.$$

Since $\sum\limits_{l=1}^{n} \mathbf{e}_{.l} d^{\mathbf{B}}_{lj} = \mathbf{d}^{\mathbf{B}}_{.j}$, then it follows (4.38). ■

4.3. Examples

In this subsection, we give an example to illustrate results obtained in the section.

1. Let us consider the matrix equation

$$\mathbf{AXB} = \mathbf{D}, \tag{4.45}$$

where

$$\mathbf{A} = \begin{pmatrix} 1 & i & i \\ i & -1 & -1 \\ 0 & 1 & 0 \\ -1 & 0 & -i \end{pmatrix}, \ \mathbf{B} = \begin{pmatrix} i & 1 & -i \\ -1 & i & 1 \end{pmatrix}, \ \mathbf{D} = \begin{pmatrix} 1 & i & 1 \\ i & 0 & 1 \\ 1 & i & 0 \\ 0 & 1 & i \end{pmatrix}.$$

Since $\operatorname{rank} \mathbf{A} = 2$ and $\operatorname{rank} \mathbf{B} = 1$, then we have the case (ii) of Theorem 4.9. We shall find the least squares solution of (4.45) by (4.11). Then we have

$$\mathbf{A}^*\mathbf{A} = \begin{pmatrix} 3 & 2i & 3i \\ -2i & 3 & 2 \\ -3i & 2 & 3 \end{pmatrix}, \ \mathbf{BB}^* = \begin{pmatrix} 3 & -3i \\ 3i & 3 \end{pmatrix}, \ \widetilde{\mathbf{D}} = \mathbf{A}^*\mathbf{DB}^* = \begin{pmatrix} 1 & -i \\ -i & -1 \\ -i & -1 \end{pmatrix},$$

and $\sum\limits_{\alpha\in I_{1,\,2}} |(\mathbf{BB}^*)\,{}^{\alpha}_{\alpha}| = 3+3=6$,

$$\sum_{\beta\in J_{2,\,3}} \left|(\mathbf{A}^*\mathbf{A})\,{}^{\beta}_{\beta}\right| = \det\begin{pmatrix}3 & 2i\\ -2i & 3\end{pmatrix} + \det\begin{pmatrix}3 & 2\\ 2 & 3\end{pmatrix} + \det\begin{pmatrix}3 & 3i\\ -3i & 3\end{pmatrix} = 10.$$

By (4.17), we can get

$$\mathbf{d}^{\mathbf{B}}_{.1} = \begin{pmatrix}1\\ -i\\ -i\end{pmatrix}, \quad \mathbf{d}^{\mathbf{B}}_{.2} = \begin{pmatrix}-i\\ -1\\ -1\end{pmatrix}.$$

Since $(\mathbf{A}^*\mathbf{A})_{.\,1}\left(\mathbf{d}^{\,\mathbf{B}}_{.1}\right) = \begin{pmatrix}1 & 2i & 3i\\ -i & 3 & 2\\ -i & 2 & 3\end{pmatrix}$, then finally we obtain

$$x_{11} = \frac{\sum\limits_{\beta\in J_{2,\,3}\{i\}} \left|(\mathbf{A}^*\mathbf{A})_{.\,1}\left(\mathbf{d}^{\,\mathbf{B}}_{.1}\right)\,{}^{\beta}_{\beta}\right|}{\sum\limits_{\beta\in J_{2,3}} \left|(\mathbf{A}^*\mathbf{A})^{\beta}_{\beta}\right| \sum\limits_{\alpha\in I_{1,2}} |(\mathbf{BB}^*)^{\alpha}_{\alpha}|} = \frac{\det\begin{pmatrix}1 & 2i\\ -i & 3\end{pmatrix} + \det\begin{pmatrix}1 & 3i\\ -i & 3\end{pmatrix}}{60} = -\frac{1}{60}.$$

Similarly,

$$x_{12} = \frac{\det\begin{pmatrix}-i & 2i\\ -1 & 3\end{pmatrix} + \det\begin{pmatrix}-i & 3i\\ -1 & 3\end{pmatrix}}{60} = -\frac{i}{60},$$

$$x_{21} = \frac{\det\begin{pmatrix}3 & 1\\ -2i & -i\end{pmatrix} + \det\begin{pmatrix}-i & 2\\ -i & 3\end{pmatrix}}{60} = -\frac{2i}{60},$$

$$x_{22} = \frac{\det\begin{pmatrix}3 & -i\\ -2i & -1\end{pmatrix} + \det\begin{pmatrix}-1 & 2\\ -1 & 3\end{pmatrix}}{60} = -\frac{2}{60},$$

$$x_{31} = \frac{\det\begin{pmatrix}3 & 1\\ -3i & -i\end{pmatrix} + \det\begin{pmatrix}3 & -i\\ 2 & -i\end{pmatrix}}{60} = -\frac{i}{60},$$

$$x_{32} = \frac{\det\begin{pmatrix}3 & -i\\ -3i & -1\end{pmatrix} + \det\begin{pmatrix}3 & -1\\ 2 & -1\end{pmatrix}}{60} = -\frac{1}{60}.$$

2. Let us consider the matrix equation (4.45), where

$$\mathbf{A} = \begin{pmatrix}2 & 0 & 0\\ -i & i & i\\ -i & -i & -i\end{pmatrix},\ \mathbf{B} = \begin{pmatrix}1 & -1 & 1\\ i & -i & i\\ -1 & 1 & 2\end{pmatrix},\ \mathbf{D} = \begin{pmatrix}1 & i & 1\\ i & 0 & 1\\ 1 & i & 0\end{pmatrix}.$$

We shall find the Drazin inverse solution of (4.45) by (4.11). We obtain

$$\mathbf{A}^2 = \begin{pmatrix}4 & 0 & 0\\ 2-2i & 0 & 0\\ -2-2i & 0 & 0\end{pmatrix},\ \mathbf{A}^3 = \begin{pmatrix}8 & 0 & 0\\ 4-4i & 0 & 0\\ -4-4i & 0 & 0\end{pmatrix},$$

$$\mathbf{B}^2 = \begin{pmatrix} -i & i & 3-i \\ 1 & -1 & 1+3i \\ -3+i & 3-i & 3+i \end{pmatrix}.$$

Since $\operatorname{rank}\mathbf{A} = 2$ and $\operatorname{rank}\mathbf{A}^2 = \operatorname{rank}\mathbf{A}^2 = 1$, then $k_1 = \operatorname{Ind}\mathbf{A} = 2$ and $r_1 = 1$. Since $\operatorname{rank}\mathbf{B} = \operatorname{rank}\mathbf{B}^2 = 2$, then $k_2 = \operatorname{Ind}\mathbf{B} = 1$ and $r_2 = 2$. Then we have

$$\widetilde{\mathbf{D}} = \mathbf{A}^2\mathbf{D}\mathbf{B} = \begin{pmatrix} -4 & 4 & 8 \\ -2+2i & 2-2i & 4-4i \\ 2+2i & -2-2i & -4-4i \end{pmatrix},$$

and $\sum\limits_{\beta\in J_{1,3}} \left|\left(\mathbf{A}^3\right)_{\beta}^{\beta}\right| = 8+0+0 = 8,$

$$\sum_{\alpha\in I_{2,3}} \left|\left(\mathbf{B}^2\right)_{\alpha}^{\alpha}\right| =$$
$$\det\begin{pmatrix} -i & i \\ 1 & -1 \end{pmatrix} + \det\begin{pmatrix} -1 & 1+3i \\ 3-i & 3+i \end{pmatrix} + \det\begin{pmatrix} -i & 3-i \\ -3+i & 3+i \end{pmatrix} =$$
$$0 + (-9-9i) + (9-9i) = -18i.$$

By (4.13), we can get

$$\mathbf{d}_{.1}^{\mathbf{B}} = \begin{pmatrix} 12-12i \\ -12i \\ -12 \end{pmatrix},\ \mathbf{d}_{.2}^{\mathbf{B}} = \begin{pmatrix} -12+12i \\ 12i \\ 12 \end{pmatrix},\ \mathbf{d}_{.3}^{\mathbf{B}} = \begin{pmatrix} 8 \\ -12-12i \\ -12+12i \end{pmatrix}.$$

Since $\mathbf{A}^3_{.1}\left(\mathbf{d}_{.1}^{\mathbf{B}}\right) = \begin{pmatrix} 12-12i & 0 & 0 \\ -12i & 0 & 0 \\ -12 & 0 & 0 \end{pmatrix}$, then finally we obtain

$$x_{11} = \frac{\sum\limits_{\beta\in J_{1,3}\{1\}} \left|\mathbf{A}^3_{.1}\left(\mathbf{d}_{.1}^{\mathbf{B}}\right)_{\beta}^{\beta}\right|}{\sum\limits_{\beta\in J_{1,3}} \left|\left(\mathbf{A}^3\right)_{\beta}^{\beta}\right| \sum\limits_{\alpha\in I_{2,3}} \left|\left(\mathbf{B}^2\right)_{\alpha}^{\alpha}\right|} = \frac{12-12i}{8\cdot(-18i)} = \frac{1+i}{12}.$$

Similarly,

$$x_{12} = \frac{-12+12i}{8\cdot(-18i)} = \frac{-1-i}{12},\quad x_{13} = \frac{8}{8\cdot(-18i)} = \frac{i}{18},$$

$$x_{21} = \frac{-12i}{8\cdot(-18i)} = \frac{1}{12},\ x_{22} = \frac{12i}{8\cdot(-18i)} = -\frac{1}{12},\ x_{23} = \frac{-12-12i}{8\cdot(-18i)} = \frac{1-i}{12},$$

$$x_{31} = \frac{12}{8\cdot(-18i)} = -\frac{i}{12},\ x_{32} = \frac{-12}{8\cdot(-18i)} = \frac{i}{12}.\ x_{33} = \frac{-12+12i}{8\cdot(-18i)} = \frac{-1-i}{12}.$$

Then

$$\mathbf{X} = \begin{pmatrix} \frac{1+i}{12} & \frac{-1-i}{12} & \frac{i}{18} \\ \frac{1}{12} & -\frac{1}{12} & \frac{1-i}{12} \\ -\frac{i}{12} & \frac{i}{12} & \frac{-1-i}{12} \end{pmatrix}$$

is the Drazin inverse solution of (4.45).

5. An Application of the Determinantal Representations of the Drazin Inverse to Some Differential Matrix Equations

In this section we demonstrate an application of the determinantal representations (2.20) and (2.21) of the Drazin inverse to solutions of the following differential matrix equations, $\mathbf{X}' + \mathbf{AX} = \mathbf{B}$ and $\mathbf{X}' + \mathbf{XA} = \mathbf{B}$, where the matrix $\mathbf{A}$ is singular.

Consider the matrix differential equation

$$\mathbf{X}' + \mathbf{AX} = \mathbf{B} \tag{5.1}$$

where $\mathbf{A} \in \mathbb{C}^{n\times n}$, $\mathbf{B} \in \mathbb{C}^{n\times n}$ are given, $\mathbf{X} \in \mathbb{C}^{n\times n}$ is unknown. It's well-known that the general solution of (5.1) is found to be

$$\mathbf{X}(t) = \exp^{-\mathbf{A}t}\left(\int \exp^{\mathbf{A}t} dt\right)\mathbf{B}$$

If $\mathbf{A}$ is invertible, then

$$\int \exp^{\mathbf{A}t} dt = \mathbf{A}^{-1}\exp^{\mathbf{A}t} + \mathbf{G},$$

where $\mathbf{G}$ is an arbitrary $n \times n$ matrix. If $\mathbf{A}$ is singular, then the following theorem gives an answer.

Theorem 5.1. *([63], Theorem 1) If $\mathbf{A}$ has index k, then*

$$\int \exp^{\mathbf{A}t} dt = \mathbf{A}^{D}\exp^{\mathbf{A}t} + (\mathbf{I} - \mathbf{AA}^{D})t\left[\mathbf{I} + \frac{\mathbf{A}}{2}t + \frac{\mathbf{A}^2}{3!}t^2 + \ldots + \frac{\mathbf{A}^{k-1}}{k!}t^{k-1}\right] + \mathbf{G}.$$

Using Theorem 5.1 and the power series expansion of $\exp^{-\mathbf{A}t}$, we get an explicit form for a general solution of (5.1)

$$\mathbf{X}(t) = \\ \left\{\mathbf{A}^{D} + (\mathbf{I} - \mathbf{AA}^{D})t\left(\mathbf{I} - \tfrac{\mathbf{A}}{2}t + \tfrac{\mathbf{A}^2}{3!}t^2 - \ldots(-1)^{k-1}\tfrac{\mathbf{A}^{k-1}}{k!}t^{k-1}\right) + \mathbf{G}\right\}\mathbf{B}.$$

If we put $\mathbf{G} = \mathbf{0}$, then we obtain the following partial solution of (5.1),

$$\begin{array}{c} \mathbf{X}(t) = \mathbf{A}^{D}\mathbf{B} + (\mathbf{B} - \mathbf{A}^{D}\mathbf{AB})t - \frac{1}{2}(\mathbf{AB} - \mathbf{A}^{D}\mathbf{A}^2\mathbf{B})t^2 + \ldots \\ \frac{(-1)^{k-1}}{k!}(\mathbf{A}^{k-1}\mathbf{B} - \mathbf{A}^{D}\mathbf{A}^{k}\mathbf{B})t^{k}. \end{array} \tag{5.2}$$

Denote $\mathbf{A}^{l}\mathbf{B} =: \widehat{\mathbf{B}}^{(l)} = (\widehat{b}_{ij}^{(l)}) \in \mathbb{C}^{n\times n}$ for all $l = \overline{1, 2k}$.

Theorem 5.2. *The partial solution (5.2), $\mathbf{X}(t) = (x_{ij})$, possess the following determinantal representation,*

$$\begin{array}{l} x_{ij} = \frac{\sum\limits_{\beta \in J_{r,\,n}\{i\}} \left|\left(\mathbf{A}^{k+1}_{.\,i}\left(\widehat{\mathbf{b}}^{(k)}_{.j}\right)\right)^{\beta}_{\beta}\right|}{\sum\limits_{\beta \in J_{r,\,n}} \left|\left(\mathbf{A}^{k+1}\right)^{\beta}_{\beta}\right|} + \left(b_{ij} - \frac{\sum\limits_{\beta \in J_{r,\,n}\{i\}} \left|\left(\mathbf{A}^{k+1}_{.\,i}\left(\widehat{\mathbf{b}}^{(k+1)}_{.j}\right)\right)^{\beta}_{\beta}\right|}{\sum\limits_{\beta \in J_{r,\,n}} \left|\left(\mathbf{A}^{k+1}\right)^{\beta}_{\beta}\right|}\right)t \\ -\frac{1}{2}\left(\widehat{b}^{(1)}_{ij} - \frac{\sum\limits_{\beta \in J_{r,\,n}\{i\}} \left|\left(\mathbf{A}^{k+1}_{.\,i}\left(\widehat{\mathbf{b}}^{(k+2)}_{.j}\right)\right)^{\beta}_{\beta}\right|}{\sum\limits_{\beta \in J_{r,\,n}} \left|\left(\mathbf{A}^{k+1}\right)^{\beta}_{\beta}\right|}\right)t^2 + \ldots \\ \frac{(-1)^{k}}{k!}\left(\widehat{b}^{(k-1)}_{ij} - \frac{\sum\limits_{\beta \in J_{r,\,n}\{i\}} \left|\left(\mathbf{A}^{k+1}_{.\,i}\left(\widehat{\mathbf{b}}^{(2k)}_{.j}\right)\right)^{\beta}_{\beta}\right|}{\sum\limits_{\beta \in J_{r,\,n}} \left|\left(\mathbf{A}^{k+1}\right)^{\beta}_{\beta}\right|}\right)t^{k} \end{array} \tag{5.3}$$

for all $i, j = \overline{1, n}$.

Proof. Using the determinantal representation of the identity $\mathbf{A}^D\mathbf{A}$ (2.27), we obtain the following determinantal representation of the matrix $\mathbf{A}^D\mathbf{A}^m\mathbf{B} := (y_{ij})$,

$$y_{ij} = \sum_{s=1}^{n} p_{is} \sum_{t=1}^{n} a_{st}^{(m-1)} b_{tj} = \sum_{\beta \in J_{r,n}\{i\}} \frac{\sum_{s=1}^{n} \left| \left(\mathbf{A}^{k+1}_{.i} \left(\mathbf{a}_{.s}^{(k+1)} \right) \right)^{\beta}_{\beta} \right| \cdot \sum_{t=1}^{n} a_{st}^{(m-1)} b_{tj}}{\sum_{\beta \in J_{r,n}} \left| \left(\mathbf{A}^{k+1} \right)^{\beta}_{\beta} \right|} =$$

$$\sum_{\beta \in J_{r,n}\{i\}} \frac{\sum_{t=1}^{n} \left| \left(\mathbf{A}^{k+1}_{.i} \left(\mathbf{a}_{.t}^{(k+m)} \right) \right)^{\beta}_{\beta} \right| \cdot b_{tj}}{\sum_{\beta \in J_{r,n}} \left| \left(\mathbf{A}^{k+1} \right)^{\beta}_{\beta} \right|} = \frac{\sum_{\beta \in J_{r,n}\{i\}} \left| \left(\mathbf{A}^{k+1}_{.i} \left(\widehat{\mathbf{b}}_{.j}^{(k+m)} \right) \right)^{\beta}_{\beta} \right|}{\sum_{\beta \in J_{r,n}} \left| \left(\mathbf{A}^{k+1} \right)^{\beta}_{\beta} \right|}$$

for all $i, j = \overline{1, n}$ and $m = \overline{1, k}$. From this and the determinantal representation of the Drazin inverse solution (4.34) and the identity (2.27) it follows (5.3). ∎

Corollary 5.3. *If* $Ind\mathbf{A} = 1$, *then the partial solution of (5.1),*

$$\mathbf{X}(t) = (x_{ij}) = \mathbf{A}^g\mathbf{B} + (\mathbf{B} - \mathbf{A}^g\mathbf{A}\mathbf{B})t,$$

possess the following determinantal representation

$$x_{ij} = \frac{\sum_{\beta \in J_{r,n}\{i\}} \left| \left(\mathbf{A}^2_{.i} \left(\widehat{\mathbf{b}}_{.j}^{(1)} \right) \right)^{\beta}_{\beta} \right|}{\sum_{\beta \in J_{r,n}} \left| \left(\mathbf{A}^2 \right)^{\beta}_{\beta} \right|} + \left(b_{ij} - \frac{\sum_{\beta \in J_{r,n}\{i\}} \left| \left(\mathbf{A}^2_{.i} \left(\widehat{\mathbf{b}}_{.j}^{(2)} \right) \right)^{\beta}_{\beta} \right|}{\sum_{\beta \in J_{r,n}} \left| \left(\mathbf{A}^2 \right)^{\beta}_{\beta} \right|} \right) t. \tag{5.4}$$

for all $i, j = \overline{1, n}$.

Consider the matrix differential equation

$$\mathbf{X}' + \mathbf{X}\mathbf{A} = \mathbf{B} \tag{5.5}$$

where $\mathbf{A} \in \mathbb{C}^{n\times n}$, $\mathbf{B} \subset \mathbb{C}^{n\times n}$ are given, $\mathbf{X} \in \mathbb{C}^{n\times n}$ is unknown. The general solution of (5.5) is found to be

$$\mathbf{X}(t) = \mathbf{B} \exp^{-\mathbf{A}t} \left(\int \exp^{\mathbf{A}t} dt \right)$$

If $\mathbf{A}$ is singular, then an explicit form for a general solution of (5.5) is

$$\mathbf{X}(t) =$$
$$\mathbf{B} \left\{ \mathbf{A}^D + (\mathbf{I} - \mathbf{A}\mathbf{A}^D)t \left(\mathbf{I} - \frac{\mathbf{A}}{2}t + \frac{\mathbf{A}^2}{3!}t^2 + ...(-1)^{k-1}\frac{\mathbf{A}^{k-1}}{k!}t^{k-1} \right) + \mathbf{G} \right\}.$$

If we put $\mathbf{G} = \mathbf{0}$, then we obtain the following partial solution of (5.5),

$$\begin{aligned} \mathbf{X}(t) = \mathbf{B}\mathbf{A}^D + (\mathbf{B} - \mathbf{B}\mathbf{A}\mathbf{A}^D)t - \tfrac{1}{2}(\mathbf{B}\mathbf{A} - \mathbf{B}\mathbf{A}^2\mathbf{A}^D)t^2 + ... \\ \tfrac{(-1)^{k-1}}{k!}(\mathbf{B}\mathbf{A}^{k-1} - \mathbf{B}\mathbf{A}^k\mathbf{A}^D)t^k. \end{aligned} \tag{5.6}$$

Denote $\mathbf{B}\mathbf{A}^l =: \check{\mathbf{B}}^{(l)} = (\check{b}_{ij}^{(l)}) \in \mathbb{C}^{n\times n}$ for all $l = \overline{1, 2k}$. Using the determinantal representation of the Drazin inverse solution (4.36), the group inverse (2.25) and the identity (2.26) we evidently obtain the following theorem.

Theorem 5.4. *The partial solution (5.6), $\mathbf{X}(t) = (x_{ij})$, possess the following determinantal representation,*

$$x_{ij} = \frac{\sum\limits_{\alpha \in I_{r,n}\{j\}} \left|\left(\mathbf{A}^{k+1}_{j.}\left(\check{\mathbf{b}}^{(k)}_{.i}\right)\right)^{\alpha}_{\alpha}\right|}{\sum\limits_{\alpha \in I_{r,n}} \left|\left(\mathbf{A}^{k+1}\right)^{\alpha}_{\alpha}\right|} + \left(b_{ij} - \frac{\sum\limits_{\alpha \in I_{r,n}\{j\}} \left|\left(\mathbf{A}^{k+1}_{j.}\left(\check{\mathbf{b}}^{(k+1)}_{i.}\right)\right)^{\alpha}_{\alpha}\right|}{\sum\limits_{\alpha \in I_{r,n}} \left|\left(\mathbf{A}^{k+1}\right)^{\alpha}_{\alpha}\right|}\right) t$$
$$-\frac{1}{2}\left(\check{b}^{(1)}_{ij} - \frac{\sum\limits_{\alpha \in I_{r,n}\{j\}} \left|\left(\mathbf{A}^{k+1}_{j.}\left(\check{\mathbf{b}}^{(k+2)}_{i.}\right)\right)^{\alpha}_{\alpha}\right|}{\sum\limits_{\alpha \in I_{r,n}} \left|\left(\mathbf{A}^{k+1}\right)^{\alpha}_{\alpha}\right|}\right) t^2 + \ldots$$
$$\frac{(-1)^k}{k!}\left(\check{b}^{(k-1)}_{ij} - \frac{\sum\limits_{\alpha \in I_{r,n}\{j\}} \left|\left(\mathbf{A}^{k+1}_{j.}\left(\check{\mathbf{b}}^{(2k)}_{i.}\right)\right)^{\alpha}_{\alpha}\right|}{\sum\limits_{\alpha \in I_{r,n}} \left|\left(\mathbf{A}^{k+1}\right)^{\alpha}_{\alpha}\right|}\right) t^k$$

for all $i, j = \overline{1, n}$.

Corollary 5.5. *If $Ind\mathbf{A} = 1$, then the partial solution of (5.5),*

$$\mathbf{X}(t) = (x_{ij}) = \mathbf{B}\mathbf{A}^g + (\mathbf{B} - \mathbf{B}\mathbf{A}\mathbf{A}^g)t,$$

possess the following determinantal representation

$$x_{ij} = \frac{\sum\limits_{\alpha \in I_{r,n}\{j\}} \left|\left(\mathbf{A}^2_{j.}\left(\widehat{\mathbf{b}}^{(1)}_{i.}\right)\right)^{\alpha}_{\alpha}\right|}{\sum\limits_{\alpha \in I_{r,n}} \left|\left(\mathbf{A}^2\right)^{\alpha}_{\alpha}\right|} + \left(b_{ij} - \frac{\sum\limits_{\alpha \in I_{r,n}\{j\}} \left|\left(\mathbf{A}^2_{j.}\left(\widehat{\mathbf{b}}^{(2)}_{i.}\right)\right)^{\alpha}_{\alpha}\right|}{\sum\limits_{\alpha \in I_{r,n}} \left|\left(\mathbf{A}^2\right)^{\alpha}_{\alpha}\right|}\right) t.$$

for all $i, j = \overline{1, n}$.

5.1. Example

1. Let us consider the differential matrix equation

$$\mathbf{X}' + \mathbf{A}\mathbf{X} = \mathbf{B}, \tag{5.7}$$

where

$$\mathbf{A} = \begin{pmatrix} 1 & -1 & 1 \\ i & -i & i \\ -1 & 1 & 2 \end{pmatrix}, \ \mathbf{B} = \begin{pmatrix} 1 & i & 1 \\ i & 0 & 1 \\ 1 & i & 0 \end{pmatrix}.$$

Since $\operatorname{rank} \mathbf{A} = \operatorname{rank} \mathbf{A}^2 = 2$, then $k = \operatorname{Ind} \mathbf{A} = 1$ and $r = 2$. The matrix $\mathbf{A}$ is the group inverse. We shall find the partial solution of (5.7) by (5.4). We have

$$\mathbf{A}^2 = \begin{pmatrix} -i & i & 3-i \\ 1 & -1 & 1+3i \\ -3+i & 3-i & 3+i \end{pmatrix}, \ \widehat{\mathbf{B}}^{(1)} = \mathbf{A}\mathbf{B} = \begin{pmatrix} 2-i & 2i & 0 \\ 1+2i & -2 & 0 \\ 1+i & i & 0 \end{pmatrix},$$

$$\widehat{\mathbf{B}}^{(2)} = \mathbf{A}^2\mathbf{B} = \begin{pmatrix} 2-2i & 2+3i & 0 \\ 2+2i & -3+2i & 0 \\ 1+5i & -2 & 0 \end{pmatrix}.$$

and

$$\sum_{\alpha \in J_{2,3}} \left| \left(\mathbf{A}^2\right)_{\beta}^{\beta} \right| =$$
$$\det\begin{pmatrix} -i & i \\ 1 & -1 \end{pmatrix} + \det\begin{pmatrix} -1 & 1+3i \\ 3-i & 3+i \end{pmatrix} + \det\begin{pmatrix} -i & 3-i \\ -3+i & 3+i \end{pmatrix} =$$
$$0 + (-9-9i) + (9-9i) = -18i.$$

Since $\left(\mathbf{A}^2\right)_{.1}\left(\widehat{\mathbf{b}}_{.1}^{(1)}\right) = \begin{pmatrix} 2-i & i & 3-i \\ 1+2i & -1 & 1+3i \\ 1+i & 3-i & 3+i \end{pmatrix}$ and

$$\left(\mathbf{A}^2\right)_{.1}\left(\widehat{\mathbf{b}}_{.1}^{(2)}\right) = \begin{pmatrix} 2-2i & i & 3-i \\ 2+2i & -1 & 1+3i \\ 1+5i & 3-i & 3+i \end{pmatrix},$$

then finally we obtain

$$x_{11} = \frac{\sum\limits_{\beta \in J_{2,3}\{1\}} \left| \left(\mathbf{A}^2_{.1}\left(\widehat{\mathbf{b}}_{.1}^{(1)}\right)\right)_{\beta}^{\beta} \right|}{\sum\limits_{\beta \in J_{2,3}} \left| \left(\mathbf{A}^2\right)_{\beta}^{\beta} \right|} + \left(b_{11} - \frac{\sum\limits_{\beta \in J_{2,3}\{1\}} \left| \left(\mathbf{A}^2_{.1}\left(\widehat{\mathbf{b}}_{.1}^{(2)}\right)\right)_{\beta}^{\beta} \right|}{\sum\limits_{\beta \in J_{2,3}} \left| \left(\mathbf{A}^2\right)_{\beta}^{\beta} \right|} \right) t =$$
$$\tfrac{3-3i}{-18i} + \left(1 - \tfrac{-18i}{-18i}\right) t = \tfrac{1+i}{6}.$$

Similarly,

$$x_{12} = \frac{-3+3i}{-18i} + \left(i - \frac{9+9i}{-18i}\right) t = \frac{-1-i}{6} + \frac{1+i}{2} t, \; x_{13} = 0 + (1-0)\,t = t,$$

$$x_{21} = \frac{3+3i}{-18i} + \left(i - \frac{-18}{-18i}\right) t = \frac{-1+i}{6},$$

$$x_{22} = \frac{-3-3i}{-18i} + \left(0 - \frac{-9+9i}{-18i}\right) t = \frac{1-i}{6} + \frac{1+i}{2} t, \; x_{23} = 0 + (1-0)\,t = t,$$

$$x_{31} = \frac{-12i}{-18i} + \left(1 - \frac{-18i}{-18i}\right) t = \frac{2}{3},$$

$$x_{32} = \frac{9+3i}{-18i} + \left(i - \frac{-18}{-18i}\right) t = \frac{-1+3i}{6}, \; x_{33} = 0 + (0-0)\,t = 0.$$

Then

$$\mathbf{X} = \frac{1}{6}\begin{pmatrix} 1+i & -1-i+(3+3i)t & t \\ -1+i & 1-i+(3+3i)t & t \\ 4 & -1+3i & 0 \end{pmatrix}$$

is the partial solution of (5.7) .

6. Conclusion

From student years it is well known that Cramer's rule may only be used when the system is square and the coefficient matrix is invertible. In this chapter we are considered various cases of Cramer's rule for generalized inverse solutions of systems of linear equations and matrix equations when the coefficient matrix is not square or non-invertible. The results of this chapter have practical and theoretical importance because they give an explicit representation of an individual component of solutions independently of all other components. Also the results of this chapter can be extended to matrices over rings (and now this is done in the quaternion skew field), to polynomial matrices, etc.

References

[1] E. H. Moore, On the reciprocal of the general algebraic matrix, *Bulletin of the American Mathematical Society* 26 (9) (1920) 394–395.

[2] A. Bjerhammar, Application of calculus of matrices to method of least squares; with special references to geodetic calculations, *Trans. Roy. Inst. Tech. Stockholm* 49 (1951) 1–86.

[3] R. Penrose, A generalized inverse for matrices, *Proc. Camb. Philos. Soc.* 51 (1955) 406–413.

[4] M. P.Drazin, Pseudo-inverses in associative rings and semigroups, *The American Mathematical Monthly* 65 (7) (1958) 506–514.

[5] K.M. Prasad, R.B. Bapat, A note of the Khatri inverse, *Sankhya: Indian J. Stat.* 54 (1992) 291–295.

[6] I.I. Kyrchei, Analogs of the adjoint matrix for generalized inverses and corresponding Cramer rules, *Linear and Multilinear Algebra* 56(4) (2008) 453–469.

[7] Ivan Kyrchei, Analogs of Cramer's rule for the minimum norm least squares solutions of some matrix equations, *Applied Mathematics and Computation* 218 (2012) 6375–6384.

[8] Ivan Kyrchei, Explicit formulas for determinantal representations of the Drazin inverse solutions of some matrix and differential matrix equations, *Applied Mathematics and Computation* 219 (2013) 1576–1589.

[9] Ivan I. Kyrchei, The theory of the column and row determinants in a quaternion linear algebra, *Advances in Mathematics Research* 15, pp. 301–359. Nova Sci. Publ., New York, 2012.

[10] I.I. Kyrchei Cramer's rule for some quaternion matrix equations, *Applied Mathematics and Computation* 217(5) (2010) 2024–2030.

[11] I.I. Kyrchei, Determinantal representation of the MoorePenrose inverse matrix over the quaternion skew field, *Journal of Mathematical Sciences* 180(1) (2012) 23–33

[12] I.I. Kyrchei Determinantal representations of the Moore-Penrose inverse over the quaternion skew field and corresponding Cramer's rules, *Linear and Multilinear Algebra* 59(4) (2011), pp. 413–431.

[13] Ivan Kyrchei, Explicit representation formulas for the minimum norm least squares solutions of some quaternion matrix equations, *Linear Algebra and Its Applications* 438(1) (2013) 136–152.

[14] Ivan Kyrchei, Determinantal representations of the Drazin inverse over the quaternion skew field with applications to some matrix equations, *Applied Mathematics and Computation* 238 (2014) 193–207.

[15] X. Liu, G. Zhu, G. Zhou, Y. Yu, An analog of the adjugate matrix for the outer inverse $\mathbf{A}^{(2)}_{T,S}$, *Mathematical Problems in Engineering*, Volume 2012, Article ID 591256, 14 pages.

[16] R. B. Bapat, K. P. S. Bhaskara, K. Manjunatha Prasad, Generalized inverses over integral domains, *Linear Algebra and Its Applications* 140 (1990) 181–196.

[17] A. Ben-Israel, Generalized inverses of marices: a perspective of the work of Penrose, *Math. Proc. Camb. Phil. Soc.* 100 (1986) 401–425.

[18] R. Gabriel, Das verallgemeinerte inverse eineer matrix, deren elemente einem beliebigen Körper angehören, *J.Reine angew math.* 234 (1967) 107–122.

[19] P. S. Stanimirovic', General determinantal representation of pseudoinverses of matrices, *Mat. Vesnik* 48 (1996) 1–9.

[20] A. Ben-Israel, On matrices of index zero or one, *SIAM J. Appl. Math.* 17 (1969) 111–1121.

[21] R. A. Horn, C. R. Johnson, *Matrix Analysis.* Cambridge etc., Cambridge University Press, 1985.

[22] P. Lancaster and M. Tismenitsky, *Theory of matrices*, Acad. Press., New York, 1969.

[23] C.F. Van Loan, Generalizing the singular value decomposition, *SIAM J. Numer. Anal.* 13 (1976) 76–83.

[24] Y. Wei, H. Wu, The representation and approximation for the weighted Moore-Penrose inverse, *Applied Mathematics and Computation* 121 (2001) 17–28.

[25] Stephen L. Campbell, Carl D. Meyer Jr., Weak Drazin inverses, *Linear Algebra and Its Applications.* 20(2) (1978) 167–178.

[26] P.S. Stanimirovic', D.S. Djordjevic', Full-rank and determinantal representation of the Drazin inverse, *Linear Algebra and Its Applications* 311 (2000) 131–151.

[27] Carl D. Meyer Jr. , Limits and the index of a square matrix *SIAM J. Appl. Math.* 26(3) (1974) 506–515.

[28] R. E. Cline, T. N. E. Greville, A Drazin inverse for rectangular matrices, *Linear Algebra and Its Applications* 29 (1980) 53–62.

[29] Y. Wei, Integral representation of the W-weighted Drazin inverse, *Applied Mathematics and Computation* 144 (2003) 3–10.

[30] Y. Wei, C.-W. Woo, T. Lei, A note on the perturbation of the W- weighted Drazin Inverse, *Applied Mathematics and Computation* 149 (2004) 423–430.

[31] Y. Wei, A characterization for the W -weighted Drazin inverse and a Cramer rule for the W - weighted Drazin inverse solution, *Applied Mathematics and Computation* 125 (2002) 303–310.

[32] Zeyad Abdel Aziz Al Zhour, Adem Kiliçman, Malik Hj. Abu Hassa, New representations for weighted Drazin inverse of matrices, *Int. Journal of Math. Analysis* 1(15) (2007) 697–708

[33] S.M. Robinson, A short proof of Cramer's rule, *Math. Mag.* 43 (1970) 94–95.

[34] A. Ben-Israel, A Cramer rule for least-norm solutions of consistent linear equations, *Linear Algebra and Its Applications* 43 (1982) 223–226.

[35] G.C. Verghese, A Cramer rule for least-norm least-square-error solution of inconsistent linear equations, *Linear Algebra and Its Applications* 48 (1982) 315–316.

[36] H. J. Werner, On extension of Cramer's rule for solutions of restricted linear systems, *Linear and Multilinear Algebra* 15 (1984) 319–330.

[37] Y. Chen, A Cramer rule for solution of the general restricted linear equation, *Linear and Multilinear Algebra* 34 (1993) 177–186

[38] J. Ji, Explicit expressions of the generalized inverses and condensed Cramer rules, *Linear Algebra and Its Applications* 404 (2005) 183–192.

[39] G. Wang, A Cramer rule for minimum-norm (T) least-squares (S) solution of inconsistent linear equations, *Linear Algebra and Its Applications* 74 (1986) 213–218.

[40] G. Wang, A Cramer rule for finding the solution of a class of singular equations, *Linear Algebra and Its Applications* 116 (1989) 27–34.

[41] G. Chen, X. Chen. A new splitting for singular linear system and Drazin inverse. *J. East China Norm. Univ. Natur. Sci. Ed.* 3 (1996) 12–18.

[42] A. Sidi. A unified approach to Krylov subspace methods for the Drazin-inverse solution of singular nonsymmetric linear systems. *Linear Algebra and Its Applications* 298 (1999) 99–113.

[43] Y. M. Wei, H. B. Wu, Additional results on index splittings for Drazin inverse solutions of singular linear systems, *The Electronic Journal of Linear Algebra* 8 (2001) 83–93.

[44] P.S. Stanimirovic. A representation of the minimal P-norm solution. *Novi Sad J. Math.* 3(1) (2000) 177–183.

[45] H. Dai, On the symmetric solution of linear matrix equation, *Linear Algebra and Its Applications* 131 (1990) 1–7.

[46] M. Dehghan, M. Hajarian, The reflexive and anti-reflexive solutions of a linear matrix equation and systems of matrix equations, *Rocky Mountain J. Math.* 40 (2010) 825–848.

[47] F.J. Henk Don, On the symmetric solutions of a linear matrix equation, *Linear Algebra and Its Applications* 93 (1987) 1–7.

[48] Z.Y. Peng and X.Y. Hu, The generalized reflexive solutions of the matrix equations AX = D and AXB = D, *Numer. Math.* 5 (2003) 94–98.

[49] W.J. Vetter, Vector structures and solutions of linear matrix equations, *Linear Algebra and Its Applications.* 9 (1975) 181–188.

[50] C.G. Khatri, S.K. Mitra, Hermitian and nonnegative definite solutions of linear matrix equations, *SIAM J. Appl. Math.* 31 (1976) 578–585.

[51] Q.W. Wang, Bisymmetric and centrosymmetric solutions to systems of real quaternion matrix equations, *Comput. Math. Appl.* 49 (2005) 641–650.

[52] Q.W. Wang, The general solution to a system of real quaternion matrix equations, *Comput. Math. Appl.* 49 (56) (2005) 665–675.

[53] Yao-tang Li, Wen-jing Wu, Symmetric and skew-antisymmetric solutions to systems of real quaternion matrix equations, *Comput. Math. Appl.* 55 (2008) 1142–1147.

[54] Y. H. Liu, Ranks of least squares solutions of the matrix equation AXB=C, *Comput. Math. Appl.* 55 (2008) 1270–1278.

[55] Q.W. Wang, S. W. Yu, Extreme ranks of real matrices in solution of the quaternion matrix equation AXB=C with applications, *Algebra Colloquium*, 17 (2) (2010) 345–360.

[56] G. Wang, S. Qiao, Solving constrained matrix equations and Cramer rule, *Applied Mathematics and Computation* 159 (2004) 333–340.

[57] Y. Qiu, A. Wang, Least-squares solutions to the equations AX = B, XC = D with some constraints, *Applied Mathematics and Computation* 204 (2008) 872–880.

[58] Y. Yuan, Least-squares solutions to the matrix equations AX = B and XC = D, *Applied Mathematics and Computation* 216 (2010) 3120-3125.

[59] G. Wang, Z. Xu, Solving a kind of restricted matrix equations and Cramer rule *Applied Mathematics and Computation* 162 (2005) 329–338.

[60] C. Gu, G. Wang, Condensed Cramer rule for solving restricted matrix equations, *Applied Mathematics and Computation* 183 (2006) 301–306

[61] G.J. Song, Q.W. Wang, H.X. Chang, Cramer rule for the unique solution of restricted matrix equations over the quaternion skew field, *Comput. Math. Appl.* 61 (2011) 1576–1589.

[62] Z.Xu, G. Wang, On extensions of Cramer's rule for solutions of restricted matrix equtions, *Journal of Lanzhou University* 42 (3) (2006) 96–100.

[63] S. L. Campbell, C. D. Meyer, JR. and N. J. Rose, Applications of the Drazin inverse to linear systems of differential equations with singular constant coefficients, *SIAM J. Appl. Math.* 31 (1976) 411–425.

In: Advances in Linear Algebra Research
Editor: Ivan Kyrchei, pp. 133-149
ISBN: 978-1-63463-565-3

Chapter 4

FEEDBACK ACTIONS ON LINEAR SYSTEMS OVER VON NEUMANN REGULAR RINGS

Andrés Sáez-Schwedt*
Departamento de Matemáticas, Universidad de León,
Campus de Vegazana, León, Spain

Abstract

Many physical systems in science and engineering can be described at time t in terms of an n-dimensional state vector $x(t)$ and an m-dimensional input vector $u(t)$, governed by an evolution equation of the form $x'(t) = A \cdot x(t) + B \cdot u(t)$, if the time is continuous, or $x(t+1) = A \cdot x(t) + B \cdot u(t)$ in the discrete case. Thus, the system is completely described by the pair of matrices (A, B) of sizes $n \times n$ and $n \times m$ respectively.

In two instances feedback is used to modify the structure of a given system (A, B): first, A can be replaced by $A + BF$, with some characteristic polynomial that ensures stability of the new system $(A + BF, B)$; and second, combining changes of bases with a feedback action $A \mapsto A + BF$ one obtains an equivalent system with a simpler structure.

Given a system (A, B), let $\mathcal{R}(A, B)$ denote the set of states reachable at finite time when starting with initial condition $x(0) = 0$ and varying $u(t)$, i.e., $\mathcal{R}(A, B)$ is the right image of the matrix $[B|AB|A^2B|\cdots]$. Also, let $Pols(A, B)$ denote the set of characteristic polynomials of all possible matrices $A + BF$, as F varies.

Classically, (A, B) have their entries in the field of real or complex numbers, but the concept of discrete-time system is generalized to matrix pairs with coefficients in an arbitrary commutative ring R. Therefore, techniques from Linear Algebra over commutative rings are needed.

In this chapter, the following problems are studied and solved when R is a commutative von Neumann regular ring:

- A canonical form is obtained for the feedback equivalence of systems (combination of basis changes with a feedback action).

*E-mail address: asaes@unileon.es

- Given a system (A, B), it is proved that there exist a matrix F and a vector u such that the single-input system $(A+BF, Bu)$ has the same reachable states and the same assignable polynomials as the original system, i.e., $\mathcal{R}(A+BF, Bu) = \mathcal{R}(A, B)$ and $Pols(A+BF, Bu) = Pols(A, B)$.

Keywords: Systems over commutative rings, commutative von Neumann regular ring, feedback classification, polynomials assignable by state feedback

AMS Subject Classification: 93B25, 93B52, 13F99

1. Introduction

The theory of systems over rings had its origins in the 1970s, in order to cover with situations where it is too restrictive to consider systems with real or complex coefficients. The motivation for this approach is explained in [1, 2, 3], and the necessary Linear Algebra over commutative rings is nicely covered in [4].

Let R be a commutative ring with 1. An m-input, n-dimensional *linear control system* over R, or shortly a system of size (n, m) over R, is a pair of matrices (A, B) of sizes $n \times n$ and $n \times m$, respectively, with coefficients in R. The following control process is associated with (A, B):

$$x(t+1) = A \cdot x(t) + B \cdot u(t),$$

where the vector $x(t) \in R^n$ is the *state* of the system at time t, and $u(t) \in R^m$ is the *input* at time t. If we put the initial condition $x(0) = 0$ and choose arbitrary input vectors $u(1), u(2), \ldots$, then the following states are obtained:

$$\begin{aligned} x(1) &= A \cdot x(0) + B \cdot u(1) = B \cdot u(1) \\ x(2) &= A \cdot x(1) + B \cdot u(2) = AB \cdot u(1) + B \cdot u(2) \\ x(3) &= A \cdot x(2) + B \cdot u(3) = A^2B \cdot u(1) + AB \cdot u(2) + B \cdot u(3) \\ &\ldots \end{aligned}$$

This means that the set of states which are reachable at some finite time, when starting from the origin $x(0) = 0$ and choosing appropriate inputs $(u(t))_{t\geq 1}$, is a submodule of R^n which can be defined as the image of the infinite matrix $[B|AB|A^2B|\cdots]$. Note that, by Cayley-Hamilton's theorem, A^n and higher powers of A are linear combinations of $A, A^2, \ldots, A^{n-1}$ and I, the $n \times n$ identity matrix, therefore it suffices to consider the image of the so called reachability matrix $A^*B := [B|AB|A^2B|\cdots|A^{n-1}B]$.

This module of reachable states has different notations in the literature, among others: $Reach(A, B)$ [3], $N_n^{(A,B)}$ [5, 6], $<A|B>$ [7], $im(A^*B)$ [8] or $\mathcal{R}(A, B)$ [9]. The chosen notation for this chapter will be $\mathcal{R}(A, B)$, and in a few places $im(A^*B)$. With these notations, the system (A, B) is called *reachable* if $\mathcal{R}(A, B) = R^n$.

In Control Theory, if (A, B) have real coefficients, then the (complex) eigenvalues of A (which are the poles of the transfer function) are important to guarantee the stability of the system. Such stability depends on the positions of the poles in the complex plane (see [3]). Therefore, it is customary to replace the "open-loop" system by a "closed-loop" system, that is, a system in which the state is "fed back" as an input. The input is taken as $u(t) = w(t) + Fx(t)$, where $w(t)$ can be interpreted as an uncontrolled input (e.g., a disturbance or perturbation) and $Fx(t)$ is a new term that depends linearly on the state. Then, the equation $x(t+1) = Ax(t) + B(w(t) + Fx(t))$ is that of a new system with state $x(t)$, input $w(t)$ and matrices $(A + BF, B)$. Within this context, a matrix F is looked for, such that $A + BF$ has some prescribed characteristic polynomial.

When applying state feedback to a given system (A, B), it is interesting to describe exactly which polynomials arise as characteristic polynomials of $A+BF$, for some F. With this in mind, in [10, 11] the notation $Pols(A, B)$ is used to refer to the set of polynomials assignable by feedback to the system (A, B), and consists of all characteristic polynomials $\chi(A + BF) = det(xI - A - BF)$, as F ranges over all $m \times n$ matrices with entries in R.

In order to make this introduction self-contained and historically complete, we recall some wellknown properties of systems over rings (see e.g., [1, 2]). The system (A, B) is called (i) *pole assignable* if given arbitrary scalars $a_1, \ldots, a_n$ in R there exists a matrix F such that $\chi(A + BF) = (x - a_1)\cdots(x - a_n)$; (ii) *coefficient assignable* if for any monic polynomial $f(x)$ of degree n there exists a matrix F such that $\chi(A + BF) = f(x)$; and (iii) *feedback cyclizable* if there exist a matrix F and a vector u such that $(A + BF, Bu)$ is reachable. It is known that feedback cyclizability $\Rightarrow$ coefficient assignability $\Rightarrow$ pole assignability $\Rightarrow$ reachability, and that all properties are equivalent when R is a field.

Rings for which some of the converse implications hold are denoted as follows: R is a PA ring (resp. CA ring) (resp. FC ring) if reachable systems over R are pole assignable (resp. coefficient assignable) (resp. feedback cyclizable). Then, it is immediate that $FC \Rightarrow CA \Rightarrow PA$, and it can be seen that the implications are strict.

Why is the feedback cyclization property so important? Suppose that we are given a system (A, B) of size (n, m), and we want to assign to it a certain polynomial $f(x)$. This means that we have to find an $m \times n$ matrix K such that $\chi(A + BK) = f(x)$, i.e., we are faced with a problem with mn degrees of freedom. But, if we are lucky to find F, u such that $(A + BF, Bu)$ is reachable, then, taking into account that for single-input systems reachability implies coefficient assignability [1, Theorem 3.2], there exists certainly a $1 \times n$ matrix K such that $\chi(A + BF + BuK) = f(x)$, i.e., we have reduced our problem to one with n degrees of freedom.

Various attempts have been made to extend the properties PA, CA, FC to non reachable systems. In [8, 12, 13], the notion of *residual rank* is used: a system (A, B) over R has residual rank r, and we write $res.rk(A, B) = r$, if r is the greatest integer such that the ideal of R generated by the $r \times r$ minors of the reachability matrix A^*B is the whole ring R. Reachable systems correspond to the case of maximum possible residual rank: $res.rk(A, B) = n$. Now, consider a system (A, B) over R with $res.rk(A, B) = r$. Then, (A, B) is said to be (i) pole assignable if, given arbitrary scalars $a_1, \ldots, a_r$, there exists a matrix F such that $\chi(A + BF)$ is a multiple of $(x - a_1)\cdots(x - a_r)$; (ii) coefficient assignable if, given any monic polynomial $f(x)$ of degree r, there exists F such that $\chi(A + BF)$ is a multiple of $f(x)$; and (iii) feedback cyclizable if there exist F, u such that $res.rk(A + BF, Bu) = res.rk(A, B)$.

With these notations, R is called a PS ring (resp. strong CA ring) (resp. strong FC ring) if any system over R is pole assignable (resp. coefficient assignable) (resp. feedback cyclizable). As one would expect, the relation among the strong forms of PA, CA, FC

and the usual forms is the following:

$$\begin{array}{ccccc} \text{strong } FC & \Rightarrow & \text{strong } CA & \Rightarrow & PS \\ \Downarrow & & \Downarrow & & \Downarrow \\ FC & \Rightarrow & CA & \Rightarrow & PA \end{array}$$

Although these strong versions of PA, CA and FC allow to study non reachable systems, they are only interesting for systems (A, B) with residual rank at least one, which is equivalent to requiring that the ideal generated by all the entries of B is the whole ring R. If this is not the case, then (A, B) trivially satisfies the feedback cyclization property, and hence all the other properties: indeed, if $res.rk(A, B) = 0$, then for any F, u one has $res.rk(A + BF, Bu) = 0$ (the ideal generated by the entries of the vector Bu is contained in the ideal generated by the entries of B, and thus is not R).

A further step was made in [6, 11], allowing to study arbitrary non reachable systems. Observe that for any matrix $F \in R^{m\times n}$ and any vector $u \in R^m$, it is immediate that $\mathcal{R}(A + BF, Bu) \subseteq \mathcal{R}(A, B)$ (see e.g., [6] or [3]) and $Pols(A + BF, Bu) \subseteq Pols(A, B)$ (because a matrix of the form $(A + BF) + (Bu)F'$ can always be written as $A + BF''$, with $F'' = F + uF'$). An interesting question is: when do the reverse inclusions hold? Or equivalently, when are the following problems solvable?

$$\begin{array}{rcl} (Reach) & : & \text{Find } F, u \text{ such that } \mathcal{R}(A + BF, Bu) = \mathcal{R}(A, B) \\ (Pols) & : & \text{Find } F, u \text{ such that } Pols(A + BF, Bu) = Pols(A, B) \end{array}$$

Definition 1. *(Properties Reach and Pols)* A ring R is said to satisfy $Reach$ (resp. $Pols$) if all systems over R satisfy $Reach$ (resp. $Pols$).

It is clear that $Reach$ implies the strong FC property: if $\mathcal{R}(A+BF, Bu) = \mathcal{R}(A, B)$, then the matrices A^*B and $(A + BF)^*(Bu)$ have the same image, hence the same ideals of minors, and so $res.rk(A + BF, Bu) = res.rk(A, B)$. In [6, Theorem 4], it is proved that a ring R satisfies $Reach$ (called in that reference Problem (1)) if and only if R is von Neumann regular, i.e., for any a in R there exists x such that $a^2x = a$. Also, the condition $Reach$ is strictly stronger than strong FC, as can be seen in [8, Proposition 3.4], where various examples of strong FC rings are given, which are not von Neumann regular. Then, in [11, Theorem 2], it is proved that von Neumann regular rings satisfy $Pols$, and in fact, both problems $Reach$ and $Pols$ are solved simultaneously: there exist a matrix F and a vector u such that $\mathcal{R}(A + BF, Bu) = \mathcal{R}(A, B)$ and $Pols(A + BF, Bu) = Pols(A, B)$.

This means that starting from an arbitrary multi-input system (A, B), we obtain a single-input system $(A + BF, Bu)$ which has the same module of reachable states and the same set of assignable polynomials as the original system (A, B), a reduction of great importance in systems theory.

One of the main tools in proving the previous results is the use of *feedback equivalence*. Two systems (A, B) and (A', B') of size (n, m) over R are feedback equivalent if there exist invertible matrices P, Q and a feedback matrix K such that $A' = PAP^{-1} + PBK$ and $B' = PBQ$. Any feedback transformation is a combination of the following three types of row and column operations: (i) $(A, B) \mapsto (PAP^{-1}, PB)$, corresponding to a change of basis in R^n, (ii) $(A, B) \mapsto (A, BQ)$, which applies column operations on B, and

(iii) $(A, B) \mapsto (A + BK, B)$, a closed loop as explained before, which can be interpreted as adding to each column of A a linear combination of columns of B.

Note that equivalence and similarity of matrices over R are particular cases of feedback equivalence. Indeed, if (A, B) and (A', B') are feedback equivalent, then B, B' are equivalent ($B' = PBQ$, with P, Q invertible), while the equivalence of $(A, 0), (A', 0)$ implies similarity of A, A' (i.e., $A' = PAP^{-1}$). Therefore, a canonical form under feedback equivalence only makes sense if a canonical form is known for equivalence (i.e., the ring is an elementary divisor ring and a certain Smith form is known), and also a canonical form for similarity of matrices is known.

The effect of feedback equivalence on the module of reachable states is the following: $\mathcal{R}(A', B') = \phi_P(\mathcal{R}(A, B))$ [5, Lemma 2.1], where ϕ_P is the isomorphism of R^n with matrix P in the standard bases. On the other side, feedback has no effect on the set of assignable polynomials: $Pols(A', B') = Pols(A, B)$, as can be easily proved, see e.g., [10].

In this chapter, after recalling in Section 2 what is known for systems over fields, Section 3 is devoted to explaining how to derive a feedback canonical form for the feedback equivalence of systems over a von Neumann regular ring, and in Section 4 this canonical form is of fundamental importance to prove the simultaneous solution of $Reach$ and $Pols$.

2. Preliminaries: What is Known for Fields?

For a reachable system (A, B) of size (n, m) over a field k, our problems $Reach$ and $Pols$ are trivially solved by the wellknown Heymann's Lemma [14]: there exist a matrix F and a vector u of sizes $m \times n$ and $m \times u$ respectively such that the single-input system $(A + BF, Bu)$ is reachable. This means that $\mathcal{R}(A, B)$ and $\mathcal{R}(A + BF, Bu)$ are both equal to k^n, therefore problem $Reach$ is solved. But reachable systems over fields are always coefficient assignable [1], from which it follows that $Pols(A, B)$ and $Pols(A + BF, Bu)$ are both equal to the set of all monic polynomials of degree n over k, hence the condition $Pols$ is solved, by means of the same matrix F and vector u which solve $Reach$.

Also, a canonical form is known for reachable systems over a field k, see [15] or [5]. For any positive integer i, let the $i \times i$ matrix A_i denote one of the companion matrices of the polynomial x^i (with zeroes on the main diagonal, and ones immediately below), and let b_i be the first basic vector of k^i. Then, given a reachable system (A, B) of size (n, m), there exist positive integers $k_1 \geq \cdots \geq k_s$, with sum n, such that (A, B) is feedback equivalent to a canonical form $(\hat{A}, \hat{B})$, such that $\hat{A}$ is formed with the blocks $A_{k_1}, \ldots, A_{k_s}$, and $\hat{B}$ has blocks $b_{k_1}, \ldots, b_{k_s}$, completed with zero columns. The elements k_i are called the Kronocker indices. For example, the canonical form associated to a reachable system of size $(7, 4)$ with indices $4, 2, 1$ is given by:

$$\hat{A} = \left[\begin{array}{cccc|cc|c} 0 & 0 & 0 & 0 & 0 & 0 & 0 \\ 1 & 0 & 0 & 0 & 0 & 0 & 0 \\ 0 & 1 & 0 & 0 & 0 & 0 & 0 \\ 0 & 0 & 1 & 0 & 0 & 0 & 0 \\ \hline 0 & 0 & 0 & 0 & 0 & 0 & 0 \\ 0 & 0 & 0 & 0 & 1 & 0 & 0 \\ \hline 0 & 0 & 0 & 0 & 0 & 0 & 0 \end{array}\right], \hat{B} = \left[\begin{array}{c|c|c|c} 1 & 0 & 0 & 0 \\ 0 & 0 & 0 & 0 \\ 0 & 0 & 0 & 0 \\ 0 & 0 & 0 & 0 \\ \hline 0 & 1 & 0 & 0 \\ 0 & 0 & 0 & 0 \\ \hline 0 & 0 & 1 & 0 \end{array}\right].$$

For non reachable systems over fields, the situation is solved via the so called Kalman decomposition [3, Lemma 3.3.3]. Concretely, let (A, B) be a system of size (n, m) over a field k. Then, there exists an invertible matrix P such that

$$PAP^{-1} = \begin{bmatrix} A_1 & A_2 \\ 0 & A_3 \end{bmatrix}, PB = \begin{bmatrix} B_1 \\ 0 \end{bmatrix}, \tag{1}$$

where A_1 is $r \times r$, B_1 is $r \times m$, and the m-input r-dimensional system (A_1, B_1) is reachable. Consequently, one can apply Heymann's Lemma to (A_1, B_1), taking F_1 and u such that $(A_1 + B_1F_1, B_1u)$ is reachable. Now, defining $F = [F_1, 0]$, one can see that $\mathcal{R}(A + BF, Bu) = \mathcal{R}(A, B)$ (this is easy) and $Pols(A + BF, Bu) = Pols(A, B)$ (for this, one needs the Pole-Shifting Theorem [3, Theorem 13], which states that the polynomials assignable to (A, B) are exactly all monic r-degree polynomials, multiplied by the fixed polynomial $\chi(A_3)$). Although not explicitely stated, is is clear from the results of [3, Chapter 5] that $Pols(A, B) = Pols(A_1, B_1) \cdot \chi(A_3)$, and also $Pols(A + BF, Bu) = Pols(A_1 + B_1F_1, B_1u) \cdot \chi(A_3)$, and in fact multiplication by $\chi(A_3)$ induces a bijection between the corresponding sets $Pols(.)$, this will be explained in detail in Lemma 5. From this and the fact that $Pols(A_1 + B_1F_1, B_1u) = Pols(A_1, B_1)$, it follows that $Pols(A + BF, Bu) = Pols(A, B)$, and so $Pols$ has been solved, by means of the same F, u which solve $Reach$. In this situation, a canonical form for (A, B) is obtained combining the Brunovsky canonical form of the reachable system (A_1, B_1), together with a canonical form for A_3 under similarity.

If R is a finite product of fields, then it is a regular ring. This is the case, for example, if $R = \mathbb{Z}/(d\mathbb{Z})$, where d is a squarefree integer, i.e., not divisible by the square of a prime. In this case, the simultaneous solution of $Reach$ and $Pols$ for systems over fields can be immediately extended to a finite product of fields, without needing any additional properties of regular rings.

The key fact when working with regular rings will be that, even if there are infinitely many maximal ideales (i.e., R has infinitely many residue fields), any system can be split into a *finite* famillly of systems, each of which *behaves like* a system over a field. Concretely, each of these systems will satisfy a Kalman-type decomposition, with the reachable part in Brunovsky canonical form.

3. Derivation of a Canonical Form

In this section, we will outline the construction made in [9, Theorem 5] to obtain a canonical form for the feedback equivalence of systems over von Neumann regular rings. Let R be a von Neumann regular ring. Many important properties of this class of rings can be found in [4, 16, 17, 18], among others: R has Krull dimension zero (all prime ideals are maximal) and no nonzero nilpotents, or any finitely generated ideal is principal and generated by an idempotent.

Examples of commutative von Neumann regular rings are fields, sub-rings of arbitrary products of fields, and typically many rings of continuous functions (e.g., function rings over P-spaces, see [19, Theorem 14.29]), and semi-simple finite rings. The importance of regular rings in systems theory is discussed, among other references, in [6, 9, 20].

Let us see how we can take profit of certain conditions of R. The first property of interest is that R is an elementary divisor ring [16, Theorem 11]: given a matrix B, there exist invertible matrices P, Q such that

$$PBQ = \left[\begin{array}{ccc|c} d_1 & & & \\ & \ddots & & 0 \\ & & d_r & \\ \hline & 0 & & 0 \end{array}\right]$$

with zero blocks of appropriate sizes and diagonal elements each one dividing the next one: $d_1|d_2|\cdots|d_r$. Moreover, since each element of a regular ring is the product of a unit with an idempotent [16, Lemma 10], after multiplying P or Q with appropriate units one can assume that all d_i's are idempotent. This means that if the system (A, B) is replaced by the feedback equivalent system (PAP^{-1}, PBQ), the second matrix of the system has the above diagonal form, and so one can assume from the beginning that -up to feedback equivalence- B itself has this form.

For example, take the ring $R = \mathbb{Z}/210\mathbb{Z}$, a regular ring because $210 = 2 \cdot 3 \cdot 5 \cdot 7$ is squarefree, and consider the divisibility conditions $2|2 \cdot 3|2 \cdot 3 \cdot 5$. The corresponding idempotents are respectively $\{d_1 = 106, d_2 = 36, d_3 = 120\}$ (for instance, $2 = 107 \cdot 106$, with 107 a unit and 106 idempotent, $2 \cdot 3 = 41 \cdot 36$, etc.). So, a typical matrix B could be:

$$B = \begin{bmatrix} 106 & 0 & 0 \\ 0 & 36 & 0 \\ 0 & 0 & 120 \\ 0 & 0 & 0 \end{bmatrix}$$

The second property which we will use is the following simple trick with idempotents. If d is an idempotent of the ring R (for this R does not need to be regular), one has a direct sum decomposition $R = dR \oplus (1-d)R$. Note that $1-d$ is also idempotent ($(1-d)^2 = 1 - 2d + d^2 = 1 - 2d + d = 1 - d$) and the idempotents $\{d, 1-d\}$ are orthogonal to each other: $d(1-d) = d - d^2 = 0$. Given any element a of R, it is easy to add to a some multiple of d and obtain an element orthogonal to d: indeed, the identity $a = da + (1-d)a$ means that $a - ad$ is a multiple of $1-d$, and thus is orthogonal to d.

Now, observe that the feedback operation which consists of replacing A by $A+BF$ can be interpreted as adding multiples of columns of B to the columns of A. If a_{ij} denotes the element of A in position (i, j), it is clear that $a_{ij} - d_i a_{ij}$ is orthogonal to d_i, and because of the divisibility conditions, this also implies orthogonality with all $d_{i'}$ for $i' > i$. Since for all i, j the replacement $a_{ij} \mapsto a_{ij} - d_i a_{ij}$ can be achieved with a suitable feedback operation of the form $A \mapsto A + BK$, the final result is that we can assume that d_i is orthogonal to the i-th row of A, and to all the preceeding rows.

At this point, we can assume that B is diagonal with diagonal elements $\{d_i\}$, all idempotent and each one dividing the following one, and that d_i is orthogonal to the rows $1, \ldots, i$ of A.

A typical system in this form could be for example:

$$A = \begin{bmatrix} 105 & 105 & 0 & 105 \\ 35 & 0 & 105 & 140 \\ 14 & 112 & 7 & 126 \\ 173 & 191 & 162 & 147 \end{bmatrix}, B = \begin{bmatrix} 106 & 0 & 0 \\ 0 & 36 & 0 \\ 0 & 0 & 120 \\ 0 & 0 & 0 \end{bmatrix}$$

The next trick is a standard argument from Commutative Algebra. Starting from a finite number of idempotent elements $d_1, \ldots, d_r$, one can derive a finite family of pairwise orthogonal idempotens and with sum 1. The construction is as follows:

$$\begin{cases} e_0 &= 1 - d_1 \\ e_1 &= d_1(1 - d_2) \\ & \vdots \\ e_{r-1} &= d_{r-1}(1 - d_r) \\ e_r &= d_r \end{cases}$$

In this case, one has $R = e_1 R \oplus \cdots \oplus e_r R$, a direct sum of rings (each $e_i R$ is commutative and with unit element e_i), and the feedback classification of (A, B) over R is reduced to the classification of each $(e_i A, e_i B)$ over $e_i R$.

Applying this procedure to $\{d_1 = 106, d_2 = 36, d_3 = 120\}$ yields the orthogonal family $\{e_0 = 105, e_1 = 70, e_2 = 126, e_3 = 120\}$, with sum 1. Accordingly, one has to consider these four systems:

$$105 \cdot A = \begin{bmatrix} 105 & 105 & 0 & 105 \\ 105 & 0 & 105 & 0 \\ 0 & 0 & 105 & 0 \\ 105 & 105 & 0 & 105 \end{bmatrix}, \quad 105 \cdot B = \begin{bmatrix} 0 & 0 & 0 \\ 0 & 0 & 0 \\ 0 & 0 & 0 \\ 0 & 0 & 0 \end{bmatrix}$$

$$70 \cdot A = \begin{bmatrix} 0 & 0 & 0 & 0 \\ 140 & 0 & 0 & 140 \\ 140 & 70 & 70 & 0 \\ 140 & 140 & 0 & 0 \end{bmatrix}, \quad 70 \cdot B = \begin{bmatrix} 70 & 0 & 0 \\ 0 & 0 & 0 \\ 0 & 0 & 0 \\ 0 & 0 & 0 \end{bmatrix}$$

$$126 \cdot A = \begin{bmatrix} 0 & 0 & 0 & 0 \\ 0 & 0 & 0 & 0 \\ 84 & 42 & 42 & 126 \\ 168 & 126 & 42 & 42 \end{bmatrix}, \quad 126 \cdot B = \begin{bmatrix} 126 & 0 & 0 \\ 0 & 126 & 0 \\ 0 & 0 & 0 \\ 0 & 0 & 0 \end{bmatrix}$$

$$120 \cdot A = \begin{bmatrix} 0 & 0 & 0 & 0 \\ 0 & 0 & 0 & 0 \\ 0 & 0 & 0 & 0 \\ 180 & 30 & 120 & 0 \end{bmatrix}, \quad 120 \cdot B = \begin{bmatrix} 120 & 0 & 0 \\ 0 & 120 & 0 \\ 0 & 0 & 120 \\ 0 & 0 & 0 \end{bmatrix}$$

Observe the identity blocks in the right matrices, and the zero blocks in the left. The first system $(e_0 A, e_0 B)$ has second matrix zero, so its feedback classification is equivalent to the classification of $e_0 A$ by similarity. All other systems have the following block structure

$$(e_i A, e_i B) = \left(\left[\begin{array}{c|c} 0 & 0 \\ \hline e_i B_i & e_i A_i \end{array} \right], \left[\begin{array}{c|c} e_i I & 0 \\ \hline 0 & 0 \end{array} \right] \right),$$

where e_iI is an identity matrix in the ring e_iR with identity e_i. In this situation, since the system (e_iA_i, e_iB_i) has size smaller than (A, B), one can repeat the procedure recursively.

Suppose we are given a system of this form:

$$A = \begin{bmatrix} 0 & 0 \\ B_1 & A_1 \end{bmatrix}, B = \begin{bmatrix} I & 0 \\ 0 & 0 \end{bmatrix}$$

If a canonical form is known for the smaller system (A_1, B_1), then it can be "copied and pasted" into (A, B). Indeed, let P_1, Q_1, K_1 be matrices which transform (A_1, B_1) into a certain canonical form $(\hat{A}_1, \hat{B}_1) = (P_1A_1P_1^{-1} + P_1B_1K_1, P_1B_1Q_1)$. Considering the matrices

$$P = \begin{bmatrix} Q_1^{-1} & -Q_1^{-1}K_1P_1 \\ 0 & P_1 \end{bmatrix}, P^{-1} = \begin{bmatrix} Q_1 & K_1 \\ 0 & P_1^{-1} \end{bmatrix}, Q = \begin{bmatrix} Q_1 & 0 \\ 0 & I \end{bmatrix},$$

after operating it can be seen that

$$PAP^{-1} = \begin{bmatrix} * & * \\ \underbrace{P_1B_1Q_1}_{\hat{B}1} & \underbrace{P_1A_1P_1^{-1} + P_1B_1K_1}_{\hat{A}_1} \end{bmatrix}, PBQ = \begin{bmatrix} I & 0 \\ 0 & 0 \end{bmatrix}.$$

A further feedback replaces the $*$'s with zeroes. Finally, this type of recursive "copypasting" of canonical forms allows to prove the following result.

Theorem 2. *Let $\Sigma = (A, B)$ be a system of size (n, m) over a regular ring R. Then, there exists a finite family of idempotents $e_1, \ldots, e_k$, pairwise orthogonal and with sum 1, such that for each i, the system (e_iA, e_iB) over the regular ring e_iR is feedback equivalent to a system with the block form:*

$$(\hat{A}_i, \hat{B}_i) = \left(\begin{bmatrix} e_iA_i & 0 \\ 0 & e_iC_i \end{bmatrix}, \begin{bmatrix} e_iB_i \\ 0 \end{bmatrix} \right),$$

where the pair (e_iA_i, e_iB_i) is reachable and in Brunovsky canonical form over e_iR.

Proof. This is proved in detail in [9, Theorem 5], following the techniques previously outlined. □

4. Simultaneous Solution to Reach and Pols

This section will be devoted to proving [11, Theorem 2], which we state here.

Theorem 3. *Let (A, B) be a system of size (n, m) over a commutative von Neumann regular R. Then, the following conditions hold:*

(P1) Given a matrix F and a vector u, the system (A, B) satisfies $Reach$ via F, u if and only if (A, B) satisfies $Pols$ via the same F, u.

(P2) There exist a matrix F and a vector u which solve problems $Reach$ and $Pols$ simultaneously for (A, B).

Proof. Let $e_1, \ldots, e_k$ be idempotents with the conditions of the last theorem. To prove the theorem, the following steps will be followed:

Step 1: prove that all reachable systems (e_iA_i, e_iB_i) satisfy (P1) and (P2).

Step 2: show that (P1) and (P2) are translated from the small size system (e_iA_i, e_iB_i) to the large size system $(\hat{A}_i, \hat{B}_i)$.

Step 3: prove that (P1) and (P2) are propagated from $(\hat{A}_i, \hat{B}_i)$ to the feedback equivalent system (e_iA, e_iB).

Step 4: prove that the solution of (P1) and (P2) in all systems (e_iA, e_iB) over e_iR can be extended to a global solution for (A, B) over R. □

We start by solving Step 1.

Lemma 4. *Let (A, B) be a feedback cyclizable system of size (n, m) over a ring R. Then, (P1) and (P2) hold for (A, B).*

Proof. This is [11, Lemma 3]. We include the proof here, in order to clarify the relation of $Reach$ and $Pols$ with the feedback cyclization and coefficient assignability properties.

First, note that if (A, B) is feedback cyclizable, by [1] it is also reachable ($\mathcal{R}(A, B) = R^n$) and coefficient assignable ($Pols(A, B)$ is equal to all monic polynomials of degree n).

Now, it is clear that solving $Reach$ for (A, B) via a matrix F and a vector u is equivalent to $(A + BF, Bu)$ being reachable, and solving $Pols$ for (A, B) via F, u is equivalent to $(A+BF, Bu)$ being coefficient assignable. Since reachability and coefficient assignability are equivalent properties for the single-input system $(A + BF, Bu)$ [1, Theorem 3.2], it follows that (P1) holds.

Finally, by the feedback cyclization property there exist a matrix F and a vector u with $(A + BF, Bu)$ reachable, hence F, u solve both $Reach$ and $Pols$ for (A, B), and (P2) holds. □

With the notations of Theorem 3, for each $i = 1, \ldots, k$ we have that e_iR is a commutative ring with unit element e_i, and since R is von Neumann regular, then so is e_iR. But regular rings are zero-dimensional and hence they are FC- rings (see [21]), therefore reachable systems are always feedback cyclizable, and hence Step 1 is solved.

Next lemma shows how a variant of the classical Kalman controllability decomposition [3, Lemma 3.3.3] allows a solution of Step 2. We had to adapt to the language of rings and modules what was done in [3, Chap. 5] for fields and vector spaces. Also, we will present our results in a more general context than it is strictly necessary to continue with the proof of Theorem 3. Concretely, we will not require the Kalman decomposition to have a reachable part.

Lemma 5. *Let (A, B) be a system of size (n, m) over a ring R given by:*

$$A = \begin{bmatrix} A_1 & A_2 \\ 0 & A_3 \end{bmatrix}, \quad B = \begin{bmatrix} B_1 \\ 0 \end{bmatrix}.$$

with (A_1, B_1) (not necessarily reachable) of size (r, m) ($1 \leq r \leq n$). If (A_1, B_1) satisfies (P1) and (P2), then the same is true for (A, B).

Proof. This is proved in [11, Lemma 4]. First, observe that the limit case $r = n$ is trivial because $A = A_1$ and $B = B_1$.

Assume $r \leq n-1$. For any $m \times n$ matrix $F = [F_1 \ *]$, where F_1 is $m \times r$, one has

$$A + BF = \begin{bmatrix} A_1 + B_1F_1 & * \\ 0 & A_3 \end{bmatrix}, \quad \chi(A+BF) = \underbrace{\chi(A_1 + B_1F_1) \cdot \chi(A_3)}_{=\varphi(\chi(A_1+B_1F_1))}, \tag{2}$$

where $\varphi : R[x] \to R[x]$ is the map given by $\varphi(f(x)) = \chi(A_3)f(x)$. By making F and F_1 vary and noting that φ is injective (because $\chi(A_3)$ is monic), we see that

$$\varphi \text{ gives a bijection between } Pols(A_1, B_1) \text{ and } Pols(A, B) \tag{3}$$

This is actually a generalization of the Pole-Shifting Theorem [3, Theorem 13]. Now, for $F = [F_1 \ *]$ as before and $u = u_1 \in R^m$, by a similar reasoning one has that

$$\varphi \text{ gives a bijection between } Pols(A_1 + B_1F_1, B_1u_1) \text{ and } Pols(A + BF, Bu) \tag{4}$$

Combining (3) and (4), it is clear that

$$F, u \text{ solve } Pols \text{ for } (A, B) \text{ if and only if } F_1, u_1 \text{ solve } Pols \text{ for } (A_1, B_1) \tag{5}$$

On the other hand, because of the block form of (A, B), one has:

$$im(A^*B) = im\left(\begin{bmatrix} B_1 & A_1B_1 & \cdots & A_1^{n-1}B_1 \\ 0 & 0 & \cdots & 0 \end{bmatrix}\right) = im\left(\begin{bmatrix} A_1{}^*B_1 \\ 0 \end{bmatrix}\right), \tag{6}$$

where the last equality holds by Cayley-Hamilton. Since $(A+BF, Bu)$ has the same block structure as (A, B), it follows that

$$im((A+BF)^*(Bu)) = im\left(\begin{bmatrix} (A_1 + B_1F_1)^*(B_1u_1) \\ 0 \end{bmatrix}\right), \tag{7}$$

and as an immediate consequence (recall that $\mathcal{R}(A, B) = im(A^*B)$) we have that

$$F, u \text{ solve } Reach \text{ for } (A, B) \text{ if and only if } F_1, u_1 \text{ solve } Reach \text{ for } (A_1, B_1) \tag{8}$$

From (5) and (8) we see that (P2) and (P3) are extended from (A_1, B_1) to (A, B). □

Note that although the limit case $r = 0$ is not covered by the previous lemma, this situation corresponds to $B = 0$, and the properties (P1), (P2) hold trivially for (A, B): indeed, $Pols(A, B) - \chi(A)$ and $\mathcal{R}(A, B)$ are both zero, and any pair F, u solves $Reach$ and $Pols$. In any case, Step 2 is proved and we can proceed with Step 3.

As usual in most systems theory problems, the feedback equivalence of systems preserves the properties of interest.

Lemma 6. *Let (A, B) and (A', B') be feedback equivalent systems over a commutative ring R (i.e., there exist invertible matrices P, Q and a feedback matrix K such that $A' = PAP^{-1} + PBK$ and $B' = PBQ$). If (P1) and (P2) hold for (A, B), then the same is true for (A', B').*

Proof. See [11, Lemma 5]. □

Now, Step 3 is complete, which moves as right along to Step 4.

Consider the following situation: let R be a ring and $e_1, \ldots, e_k$ pairwise orthogonal idempotents with sum 1. Then, R can be identified with $\bigoplus_{i=1}^k e_i R$, where each $e_i R$ can be viewed as an ideal within R or as a commutative ring with unit element e_i. If in addition R is regular, then all $e_i R$ are also regular. The following lemma collects the properties of such a decomposition which we will need to complete the proof of Theorem 3.

Lemma 7. *(Idempotent decomposition) Let $R = \bigoplus_{i=1}^k e_i R$ be a regular ring decomposed as above. Then, for any system (A, B) over R one has:*

(i) $\mathcal{R}(A, B) = \bigoplus_{i=1}^k \mathcal{R}(e_i A, e_i B)$.

(ii) $Pols(A, B) = \bigoplus_{i=1}^k Pols(e_i A, e_i B)$.

(iii) *(A, B) satisfies Reach (resp. $Pols$) over R via F, u if and only if $(e_i A, e_i B)$ satisfies Reach (resp. $Pols$) over $e_i R$ via $e_i F, e_i u$ for each i.*

(iv) *If for all i $(e_i A, e_i B)$ satisfies (P1) and (P2), then the same holds for (A, B).*

Proof. See [11, Lemma 6]. The proof is reproduced here, in order to show the importance of working with idempotents. (i) and (iii) ($Reach$) are immediate, because they involve equalities of images of matrices, which can be ckecked in each component.

In order to prove (ii) and (iii) ($Pols$), note that: [1]

$$\underbrace{\chi(A)}_{\text{over } R} = |xI - A| = \sum_{i=1}^{k} e_i |xI - A| = \sum_{i=1}^{k} |xe_i I - e_i A| = \sum_{i=1}^{k} \underbrace{\chi(e_i A)}_{\text{over } e_i R}, \tag{9}$$

because $|eA| = e|A|$ for any square matrix A and any idempotent scalar e, and characteristic polynomials over $e_i R$ are computed by using the identity matrix over such a ring, i.e., the matrix $e_i I$. From the identification of Eq. (9), it follows easily that (ii) and (iii) ($Pols$) hold.

(iv) The propagation of properties (P1) and (P2) is an immediate consequence of (iii). □

This finishes Step 4, hence the proof of Theorem 3 is complete.

The following simple examples shows how problems $Reach$ and $Pols$ are solved simultaneously.

Example 8. Consider the regular ring $R = \mathbb{Z}/6\mathbb{Z}$ and the matrices:

$$A = \begin{bmatrix} 0 & 0 \\ 0 & 1 \end{bmatrix}, B = \begin{bmatrix} 1 & 0 \\ 0 & 3 \end{bmatrix}, F = \begin{bmatrix} 0 & 0 \\ 0 & 0 \end{bmatrix}, u = \begin{bmatrix} 1 \\ 1 \end{bmatrix}.$$

Then, one has:

$$\begin{array}{rcccl} \mathcal{R}(A + BF, Bu) & = & \mathcal{R}(A, B) & = & R \oplus 3R \\ Pols(A + BF, Bu) & = & Pols(A, B) & = & x^2 + 5 + k_1 \cdot (x + 2) + k_2 \cdot 3, \end{array}$$

for arbitrary k_1, k_2 in R.

[1] Here, the determinant of a sum of matrices happens to be equal to the sum of determinants, but please, kids, don't do this at home, this is an exceptional consequence of the behaviour of idempotents.

5. Other Techniques and Future Work

A technique commonly used in systems theory is to consider properties which are preserved by only a part of the feedback transformations, in such a way that recursive proofs can be done, precisely because of the restricted feedback operations. This will be illustrated with the solution given in [6] to the problem *Reach*.

Let (A, B) be a system, with B partitioned as $B = [b_1|B_2]$, being b_1 the first column of B, and consider the following problem:

$$\text{Find } X, y \text{ such that } \mathcal{R}(A + B_2X, b_1 + B_2y) = \mathcal{R}(A, B) \tag{10}$$

This condition trivially implies *Reach*, by putting $F = \begin{bmatrix} 0 \\ X \end{bmatrix}, u = \begin{bmatrix} 1 \\ y \end{bmatrix}$.

It is proved in [6, Lemma 2] that this strong form of *Reach* is preserved by a special subset of feedback transformations. Concretely, if $(A', B') = (PAP^{-1} + PBK, PBQ)$, with Q of the form $\begin{bmatrix} q_1 & 0 \\ q_2 & Q_3 \end{bmatrix}$, then (A, B) satisfies the above condition if and only if (A', B') does.

Then, in [6, Theorem 4] it is proved that -up to feedback transformations with Q of the special form indicated before- any system (A, B) of size (n, m) can be taken of the form

$$A = \begin{bmatrix} a_{11} & a_{12} \\ a_{21} & A_{22} \end{bmatrix}, B = \left[\begin{array}{c|c} d_1 & b_{12} \\ 0 & B_{22} \end{array} \right] = [b_1|B_2],$$

with $A_{22} \in R^{(n-1)\times(n-1)}$, $B_{22} \in R^{(n-1)\times(m-1)}$, and the remaining blocks of appropriate sizes. For systems with this block structure, there is a relation between the reachable module of (A, B) and that of the system $(A_{22}, [a_{21}|B_{22}])$ of size $(n-1, m)$, see [6, Lemma 3]:

$$\mathcal{R}(A, B) = dR \oplus d\mathcal{R}(A_{22}, [a_{21}|B_{22}]). \tag{11}$$

Applying the induction hypothesis to the system $(A_{22}, [a_{21}|B_{22}])$, there exist matrices X_1, y_1 such that $\mathcal{R}(A_{22} + B_{22}X_1, a_{21} + B_{22}y_1) = \mathcal{R}(A_{22}, [a_{21}|B_{22}])$. Afterwards, if we define $X = [y_1\ X_1]$, after operating one has:

$$A + B_2X = \begin{bmatrix} * & * \\ a_{21} + B_{22}y_1 & A_{22} + B_{22}X_1 \end{bmatrix}, b_1 = \begin{bmatrix} d_1 \\ 0 \end{bmatrix}.$$

Finally, applying the decomposition (11) it follows that (A, B) satisfies the condition (10) by means of $X = [y_1\ X_1]$ and $y = 0$. Note that if Q had not contained this zero block, this type of inductive proof would not have been possible.

Another interesting technique, specially suited to extend results from fields to von Neumann regular rings, is the one used in [20, Theorem 4.1], where it is proved that two systems are equivalent over a regular ring R if and only if they are equivalent over the residue field $R/\mathfrak{m}$, for all maximal ideals $\mathfrak{m}$ of R. As an illustration of this, we will reprove the fact that von Neumann regular rings satisfy *Reach*.

Given a system (A, B) over a regular ring R, the existence of a matrix F and a vector u satisfying $\mathcal{R}(A + BF, Bu) = \mathcal{R}(A, B)$ is equivalent to saying that the reachability matrices A^*B and $(A+BF)^*(Bu)$ have the same image, which means the existence of matrices X, Y of appropriate sizes such that $((A+BF)^*Bu)\cdot X = A^*B$ and $(A^*B)\cdot Y = (A+BF)^*(Bu)$, which we write as

$$\begin{cases} ((A+BF)^*Bu)\cdot X - A^*B &= 0 \\ (A^*B)\cdot Y - (A+BF)^*(Bu) &= 0 \end{cases} \tag{12}$$

Note that this is a collection of $2mn^2$ equalities, one for each entry of two $n \times nm$ matrix equalities. We will (re)prove that there exist matrices F, u, X, Y satisfying (12).

For each maximal ideal $\mathfrak{m}$ of R, the system $(A(\mathfrak{m}), B(\mathfrak{m}))$ over the field $R/\mathfrak{m}$ satisfies $Reach$, i.e., there exist matrices $F(\mathfrak{m}), u(\mathfrak{m})$ with coefficients in $R/\mathfrak{m}$ such that

$$\mathcal{R}(A(\mathfrak{m}) + B(\mathfrak{m})F(\mathfrak{m}), B(\mathfrak{m})u(\mathfrak{m})) = \mathcal{R}(A(\mathfrak{m}), B(\mathfrak{m}))$$

Since projection modulo $\mathfrak{m}$ is surjective, one can take matrices $F^{\mathfrak{m}}, u^{\mathfrak{m}}, X^{\mathfrak{m}}, Y^{\mathfrak{m}}$ over R which reduced modulo $\mathfrak{m}$ are precisely $F(\mathfrak{m}), u(\mathfrak{m}), X(\mathfrak{m}), Y(\mathfrak{m})$. Reasoning like in (12), the following congruences modulo $\mathfrak{m}$ hold:

$$\begin{cases} ((A+BF^{\mathfrak{m}})^*Bu^{\mathfrak{m}})\cdot X^{\mathfrak{m}} - A^*B &\equiv 0 \pmod{\mathfrak{m}} \\ (A^*B)\cdot Y^{\mathfrak{m}} - (A+BF^{\mathfrak{m}})^*(Bu^{\mathfrak{m}}) &\equiv 0 \pmod{\mathfrak{m}} \end{cases} \tag{13}$$

At this point, we recall that every localization of a von Neumann regular ring at a prime (i.e., maximal) ideal is a field (see e.g., [17] or [18]). In fact, there is a canonical ismorphism between the local ring $R_{\mathfrak{m}}$ and the residue field $R/\mathfrak{m}$, the correspondence being given by $\frac{a}{1} \mapsto a + \mathfrak{m}$. This means that an element a belongs to $\mathfrak{m}$ (i.e., it is zero in $R/\mathfrak{m}$) if and only if it is zero in the ring of fractions $R_{\mathfrak{m}}$, i.e., there exists an element $t \notin \mathfrak{m}$ such that $ta = 0$, a condition in general much stronger than just saying "a lies in $\mathfrak{m}$".

Applying this fact to (13), there exists a finite family of elements $s_i \notin \mathfrak{m}$ (one for each of the $2mn^2$ congruences) which transform the congruences "something belongs to $\mathfrak{m}$" into equalities "something equals zero". Now, if we define $s^{\mathfrak{m}} = \prod_{i=1}^{2mn^2} s_i$, it follows that $s^{\mathfrak{m}} \notin \mathfrak{m}$ and

$$\begin{cases} s^{\mathfrak{m}}\left(((A+BF^{\mathfrak{m}})^*Bu^{\mathfrak{m}})\cdot X^{\mathfrak{m}} - A^*B\right) &= 0 \\ s^{\mathfrak{m}}\left((A^*B)\cdot Y^{\mathfrak{m}} - (A+BF^{\mathfrak{m}})^*(Bu^{\mathfrak{m}})\right) &= 0 \end{cases} \tag{14}$$

Moreover, knowing that any element of a regular ring is the product of a unit with an idempotent [16, Lemma 10], $s^{\mathfrak{m}}$ can be assumed to be idempotent.

The condition $s^{\mathfrak{m}} \notin \mathfrak{m}$ ensures that none of the elements $s^{\mathfrak{m}}$ belong to all maximal ideals of R, which means that the (possibly infinite) collection $\{s^{\mathfrak{m}}\}$ generates an ideal which is the whole ring R. Choosing a finite linear combination which sums up 1:

$$\alpha_1 s^{\mathfrak{m}_1} + \cdots + \alpha_r s^{\mathfrak{m}_r} = 1,$$

one can say that $s^{\mathfrak{m}_1}, \ldots, s^{\mathfrak{m}_r}$ span R. This is the crucial step: we have extracted a finite subset of something possibly infinite.

Now, starting from $s^{\mathfrak{m}_1}, \ldots, s^{\mathfrak{m}_r}$, the following new family is constructed:

$$\begin{cases} e_1 = s^{\mathfrak{m}_1} \\ e_2 = (1 - s^{\mathfrak{m}_1})s^{\mathfrak{m}_2} \\ \vdots \\ e_r = (1 - s^{\mathfrak{m}_1})\cdots(1 - s^{\mathfrak{m}_{r-1}})s^{\mathfrak{m}_r} \end{cases}$$

It is clear that each e_t is idempotent (it is a product of idempotents), and that the e_t's are pairwise orthogonal: indeed, if $t < l$, then e_t is orthogonal to e_l because e_t has a factor $s^{\mathfrak{m}_t}$ and e_l has a factor $1 - s^{\mathfrak{m}_t}$, and $s^{\mathfrak{m}_t}$ is orthogonal to $1 - s^{\mathfrak{m}_t}$.

In addition, the elements e_t generate R, an immediate consequence of the fact that a product ab belongs to a prime ideal if and only if one of a, b belongs to it. Indeed, if $\mathfrak{m}$ were a maximal ideal containing all the e_t's, then in the expression of e_1 we see that $\mathfrak{m}$ contains $s^{\mathfrak{m}_1}$, hence it does not contain $1 - s^{\mathfrak{m}_1}$, therefore in the expression of e_2 we deduce that $\mathfrak{m}$ contains $s^{\mathfrak{m}_2}$, and so on. Continuing in this manner, one arrives at an equation saying that $\mathfrak{m}$ contains a product of r factors without containing any of the first $r-1$ factors, so it must contain the last factor, which is $s^{\mathfrak{m}_r}$, hence $\mathfrak{m}$ contains $s^{\mathfrak{m}_1}, \ldots, s^{\mathfrak{m}_r}$, contradicting the fact that these elements generate R.

Now, define the following matrices:

$$F = \sum_{t=1}^{r} e_t F^{\mathfrak{m}_t},\ u = \sum_{t=1}^{r} e_t u^{\mathfrak{m}_t},\ X = \sum_{t=1}^{r} e_t X^{\mathfrak{m}_t},\ Y = \sum_{t=1}^{r} e_t Y^{\mathfrak{m}_t},$$

which are the candidates to be a solution of (12). Once we know that $e_1, \ldots, e_r$ generate R, it is clear that a matrix with coefficients in R is zero if and only if it is annihilated by all of the e_t's. Therefore, in order to prove (12) we have to check that for all t one has:

$$\begin{cases} e_t\left(((A+BF)^*Bu)\cdot X - A^*B\right) & = & 0 \\ e_t\left((A^*B)\cdot Y - (A+BF)^*(Bu)\right) & = & 0 \end{cases}$$

Note that, in virtue of the orthogonality relations among the e_t's, one has that $e_t F = e_t F^{\mathfrak{m}_t}$, $e_t u = e_t u^{\mathfrak{m}_t}$, $e_t X = e_t X^{\mathfrak{m}_t}$, $e_t Y = e_t Y^{\mathfrak{m}_t}$. Using this and the fact that any power of e_t is equal to e_t, it follows that the last equation is equivalent to:

$$\begin{cases} e_t\left(((A+BF^{\mathfrak{m}_t})^*Bu^{\mathfrak{m}_t})\cdot X^{\mathfrak{m}_t} - A^*B\right) & = & 0 \\ e_t\left((A^*B)\cdot Y^{\mathfrak{m}_t} - (A+BF^{\mathfrak{m}_t})^*(Bu^{\mathfrak{m}_t})\right) & = & 0 \end{cases}$$

But e_t is a multiple of $s^{\mathfrak{m}_t}$, hence the above equality is true, because of Eq. (14), in the case $\mathfrak{m} = \mathfrak{m}_t$, and so F, u, X, Y give a solution to (12). This completes the proof that von Neumann regular rings satisfy *Reach*. With an analogous procedure, it can also be (re)proved that *Pols* is also solved, with the same F, u used to solve *Reach*.

6. Conclusion

In this chapter we have studied a very interesting application of Linear Algebra over commutative rings to systems theory, when the underlying coefficient ring is von Neumann regular. While Linear Algebra over rings is generally very complex, in the case of regular rings we have seen that certain idempotent decompositions make life easier, and allow us to extend various results from systems theory over fields, to systems over von Neumann regular rings. Concretely, we have presented a canonical form for the feedback equivalence of systems, and a solution to the problems of coefficient assignability and feedback cyclization, in a very strong form which is valid for arbitrary non reachable systems.

References

[1] J.W. Brewer, J.W. Bunce, F.S.Van Vleck, Linear systems over commutative rings, *Lecture Notes in Pure and Applied Mathematics* 104, Marcel Dekker Inc., 1986.

[2] R. Bumby, E.D. Sontag, H.J. Sussmann, W.V. Vasconcelos, Remarks on the pole-shifting problem over rings. *J. Pure Appl. Algebra* 20 (1981) 113–127.

[3] E.D. Sontag, *Mathematical Control Theory: Deterministic Finite Dimensional Systems*; Springer, 1990.

[4] B.R. McDonald, Linear Algebra over Commutative Rings, *Lecture Notes in Pure and Applied Mathematics* 87, Marcel Dekker Inc., 1984.

[5] J.A. Hermida-Alonso, M.P. Pérez, T. Sánchez-Giralda, Brunovsky's canonical form for linear dynamical systems over commutative rings. *Linear Algebra Appl.* 233 (1996) 131–147.

[6] A. Sáez-Schwedt, Cyclic accesibility of reachable states characterizes von Neumann regular rings, *Linear Algebra Appl.* 2010, 433, 1187–1193.

[7] J.W. Brewer, W. Schmale, (A, B)-cyclic submodules. *Linear Algebra Appl.* 301 (1999) 63–80.

[8] A. Sáez-Schwedt, T. Sánchez-Giralda, Strong feedback cyclization for systems over rings. *Systems & Control Lett.* 57 (2008) 1–7.

[9] A. Sáez-Schwedt, W. Schmale, Feedback classification of linear systems over von Neumann regular rings, *Linear Algebra Appl.* (438) 2013 1852–1862.

[10] M. Carriegos, A. DeFrancisco-Iribarren, R. Santamaría-Sánchez, The effective calculation of all assignable polynomials to a single-input system over a Bézout domain, *IEEE Trans. Automat. Control* 55 (2010) 222–225.

[11] A. Sáez-Schwedt, Assignable polynomials to linear systems over von Neumann regular rings, *Linear Algebra Appl.* (2014) in press, http://dx.doi.org/10.1016/j.laa.2014.04.29

[12] M. Carriegos, J.A. Hermida-Alonso, T. Sánchez-Giralda, Pole-shifting for linear systems over commutative rings, *Linear Algebra Appl.* 346 (2002) 97–107.

[13] A. Sáez-Schwedt, T. Sánchez-Giralda, Coefficient assignability and a block decomposition for systems over rings, *Linear Algebra Appl.* 429 (2008) 1277–1287.

[14] M. Heymann, Comments "On pole assignment in muti-input controllable linear systems", *IEEE Trans. Automat. Control* 13 (1968) 748–749.

[15] P.A. Brunovsky, A classification of linear controllable systems, *Kybernetika* 3 (1970) 173–187.

[16] L. Gillman, M. Henriksen, Some remarks about elementary divisor rings, *Trans. Amer. Math. Soc.* 82 (1956) 362–365.

[17] R. Gilmer, Background and preliminaries on zero-dimensional rings. In: D.F. Anderson, D.E. Dobbs (Eds.) *Zero-dimensional Commutative Rings, Lecture Notes in Pure and Applied Mathematics* 171, pp. 1–25. Marcel Dekker Inc., 1995.

[18] K. Goodearl, *Von Neumann Regular Rings*, Pitman, 1979.

[19] L. Gillman, M. Jerison, Rings of continuous functions, *Grad. Texts in Math.* 43; Springer, 1976.

[20] M. Carriegos, J.A. Hermida-Alonso, T. Sánchez-Giralda, The pointwise feedback relation for linear dynamical systems, *Linear Algebra Appl.* 279 (1998) 119-134.

[21] J.W. Brewer, D. Katz, W. Ullery, Pole assignability in polynomial rings, power series rings and Prüfer domains, *J. Algebra* 106 (1987) 265–286.

In: Advances in Linear Algebra Research
Editor: Ivan Kyrchei, pp. 151-183
ISBN: 978-1-63463-565-3

Chapter 5

How to Characterize Properties of General Hermitian Quadratic Matrix-Valued Functions by Rank and Inertia

Yongge Tian*
China Economics and Management Academy,
Central University of Finance and Economics, Beijing, China

Abstract

This chapter gives a comprehensive investigation to behaviors of a general Hermitian quadratic matrix-valued function

$$\phi(X) = (\,AXB + C\,)M(\,AXB + C)^{*} + D$$

by using ranks and inertias of matrices. The author first establishes a group of analytical formulas for calculating the global maximal and minimal ranks and inertias of $\phi(X)$. Based on the formulas, the author derives necessary and sufficient conditions for $\phi(X)$ to be a positive definite, positive semi-definite, negative definite, negative semi-definite function, respectively, and then solves two optimization problems of finding two matrices $\widehat{X}$ or $\widetilde{X}$ such that $\phi(X) \succcurlyeq \phi(\widehat{X})$ and $\phi(X) \preccurlyeq \phi(\widetilde{X})$ hold for all X, respectively. As extensions, the author considers definiteness and optimization problems in the Löwner sense of the following two types of multiple Hermitian quadratic matrix-valued function

$$\phi(\,X_1, \ldots, X_k\,) = \left(\sum_{i=1}^{k} A_iX_iB_i + C \right) M \left(\sum_{i=1}^{k} A_iX_iB_i + C \right)^{*} + D,$$

$$\psi(\,X_1, \ldots, X_k\,) = \sum_{i=1}^{k} (\,A_iX_iB_i + C_i\,)M_i(\,A_iX_iB_i + C_i\,)^{*} + D.$$

Some open problems on algebraic properties of these matrix-valued functions are mentioned at the end of the chapter.

*E-mail address: yongge.tian@gmail.com

Keywords: Matrix-valued function; matrix equation; matrix inequality; Löwner partial ordering; rank; inertia

AMS Subject Classification: 15A24; 15B57; 47A62; 65K10

1. Introduction

A matrix-valued function is a map between two matrix spaces $\mathbb{C}^{m\times n}$ and $\mathbb{C}^{p\times q}$, which can generally be written as

$$Y = f(X) \text{ for } Y \in \mathbb{C}^{m\times n} \text{ and } X \in \mathbb{C}^{p\times q}, \tag{1}$$

or briefly, $f : \mathbb{C}^{p\times q} \to \mathbb{C}^{m\times n}$, where $\mathbb{C}^{m\times n}$ denotes the set of all $m \times n$ complex matrices. If $f^*(X) = f(X)$ for all X in (1), then it is said to be Hermitian. Mappings between matrix spaces can be constructed arbitrarily from ordinary operations of given matrices and variable matrices, but linear and quadratic matrix-valued functions are widely used and are extensively studied from theoretical and applied points of view. It is extremely difficult at any case to give a comprehensive investigation and obtain analytical results to a given matrix-valued function because of non-commutativity of matrix algebra.

One of the fundamental quadratic cases of (1) is the following Hermitian quadratic matrix-valued function (HQMVF):

$$\phi(X) = (\,AXB + C\,)M(\,AXB + C)^* + D, \tag{2}$$

where $A \in \mathbb{C}^{n\times p}$, $B \in \mathbb{C}^{m\times q}$, $C \in \mathbb{C}^{n\times q}$, $D \in \mathbb{C}_{\mathrm{H}}^{n}$ and $M \in \mathbb{C}_{\mathrm{H}}^{q}$ are given, and $X \in \mathbb{C}^{p\times m}$ is a variable matrix. We may treat (2) as a combination $\phi = \tau\circ\rho$ of the following two simple linear and quadratic Hermitian matrix-valued functions:

$$\rho : X \to AXB + C, \quad \tau : Y \to YMY^* + D. \tag{3}$$

Quadratic objects are cornerstones of mathematics and are prominent subjects of study in many fields of mathematics. As a natural extension of ordinary quadratic functions, quadratic matrix-valued functions (including quadratic forms as their special cases) occur widely in both theoretical and applied fields of sciences. Many contributions on matrix-valued functions and their behaviors can be found in the literature; see, e.g., [1–12]. As an extension, we also consider the following two types of general quadratic matrix-valued function

$$\phi(\,X_1,\ldots,X_k\,) = \left(\sum_{i=1}^{k} A_iX_iB_i + C\,\right)M\left(\sum_{i=1}^{k} A_iX_iB_i + C\,\right)^* + D, \tag{4}$$

$$\psi(\,X_1,\ldots,X_k\,) = \sum_{i=1}^{k} (A_iX_iB_i + C_i)M_i(A_iX_iB_i + C_i)^* + D, \tag{5}$$

where A_i, B_i, C_i, C, D, M_i and M are given matrices with M_i, M and D Hermitian, X_i is a variable matrix, $i = 1,\ldots,k$.

The rank and inertia of a Hermitian matrix are two generic concepts in matrix theory for describing the dimension of the row or column vector space and the sign distribution of the real eigenvalues of the matrix, which are well understood and are easy to compute by the well-known elementary or Hermitian congruent matrix operations. These two quantities play an essential role in characterizing algebraic properties of matrices and relations between two matrices. These integer indices occur only in finite-dimensional algebras, which are not replaceable and cannot be approximated by other continuous quantities. Because the rank and inertia of a matrix can only take finite integers between 0 and the dimensions of the matrix, it is not hard to give upper and lower bounds for ranks and inertias of matrices, and further the global maximal and minimal values of rank and inertia of a Hermitian matrix-valued function do exist no matter what decision domains of variable matrices are given. In this case, it is really expected to establish certain analytical formulas for calculating the maximal and minimal ranks and inertias of a given matrix-valued function from theoretical and applied points of view. However, due to the integer property of rank and inertia of matrix, inexact or approximate values of maximal and minimal ranks and inertias of matrices are less valuable. In other words, we cannot use approximation methods to find the maximal and minimal possible ranks and inertias of matrix-valued function. This fact means that methods of solving matrix rank and inertia optimization problems are not consistent with any of the ordinary continuous and discrete problems in optimization theory.

Precisely, matrix rank and inertia optimization problems are a class of discontinuous optimization problems, in which decision variables are matrices running over certain matrix sets, while the ranks and inertias of the variable matrices are some integer-valued objective functions. It has been known that matrix rank and inertia optimization problems are NP-hard in general due to discontinuity and combinational nature of ranks and inertias of matrices and algebraic structures of the given matrix sets. However, it is really lucky that we can establish analytical formulas for calculating extremal ranks and inertias of matrix-valued functions for some special feasible matrix sets by using various expansion formulas for ranks and inertias of matrices and some tricky matrix operations. The present author has been working on this topic with great effort in the past decades by proving a huge amount of new formulas for calculating ranks and inertias of matrices, and giving analytical solutions to many matrix rank and inertia optimization problems.

The main purpose of this chapter is to establish a unified theory on optimizations of ranks and inertias of HQMVFs, as well as partial orderings of HQMVFs by using pure algebraic operations of matrices. We also present many consequences on behaviors of HQMVFs, and show their applications in solving Hermitian quadratic matrix equations and establishing Hermitian quadratic matrix inequalities.

Before proceeding, we introduce the notation to the reader and explain its usage in this chapter.

$\mathbb{C}^{m\times n}$ denotes the set of all $m \times n$ complex matrices;

$\mathbb{C}_{\mathrm{H}}^{m}$ denotes the set of all $m \times m$ Hermitian complex matrices;

A^*, $r(A)$ and $\mathscr{R}(A)$ stand for the conjugate transpose, rank and range (column space) of a matrix $A \in \mathbb{C}^{m\times n}$, respectively;

I_m denotes the identity matrix of order m;

$[\,A,\; B\,]$ denotes a row block matrix consisting of A and B;

the Moore–Penrose inverse of $A \in \mathbb{C}^{m\times n}$, denoted by $A^{\dagger}$, is defined to be the unique solution $X \in \mathbb{C}^{n\times m}$ satisfying the four matrix equations $AXA = A$, $XAX = X$,

$(AX)^* = AX$ and $(XA)^* = XA$;

the symbols E_A and F_A stand for $E_A = I_m - AA^\dagger$ and $F_A = I_n - A^\dagger A$, which satisfy $E_{A^*} = F_A$ and $F_{A^*} = E_A$;

an $X \in \mathbb{C}^{n\times m}$ is called a g-inverse of $A \in \mathbb{C}^{m\times n}$, denoted by A^-, if it satisfies $AXA = A$;

an $X \in \mathbb{C}_{\mathrm{H}}^{m}$ is called a Hermitian g-inverse of $A \in \mathbb{C}_{\mathrm{H}}^{m}$, denoted by $A^\sim$, if it satisfies $AXA = A$; called a reflexive Hermitian g-inverse of $A \in \mathbb{C}_{\mathrm{H}}^{m}$, denoted by $A_r^\sim$, if it satisfies $AXA = A$ and $XAX = X$;

$i_+(A)$ and $i_-(A)$, called the partial inertia of $A \in \mathbb{C}_{\mathrm{H}}^{m}$, are defined to be the number of the positive and negative eigenvalues of A counted with multiplicities, respectively, which satisfy $r(A) = i_+(A) + i_-(A)$;
$A \succ 0$, $A \succcurlyeq 0$, $A \prec 0$ and $A \preccurlyeq 0$ mean that A is a Hermitian positive definite, positive semi-definite, negative definite, negative semi-definite matrix, respectively;

two $A, B \in \mathbb{C}_{\mathrm{H}}^{m}$ are said to satisfy the inequalities $A \succ B$, $A \succcurlyeq B$, $A \prec B$ and $A \preccurlyeq B$ in the Löwner partial ordering if $A - B$ is positive definite, positive semi-definite, negative definite and negative semi-definite, respectively.

In what follows, we take the rank and inertia of $\phi(X)$ in (2) as integer-valued objective functions, and solve the following optimization problem.

Problem 1. For $\phi(X)$ in (2), establish analytical formulas for calculating the following global extremal ranks and inertias

$$\max_{X\in\mathbb{C}^{p\times m}} r[\phi(X)], \quad \min_{X\in\mathbb{C}^{p\times m}} r[\phi(X)], \quad \max_{X\in\mathbb{C}^{p\times m}} i_\pm[\phi(X)], \quad \min_{X\in\mathbb{C}^{p\times m}} i_\pm[\phi(X)]. \tag{6}$$

Based on the analytical formulas, the author then solves the following two problems on equalities and inequalities of $\phi(X)$.

Problem 2. For $\phi(X)$ in (2),

(i) establish necessary and sufficient conditions for the existence of an $X \in \mathbb{C}^{p\times m}$ such that

$$\phi(X) = 0, \text{ i.e., } (\,AXB + C\,)M(\,AXB + C)^* + D = 0; \tag{7}$$

(ii) establish necessary and sufficient conditions for the following inequalities

$$\phi(X) \succ 0, \ \ \phi(X) \succcurlyeq 0, \ \ \phi(X) \prec 0, \ \ \phi(X) \preccurlyeq 0 \tag{8}$$

to hold for an $X \in \mathbb{C}^{p\times m}$, respectively;

(iii) establish necessary and sufficient conditions for

$$\phi(X) \succ 0, \ \ \phi(X) \succcurlyeq 0, \ \ \phi(X) \prec 0, \ \ \phi(X) \preccurlyeq 0 \ \text{ for all } X \in \mathbb{C}^{p\times m} \tag{9}$$

to hold, respectively, namely, to give identifying conditions for $\phi(X)$ to be positive definite, positive semi-definite, negative definite, negative semi-definite function, respectively.

Problem 3. For $\phi(X)$ in (2), establish necessary and sufficient conditions for the existence of $\widehat{X}$, $\widetilde{X} \in \mathbb{C}^{p\times m}$ such that

$$\phi(X) \succcurlyeq \phi(\widehat{X}) \text{ and } \phi(X) \preccurlyeq \phi(\widetilde{X}) \tag{10}$$

hold for all $X \in \mathbb{C}^{p\times m}$, respectively, and derive analytical expressions of the two matrices $\widehat{X}$ and $\widetilde{X}$.

It should be pointed out that the well-known Lagrangian method in classic analysis is not available for solving (10) because the optimal criteria in (10) are defined from the Löwner partial ordering instead of traces or norms of matrices. In this case, we use rank and inertia formulas of matrices and regular operations of the given matrices and their generalized inverses to establish a standard algebraic method for deriving analytical solutions to these two optimization problems.

2. Preliminaries

Lemma 4. *Let $\mathcal{S}$ be a matrix subset in $\mathbb{C}^{m\times n}$, and $\mathcal{H}$ be a matrix subset in $\mathbb{C}_{\mathrm{H}}^{m}$. Then the following hold.*

(a) *Under $m = n$, $\mathcal{S}$ has a nonsingular matrix if and only if $\max_{X\in\mathcal{S}} r(X) = m$.*

(b) *Under $m = n$, all $X \in \mathcal{S}$ are nonsingular if and only if $\min_{X\in\mathcal{S}} r(X) = m$.*

(c) *$0 \in \mathcal{S}$ if and only if $\min_{X\in\mathcal{S}} r(X) = 0$.*

(d) *$\mathcal{S} = \{0\}$ if and only if $\max_{X\in\mathcal{S}} r(X) = 0$.*

(e) *$\mathcal{H}$ has a matrix $X \succ 0$ ($X \prec 0$) if and only if*

$$\max_{X\in\mathcal{H}} i_{+}(X) = m \quad \left(\max_{X\in\mathcal{H}} i_{-}(X) = m \right).$$

(f) *All $X \in \mathcal{H}$ satisfy $X \succ 0$ ($X \prec 0$), namely, $\mathcal{H}$ is a subset of the cone of positive definite matrices (negative definite matrices), if and only if*

$$\min_{X\in\mathcal{H}} i_{+}(X) = m \quad \left(\min_{X\in\mathcal{H}} i_{-}(X) = m \right).$$

(g) *$\mathcal{H}$ has a matrix $X \succcurlyeq 0$ ($X \preccurlyeq 0$) if and only if*

$$\min_{X\in\mathcal{H}} i_{-}(X) = 0 \quad \left(\min_{X\in\mathcal{H}} i_{+}(X) = 0 \right).$$

(h) *All $X \in \mathcal{H}$ satisfy $X \succcurlyeq 0$ ($X \preccurlyeq 0$), namely, $\mathcal{H}$ is a subset of the cone of positive semi-definite matrices (negative semi-definite matrices), if and only if*

$$\max_{X\in\mathcal{H}} i_{-}(X) = 0 \quad \left(\max_{X\in\mathcal{H}} i_{+}(X) = 0 \right).$$

In order to simplify block matrices, we use the following three types of elementary block matrix operation (EBMO, for short):

(i) interchange two block rows (columns) in a block matrix;

(ii) multiply a block row (column) by a nonsingular matrix from the left-hand (right-hand) side in a block matrix;

(iii) add a block row (column) multiplied by a matrix from the left-hand (right-hand) side to another block row (column);

and use the following three types of elementary block congruence matrix operation (EBCMO, for short) for a block Hermitian matrix with the same row and column partition:

(i) interchange ith and jth block rows, while interchange ith and jth block columns in the block Hermitian matrix;

(ii) multiply ith block row by a nonsingular matrix P from the left-hand side, while multiply ith block column by P^* from the right-hand side;

(iii) add ith block row multiplied by a matrix P from the left-hand side to jth block row, while add ith block column multiplied by P^* from the right-hand side to jth block column.

It is obvious that EBMOs do not change the rank of a block matrix, while EBCMOs do not change the inertia of a block Hermitian matrix.

The question of whether a given matrix-valued function is semi-definite everywhere is ubiquitous in matrix theory and applications. Lemma 4(e)–(h) assert that if certain explicit formulas for calculating the global maximal and minimal inertias of Hermitian matrix-valued functions are established, we can use them as a quantitative tool, as demonstrated in Sections 3–6 below, to derive necessary and sufficient conditions for the matrix-valued functions to be definite or semi-definite. In addition, we are able to use these inertia formulas to establish various matrix inequalities in the Löwner sense, and to solve many matrix optimization problems in the Löwner sense.

The Results in the following lemma are well known.

Lemma 5. *Let $A \in \mathbb{C}_{\rm H}^{m}$, $B \in \mathbb{C}_{\rm H}^{n}$, $Q \in \mathbb{C}^{m\times n}$, and $P \in \mathbb{C}^{p\times m}$ with $r(P) = m$. Then, the following rank and inertia formulas hold*

$$i_{\pm}(PAP^*) = i_{\pm}(A), \tag{11}$$

$$i_{\pm}(\lambda A) = \begin{cases} i_{\pm}(A) & if\ \lambda > 0 \\ i_{\mp}(A) & if\ \lambda < 0 \end{cases}, \tag{12}$$

$$i_{\pm}\begin{bmatrix} A & 0 \\ 0 & B \end{bmatrix} = i_{\pm}(A) + i_{\pm}(B), \tag{13}$$

$$i_{+}\begin{bmatrix} 0 & Q \\ Q^* & 0 \end{bmatrix} = i_{-}\begin{bmatrix} 0 & Q \\ Q^* & 0 \end{bmatrix} = r(Q). \tag{14}$$

Lemma 6. *([13]) Let $A \in \mathbb{C}_{\rm H}^{m}$, $B \in \mathbb{C}^{m\times n}$, $D \in \mathbb{C}_{\rm H}^{n}$, and let*

$$M_1 = \begin{bmatrix} A & B \\ B^* & 0 \end{bmatrix}, \quad M_2 = \begin{bmatrix} A & B \\ B^* & D \end{bmatrix}.$$

Then, the following expansion formulas for ranks and inertias of matrices hold

$$i_{\pm}(M_1) = r(B) + i_{\pm}(E_B A E_B), \tag{15}$$
$$r(M_1) = 2r(B) + r(E_B A E_B), \tag{16}$$
$$i_{\pm}(M_2) = i_{\pm}(A) + i_{\pm}\begin{bmatrix} 0 & E_A B \\ B^* E_A & D - B^* A^{\dagger} B \end{bmatrix}, \tag{17}$$
$$r(M_2) = r(A) + r\begin{bmatrix} 0 & E_A B \\ B^* E_A & D - B^* A^{\dagger} B \end{bmatrix}. \tag{18}$$

In particular, the following hold.

(a) *If* $A \succcurlyeq 0$, *then*

$$i_{+}(M_1) = r[\, A,\, B\,], \;\; i_{-}(M_1) = r(B), \;\; r(M_1) = r[\, A,\, B\,] + r(B). \tag{19}$$

(b) *If* $A \preccurlyeq 0$, *then*

$$i_{+}(M_1) = r(B), \;\; i_{-}(M_1) = r[\, A,\, B\,], \;\; r(M_1) = r[\, A,\, B\,] + r(B). \tag{20}$$

(c) *If* $\mathscr{R}(B) \subseteq \mathscr{R}(A)$, *then*

$$i_{\pm}(M_2) = i_{\pm}(A) + i_{\pm}(\, D - B^* A^{\dagger} B\,), \;\; r(M_2) = r(A) + r(\, D - B^* A^{\dagger} B\,). \tag{21}$$

(d) $r(M_2) = r(A) \Leftrightarrow \mathscr{R}(B) \subseteq \mathscr{R}(A)$ *and* $D = B^* A^{\dagger} B$.

(e) $M_2 \succcurlyeq 0 \Leftrightarrow A \succcurlyeq 0$, $\mathscr{R}(B) \subseteq \mathscr{R}(A)$ *and* $D - B^* A^{\dagger} B \succcurlyeq 0$.

Lemma 7. *([14]) Let* $A \in \mathbb{C}^{m\times p}$, $B \in \mathbb{C}^{q\times n}$ *and* $C \in \mathbb{C}^{m\times n}$ *be given. Then the matrix equation* $AXB = C$ *is consistent if and only if* $\mathscr{R}(C) \subseteq \mathscr{R}(A)$ *and* $\mathscr{R}(C^*) \subseteq \mathscr{R}(B^*)$, *or equivalently,* $AA^{\dagger}CB^{\dagger}B = AA^{\dagger}C = CB^{\dagger}B = C$. *In this case, the general solution can be written as*

$$X = A^{\dagger}CB^{\dagger} + F_A V_1 + V_2 E_B, \tag{22}$$

where V_1 *and* V_2 *are arbitrary matrices. In particular,* $AXB = C$ *has a unique solution if and only if*

$$r(A) = p, \;\; r(B) = q, \;\; \mathscr{R}(C) \subseteq \mathscr{R}(A), \;\; \mathscr{R}(C^*) \subseteq \mathscr{R}(B^*). \tag{23}$$

Lemma 8. *([15–17]) Let* $A \in \mathbb{C}^{m\times n}$, $B \in \mathbb{C}^{m\times p}$ *and* $C \in \mathbb{C}^{q\times n}$ *be given, and* $X \in \mathbb{C}^{p\times q}$ *be a variable matrix. Then, the global maximal and minimal ranks of* $A + BXC$ *are given by the following explicit formulas*

$$\max_{X\in\mathbb{C}^{p\times q}} r(\, A + BXC\,) = \min\left\{ r[\, A,\, B\,],\; r\begin{bmatrix} A \\ C \end{bmatrix} \right\}, \tag{24}$$
$$\min_{X\in\mathbb{C}^{p\times q}} r(\, A + BXC\,) = r[\, A,\, B\,] + r\begin{bmatrix} A \\ C \end{bmatrix} - r\begin{bmatrix} A & B \\ C & 0 \end{bmatrix}. \tag{25}$$

Lemma 9 ([18]). *Let $A \in \mathbb{C}_{\rm H}^{m}$, $B \in \mathbb{C}^{m\times n}$ and $C \in \mathbb{C}^{p\times m}$ be given, $X \in \mathbb{C}^{n\times p}$ be a variable matrix. Then, the global maximal and minimal ranks and inertias of $A + BXC + (BXC)^*$ are given by the following explicit formulas*

$$\max_{X\in\mathbb{C}^{n\times p}} r[\, A + BXC + (BXC)^* \,] = \min\left\{ r[\, A,\, B,\, C^* \,],\ r\begin{bmatrix} A & B \\ B^* & 0 \end{bmatrix},\ r\begin{bmatrix} A & C^* \\ C & 0 \end{bmatrix} \right\}, \tag{26}$$

$$\min_{X\in\mathbb{C}^{n\times p}} r[\, A + BXC + (BXC)^* \,] = 2r[\, A,\, B,\, C^* \,] + \max\{\, s_+ + s_-,\ t_+ + t_-,\ s_+ + t_-,\ s_- + t_+ \}, \tag{27}$$

$$\max_{X\in\mathbb{C}^{n\times p}} i_{\pm}[\, A + BXC + (BXC)^* \,] = \min\left\{ i_{\pm}\begin{bmatrix} A & B \\ B^* & 0 \end{bmatrix},\ i_{\pm}\begin{bmatrix} A & C^* \\ C & 0 \end{bmatrix} \right\}, \tag{28}$$

$$\min_{X\in\mathbb{C}^{n\times p}} i_{\pm}[\, A + BXC + (BXC)^* \,] = r[\, A,\, B,\, C^* \,] + \max\{\, s_{\pm},\ t_{\pm} \,\}, \tag{29}$$

where

$$s_{\pm} = i_{\pm}\begin{bmatrix} A & B \\ B^* & 0 \end{bmatrix} - r\begin{bmatrix} A & B & C^* \\ B^* & 0 & 0 \end{bmatrix},\quad t_{\pm} = i_{\pm}\begin{bmatrix} A & C^* \\ C & 0 \end{bmatrix} - r\begin{bmatrix} A & B & C^* \\ C & 0 & 0 \end{bmatrix}.$$

The matrices X that satisfy (26)–(29) are not necessarily unique, and their expressions were also given in [18] by using generalized inverses of matrices and certain simultaneous decomposition of the three given matrices A, B and C.

3. Basic Formulas

We first solve Problem 1 through a linearization method and Lemma 9. The method was adopted in the author's papers [19,20].

Theorem 10. *Let $\phi(X)$ be as given in (2), and define*

$$N_1 = \begin{bmatrix} D + CMC^* & A \\ A^* & 0 \end{bmatrix}, \qquad N_2 = \begin{bmatrix} D + CMC^* & CMB^* & A \\ A^* & 0 & 0 \end{bmatrix}, \tag{30}$$

$$N_3 = \begin{bmatrix} D + CMC^* & CMB^* \\ BMC^* & BMB^* \end{bmatrix},\quad N_4 = \begin{bmatrix} D + CMC^* & CMB^* & A \\ BMC^* & BMB^* & 0 \end{bmatrix}. \tag{31}$$

Then, the global maximal and minimal ranks and inertias of $\phi(X)$ are given by the following explicit formulas

$$\max_{X\in\mathbb{C}^{p\times m}} r[\, \phi(X) \,] = \min\{ r[\, D + CMC^*,\, CMB^*,\, A \,],\ r(N_1),\ r(N_3) \}, \tag{32}$$

$$\min_{X\in\mathbb{C}^{p\times m}} r[\, \phi(X) \,] = 2r[\, D + CMC^*,\, CMB^*,\, A \,] + \max\{\, s_1,\ s_2,\ s_3,\ s_4 \,\}, \tag{33}$$

$$\max_{X\in\mathbb{C}^{p\times m}} i_{\pm}[\, \phi(X) \,] = \min\{ i_{\pm}(N_1),\ i_{\pm}(N_3) \}, \tag{34}$$

$$\min_{X\in\mathbb{C}^{p\times m}} i_{\pm}[\, \phi(X) \,] = r[\, D + CMC^*,\, CMB^*,\, A \,] + \max\{ i_{\pm}(N_1) - r(N_2),\ i_{\pm}(N_3) - r(N_4) \}, \tag{35}$$

where

$$s_1 = r(N_1) - 2r(N_2), \;\; s_2 = r(N_3) - 2r(N_4),$$
$$s_3 = i_+(N_1) + i_-(N_3) - r(N_2) - r(N_4),$$
$$s_4 = i_-(N_1) + i_+(N_3) - r(N_2) - r(N_4).$$

Proof. It is easy to verify from (17) and (18) that

$$\begin{aligned} & i_\pm[\,(AXB+C)M(AXB+C)^* + D\,] \\ & = i_\pm\begin{bmatrix} -M & M(AXB+C)^* \\ (AXB+C)M & D \end{bmatrix} - i_\pm(-M), \end{aligned} \tag{36}$$

$$\begin{aligned} & r[\,(AXB+C)M(AXB+C)^* + D\,] \\ & = r\begin{bmatrix} -M & M(AXB+C)^* \\ (AXB+C)M & D \end{bmatrix} - r(M), \end{aligned} \tag{37}$$

that is, the inertia and rank of $\phi(X)$ in (2) can be calculated by those of the following linear matrix-valued function

$$\begin{aligned} \psi(X) &= \begin{bmatrix} -M & M(AXB+C)^* \\ (AXB+C)M & D \end{bmatrix} \\ &= \begin{bmatrix} -M & MC^* \\ CM & D \end{bmatrix} + \begin{bmatrix} 0 \\ A \end{bmatrix} X[\,BM,\,0\,] + \begin{bmatrix} MB^* \\ 0 \end{bmatrix} X^*[\,0,\,A^*\,]. \end{aligned} \tag{38}$$

Note from (36) and (37) that

$$\max_{X\in\mathbb{C}^{p\times m}} r[\,\phi(X)\,] = \max_{X\in\mathbb{C}^{p\times m}} r[\psi(X)] - r(A), \tag{39}$$

$$\min_{X\in\mathbb{C}^{p\times m}} r[\,\phi(X)\,] = \min_{X\in\mathbb{C}^{p\times m}} r[\psi(X)] - r(A), \tag{40}$$

$$\max_{X\in\mathbb{C}^{p\times m}} i_\pm[\,\phi(X)\,] = \max_{X\in\mathbb{C}^{p\times m}} i_\pm[\psi(X)] - i_\mp(A), \tag{41}$$

$$\min_{X\in\mathbb{C}^{p\times m}} i_\pm[\,\phi(X)\,] = \min_{X\in\mathbb{C}^{p\times m}} i_\pm[\psi(X)] - i_\mp(A). \tag{42}$$

Applying Lemma 9 to (38), we first obtain

$$\max_{X\in\mathbb{C}^{p\times m}} r[\psi(X)] = \min\{\, r(H),\;\; r(G_1),\;\; r(G_2)\,\}, \tag{43}$$

$$\min_{X\in\mathbb{C}^{p\times m}} r[\psi(X)] = 2r(H) + \max\{\, s_+ + s_-,\;\; t_+ + t_-,\;\; s_+ + t_-,\;\; s_- + t_+\}, \tag{44}$$

$$\max_{X\in\mathbb{C}^{p\times m}} i_\pm[\psi(X)] = \min\{\, i_\pm(G_1),\;\; i_\pm(G_2)\,\}, \tag{45}$$

$$\min_{X\in\mathbb{C}^{p\times m}} i_\pm[\psi(X)] = r(H) + \max\{\, s_\pm,\;\; t_\pm\,\}, \tag{46}$$

where

$$H = \begin{bmatrix} -M & MC^* & 0 & MB^* \\ CM & D & A & 0 \end{bmatrix},$$

$$G_1 = \begin{bmatrix} -M & MC^* & 0 \\ CM & D & A \\ 0 & A^* & 0 \end{bmatrix}, \; G_2 = \begin{bmatrix} -M & MC^* & MB^* \\ CM & D & 0 \\ BM & 0 & 0 \end{bmatrix},$$

$$H_1 = \begin{bmatrix} -M & MC^* & 0 & MB^* \\ CM & D & A & 0 \\ 0 & A^* & 0 & 0 \end{bmatrix}, \quad H_2 = \begin{bmatrix} -M & MC^* & MB^* & 0 \\ CM & D & 0 & A \\ BM & 0 & 0 & 0 \end{bmatrix},$$

and

$$s_{\pm} = i_{\pm}(G_1) - r(H_1), \quad t_{\pm} = i_{\pm}(G_2) - r(H_2).$$

It is easy to derive from Lemmas 5 and 6, elementary matrix operations and congruence matrix operations that

$$r(H) = r(M) + r[\, D + CMC^*,\, CMB^*,\, A\,], \tag{47}$$

$$r(H_1) = r(M) + r\begin{bmatrix} D + CMC^* & CMB^* & A \\ A^* & 0 & 0 \end{bmatrix} = r(M) + r(N_2), \tag{48}$$

$$r(H_2) = r(M) + r\begin{bmatrix} D + CMC^* & CMB^* & A \\ BMC^* & BMB^* & 0 \end{bmatrix} = r(M) + r(N_4), \tag{49}$$

$$i_{\pm}(G_1) = i_{\mp}(M) + i_{\pm}\begin{bmatrix} C + CMC^* & A \\ A^* & 0 \end{bmatrix} = i_{\mp}(M) + i_{\pm}(N_1), \tag{50}$$

$$i_{\pm}(G_2) = i_{\mp}(M) + i_{\pm}\begin{bmatrix} D + CMC^* & CMB^* \\ BMC^* & BMB^* \end{bmatrix} = i_{\mp}(M) + i_{\pm}(N_3). \tag{51}$$

Hence,

$$r(G_1) = r(M) + r(N_1), \quad r(G_2) = r(M) + r(N_3), \tag{52}$$

$$s_{\pm} = i_{\pm}(G_1) - r(H_1) = i_{\pm}(N_1) - r(N_2) - i_{\pm}(M), \tag{53}$$

$$t_{\pm} = i_{\pm}(G_2) - r(H_2) = i_{\pm}(N_3) - r(N_4) - i_{\pm}(M). \tag{54}$$

Substituting (47)–(54) into (43)–(46), and then (43)–(46) into (39)–(42), we obtain (32)–(35). ■

Without loss of generality, we assume in what follows that both $A \neq 0$ and $BMB^* \neq 0$ in (2). Applying Lemma 4 to (32)–(35), we obtain the following results.

Corollary 11. *Let $\phi(X)$ be as given in* (2), *N_1 and N_3 be the matrices of* (30) *and* (31), *and let $J = D + CMC^*$. Then, the following hold.*

(a) *There exists an $X \in \mathbb{C}^{p\times m}$ such that $\phi(X)$ is nonsingular if and only if $r[\, J, CMB^*,\, A\,] = n$, $r(N_1) \geqslant n$ and $r(N_3) \geqslant n$.*

(b) *$\phi(X)$ is nonsingular for all $X \in \mathbb{C}^{p\times m}$ if and only if $r(\, D + CMC^*\,) = n$, and one of the following four conditions holds*

  (i) *$BMC^* J^{-1} A = 0$ and $A^* J^{-1} A = 0$.*

  (ii) *$BMC^* J^{-1} A = 0$ and $BMC^* J^{-1} CMB^* = BMB^*$.*

  (iii) *$A^* J^{-1} A \succcurlyeq 0$, $BMB^* - BMC^* J^{-1} CMB^* \succcurlyeq 0$, $\mathscr{R}(A^* J^{-1} CMB^*) \subseteq \mathscr{R}(A^* J^{-1} A)$, and $\mathscr{R}(BMC^* J^{-1} A) \subseteq \mathscr{R}(\, BMB^* - BMC^* J^{-1} CMB^*\,)$.*

  (iv) *$A^* J^{-1} A \preccurlyeq 0$, $BMB^* - BMC^* J^{-1} CMB^* \preccurlyeq 0$, $\mathscr{R}(A^* J^{-1} CMB^*) \subseteq \mathscr{R}(A^* J^{-1} A)$ and $\mathscr{R}(BMC^* J^{-1} A) \subseteq \mathscr{R}(\, BMB^* - BMC^* J^{-1} CMB^*\,)$.*

Proof. Setting the right-hand side of (32) equal to n directly leads to the result in (a). Under the condition $r(J) = n$, (33) reduces to

$$\min_{X \in \mathbb{C}^{p\times m}} r[\,\phi(X)\,] = 2n + \max\{\, s_1,\ \ s_2,\ \ s_3,\ \ s_4 \,\}, \tag{55}$$

where

$$\begin{aligned}
s_1 =\,& r(A^*J^{-1}A) - 2r[\,A^*J^{-1}CMB^*,\ A^*J^{-1}A\,] - n,\\
s_2 =\,& r(BMB^* - BMC^*J^{-1}CMB^*) - 2r[BMB^* - BMC^*J^{-1}CMB^*,\ BMC^*J^{-1}A]\\
& -n,\\
s_3 =\,& i_-(A^*J^{-1}A) + i_-(\,BMB^* - BMC^*J^{-1}CMB^*\,) - r[\,A^*J^{-1}CMB^*,\ A^*J^{-1}A\,]\\
& -r[\,BMB^* - BMC^*J^{-1}CMB^*,\ BMC^*J^{-1}A\,] - n,\\
s_4 =\,& i_+(A^*J^{-1}A) + i_+(\,BMB^* - BMC^*J^{-1}CMB^*\,) - r[\,A^*J^{-1}CMB^*,\ A^*J^{-1}A\,]\\
& -r[\,BMB^* - BMC^*J^{-1}CMB^*,\ BMC^*J^{-1}A\,] - n.
\end{aligned}$$

Setting (55) equal to n, we see that $\phi(X)$ is nonsingular for all $X \in \mathbb{C}^{p\times m}$ if and only if $r(D + CMC^*) = n$, and one of the following four rank equalities holds

(i) $r(A^*J^{-1}A) = 2r[\,A^*J^{-1}CMB^*,\ A^*J^{-1}A\,]$;

(ii) $r(BMB^* - BMC^*J^{-1}CMB^*) = 2r[BMB^* - BMC^*J^{-1}CMB^*,\ BMC^*J^{-1}A]$;

(iii) $i_-(A^*J^{-1}A) + i_-[\,BMB^* - BMC^*J^{-1}CMB^*\,] = r[\,A^*J^{-1}CMB^*,\ A^*J^{-1}A\,] + r[\,BMB^* - BMC^*J^{-1}CMB^*,\ BMC^*J^{-1}A\,]$;

(iv) $i_+(A^*J^{-1}A) + i_+(\,BMB^* - BMC^*J^{-1}CMB^*\,) = r[\,A^*J^{-1}CMB^*,\ A^*J^{-1}A\,] + r[\,BMB^* - BMC^*J^{-1}CMB^*,\ BMC^*J^{-1}A\,] - n$;

which are further equivalent to the result in (b) by comparing both sides of the four rank equalities. ■

The following two corollaries can be shown by a similar approach.

Corollary 12. *Let $\phi(X)$ be as given in (2), and let N_1 and N_3 be the matrices of (30) and (31). Then, the following hold.*

(a) *There exists an $X \in \mathbb{C}^{p\times m}$ such that $\phi(X) = 0$, namely, the matrix equation in (7) is consistent, if and only if*

$$\begin{aligned}
&\mathscr{R}(\,D + CMC^*\,) \subseteq \mathscr{R}[\,A,\ CMB^*\,], \quad r(N_1) = 2r(A),\\
&2r[\,A,\ CMB^*\,] + r(N_3) - 2r(N_4) \leqslant 0,\\
&r[\,A,\ CMB^*\,] + i_+(N_3) - r(N_4) \leqslant 0,\\
&r[\,A,\ CMB^*\,] + i_-(N_3) - r(N_4) \leqslant 0.
\end{aligned}$$

(b) $\phi(X) = 0$ *holds for all* $X \in \mathbb{C}^{p\times m}$ *if and only if* $[\,D + CMC^*,\ CMB^*,\ A\,] = 0$, *or*

$$\begin{bmatrix} D + CMC^* & A \\ A^* & 0 \end{bmatrix} = 0, \quad \text{or} \quad \begin{bmatrix} D + CMC^* & CMB^* \\ BMC^* & BMB^* \end{bmatrix} = 0.$$

Corollary 13. *Let $\phi(X)$ be as given in* (2), *and let N_1 and N_3 be the matrices of* (30) *and* (31). *Then, the following hold.*

(a) *There exists an $X \in \mathbb{C}^{p\times m}$ such that $\phi(X) \succ 0$, namely, the matrix inequality is feasible, if and only if*

$$i_+(N_1) = n \text{ and } i_+(N_3) \geqslant n, \quad \text{or} \quad i_+(N_1) \geqslant n \text{ and } i_+(N_3) = n.$$

(b) *There exists an $X \in \mathbb{C}^{p\times m}$ such that $\phi(X) \prec 0$, the matrix inequality is feasible, if and only if*

$$i_-(N_1) = n \text{ and } i_-(N_3) \geqslant n, \quad \text{or} \quad i_-(N_1) \geqslant n \text{ and } i_-(N_3) = n.$$

(c) *$\phi(X) \succ 0$ holds for all $X \in \mathbb{C}^{p\times m}$, namely, $\phi(X)$ is a completely positive matrix-valued function, if and only if*

$$D + CMC^* \succ 0, \ N_3 \succcurlyeq 0, \ \mathscr{R}\begin{bmatrix} A \\ 0 \end{bmatrix} \subseteq \mathscr{R}(N_3).$$

(d) *$\phi(X) \prec 0$ holds for all $X \in \mathbb{C}^{p\times m}$ namely, $\phi(X)$ is a completely negative matrix-valued function, if and only if*

$$D + CMC^* \prec 0, \ N_3 \preccurlyeq 0, \ \mathscr{R}\begin{bmatrix} A \\ 0 \end{bmatrix} \subseteq \mathscr{R}(N_3).$$

(e) *There exists an $X \in \mathbb{C}^{p\times m}$ such that $\phi(X) \succcurlyeq 0$, namely, the matrix inequality is feasible, if and only if*

$$r[\, D + CMC^*, \ CMB^*, \ A\,] + i_-(N_1) \leqslant r(N_2)$$

and

$$r[\, D + CMC^*, \ CMB^*, \ A\,] + i_-(N_3) \leqslant r(N_4).$$

(f) *There exists an $X \in \mathbb{C}^{p\times m}$ such that $\phi(X) \preccurlyeq 0$, namely, the matrix inequality is feasible, if and only if*

$$r[\, D + CMC^*, \ CMB^*, \ A\,] + i_+(N_1) \leqslant r(N_2)$$

and

$$r[\, D + CMC^*, \ CMB^*, \ A\,] + i_+(N_3) \leqslant r(N_4).$$

(g) *$\phi(X) \succcurlyeq 0$ holds for all $X \in \mathbb{C}^{p\times m}$, namely, $\phi(X)$ is a positive matrix-valued function, if and only if $N_3 \succcurlyeq 0$.*

(h) *$\phi(X) \preccurlyeq 0$ holds for all $X \in \mathbb{C}^{p\times m}$, namely, $\phi(X)$ is a negative matrix-valued function, if and only if $N_3 \preccurlyeq 0$.*

Two special cases of Theorem 11 are presented below.

Corollary 14. *Let*

$$\phi(X) = (\,AXB + C\,)(\,AXB + C)^* - DD^*, \tag{56}$$

and define

$$N_1 = \begin{bmatrix} CC^* - DD^* & A \\ A^* & 0 \end{bmatrix}, \quad N_2 = \begin{bmatrix} CC^* - DD^* & CB^* & A \\ A^* & 0 & 0 \end{bmatrix}, \tag{57}$$

$$N_3 = \begin{bmatrix} CC^* - DD^* & CB^* \\ BC^* & BB^* \end{bmatrix}, \quad N_4 = \begin{bmatrix} CC^* - DD^* & CB^* & A \\ BC^* & BB^* & 0 \end{bmatrix}. \tag{58}$$

Then, the global maximal and minimal ranks and inertias of $\phi(X)$ are given by

$$\max_{X \in \mathbb{C}^{p\times m}} r[\,\phi(X)\,] = \min\{\,r[\,CC^* - DD^*,\, CB^*,\, A\,],\ r(N_1),\ r(N_3)\,\}, \tag{59}$$

$$\min_{X \in \mathbb{C}^{p\times m}} r[\,\phi(X)\,] = 2r[\,CC^* - DD^*,\, CB^*,\, A\,] + \max\{\,s_1,\ s_2,\ s_3,\ s_4\,\}, \tag{60}$$

$$\max_{X \in \mathbb{C}^{p\times m}} i_{\pm}[\,\phi(X)\,] = \min\{\,i_{\pm}(N_1),\ i_{\pm}(N_3)\,\}, \tag{61}$$

$$\begin{aligned} \min_{X \in \mathbb{C}^{p\times m}} i_{\pm}[\,\phi(X)\,] &= r[\,CC^* - DD^*,\, CB^*,\, A\,] \\ &\quad + \max\{i_{\pm}(N_1) - r(N_2),\ i_{\pm}(N_3) - r(N_4)\}, \end{aligned} \tag{62}$$

where

$$\begin{aligned} s_1 &= r(N_1) - 2r(N_2),\ \ s_2 = r(N_3) - 2r(N_4), \\ s_3 &= i_+(N_1) + i_-(N_3) - r(N_2) - r(N_4), \\ s_4 &= i_-(N_1) + i_+(N_3) - r(N_2) - r(N_4). \end{aligned}$$

Let

$$\phi(X) = (\,AXB + C\,)(\,AXB + C)^* - I_n. \tag{63}$$

Then $\phi(X) = 0$ means that the rows of $AXB + C$ are orthogonal each other. Further, if $AXB + C$ is square, $\phi(X) = 0$ means that $AXB + C$ is unitary.

Corollary 15. *Let $\phi(X)$ be as given in* (63), *and define*

$$N_1 = \begin{bmatrix} CC^* - I_n & A \\ A^* & 0 \end{bmatrix}, \quad N_2 = \begin{bmatrix} CC^* - I_n & CB^* & A \\ A^* & 0 & 0 \end{bmatrix}, \tag{64}$$

$$N_3 = \begin{bmatrix} CC^* - I_n & CB^* \\ BC^* & BB^* \end{bmatrix}, \quad N_4 = \begin{bmatrix} CC^* - I_n & CB^* & A \\ BC^* & BB^* & 0 \end{bmatrix}. \tag{65}$$

Then, the global maximal and minimal ranks and inertias of $\phi(X)$ are given by

$$\max_{X \in \mathbb{C}^{p\times m}} r[\,\phi(X)\,] = \min\{\,r[\,CC^* - I_n,\, CB^*,\, A\,],\ r(N_1),\ r(N_3)\,\}, \tag{66}$$

$$\min_{X \in \mathbb{C}^{p\times m}} r[\,\phi(X)\,] = 2r[\,CC^* - I_n,\, CB^*,\, A\,] + \max\{\,s_1,\ s_2,\ s_3,\ s_4\,\}, \tag{67}$$

$$\max_{X \in \mathbb{C}^{p\times m}} i_{\pm}[\,\phi(X)\,] = \min\{\,i_{\pm}(N_1),\ i_{\pm}(N_3)\,\}, \tag{68}$$

$$\begin{aligned} \min_{X \in \mathbb{C}^{p\times m}} i_{\pm}[\,\phi(X)\,] &= r[\,CC^* - I_n,\, CB^*,\, A\,] \\ &\quad + \max\{\,i_{\pm}(N_1) - r(N_2),\ i_{\pm}(N_3) - r(N_4)\}, \end{aligned} \tag{69}$$

where

$$s_1 = r(N_1) - 2r(N_2), \;\; s_2 = r(N_3) - 2r(N_4),$$
$$s_3 = i_+(N_1) + i_-(N_3) - r(N_2) - r(N_4),$$
$$s_4 = i_-(N_1) + i_+(N_3) - r(N_2) - r(N_4).$$

Whether a given function is null or positive or nonnegative everywhere is a fundamental consideration when characterizing behaviors of the function. It was realized in matrix theory that the determination of the definiteness or semi-definiteness of a general matrix-valued function is NP-hard. Corollaries 11–13, however, show that we are really able to characterize the definiteness and semi-definiteness of (2) by using some ordinary and elementary methods. These results set up a criterion for characterizing definiteness and semi-definiteness of nonlinear matrix-valued functions, and will prompt more investigations on this challenging topic.

Recall that a Hermitian matrix A can uniquely be decomposed as the difference of two disjoint Hermitian positive semi-definite definite matrices

$$A = A_1 - A_2, \;\; A_1A_2 = A_2A_1 = 0, \;\; A_1 \succcurlyeq 0, \;\; A_2 \succcurlyeq 0. \tag{70}$$

Applying this assertion to (2), we obtain the following result.

Corollary 16. *Let $\phi(X)$ be as given in* (2). *Then, $\phi(X)$ can always be decomposed as*

$$\phi(X) = \phi_1(X) - \phi_2(X), \tag{71}$$

where

$$\phi_1(X) = (\,AXB + C\,)M_1(\,AXB + C)^* + D_1,$$
$$\phi_2(X) = (\,AXB + C\,)M_2(\,AXB + C)^* + D_2$$

satisfy

$$\phi_1(X) \succcurlyeq 0 \;\; and \;\; \phi_2(X) \succcurlyeq 0 \tag{72}$$

for all $X \in \mathbb{C}^{p\times m}$.

Proof. Note from (70) that the two Hermitian matrices D and M in (2) can uniquely be decomposed as

$$D = D_1 - D_2, \quad D_1D_2 = D_2D_1 = 0, \quad D_1 \succcurlyeq 0, \quad D_2 \succcurlyeq 0,$$
$$M = M_1 - M_2, \;\; M_1M_2 = M_2M_1 = 0, \;\; M_1 \succcurlyeq 0, \;\; M_2 \succcurlyeq 0.$$

So that both $\phi_1(X)$ and $\phi_2(X)$ in (71) are positive matrix-valued functions. ■

Corollary 17. *Let* $\phi(X)$ *be as given in* (2), *and suppose that* $AXB+C=0$ *has a solution, i.e.,* $\mathscr{R}(C)\subseteq\mathscr{R}(A)$ *and* $\mathscr{R}(C^*)\subseteq\mathscr{R}(B^*)$, *and let* $N=\begin{bmatrix} D & A \\ A^* & 0 \end{bmatrix}$. *Then,*

$$\max_{X\in\mathbb{C}^{p\times m}} r[\,\phi(X)\,] = \min\{r[\,A,\,D\,],\ \ r(D)+r(BMB^*)\}, \tag{73}$$

$$\min_{X\in\mathbb{C}^{p\times m}} r[\,\phi(X)\,] = \max\{2r[\,A,\,D\,]-r(N),\ \ r(D)-r(BMB^*),$$
$$r[\,A,\,D\,]+i_-(D)-i_+(BMB^*)-i_-(N),$$
$$r[\,A,\,D\,]+i_+(D)-i_-(BMB^*)-i_+(N)\}, \tag{74}$$

$$\max_{X\in\mathbb{C}^{p\times m}} i_\pm[\,\phi(X)\,] = \min\{i_\pm(N),\ \ i_\pm(D)+i_\pm(BMB^*)\}, \tag{75}$$

$$\min_{X\in\mathbb{C}^{p\times m}} i_\pm[\,\phi(X)\,] = \max\{\,r[\,A,\,D\,]-i_\mp(N),\ \ i_\pm(D)-i_\mp(BMB^*)\}. \tag{76}$$

We next solve the two Hermitian quadratic optimization problems in (10), where the two matrices $\phi(\widehat{X})$ and $\phi(\widetilde{X})$ are called the global maximal and minimal matrices of $\phi(X)$ in (2) in the Löwner partial ordering, respectively.

Corollary 18. *Let* $\phi(X)$ *be as given in* (2), *and let* $N=\begin{bmatrix} D+CMC^* & CMB^* \\ BMC^* & BMB^* \end{bmatrix}$. *Then, there exists an* $\widehat{X}\in\mathbb{C}^{p\times m}$ *such that*

$$\phi(X)\succcurlyeq\phi(\widehat{X}) \tag{77}$$

holds for all $X\in\mathbb{C}^{p\times m}$ *if and only if*

$$BMB^*\succcurlyeq 0,\ \ \mathscr{R}(CMB^*)\subseteq\mathscr{R}(A),\ \ \mathscr{R}(BMC^*)\subseteq\mathscr{R}(BMB^*). \tag{78}$$

In this case, the following hold.

(a) *The matrix* $\widehat{X}\in\mathbb{C}^{p\times m}$ *satisfying* (77) *is the solution of the linear matrix equation*

$$A\widehat{X}BMB^*+CMB^*=0. \tag{79}$$

Correspondingly,

$$\widehat{X}=-A^\dagger CMB^*(BMB^*)^\dagger+F_AV_1+V_2E_{BMB^*}, \tag{80}$$

$$\phi(\widehat{X})=D+CMC^*-CMB^*(BMB^*)^\dagger BMC^*, \tag{81}$$

$$\phi(X)-\phi(\widehat{X})=(\,AXB+C\,)MB^*(BMB^*)^\dagger BM(\,AXB+C\,)^*, \tag{82}$$

where V_1 *and* V_2 *are arbitrary matrices.*

(b) *The ranks and inertias of* $\phi(\widehat{X})$ *and* $\phi(X)-\phi(\widehat{X})$ *are given by*

$$i_+[\,\phi(\widehat{X})\,]=i_+(N)-r(BMB^*),\ i_-[\,\phi(\widehat{X})\,]=i_-(N), \tag{83}$$

$$r[\,\phi(\widehat{X})\,]=r(N)-r(BMB^*), \tag{84}$$

$$i_+[\,\phi(X)-\phi(\widehat{X})\,]=r[\,\phi(X)-\phi(\widehat{X})\,]=r(\,AXBMB^*+CMB^*\,). \tag{85}$$

(c) *The matrix* $\widehat{X} \in \mathbb{C}^{p\times m}$ *satisfying* (77) *is unique if and only if*

$$r(A) = p, \ \ \mathscr{R}(CMB^*) \subseteq \mathscr{R}(A), \ \ BMB^* \succ 0; \tag{86}$$

under this condition,

$$\widehat{X} = -A^{\dagger}CMB^*(BMB^*)^{-1}, \tag{87}$$
$$\phi(\widehat{X}) = D + CMC^* - CMB^*(BMB^*)^{-1}BMC^*, \tag{88}$$
$$\phi(X) - \phi(\widehat{X}) = (\,AXB + C\,)MB^*(BMB^*)^{-1}BM(\,AXB + C\,)^*. \tag{89}$$

(d) $\widehat{X} = 0$ *is a solution of* (77) *if and only if* $BMB^* \succcurlyeq 0$ *and* $CMB^* = 0$. *In this case,* $\phi(0) = D + CMC^*$.

(e) $\widehat{X} = 0$ *is the unique solution of* (77) *if and only if* $r(A) = p$, $CMB^* = 0$ *and* $BMB^* \succ 0$. *In this case,* $\phi(0) = D + CMC^*$.

(f) *There exists an* $\widehat{X} \in \mathbb{C}^{p\times m}$ *such that*

$$\phi(X) \succcurlyeq \phi(\widehat{X}) \succcurlyeq 0 \tag{90}$$

holds for all $X \in \mathbb{C}^{p\times m}$ *if and only if*

$$\mathscr{R}(CMB^*) \subseteq \mathscr{R}(A) \ \ \text{and} \ \ N \succcurlyeq 0. \tag{91}$$

In this case, the matrix $\widehat{X} \in \mathbb{C}^{p\times m}$ *satisfying* (90) *is unique if and only if*

$$r(A) = p, \ \ \mathscr{R}(CMB^*) \subseteq \mathscr{R}(A), \ \ BMB^* \succ 0, \ \ N \succcurlyeq 0. \tag{92}$$

Proof. Let

$$\psi(X) = \phi(X) - \phi(\widehat{X}) = (\,AXB + C\,)M(\,AXB + C)^* - \left(A\widehat{X}B + C\right)M\left(A\widehat{X}B + C\right)^*.$$

Then, $\phi(X) \succcurlyeq \phi(\widehat{X})$ is equivalent to $\psi(X) \succcurlyeq 0$. Under $A \neq 0$, we see from Corollary 13(g) that $\psi(X) \succcurlyeq 0$ holds for all $X \in \mathbb{C}^{p\times m}$ if and only if

$$\begin{bmatrix} CMC^* - \left(A\widehat{X}B + C\right)M\left(A\widehat{X}B + C\right)^* & CMB^* \\ BMC^* & BMB^* \end{bmatrix} \succcurlyeq 0, \tag{93}$$

which, by Lemma 6(e), is further equivalent to

$$BMB^* \succcurlyeq 0, \ \ \mathscr{R}(BMC^*) \subseteq \mathscr{R}(BMB^*), \tag{94}$$
$$CMC^* - \left(A\widehat{X}B + C\right)M\left(A\widehat{X}B + C\right)^* - CMB^*(BMB^*)^{\dagger}BMC^* \succcurlyeq 0. \tag{95}$$

In this case, it is easy to verify

$$\begin{aligned} & CMC^* - \left(A\widehat{X}B + C\right)M\left(\,A\widehat{X}B + C\right)^* - CMB^*(BMB^*)^{\dagger}BMC^* \\ &= -(\,A\widehat{X}BMB^* + CMB^*\,)(BMB^*)^{\dagger}(\,A\widehat{X}BMB^* + CMB^*\,)^*, \end{aligned} \tag{96}$$

and therefore, the inequality in (95) is equivalent to $CMB^* + A\widehat{X}BMB^* = 0$. By Lemma 7, this matrix equation is solvable if and only if $\mathscr{R}(CMB^*) \subseteq \mathscr{R}(A)$ and $\mathscr{R}(BMC^*) \subseteq \mathscr{R}(BMB^*)$. In this case, the general solution of the equation is (80) by Lemma 7, and (95) becomes

$$CMC^* - \left(A\widehat{X}B + C \right) M \left(A\widehat{X}B + C \right)^* - CMB^*(BMB^*)^{\dagger}BMC^* = 0.$$

Thus (81) and (82) follow. The results in (b)–(f) follow from (a). Their derivations are left to the reader. ∎

The following corollary can be shown similarly.

Corollary 19. *Let $\phi(X)$ be as given in* (2), *and let* $N = \begin{bmatrix} D + CMC^* & CMB^* \\ BMC^* & BMB^* \end{bmatrix}$. *Then, there exists an $\widetilde{X} \in \mathbb{C}^{p\times m}$ such that*

$$\phi(X) \preccurlyeq \phi(\widetilde{X}) \tag{97}$$

holds for all $X \in \mathbb{C}^{p\times m}$ if and only if

$$BMB^* \preccurlyeq 0, \;\; \mathscr{R}(CMB^*) \subseteq \mathscr{R}(A), \;\; \mathscr{R}(BMC^*) \subseteq \mathscr{R}(BMB^*). \tag{98}$$

In this case, the following hold.

(a) *The matrix $\widetilde{X}$ satisfying* (97) *is the solution of the linear matrix equation*

$$A\widetilde{X}BMB^* + CMB^* = 0. \tag{99}$$

Correspondingly,

$$\widetilde{X} = -A^{\dagger}CMB^*(BMB^*)^{\dagger} + F_AV_1 + V_2E_{BMB^*}, \tag{100}$$
$$\phi(\widetilde{X}) = D + CMC^* - CMB^*(BMB^*)^{\dagger}BMC^*, \tag{101}$$
$$\phi(X) - \phi(\widetilde{X}) = (\, AXB + C\,)MB^*(BMB^*)^{\dagger}BM(\, AXB + C\,)^*, \tag{102}$$

where V_1 and V_2 are arbitrary matrices.

(b) *The ranks and inertias of $\phi(\widetilde{X})$ and $\phi(X) - \phi(\widetilde{X})$ are given by*

$$i_+[\,\phi(\widetilde{X})\,] = i_+(N), \; i_-[\,\phi(\widetilde{X})\,] = i_-(N) - r(BMB^*), \tag{103}$$
$$r[\,\phi(\widehat{X})\,] = r(N) - r(BMB^*), \tag{104}$$
$$i_-[\,\phi(X) - \phi(\widetilde{X})\,] = r[\,\phi(X) - \phi(\widetilde{X})\,] = r(\, AXBMB^* + CMB^*\,). \tag{105}$$

(c) *The matrix $\widetilde{X} \in \mathbb{C}^{p\times m}$ satisfying* (97) *is unique if and only if*

$$r(A) = p, \;\; \mathscr{R}(CMB^*) \subseteq \mathscr{R}(A), \;\; BMB^* \prec 0. \tag{106}$$

In this case,

$$\widetilde{X} = -A^{\dagger}CMB^*(BMB^*)^{-1}, \tag{107}$$
$$\phi(\widetilde{X}) = D + CMC^* - CMB^*(BMB^*)^{-1}BMC^*, \tag{108}$$
$$\phi(X) - \phi(\widetilde{X}) = (\, AXB + C\,)MB^*(BMB^*)^{-1}BM(\, AXB + C\,)^*. \tag{109}$$

(d) $\widetilde{X} = 0$ *is a solution of* (97) *if and only if* $BMB^* \preccurlyeq 0$ *and* $CMB^* = 0$. *In this case,* $\phi(0) = D + CMC^*$.

(e) $\widetilde{X} = 0$ *is the unique solution of* (97) *if and only if* $r(A) = p$, $CMB^* = 0$ *and* $BMB^* \prec 0$. *In this case,* $\phi(0) = D + CMC^*$.

(f) *There exists an* $\widetilde{X} \in \mathbb{C}^{p\times m}$ *such that*

$$\phi(X) \preccurlyeq \phi(\widetilde{X}) \preccurlyeq 0 \tag{110}$$

holds for all $X \in \mathbb{C}^{p\times m}$ *if and only if*

$$\mathscr{R}(CMB^*) \subseteq \mathscr{R}(A) \ \ and \ \ N \preccurlyeq 0. \tag{111}$$

In this case, the matrix $\widetilde{X} \in \mathbb{C}^{p\times m}$ *satisfying* (110) *is unique if and only if*

$$r(A) = p, \ \ \mathscr{R}(CMB^*) \subseteq \mathscr{R}(A), \ \ BMB^* \prec 0, \ \ N \preccurlyeq 0. \tag{112}$$

4. The Convexity and Concavity of Hermitian Matrix-Valued Function

As usual, the matrix-valued function $\phi(X)$ in (2) is said to be convex if and only if

$$\phi\left(\frac{1}{2}X_1 + \frac{1}{2}X_2\right) \preccurlyeq \frac{1}{2}\phi(X_1) + \frac{1}{2}\phi(X_2) \tag{113}$$

holds for all $X_1, \ X_2 \in \mathbb{C}^{p\times m}$; said to be concave if and only if

$$\phi\left(\frac{1}{2}X_1 + \frac{1}{2}X_2\right) \succcurlyeq \frac{1}{2}\phi(X_1) + \frac{1}{2}\phi(X_2) \tag{114}$$

holds for all $X_1, \ X_2 \in \mathbb{C}^{p\times m}$. It is easy to verify that

$$\phi\left(\frac{1}{2}X_1 + \frac{1}{2}X_2\right) - \frac{1}{2}\phi(X_1) - \frac{1}{2}\phi(X_2) = -\frac{1}{4}A(X_1 - X_2)BMB^*(X_1 - X_2)^*A^*, \tag{115}$$

which is a special case of (2) as well. Applying Theorem 10 to (115), we obtain the following result.

Theorem 20. *Let $\phi(X)$ be as given in (2) with $A \neq 0$ and $BMB^* \neq 0$. Then,*

$$\max_{X_1 \neq X_2,\, X_1,\, X_2 \in \mathbb{C}^{p \times m}} r\left[\phi\left(\frac{1}{2}X_1 + \frac{1}{2}X_2\right) - \frac{1}{2}\phi(X_1) - \frac{1}{2}\phi(X_2)\right] = \min\{r(A),\ r(BMB^*)\}, \tag{116}$$

$$\min_{X_1 \neq X_2,\, X_1,\, X_2 \in \mathbb{C}^{p \times m}} r\left[\phi\left(\frac{1}{2}X_1 + \frac{1}{2}X_2\right) - \frac{1}{2}\phi(X_1) - \frac{1}{2}\phi(X_2)\right] = \begin{cases} 1 & BMB^* \succ 0 \text{ and } r(A) = p \\ 1 & BMB^* \prec 0 \text{ and } r(A) = p \\ 0 & \text{otherwise} \end{cases}, \tag{117}$$

$$\max_{X_1 \neq X_2,\, X_1,\, X_2 \in \mathbb{C}^{p \times m}} i_+\left[\phi\left(\frac{1}{2}X_1 + \frac{1}{2}X_2\right) - \frac{1}{2}\phi(X_1) - \frac{1}{2}\phi(X_2)\right] = \min\{r(A),\ i_-(BMB^*)\}, \tag{118}$$

$$\max_{X_1 \neq X_2,\, X_1,\, X_2 \in \mathbb{C}^{p \times m}} i_-\left[\phi\left(\frac{1}{2}X_1 + \frac{1}{2}X_2\right) - \frac{1}{2}\phi(X_1) - \frac{1}{2}\phi(X_2)\right] = \min\{r(A),\ i_+(BMB^*)\}, \tag{119}$$

$$\min_{X_1 \neq X_2,\, X_1,\, X_2 \in \mathbb{C}^{p \times m}} i_+\left[\phi\left(\frac{1}{2}X_1 + \frac{1}{2}X_2\right) - \frac{1}{2}\phi(X_1) - \frac{1}{2}\phi(X_2)\right] = \begin{cases} 1 & BMB^* \prec 0 \text{ and } r(A) = p \\ 0 & BMB^* \not\prec 0 \text{ or } r(A) < p \end{cases}, \tag{120}$$

$$\min_{X_1 \neq X_2,\, X_1,\, X_2 \in \mathbb{C}^{p \times m}} i_-\left[\phi\left(\frac{1}{2}X_1 + \frac{1}{2}X_2\right) - \frac{1}{2}\phi(X_1) - \frac{1}{2}\phi(X_2)\right] = \begin{cases} 1 & BMB^* \succ 0 \text{ and } r(A) = p \\ 0 & BMB^* \not\succ 0 \text{ or } r(A) < p \end{cases}. \tag{121}$$

In consequence, the following hold.

(a) *There exist $X_1,\ X_2 \in \mathbb{C}^{p \times m}$ with $X_1 \neq X_2$ such that $\phi\left(\frac{1}{2}X_1 + \frac{1}{2}X_2\right) - \frac{1}{2}\phi(X_1) - \frac{1}{2}\phi(X_2)$ is nonsingular if and only if both $r(A) = n$ and $r(BMB^*) \succcurlyeq n$.*

(b) *There exist $X_1,\ X_2 \in \mathbb{C}^{p \times m}$ with $X_1 \neq X_2$ such that $\phi\left(\frac{1}{2}X_1 + \frac{1}{2}X_2\right) = \frac{1}{2}\phi(X_1) + \frac{1}{2}\phi(X_2)$ if and only if $BMB^* \not\succ 0$ and $BMB^* \not\prec 0$, or $r(A) < p$.*

(c) *There exist $X_1,\ X_2 \in \mathbb{C}^{p \times m}$ with $X_1 \neq X_2$ such that $\phi\left(\frac{1}{2}X_1 + \frac{1}{2}X_2\right) \succ \frac{1}{2}\phi(X_1) + \frac{1}{2}\phi(X_2)$ if and only if both $BMB^* \prec 0$ and $r(A) = n$.*

(d) *There exist $X_1,\ X_2 \in \mathbb{C}^{p \times m}$ with $X_1 \neq X_2$ such that $\phi\left(\frac{1}{2}X_1 + \frac{1}{2}X_2\right) \prec \frac{1}{2}\phi(X_1) + \frac{1}{2}\phi(X_2)$ if and only if both $BMB^* \succ 0$ and $r(A) = n$.*

(e) *There exist $X_1,\ X_2 \in \mathbb{C}^{p \times m}$ with $X_1 \neq X_2$ such that $\phi\left(\frac{1}{2}X_1 + \frac{1}{2}X_2\right) \succcurlyeq \frac{1}{2}\phi(X_1) + \frac{1}{2}\phi(X_2)$ if and only if either $BMB^* \not\succ 0$ or $r(A) < p$.*

(f) *There exist $X_1,\ X_2 \in \mathbb{C}^{p \times m}$ with $X_1 \neq X_2$ such that $\phi\left(\frac{1}{2}X_1 + \frac{1}{2}X_2\right) \preccurlyeq \frac{1}{2}\phi(X_1) + \frac{1}{2}\phi(X_2)$ if and only if either $BMB^* \not\prec 0$ or $r(A) < p$.*

(g) *$\phi\left(\frac{1}{2}X_1 + \frac{1}{2}X_2\right) \succcurlyeq \frac{1}{2}\phi(X_1) + \frac{1}{2}\phi(X_2)$ for all $X_1,\ X_2 \in \mathbb{C}^{p \times m}$ with $X_1 \neq X_2$ if and only if $BMB^* \preccurlyeq 0$.*

(h) $\phi\left(\frac{1}{2}X_1 + \frac{1}{2}X_2\right) \preccurlyeq \frac{1}{2}\phi(X_1) + \frac{1}{2}\phi(X_2)$ *for all* $X_1,\ X_2 \in \mathbb{C}^{p\times m}$ *with* $X_1 \neq X_2$ *if and only if* $BMB^* \succcurlyeq 0$.

(i) *If* $\phi(X)$ *is a positive semi-definite matrix-valued function, then* $\phi(X)$ *is convex.*

(j) *If* $\phi(X)$ *is a negative semi-definite matrix-valued function, then* $\phi(X)$ *is concave.*

5. Two Types of Multiple HQMVF

As an extension of (2), we may construct a multiple Hermitian quadratic matrix-valued function as follows

$$\phi(X_1, \ldots, X_k) = \left(\sum_{i=1}^{k} A_iX_iB_i + C\right) M \left(\sum_{i=1}^{k} A_iX_iB_i + C\right)^* + D, \tag{122}$$

where $0 \neq A_i \in \mathbb{C}^{n\times p_i}$, $B_i \in \mathbb{C}^{m_i\times q}$, $C \in \mathbb{C}^{n\times q}$, $D \in \mathbb{C}^n_{\mathrm{H}}$ and $M \in \mathbb{C}^q_{\mathrm{H}}$ are given, and $X_i \in \mathbb{C}^{p_i\times m_i}$ is a variable matrix, $i = 1, \ldots, k$. We treat it as a combined non-homogeneous linear and quadratic Hermitian matrix-valued functions $\phi = \tau \circ \psi$:

$$\psi : \mathbb{C}^{p_1\times m_1} \oplus \cdots \oplus \mathbb{C}^{p_k\times m_k} \to \mathbb{C}^{n\times q}, \quad \tau : \mathbb{C}^{n\times q} \to \mathbb{C}^n_{\mathrm{H}}.$$

Because more independent variable matrices occur in (122), the author fails to establish analytical formulas for calculating the extremal ranks and inertias of (122). In this section, we only consider the following two problems on the semi-definiteness and optimization in the Löwner sense of (122):

(i) establish necessary and sufficient conditions for $\phi(X_1, \ldots, X_k) \succcurlyeq 0$ ($\phi(X_1, \ldots, X_k) \preccurlyeq 0$) to hold for all $X_1, \ldots, X_k$;

(ii) establish necessary and sufficient conditions for the existence of $\widehat{X}_1, \ldots, \widehat{X}_k$ and $\widetilde{X}_1, \ldots, \widetilde{X}_k$ such that

$$\phi(X_1, \ldots, X_k) \succcurlyeq \phi(\widehat{X}_1, \ldots, \widehat{X}_k), \quad \phi(X_1, \ldots, X_k) \preccurlyeq \phi(\widetilde{X}_1, \ldots, \widetilde{X}_k) \tag{123}$$

hold for all $X_1, \ldots, X_k$ in the Löwner partial ordering, respectively, and give analytical expressions of $\widehat{X}_1, \ldots, \widehat{X}_k$ and $\widetilde{X}_1, \ldots, \widetilde{X}_k$.

Theorem 21. *Let* $\phi(X_1, \ldots, X_k)$ *be as given in* (122), *and define* $B^* = [B_1^*, \ldots, B_k^*]$. *Also let*

$$N = \begin{bmatrix} D + CMC^* & CMB^* \\ BMC^* & BMB^* \end{bmatrix}. \tag{124}$$

Then, the following hold.

(a) $\phi(X_1, \ldots, X_k) \succcurlyeq 0$ *for all* $X_1 \in \mathbb{C}^{p_1\times m_1}, \ldots, X_k \in \mathbb{C}^{p_k\times m_k}$ *if and only if* $N \succcurlyeq 0$.

(b) $\phi(X_1, \ldots, X_k) \preccurlyeq 0$ *for all* $X_1 \in \mathbb{C}^{p_1\times m_1}, \ldots, X_k \in \mathbb{C}^{p_k\times m_k}$ *if and only if* $N \preccurlyeq 0$.

(c) *There exist* $\widehat{X}_1, \ldots, \widehat{X}_k$ *such that*

$$\phi(X_1, \ldots, X_k) \succcurlyeq \phi(\widehat{X}_1, \ldots, \widehat{X}_k) \tag{125}$$

holds for all $X_1 \in \mathbb{C}^{p_1 \times m_1}, \ldots, X_k \in \mathbb{C}^{p_k \times m_k}$ *if and only if*

$$BMB^* \succcurlyeq 0, \ \ \mathscr{R}(BMC^*) \subseteq \mathscr{R}(BMB^*). \tag{126}$$

In this case, the matrices $\widehat{X}_1, \ldots, \widehat{X}_k$ *are the solutions of the linear matrix equation*

$$\sum_{i=1}^{k} A_i \widehat{X}_i B_i M B^* = -CMB^*. \tag{127}$$

Correspondingly,

$$\phi(\widehat{X}_1, \ldots, \widehat{X}_k) = D + CMC^* - CMB^*(BMB^*)^{\dagger} BMC^*, \tag{128}$$

$$\phi(X_1, \ldots, X_k) - \phi(\widehat{X}_1, \ldots, \widehat{X}_k)$$
$$= \left(\sum_{i=1}^{k} A_i X_i B_i + C \right) MB^*(BMB^*)^{\dagger} BM \left(\sum_{i=1}^{k} A_i X_i B_i + C \right)^*. \tag{129}$$

(d) *There exist* $\widetilde{X}_1, \ldots, \widetilde{X}_k$ *such that*

$$\phi(X_1, \ldots, X_k) \preccurlyeq \phi(\widetilde{X}_1, \ldots, \widetilde{X}_k) \tag{130}$$

holds for all $X_1 \in \mathbb{C}^{p_1 \times m_1}, \ldots, X_k \in \mathbb{C}^{p_k \times m_k}$ *if and only if*

$$BMB^* \preccurlyeq 0, \ \ \mathscr{R}(BMC^*) \subseteq \mathscr{R}(BMB^*). \tag{131}$$

In this case, the matrices $\widetilde{X}_1, \ldots, \widetilde{X}_k$ *are the solutions of the linear matrix equation*

$$\sum_{i=1}^{k} A_i \widetilde{X}_i B_i M B^* = -CMB^*. \tag{132}$$

Correspondingly,

$$\phi(\widetilde{X}_1, \ldots, \widetilde{X}_k) = D + CMC^* - CMB^*(BMB^*)^{\dagger} BMC^*, \tag{133}$$

$$\phi(X_1, \ldots, X_k) - \phi(\widetilde{X}_1, \ldots, \widetilde{X}_k)$$
$$= \left(\sum_{i=1}^{k} A_i X_i B_i + C \right) MB^*(BMB^*)^{\dagger} BM \left(\sum_{i=1}^{k} A_i X_i B_i + C \right)^*. \tag{134}$$

Proof. Rewrite (122) as

$$\phi(X_1, \ldots, X_k)$$
$$= \left(A_1 X_1 B_1 + \sum_{i=2}^{k} A_i X_i B_i + C \right) M \left(A_1 X_1 B_1 + \sum_{i=2}^{k} A_i X_i B_i + C \right)^* + D, \tag{135}$$

and applying Corollary 18 to it, we see that $\phi(X_1,\ldots,X_k) \succcurlyeq 0$ for all $X_1 \in \mathbb{C}^{p_1\times m_1}$ if and only if

$$\begin{bmatrix} D & 0 \\ 0 & 0 \end{bmatrix} + \begin{bmatrix} \sum_{i=2}^k A_iX_iB_i + C \\ B_1 \end{bmatrix} M \begin{bmatrix} \sum_{i=2}^k A_iX_iB_i + C \\ B_1 \end{bmatrix}^* \succcurlyeq 0 \tag{136}$$

for all $X_2 \in \mathbb{C}^{p_2\times m_2},\ldots,X_k \in \mathbb{C}^{p_k\times m_k}$. Note that

$$\begin{aligned} &\begin{bmatrix} D & 0 \\ 0 & 0 \end{bmatrix} + \begin{bmatrix} \sum_{i=2}^k A_iX_iB_i + C \\ B_1 \end{bmatrix} M \begin{bmatrix} \sum_{i=2}^k A_iX_iB_i + C \\ B_1 \end{bmatrix}^* \\ &= \begin{bmatrix} D & 0 \\ 0 & 0 \end{bmatrix} + \left(\sum_{i=2}^k \begin{bmatrix} A_i \\ 0 \end{bmatrix} X_iB_i + \begin{bmatrix} C \\ B_1 \end{bmatrix}\right) M \left(\sum_{i=2}^k \begin{bmatrix} A_i \\ 0 \end{bmatrix} X_iB_i + \begin{bmatrix} C \\ B_1 \end{bmatrix}\right)^*. \end{aligned} \tag{137}$$

Applying Corollary 18, we see that this matrix is positive semi-definite for all $X_2 \in \mathbb{C}^{p_2\times m_2}$ if and only if

$$\begin{bmatrix} D & 0 & 0 \\ 0 & 0 & 0 \\ 0 & 0 & 0 \end{bmatrix} + \begin{bmatrix} \sum_{i=3}^k \begin{bmatrix} A_i \\ 0 \end{bmatrix} X_iB_i + \begin{bmatrix} C \\ B_1 \end{bmatrix} \\ B_2 \end{bmatrix} M \begin{bmatrix} \sum_{i=3}^k \begin{bmatrix} A_i \\ 0 \end{bmatrix} X_iB_i + \begin{bmatrix} C \\ B_1 \end{bmatrix} \\ B_2 \end{bmatrix}^* \succcurlyeq 0 \tag{138}$$

for all $X_3 \in \mathbb{C}^{p_3\times m_3},\ldots,X_k \in \mathbb{C}^{p_k\times m_k}$. Thus, we obtain by induction that $\phi(X_1,\ldots,X_k) \succcurlyeq 0$ for all $X_1 \in \mathbb{C}^{p_1\times m_1},\ldots,X_k \in \mathbb{C}^{p_k\times m_k}$ if and only if

$$\begin{bmatrix} D+CMC^* & CMB_1^* & \ldots & CMB_k^* \\ B_1MC^* & B_1MB_1^* & \ldots & B_1MB_k^* \\ \vdots & \vdots & \ddots & \vdots \\ B_kMC^* & B_kMB_1^* & \cdots & B_kMB_k^* \end{bmatrix} = \begin{bmatrix} D+CMC^* & CMB^* \\ BMC^* & BMB^* \end{bmatrix} \succcurlyeq 0,$$

establishing (a).

Let

$$\begin{aligned} \rho(X_1,\ldots,X_k) &= \phi(X_1,\ldots,X_k) - \phi(\widehat{X}_1,\ldots,\widehat{X}_k) \\ &= \left(\sum_{i=1}^k A_iX_iB_i + C\right) M \left(\sum_{i=1}^k A_iX_iB_i + C\right)^* \\ &\quad - \left(\sum_{i=1}^k A_i\widehat{X}_iB_i + C\right) M \left(\sum_{i=1}^k A_i\widehat{X}_iB_i + C\right)^*. \end{aligned} \tag{139}$$

Then, $\phi(X_1,\ldots,X_k) \succcurlyeq \phi(\widehat{X}_1,\ldots,\widehat{X}_k)$ for all $X_1 \in \mathbb{C}^{p_1\times m_1},\ldots,X_k \in \mathbb{C}^{p_k\times m_k}$ is equivalent to

$$\rho(X_1,\ldots,X_k) \succcurlyeq 0 \ \text{ for all } \ X_1 \in \mathbb{C}^{p_1\times m_1},\ldots,X_k \in \mathbb{C}^{p_k\times m_k}. \tag{140}$$

From (a), (140) holds if and only if

$$\begin{bmatrix} -\left(\sum_{i=1}^k A_i\widehat{X}_iB_i + C\right) M \left(\sum_{i=1}^k A_i\widehat{X}_iB_i + C\right)^* + CMC^* & CMB^* \\ BMC^* & BMB^* \end{bmatrix} \succcurlyeq 0, \tag{141}$$

which, by (93)–(95), is further equivalent to

$$BMB^* \succcurlyeq 0, \ \mathscr{R}(BMC^*) \subseteq \mathscr{R}(BMB^*), \tag{142}$$

$$CMC^* - \left(\sum_{i=1}^{k} A_i\widehat{X}_iB_i + C\right)M\left(\sum_{i=1}^{k} A_i\widehat{X}_iB_i + C\right)^* - CMB^*(BMB^*)^{\dagger}BMC^* \succcurlyeq 0. \tag{143}$$

In this case,

$$\begin{aligned} &CMC^* - \left(\sum_{i=1}^{k} A_i\widehat{X}_iB_i + C\right)M\left(\sum_{i=1}^{k} A_i\widehat{X}_iB_i + C\right)^* - CMB^*(BMB^*)^{\dagger}BMC^* \\ &= -\left(\sum_{i=1}^{k} A_i\widehat{X}_iB_i + C\right)MB^*(BMB^*)^{\dagger}BM\left(\sum_{i=1}^{k} A_i\widehat{X}_iB_i + C\right)^* \end{aligned} \tag{144}$$

holds, and therefore, (143) is equivalent to $CMB^* + \sum_{i=1}^{k} A_i\widehat{X}_iB_iMB^* = 0$. This is a general two-sided linear matrix equation involving k unknown matrices. The existence of solutions of this equation and its general solution can be derived from the Kronecker product of matrices. The details are omitted here. Result (d) can be shown similarly. ∎

Two consequences of Theorem 21 are given below.

Corollary 22. *Let*

$$\psi(X_1, \ldots, X_k) = \sum_{i=1}^{k} (A_iX_iB_i + C_i)M_i(A_iX_iB_i + C_i)^* + D, \tag{145}$$

where $0 \neq A_i \in \mathbb{C}^{n\times p_i}$, $B_i \in \mathbb{C}^{m_i\times q_i}$, $C \in \mathbb{C}^{n\times q_i}$, $D \in \mathbb{C}_{\mathrm{H}}^{n}$ *and* $M_i \in \mathbb{C}_{\mathrm{H}}^{q_i}$ *are given, and* $X_i \in \mathbb{C}^{p_i\times m_i}$ *is a variable matrix,* $i = 1, \ldots, k$. *Also define*

$$B = \mathrm{diag}(B_1, \ldots, B_k), \ C = [C_1, \ldots, C_k],$$
$$M = \mathrm{diag}(M_1, \ldots, M_k), \ N = \begin{bmatrix} D + CMC^* & CMB^* \\ BMC^* & BMB^* \end{bmatrix}.$$

Then, the following hold.

(a) $\psi(X_1, \ldots, X_k) \succcurlyeq 0$ *for all* $X_1 \in \mathbb{C}^{p_1\times m_1}, \ldots, X_k \in \mathbb{C}^{p_k\times m_k}$ *if and only if* $N \succcurlyeq 0$.

(b) $\psi(X_1, \ldots, X_k) \preccurlyeq 0$ *for all* $X_1 \in \mathbb{C}^{p_1\times m_1}, \ldots, X_k \in \mathbb{C}^{p_k\times m_k}$ *if and only if* $N \preccurlyeq 0$.

(c) *There exist* $\widehat{X}_1, \ldots, \widehat{X}_k$ *such that*

$$\psi(X_1, \ldots, X_k) \succcurlyeq \psi(\widehat{X}_1, \ldots, \widehat{X}_k) \tag{146}$$

holds for all $X_1 \in \mathbb{C}^{p_1\times m_1}, \ldots, X_k \in \mathbb{C}^{p_k\times m_k}$ *if and only if*

$$B_iM_iB_i^* \succcurlyeq 0, \ \mathscr{R}(B_iM_iC_i^*) \subseteq \mathscr{R}(B_iM_iB_i^*), \ i = 1, \ldots, k. \tag{147}$$

In this case, the matrices $\widehat{X}_1, \ldots, \widehat{X}_k$ *satisfying* (146) *are the solutions of the k linear matrix equations*

$$A_i \widehat{X}_i B_i M_i B_i^* = -C_i M_i B_i^*, \quad i = 1, \ldots, k. \tag{148}$$

Correspondingly,

$$\psi(\widehat{X}_1, \ldots, \widehat{X}_k) = D + \sum_{i=1}^{k} C_i M_i C_i^* - \sum_{i=1}^{k} C_i M_i B^* (B_i M_i B_i^*)^{\dagger} B_i M_i C_i^*, \tag{149}$$

$$\psi(X_1, \ldots, X_k) - \psi(\widehat{X}_1, \ldots, \widehat{X}_k)$$
$$= \sum_{i=1}^{k} (A_i X_i B_i + C_i) M_i B_i^* (B_i M_i B_i^*)^{\dagger} B_i M_i (A_i X_i B_i + C_i)^*. \tag{150}$$

(d) *There exist* $\widetilde{X}_1, \ldots, \widetilde{X}_k$ *such that*

$$\psi(X_1, \ldots, X_k) \preccurlyeq \psi(\widetilde{X}_1, \ldots, \widetilde{X}_k) \tag{151}$$

holds for all $X_1 \in \mathbb{C}^{p_1 \times m_1}, \ldots, X_k \in \mathbb{C}^{p_k \times m_k}$ *if and only if*

$$B_i M_i B_i^* \preccurlyeq 0, \ \mathscr{R}(B_i M_i C_i^*) \subseteq \mathscr{R}(B_i M_i B_i^*), \quad i = 1, \ldots, k. \tag{152}$$

In this case, the matrices $\widetilde{X}_1, \ldots, \widetilde{X}_k$ *satisfying* (151) *are the solutions of the k linear matrix equations*

$$A_i \widetilde{X}_i B_i M_i B_i^* = -C_i M_i B_i^*, \quad i = 1, \ldots, k. \tag{153}$$

Correspondingly,

$$\psi(\widetilde{X}_1, \ldots, \widetilde{X}_k) = D + \sum_{i=1}^{k} C_i M_i C_i^* - \sum_{i=1}^{k} C_i M_i B^* (B_i M_i B_i^*)^{\dagger} B_i M_i C_i^*, \tag{154}$$

$$\psi(X_1, \ldots, X_k) - \psi(\widetilde{X}_1, \ldots, \widetilde{X}_k)$$
$$= \sum_{i=1}^{k} (A_i X_i B_i + C_i) M_i B_i^* (B_i M_i B_i^*)^{\dagger} B_i M_i (A_i X_i B_i + C_i)^*. \tag{155}$$

Proof. Rewrite (145) as

$$\begin{aligned}
&\psi(X_1, \ldots, X_k) \\
&= [A_1 X_1 B_1 + C_1, \ldots, A_k X_k B_k + C_k] M \, [A_1 X_1 B_1 + C_1, \ldots, A_k X_k B_k + C_k]^* + D \\
&= [A_1 X_1 [B_1, \ldots, 0] + \cdots + A_k X_k [0, \ldots, B_k] + [C_1, \ldots, C_k]] M \\
&\quad \times [A_1 X_1 [B_1, \ldots, 0] + \cdots + A_k X_k [0, \ldots, B_k] + [C_1, \ldots, C_k]]^* + D,
\end{aligned} \tag{156}$$

which a special case of (122). Applying Theorem 21 to it, we obtain the result required. ∎

Corollary 23. *Let*

$$\begin{aligned}&\psi(X_1,\ldots,X_k)\\&=[A_1X_1B_1+C_1,\ldots,A_kX_kB_k+C_k]M[A_1X_1B_1+C_1,\ldots,A_kX_kB_k+C_k]^*\\&\quad+D,\end{aligned}\tag{157}$$

where $0\neq A_i\in\mathbb{C}^{n\times p_i}$, $B_i\in\mathbb{C}^{m_i\times q_i}$, $C_i\in\mathbb{C}^{n\times q_i}$, $D\in\mathbb{C}_{\mathrm{H}}^{n}$ *and* $M\in\mathbb{C}_{\mathrm{H}}^{q_1+\cdots+q_k}$ *are given, and* $X_i\in\mathbb{C}^{p_i\times m_i}$ *is variable matrix,* $i=1,\ldots,k$. *Also define*

$$B=\mathrm{diag}(B_1,\ldots,B_k)\ \ and\ \ C=[C_1,\ldots,C_k],\ \ N=\begin{bmatrix}D+CMC^*&CMB^*\\BMC^*&BMB^*\end{bmatrix}.$$

Then, the following hold.

(a) $\psi(X_1,\ldots,X_k)\succcurlyeq0$ *for all* $X_1\in\mathbb{C}^{p_1\times m_1},\ldots,X_k\in\mathbb{C}^{p_k\times m_k}$ *if and only if* $N\succcurlyeq0$.

(b) $\psi(X_1,\ldots,X_k)\preccurlyeq0$ *for all* $X_1\in\mathbb{C}^{p_1\times m_1},\ldots,X_k\in\mathbb{C}^{p_k\times m_k}$ *if and only if* $N\preccurlyeq0$.

(c) *There exist* $\widehat{X}_1,\ldots,\widehat{X}_k$ *such that*

$$\psi(X_1,\ldots,X_k)\succcurlyeq\psi(\widehat{X}_1,\ldots,\widehat{X}_k)\tag{158}$$

holds for all $X_1\in\mathbb{C}^{p_1\times m_1},\ldots,X_k\in\mathbb{C}^{p_k\times m_k}$ *if and only if*

$$BMB^*\succcurlyeq0,\ \ \mathscr{R}(BMC^*)\subseteq\mathscr{R}(BMB^*).\tag{159}$$

In this case, the matrices $\widehat{X}_1,\ldots,\widehat{X}_k$ *satisfying* (158) *are the solutions of the linear matrix equation*

$$\sum_{i=1}^{k}A_i\widehat{X}_iB_iMB^*=-CMB^*.\tag{160}$$

Correspondingly,

$$\psi(\widehat{X}_1,\ldots,\widehat{X}_k)=D+CMC^*-CMB^*(BMB^*)^{\dagger}BMC^*,\tag{161}$$

$$\begin{aligned}&\psi(X_1,\ldots,X_k)-\psi(\widehat{X}_1,\ldots,\widehat{X}_k)\\&=[A_1X_1B_1+C_1,\ldots,A_kX_kB_k+C_k]MB^*(BMB^*)^{\dagger}BM\\&\quad\times[A_1X_1B_1+C_1,\ldots,A_kX_kB_k+C_k]^*.\end{aligned}\tag{162}$$

(d) *There exist* $\widetilde{X}_1,\ldots,\widetilde{X}_k$ *such that*

$$\psi(X_1,\ldots,X_k)\preccurlyeq\psi(\widetilde{X}_1,\ldots,\widetilde{X}_k)\tag{163}$$

holds for all $X_1\in\mathbb{C}^{p_1\times m_1},\ldots,X_k\in\mathbb{C}^{p_k\times m_k}$ *if and only if*

$$BMB^*\preccurlyeq0,\ \ \mathscr{R}(BMC^*)\subseteq\mathscr{R}(BMB^*).\tag{164}$$

In this case, the matrices $\widetilde{X}_1,\ldots,\widetilde{X}_k$ *satisfying* (163) *are the solutions of the linear matrix equation*

$$\sum_{i=1}^{k}A_i\widetilde{X}_iB_iMB^*=-CMB^*.\tag{165}$$

Correspondingly,

$$\psi(\widetilde{X}_1,\ldots,\widetilde{X}_k) = D + CMC^* - CMB^*(BMB^*)^{\dagger}BMC^*, \tag{166}$$

$$\begin{aligned}&\psi(X_1,\ldots,X_k) - \psi(\widetilde{X}_1,\ldots,\widetilde{X}_k)\\ &= [\,A_1X_1B_1 + C_1,\ldots,A_kX_kB_k + C_k\,]\,MB^*(BMB^*)^{\dagger}BM\\ &\quad\times [\,A_1X_1B_1 + C_1,\ldots,A_kX_kB_k + C_k\,]^*.\end{aligned} \tag{167}$$

Proof. Rewrite (157) as

$$\begin{aligned}&\psi(X_1,\ldots,X_k)\\ &= [A_1X_1B_1 + C_1,\ldots,A_kX_kB_k + C_k]M[A_1X_1B_1 + C_1,\ldots,A_kX_kB_k + C_k]^* + D\\ &= (A_1X_1[\,B_1,\ldots,0\,] + \cdots + A_kX_k[\,0,\ldots,B_k\,] + [\,C_1,\ldots,C_k\,])M\\ &\quad\times (A_1X_1[\,B_1,\ldots,0\,] + \cdots + A_kX_k[\,0,\ldots,B_k\,] + [\,C_1,\ldots,C_k\,])^* + D,\end{aligned} \tag{168}$$

which a special case of (122). Applying Theorem 21 to it, we obtain the result required. ■

Many consequences can be derived from the results in this section. For instance,

(i) the semi-definiteness and optimization in the Löwner partial ordering of the following constrained HQMVF

$$\phi(X) = (\,AXB + C\,)M(\,AXB + C)^* + D \ \text{ s.t. } \ PXQ = R$$

can be derived;

(ii) the semi-definiteness and optimization in the Löwner partial ordering of the following matrix expressions that involve partially specified matrices

$$\begin{bmatrix} A & B \\ C & ? \end{bmatrix} M \begin{bmatrix} A & B \\ C & ? \end{bmatrix}^* + N, \ \begin{bmatrix} ? & B \\ C & ? \end{bmatrix} M \begin{bmatrix} ? & B \\ C & ? \end{bmatrix}^* + N, \ \begin{bmatrix} A & ? \\ ? & ? \end{bmatrix} M \begin{bmatrix} A & ? \\ ? & ? \end{bmatrix}^* + N$$

can be derived, in particular, necessary and sufficient conditions can be derived for the following inequalities

$$\begin{bmatrix} A & B \\ C & ? \end{bmatrix} \begin{bmatrix} A & B \\ C & ? \end{bmatrix}^* \preccurlyeq I, \ \begin{bmatrix} ? & B \\ C & ? \end{bmatrix} \begin{bmatrix} ? & B \\ C & ? \end{bmatrix}^* \preccurlyeq I, \ \begin{bmatrix} A & ? \\ ? & ? \end{bmatrix} \begin{bmatrix} A & ? \\ ? & ? \end{bmatrix}^* \preccurlyeq I$$

to always hold in the Löwner partial ordering.

6. Some Optimization Problems on the Matrix Equation $AXB = C$

Consider the following linear matrix equation

$$AXB = C, \tag{169}$$

where $A \in \mathbb{C}^{m\times n}$, $B \in \mathbb{C}^{p\times q}$ and $C \in \mathbb{C}^{m\times q}$ are given, and $X \in \mathbb{C}^{n\times p}$ is an unknown matrix. Eq. (169) is one of the best known matrix equations in matrix theory. Many papers on this equation and its applications can be found in the literature. In the Penrose's seminal paper [14], the consistency conditions and the general solution of (169) were completely derived by using generalized inverse of matrices. If (169) is not consistent, people often need to find its approximation solutions under various optimal criteria, in particular, the least-squares criterion is ubiquitously used in optimization problems which almost always admits an explicit global solution. For (169), the least-squares solution is defined to be a matrix $X \in \mathbb{C}^{n\times p}$ that minimizes the quadratic objective function

$$\mathrm{trace}[\,(C - AXB)(C - AXB)^*\,] = \mathrm{trace}[\,(C - AXB)^*(C - AXB)\,]. \tag{170}$$

The normal matrix equation corresponding to (170) is given by

$$A^*AXBB^* = A^*CB^*, \tag{171}$$

which is always consistent, and the following result is well known.

Lemma 24. *The general least-squares solution of* (169) *can be written as*

$$X = A^{\dagger}CB^{\dagger} + F_AV_1 + V_2E_B, \tag{172}$$

where $V_1,\ V_2 \in \mathbb{C}^{n\times p}$ *are arbitrary.*

Let

$$\phi_1(X) = (C - AXB)(C - AXB)^*, \quad \phi_2(X) = (C - AXB)^*(C - AXB) \tag{173}$$

be the two HQMVFs in (170). Note that

$$r[\phi_1(X)] = r[\phi_2(X)] = r(C - AXB). \tag{174}$$

Hence, we first obtain the following result from Lemma 8.

Theorem 25. *Let* $\phi_1(X)$ *and* $\phi_2(X)$ *be as given in* (173). *Then,*

$$\max_{X\in\mathbb{C}^{n\times p}} r[\phi_1(X)] = \max_{X\in\mathbb{C}^{n\times p}} r[\phi_2(X)] = \max_{X\in\mathbb{C}^{n\times p}} r(C - AXB) = \min\left\{ r[\,A,\ C\,],\ r\begin{bmatrix} B \\ C \end{bmatrix} \right\}, \tag{175}$$

$$\min_{X\in\mathbb{C}^{n\times p}} r[\phi_1(X)] = \min_{X\in\mathbb{C}^{n\times p}} r[\phi_2(X)] = \min_{X\in\mathbb{C}^{n\times p}} r(C - AXB) = r[\,A,\ C\,] + r\begin{bmatrix} B \\ C \end{bmatrix} - r\begin{bmatrix} C & A \\ B & 0 \end{bmatrix}. \tag{176}$$

Applying Theorem 10 to (173), we obtain the following result.

Theorem 26. *Let* $\phi_1(X)$ *and* $\phi_2(X)$ *be as given in* (173). *Then, the following hold.*

(a) *There exists an* $\widehat{X} \in \mathbb{C}^{n\times p}$ *such that* $\phi_1(X) \succcurlyeq \phi_1(\widehat{X})$ *holds for all* $X \in \mathbb{C}^{n\times p}$ *if and only if*

$$\mathscr{R}(CB^*) \subseteq \mathscr{R}(A). \tag{177}$$

In this case,

$$\widehat{X} = A^{\dagger}CB^{\dagger} + F_AV_1 + V_2E_B, \quad \phi_1(\widehat{X}) = CC^* - CB^{\dagger}BC^*, \tag{178}$$

where $V_1,\ V_2 \in \mathbb{C}^{n\times p}$ *are arbitrary.*

(b) *There exists an* $\widehat{X} \in \mathbb{C}^{n\times p}$ *such that* $\phi_2(X) \succcurlyeq \phi_2(\widehat{X})$ *holds for all* $X \in \mathbb{C}^{n\times p}$ *if and only if*

$$\mathscr{R}(C^*A) \subseteq \mathscr{R}(B^*). \tag{179}$$

In this case,

$$\widehat{X} = A^{\dagger}CB^{\dagger} + F_AV_1 + V_2E_B, \ \ \phi_2(\widehat{X}) = C^*C - C^*AA^{\dagger}C, \tag{180}$$

where V_1, $V_2 \in \mathbb{C}^{n\times p}$ *are arbitrary.*

Theorem 26 also motivates us to obtain the following consequence.

Theorem 27. *Let* $A \in \mathbb{C}^{m\times n}$, $B \in \mathbb{C}^{p\times q}$ *and* $C \in \mathbb{C}^{m\times q}$ *be given. Then, there always exists an* $X \in \mathbb{C}^{n\times p}$ *such that*

$$\min_{\succcurlyeq}\left\{ A^*(\,C - AXB\,)(\,C - AXB\,)^*A \,:\, X \in \mathbb{C}^{n\times p}\right\}, \tag{181}$$

$$\min_{\succcurlyeq}\left\{ B(\,C - AXB\,)^*(\,C - AXB\,)B^* \,:\, X \in \mathbb{C}^{n\times p}\right\} \tag{182}$$

hold, simultaneously, and the general solution is given by

$$\begin{aligned}
&\operatorname*{argmin}_{\succcurlyeq}\left\{ A^*(\,C - AXB\,)(\,C - AXB\,)^*A \,:\, X \in \mathbb{C}^{n\times p}\,\right\}\\
&= \operatorname*{argmin}_{\succcurlyeq}\left\{ B(\,C - AXB\,)^*(\,C - AXB\,)B^* \,:\, X \in \mathbb{C}^{n\times p}\,\right\}\\
&= \operatorname*{argmin}_{X\in\mathbb{C}^{n\times p}} \operatorname{tr}\left[\,(\,C - AXB\,)(\,C - AXB\,)^*\,\right]\\
&= A^{\dagger}CB^{\dagger} + F_AV_1 + V_2E_B,
\end{aligned} \tag{183}$$

where V_1 *and* V_2 *are arbitrary matrices, namely, the solutions of the three minimization problems in* (183) *are the same.*

For (169), the weighted least-squares solutions with respect to positive semi-define matrices M and N are defined to be matrices $X \in \mathbb{C}^{n\times p}$ that satisfy

$$\operatorname{trace}\left[\,(\,C - AXB\,)M(\,C - AXB\,)^*\,\right] = \min, \tag{184}$$

$$\operatorname{trace}\left[\,(\,C - AXB\,)^*N(\,C - AXB\,)\,\right] = \min, \tag{185}$$

respectively. In this case, the two HQMVFs in (184) and (185) are

$$\phi_1(X) = (\,C - AXB\,)M(\,C - AXB\,)^*, \ \ \phi_2(X) = (\,C - AXB\,)^*N(\,C - AXB\,). \tag{186}$$

Hence, the theory on the ranks and inertias of $\phi_1(X)$ and $\phi_2(X)$ can be established routinely.

Recall that the least-squares solution of a linear matrix equation is defined by minimizing the trace of certain HQMVF. For example, the least-squares solution of the well-known linear matrix equation

$$AXB + CYD = E, \tag{187}$$

where A, B, C, D are given, are two matrices X and Y such that

$$\begin{aligned}&\operatorname{trace}[(E - AXB - CYD)(E - AXB - CYD)^*]\\&= \operatorname{trace}[(E - AXB - CYD)^*(E - AXB - CYD)] = \min.\end{aligned}$$

Correspondingly, solutions to the Löwner partial ordering minimization problems of the two HQMVFs

$$(E - AXB - CYD)(E - AXB - CYD)^*, \quad (E - AXB - CYD)^*(E - AXB - CYD)$$

can be derived from Theorem 21.

7. Concluding Remarks

We established in this chapter a group of explicit formulas for calculating the global maximal and minimal ranks and inertias of (2) when X runs over the whole matrix space. By taking these rank and inertia formulas as quantitative tools, we characterized many algebraic properties of (2), including solvability conditions for some nonlinear matrix equations and inequalities generated from (2), and analytical solutions to the two well-known classic optimization problems on the $\phi(X)$ in the Löwner partial ordering. The results obtained and the techniques adopted for solving the matrix rank and inertia optimization problems enable us to make new extensions of some classic results on quadratic forms, Hermitian quadratic matrix equations and Hermitian quadratic matrix inequalities, and to derive many new algebraic properties of nonlinear matrix functions that can hardly be handled before. As a continuation of this work, we mention some research problems on HQMVFs for further consideration.

(i) Characterize algebraic and topological properties of generalized Stiefel manifolds composed by the collections of all matrices satisfying (32)–(35).

(ii) The difference of (2) at two given matrices $X,\ X + \Delta X \in \mathbb{C}^{p\times m}$:

$$\phi(X + \Delta X) - \phi(X)$$

is homogenous with respect to ΔX. Hence, we can add a restriction on its norm, for instance, $\|\Delta X\| = \sqrt{\operatorname{tr}[(\Delta X)(\Delta X)^*]} < \delta$. In this case, establish formulas for calculating the maximal and minimal ranks and inertias of the difference with respect to $\Delta X \neq 0$, and use them to analyze behaviors of $\phi(X)$ nearby X. Also note that any matrix $X = (x_{ij})_{p\times m}$ can be decomposed as $X = \sum_{i=1}^{p}\sum_{j=1}^{m} x_{ij}e_{ij}$. A precise analysis on the difference is to take $\Delta X = \lambda e_{ij}$ and to characterize behaviors of the difference by using the corresponding rank and inertia formulas.

(iii) Denote the real and complex parts of (2) as $\phi(X) = \phi_0(X) + i\phi_1(X)$, where $\phi_0(X)$ and $\phi_1(X)$ are two real quadratic matrix-valued functions satisfying $\phi_0^T(X) = \phi_0(X)$ and $\phi_1^T(X) = -\phi_1(X)$, where $(\cdot)^T$ means the transpose of a matrix. In this case, establish formulas for calculating the maximal and minimal ranks and inertias of $\phi_0(X)$ and $\phi_1(X)$, and use them to characterize behaviors of $\phi_0(X)$ and $\phi_1(X)$.

(iv) Partition $\phi(X)$ in (2) as

$$\phi(X) = \begin{bmatrix} \phi_{11}(X) & \phi_{12}(X) \\ \phi_{12}^*(X) & \phi_{22}(X) \end{bmatrix}.$$

In this case, establish formulas for calculating the maximal and minimal ranks and inertias of the submatrices $\phi_{11}(X)$ and $\phi_{22}(X)$ with respect to X, and utilize them to characterize behaviors of these submatrices.

(v) Most criteria related to vector and matrix optimizations are constructed via traces of matrices. An optimization theory for (2) can also be established by taking the trace of (2) as an objective function. In such a case, it would be of interest to characterize relations between the two optimization theories for (2) derived from the trace and the Löwner partial ordering.

(vi) Establish formulas for calculating the extremal ranks and inertias of

$$(AXB + C)M(AXB + C)^* + D \text{ s.t. } r(X) \leqslant k,$$

where $k \preccurlyeq \min\{p, m\}$. This rank-constrained matrix-valued function is equivalent to the following biquadratic matrix-valued function

$$(AYZB + C)M(AYZB + C)^* + D, \ Y \in \mathbb{C}^{p\times k}, \ Z \in \mathbb{C}^{k\times m}.$$

Some previous results on positive semi-definiteness of biquadratic forms can be found in [6, 12].

(vii) Establish formulas for calculating the maximal and minimal ranks and inertias of

$$(AXB + C)M(AXB + C)^* + D \text{ s.t. } PX = Q \text{ and/or } XR = S.$$

This task could be regarded as extensions of classic equality-constrained quadratic programming problems.

(viii) For two given HQMVFs

$$\phi_i(X) = (A_iXB_i + C_i)M(A_iXB_i + C_i)^* + D_i, \ i = 1, 2$$

of the same size, establish necessary and sufficient conditions for $\phi_1(X) \equiv \phi_2(X)$ to hold.

(ix) Note from (36) and (37) that the HQMVF in (2) is embed into the congruence transformation for a block Hermitian matrix consisting of the given matrices. This fact prompts us to construct some general nonlinear matrix-valued functions that can be embed in congruence transformations for block Hermitian matrices, for instance,

$$\begin{bmatrix} I_{m_1} & 0 & 0 \\ B_1X_1 & I_{m_2} & 0 \\ B_2X_2B_1X_1 & B_2X_2 & I_{m_3} \end{bmatrix} \begin{bmatrix} A_{11} & A_{12} & A_{13} \\ A_{12}^* & A_{22} & A_{23} \\ A_{13}^* & A_{23}^* & A_{33} \end{bmatrix} \begin{bmatrix} I_{m_1} & X_1^*B_1^* & X_1^*B_1^*X_2^*B_2^* \\ 0 & I_{m_2} & X_2^*B_2^* \\ 0 & 0 & I_{m_3} \end{bmatrix}$$
$$= \begin{bmatrix} * & * & * \\ * & * & * \\ * & * & \phi(X_1, X_2) \end{bmatrix},$$

where

$$\phi(X_1, X_2) = [\, B_2X_2B_1X_1,\; B_2X_2,\; I_{m_3} \,] \begin{bmatrix} A_{11} & A_{12} & A_{13} \\ A_{12}^* & A_{22} & A_{23} \\ A_{13}^* & A_{23}^* & A_{33} \end{bmatrix} \begin{bmatrix} X_1^*B_1^*X_2^*B_2^* \\ X_2^*B_2^* \\ I_{m_3} \end{bmatrix},$$

which is a special case of the following nonlinear matrix-valued function

$$\begin{aligned} &\phi(X_1,\, X_2) \\ &= (\, A_1X_1B_1 + C_1 \,)(\, A_2X_2B_2 + C_2 \,)M(\, A_2X_2B_2 + C_2)^*\,(\, A_1X_1B_1 + C_1 \,)^* \\ &\quad + D. \end{aligned}$$

In these cases, it would be of interest to establish possible formulas for calculating the extremal ranks and inertias of these nonlinear matrix-valued functions (biquadratic matrix-valued functions), in particular, to find criteria of identifying semi-definiteness of these nonlinear matrix-valued functions, and to solve the Löwner partial ordering optimization problems.

(x) Two special forms of (122) and (145) by setting $X_1 = \cdots = X_k = X$ are

$$\left(\sum_{i=1}^{k} A_iXB_i + C \right) M \left(\sum_{i=1}^{k} A_iXB_i + C \right)^* + D,$$

$$\sum_{i=1}^{k} (\, A_iXB_i + C_i \,)M_i(\, A_iXB_i + C_i \,)^* + D.$$

In this case, find criteria for the HQMVFs to be semi-definite, and solve for their global extremal matrices in the Löwner partial ordering.

(xi) Many expressions that involve matrices and their generalized inverses can be represented as quadratic matrix-valued functions, for instance,

$$D - B^*A_r^{\sim}B, \;\; A - BB^-A(BB^-)^*, \;\; A - BB^-A - A(BB^-)^* + BB^-A(BB^-)^*.$$

In these cases, it would be of interest to establish formulas for calculating the maximal and minimal ranks and inertias of these matrix expressions with respect to the reflexive Hermitian g-inverse $A_r^{\sim}$ of a Hermitian matrix A, and g-inverse B^- of B. Some recent work on the ranks and inertias of the Hermitian Schur complement $D - B^*A^{\sim}B$ and their applications was given in [18,21].

Since linear algebra is a successful theory with essential applications in most scientific fields, the methods and results in matrix theory are prototypes of many concepts and content in other advanced branches of mathematics. In particular, matrix functions, matrix equalities and matrix inequalities in the Löwner partial ordering, as well as generalized inverses of matrices were sufficiently extended to their counterparts for operators in a Hilbert space, or elements in a ring with involution, and their algebraic properties were extensively studied in the literature. In most cases, the conclusions on the complex matrices and their counterparts in general algebraic settings are analogous.

Also, note that the results in this chapter are derived from ordinary algebraic operations of the given matrices and their generalized inverses. Hence, it is no doubt that most of the conclusions in this chapter can trivially be extended to the corresponding equations and inequalities for linear operators on a Hilbert space or elements in a ring with involution.

Acknowledgments

This work was supported by National Natural Science Foundation of China (Grant No. 11271384).

References

[1] F.A. Badawi, On a quadratic matrix inequality and the corresponding algebraic Riccati equation, *Internat. J. Contr.* 6 (1982) 313–322.

[2] A. Beck, Quadratic matrix programming, *SIAM J. Optim.* 17 (2006) 1224–1238.

[3] A. Beck, Convexity properties associated with nonconvex quadratic matrix-valued functions and applications to quadratic programming, *J. Optim. Theory Appl.* 142 (2009) 1–29.

[4] V.P. Belavkin, P. Staszewski, Radon-Nikodym theorem for completely positive maps, *Reports Math. Phys.* 24 (1986) 49–55.

[5] S. Boyd, L. Vandenberghe, *Convex Optimization*, Cambridge University Press, 2004.

[6] A.P. Calderón, A note on biquadratic forms, *Linear Algebra Appl.* 7 (1973) 175–177.

[7] E. Candes, B. Recht, Exact matrix completion via convex optimization, *Found. of Comput. Math.* 9 (2009) 717–772.

[8] Y. Chabrillac, J.P. Crouzeix, Definiteness and semidefiniteness of quadratic forms revisited, *Linear Algebra Appl.* 63 (1984) 283–292.

[9] Y. Chen, Nonnegative definite matrices and their applications to matrix quadratic programming problems, *Linear Multilinear Algebra* 33 (1993) 189–201.

[10] M. Choi, Completely positive linear maps on complex matrices, *Linear Algebra Appl.* 10 (1975) 285–290.

[11] M. Choi, Positive linear maps, *Proc. Sympo. Pure Math.* 38 (1982) 583–590.

[12] M. Choi, Positive semidefinite biquadratic forms, Linear Algebra Appl. 12 (1975) 95–100.

[13] Y. Tian, Equalities and inequalities for inertias of Hermitian matrices with applications, *Linear Algebra Appl.* 433 (2010) 263–296.

[14] R. Penrose, A generalized inverse for matrices, *Proc. Cambridge Philos. Soc.* 51 (1955) 406–413.

[15] B. De Moor, G.H. Golub, The restricted singular value decomposition: properties and applications, *SIAM J. Matrix Anal. Appl.* 12 (1991) 401–425.

[16] Y. Tian, The maximal and minimal ranks of some expressions of generalized inverses of matrices, *Southeast Asian Bull. Math.* 25 (2002) 745–755.

[17] Y. Tian, S. Cheng, The maximal and minimal ranks of $A - BXC$ with applications, New York J. Math. 9 (2003) 345–362.

[18] Y. Liu, Y. Tian, Max-min problems on the ranks and inertias of the matrix expressions $A - BXC \pm (BXC)^*$ with applications, *J. Optim. Theory Appl.* 148 (2011) 593–622.

[19] Y. Tian, Solving optimization problems on ranks and inertias of some constrained nonlinear matrix functions via an algebraic linearization method, *Nonlinear Anal.* 75 (2012) 717–734.

[20] Y. Tian, Formulas for calculating the extremum ranks and inertias of a four-term quadratic matrix-valued function and their applications, *Linear Algebra Appl.* 437 (2012) 835–859.

[21] Y. Tian, On an equality and four inequalities for generalized inverses of Hermitian matrices, *Electron. J. Linear Algebra* 23 (2012) 11–42.

In: Advances in Linear Algebra Research
Editor: Ivan Kyrchei, pp. 185-237
ISBN: 978-1-63463-565-3

Chapter 6

INTRODUCTION TO THE THEORY OF TRIANGULAR MATRICES (TABLES)

Roman Zatorsky*
Precarpathian Vasyl Stefanyk National University,
Ivano-Frankivsk, Ukraine

Abstract

We consider elements of linear algebra based on triangular tables with entries in some number field and their functions, analogical to the classical notions of a matrix, determinant and permanent. Some properties are investigated and applications in various areas of mathematics are given.

Keywords: Triangular matrix, determinant, paradeterminant, permanent, parapermanent, Polya transformation, linear recurrent equations, power series

AMS Subject Classification: 15A15

1. Preface

It is difficult to overestimate the role of matrix theory and matrix method in different branches of mathematics. But in mathematics, it becomes necessary quite often to operate not only with rectangular tables, but also with tables of numbers of a different kind. Therefore, by matrices we mean tables of numbers of any form. In particular, the main "characters" of the chapter are special tables (which, for lack of a better term, will be called *triangular matrices*. As there are no ordinary triangular matrices in the chapter, there will be no misunderstanding) and their functions – paradeterminants and parapermanents, which are some analogs of the determinant and permanent of square matrices.

The paradeterminants and parapermanents of triangular matrices of the n-th order are also multilinear polynomials of the elements of these matrices, but with $\frac{n(n+1)}{2}$ variables and 2^{n-1} summands. Consequently, one could expect that the paradeterminants of the n-th

*E-mail address: romazz@rambler.ru

order are related to the determinants of special matrices of the n-th order. Theorem 5.3 validates this hypothesis.

The summation of summands in the paradeterminant is over the set of ordered partitions of an integer n into natural summands, which made it possible to write it as a triangular matrix. Accordingly, it got new properties, which are well suited for analysis of problems with linear recurrence relations and partitions.

It should also be noted that paradeterminants and determinants are in fruitful collaboration, which allows finding new properties of both functions. For instance, it gives partial solution of Polya's problem on reduction of the permanent of the square matrix to its equal determinant of the respective transformed square matrix.

The chapter is organized as follows.

Definitions of functions of triangular matrices are based on the combinatorial notions. In Section 2 we establish the combinatorial notions required for the construction of triangular matrix functions.

In Section 3, we give definitions of triangular matrices and parafunctions of triangular matrices, and operations with triangular matrices. We obtain an inverse triangular matrix and consider a paradeterminant product of triangular matrices and a scalar product of vector by parafunction.

We establish properties of parafunctions of triangular matrices In Section 4, and relations of parafunctions with other functions of matrices in Section 5.

Some theorems on applying triangular matrix calculus are considered in Section 6.

Facts set forth in Sections 2 were partly published in [1–4], in Section 3 and 4 were published in [5–9], in Sections 5 were published in [10–12], and in Sections 6 were partly published in [13, 14]

2. Combinatorial Preliminaries

Basic combinatorial notions, which give rise to the notions of the determinant and the permanent [15] of a square matrix are the notions of permutation and transversal[1]. More precisely, the determinant and the permanent of a square matrix are defined as follows[2]. Let

$$A = \begin{pmatrix} a_{11} & a_{12} & \cdots & a_{1n} \\ a_{21} & a_{22} & \cdots & a_{2n} \\ \vdots & \vdots & \cdots & \vdots \\ a_{n1} & a_{n2} & \cdots & a_{nn} \end{pmatrix}. \tag{2.1}$$

The permanent and the determinant of the square matrix (2.1) are, respectively, the numbers

$$\operatorname{per} A = \sum_{\varphi \in S_n} a_{i_1 1} a_{i_2 2} \cdot \ldots \cdot a_{i_n n},$$

[1]The transversal of a square matrix is a tuple of elements taken by one at a time from each row and each column of the matrix.

[2]The permanent can be defined for any rectangular matrix. In order to compare the definitions of determinant and permanent, however, we consider only the permanent of a square matrix.

$$\det A = \sum_{\varphi \in S_n} sign\, \varphi\, a_{i_1 1} a_{i_2 2} \cdot \ldots \cdot a_{i_n n},$$

where S_n is the set of all permutations $\varphi = \binom{1\ 2 \ldots n}{i_1 i_2 \ldots i_n}$, of order n, and sign φ is the sign of the permutation[3] φ.

Note that there is also an axiomatic approach to the definition of determinants (see, for instance, [16, 17]).

Analogous functions of triangular matrices, which are the subject of this chapter, are based on the combinatorial notions of an ordered partition of a positive integer n into positive integer summands and monotransversals [4]. In the construction of triangular matrix functions, the sets $\Xi(n)$ are also important. We turn our close attention to these and other combinatorial notions.

2.1. Multisets

In discrete mathematics, there are problems focusing on collections of objects, which may include identical ones. In this case, the language of the Cantor Set Theory presents certain difficulties and inconveniences. Since the mid-20th century, the notion of a multiset has become more and more significant in this context.

Definition 2.1. *A **multiset** A is an unordered tuple of elements of some set $[A]$. A set $[A]$ is called **a basis** of a multiset A.*

Definition 2.2. *If a multiset A consists of k_1 elements a_1, k_2 elements a_2, ..., k_n elements a_n, then it is said that this multiset has **primary specification***

$$S(A) = [a_1^{k_1}, a_2^{k_2}, \ldots, a_n^{k_n}] \tag{2.2}$$

The multiset A can then be written in its canonical form

$$A = \{a_1^{k_1}, a_2^{k_2}, \ldots, a_n^{k_n}\}. \tag{2.3}$$

*Numbers k_i, $i = 1, 2, \ldots, n$, are **exponents** of the primary specification of A.*

The primary specification of the multiset of exponents

$$\{k_1, k_2, \ldots, k_n\}$$

of the primary specification of the multiset(2.3), is called *the secondary specification* of this multiset and is denoted by

$$S(S(A)) = [[1^{\lambda_1}, 2^{\lambda_2}, \ldots, m^{\lambda_m}]],$$

and numbers λ_i, $i = 1, 2, \ldots, m$ *are exponents of the secondary specification of* A.

[3]In algebra it is proved that each permutation decomposes uniquely into a product of independent cycles. The number $n-r$, where n is the order of the permutation, and r is the number of cycles, is called the decrement *of the permutation*. The set of permutations of order n is divided into even and odd permutations. The evenness of the permutation coincides with the evenness of its decrement.

[4]See the definition 3.4 on p. 198

As the elements of a multiset are unordered, then, without losing generality, we can assume that the exponents of its primary specification satisfy the inequality

$$k_1 \geq k_2 \geq \ldots \geq k_n. \tag{2.4}$$

If

$$k_1 = k_2 = \ldots = k_n = 1,$$

then the multiset (2.3) becomes an ordinary set.

The number of all the elements of a multiset A is called its cardinality and denoted by $|A|$. It is evident that the cardinality of the multiset (2.3) equals

$$|A| = k_1 + k_2 + \ldots + k_n.$$

2.2. Partitions

There is hardly any area of knowledge where it would not be necessary to classify objects by some criteria. Classification like this usually leads to partition of some set into equivalence classes. That is precisely why much time and effort of different mathematicians have been devoted to the study of partitions.

This section covers some aspects of the theory of partition of positive integers into natural numbers. This aspect of the general partition theory has long been stimulated by a great number of problems of combinatorial and number-theoretic nature and, therefore, is well developed.

Definition 2.3. *Let Ω be the set where an associative and commutative operation $\oplus$ is defined (i.e., Ω with an operation $\oplus$ is a commutative semigroup). An m-**partition** of an element $\omega \in \Omega$ is the set*

$$\{\omega_1, \omega_2, \ldots, \omega_m\}, \ \omega_i \in \Omega,$$

the elements of which satisfy

$$\omega_1 \oplus \omega_2 \oplus \ldots \oplus \omega_m = \omega. \tag{2.5}$$

*If the order of the elements $\omega_1, \omega_2, \ldots, \omega_m$ in the partition (2.5) is important, then this m-partition is called **an ordered m-partition** of ω, or an m-**composition** of ω and is denoted by*

$$(\omega_1, \omega_2, \ldots, \omega_m).$$

Let us denote the set of all ordered and unordered m-partitions of ω by $C_m(\omega, \oplus)$ and $P_m(\omega, \oplus)$ respectively and set

$$C(\omega, \oplus) = \bigcup_{m \geq 1} C_m(\omega, \oplus) \ P(\omega, \oplus) = \bigcup_{m \geq 1} P_m(\omega, \oplus)$$

If $\Omega = \mathbb{N}$ and $\oplus$ are ordinary addition, then $C_m(n, +)$ and $P_m(n, +)$ are respectively the sets of all ordered and unordered partitions of a positive integer n into m positive integer summands, and the sets $C(n, +)$ and $P(n, +)$ are sets of all ordered and unordered partitions of the same number into any quantity of positive integer summands.

A partition of a positive integer into an unordered sum of positive integer summands may be regarded as the multiset of these summands.

Example 2.1. Let us construct the set of all ordered partitions of the number 4 into positive integer summands:

$$C(4,+) =$$
$$= \{(4),(3,1),(1,3),(2,2),(2,1,1),(1,2,1),(1,1,2),(1,1,1,1)\}.$$

$$C_2(4,+) = \{(3,1),(1,3),(2,2)\}.$$

Example 2.2.

$$P(4,+) = \{\{4\},\{3,1\},\{2,2\},\{2,1,1\},\{1,1,1,1\}\}.$$

$$P_4(9,+) = \{\{6,1,1,1\},\{5,2,1,1\},\{4,3,1,1\},\{4,2,2,1\},$$
$$\{3,3,2,1\},\{3,2,2,2\}\}.$$

Remark 2.1. In unordered partitions of a positive integer n into m positive integer summands, its summands are usually arranged in a nonascending order.

The set $C_m(n,+)$ is the set of all natural solutions of the equation

$$\alpha_1 + \ldots + \alpha_m = n, \; m \leqslant n; \tag{2.6}$$

the set $C(n,+)$ is the union of sets of all natural solutions of the equations

$$\alpha_1 + \ldots + \alpha_m = n, \; m = 1, \ldots, n;$$

the set $P_m(n,+)$ is the set of natural solutions of the system

$$\begin{cases} \alpha_1 + \alpha_2 + \ldots + \alpha_m = n, \\ \alpha_1 \geqslant \alpha_2 \geqslant \ldots \geqslant \alpha_m, \end{cases} \tag{2.7}$$

and finally, the set $P(n,+)$, with regard for zero components of a partition, is the set of all integral nonnegative solutions of the system

$$\begin{cases} \alpha_1 + \alpha_2 + \ldots + \alpha_n = n, \\ \alpha_1 \geqslant \alpha_2 \geqslant \ldots \geqslant \alpha_n, \end{cases} \tag{2.8}$$

If an unordered partition of a positive integer n consists of λ_1 ones, λ_2 twos and so on λ_n summands, which equal n, i.e., the primary specification of a partition is as follows $[1^{\lambda_1}, 2^{\lambda_2}, \ldots, n^{\lambda_n}]$, then the exponents of this specification satisfy the equation

$$\lambda_1 + 2\lambda_2 + \ldots + n\lambda_n = n \tag{2.9}$$

and the number of solutions of the system (2.8) equals the number of solutions of the equation (2.9).

But if an unordered partition consists of m summands, then the exponents of this specification satisfy the system of equations

$$\begin{cases} \lambda_1 + 2\lambda_2 + \ldots + n\lambda_n = n, \\ \lambda_1 + \lambda_2 + \ldots + \lambda_n = m, \end{cases} \tag{2.10}$$

Now the following is obvious.

Proposition 2.1. *There exists a natural bijection between the set of natural solutions of the system (2.7) and the set of integral nonnegative solutions of the system (2.10).*

Let us denote the following:

$$|C_m(n,+)| = c(n,m),\ |C(n,+)| = c(n),$$
$$|P_m(n,+)| = p(n,m),\ |P(n,+)| = p(n), \tag{2.11}$$

and we suppose

$$c(n,m) = p(n,m) = 0$$

for $n < m$.

It is obvious that

$$c(n) = \sum_{m=1}^{n} c(n,m),\ p(n) = \sum_{m=1}^{n} p(n,m). \tag{2.12}$$

Proposition 2.2. *The following equality holds*

$$c(n,m) = \binom{n-1}{m-1}. \tag{2.13}$$

Proof. Since $m \leqslant n$, then we subtract m from both members of the equation (2.6). We obtain the equation

$$(\alpha_1 - 1) + (\alpha_2 - 1) + \ldots + (\alpha_m - 1) = n - m.$$

Consequently, the bijection is established between natural solutions of the equation (2.6) and integral nonnegative solutions of the equation

$$\beta_1 + \beta_2 + \ldots + \beta_m = n - m,$$

where $\beta_i = \alpha_i - 1,\ i = 1, \ldots, m$. But the number of solutions of the latter equation equals $\binom{n-1}{m-1}$. ∎

Proposition 2.3. *The following equality holds*

$$c(n) = 2^{n-1}. \tag{2.14}$$

Proof. The validity of this proposition follows directly from the equality (2.13) and the first equality (2.12):

$$c(n) = \sum_{m=1}^{n} c(n,m) = \sum_{m=1}^{n} \binom{n-1}{m-1} = 2^{n-1}.$$

∎

Definition 2.4. *Let $\alpha = (\alpha_1, \alpha_2, \ldots, \alpha_m) \in C_m(n,+)$. A number $(n-m)$ is* ***decrement*** *of an ordered partition α. A partition with an event decrement is* ***even****, nd with an odd one —* ***odd****. To each ordered partition α, we also ascribe* ***the sign*** $(-1)^{n-m}$.

Example 2.3. *In the set*

$$C(4,+) =$$
$$= \{(4),(3,1),(1,3),(2,2),(2,1,1),(1,2,1),(1,1,2),(1,1,1,1)\}$$

the partitions

$$\{(3,1),(1,3),(2,2),(1,1,1,1)\}$$

are even, and the partitions

$$\{(4),(2,1,1),(1,2,1),(1,1,2)\}$$

are odd.

An alert reader notices that the number of even ordered partitions of the set $C(4,+)$ equals the number of its odd partitions.

The following general is true.

Proposition 2.4. *The number of even partitions of the set $C(n,+)$ equals the number of odd partitions of this set.*

Proof. From the known identity $\sum_{i=0}^{k}(-1)^i\binom{k}{i} = 0$ and the equality (2.13), it follows that

$$(-1)^{n-m}\sum_{m=1}^{n} c(n,m) = (-1)^{n-m}\sum_{m=1}^{n}\binom{n-1}{m-1}$$

$$= (-1)^{(n-1)-(m-1)}\sum_{m=1}^{n}\binom{n-1}{m-1} = (-1)^{(n-1)-(i)}\sum_{i=0}^{n-1}\binom{n-1}{i} =$$

$$= (-1)^{i}\sum_{i=0}^{n-1}\binom{n-1}{i} = 0.$$

■

2.3. Sets $\Xi(n)$

Definition 2.5. [1]. *The set $\Xi(n)$ is the set of ordered n-options*

$$\xi = (\xi(1),\xi(2),\ldots,\xi(n)) \tag{2.15}$$

from the multiset with the primary specification $\{1^1,2^2,\ldots,n^n\}$, the elements of which satisfy the condition:

1) a positive integer $\xi(j)$ satisfies

$$j \le \xi(j) \le n, \quad j = 1,2,\cdots,n;$$

2) for each $j = 1,2,\ldots,n$ the following equalities hold

$$\xi(j) = \xi(j+1) = \ldots = \xi(\xi(j)).$$

Note that for $j = n$ the inequality from the first condition is of the form $n \leq \xi(n) \leq n$, therefore $\xi(n) = n$.

Example 2.4.

$$\Xi(3) = \{(1,2,3),(2,2,3),(1,3,3),(3,3,3)\}.$$

Remark 2.2. The sets $\Xi(n)$ first appeared while finding the number of shortest paths in a Ferrer graph [11], connecting the limiting southeast point with the limiting northwest point of this graph. Nevertheless, the number of shortest paths in a Ferrer graph of the multiset $A = \{a_1^{\alpha_1}, a_2^{\alpha_2}, \ldots, a_r^{\alpha_r}\}$ was initially expressed in terms of the following formula:

$$P(\alpha_1, \alpha_2, \ldots, \alpha_r) =$$

$$= \sum_{\{s_1^{k_1},\ldots,s_l^{k_l}\}\in\Xi(r)} (-1)^{r-(\lambda_1+\ldots+\lambda_p)} \frac{\Pi_{j=1}^{l}\Pi_{i=0}^{k_j-1}(\alpha_{s_j} - k_j + i + 2)}{\Pi_{i=1}^{l} k_i!},$$

where $k_1, k_2, \ldots, k_l$ and $\lambda_1, \lambda_2, \ldots, \lambda_p$ are respectively exponents of the primary and secondary specifications (see p. 187) of the multiset $\{s_1^{k_1}, \ldots, s_l^{k_l}\}$. But it's been a long time before $\Xi(n)$ is interpreted as the set used to add summands of the paradeterminant (see Theorem 3.2 on p. 201) of some triangular matrix.

Remark 2.3. (Levitskaya A.A.) The following set may be written in its explicit form $\Xi(n)$:

$$\Xi(n) = \bigcup_{k=0}^{n-1} \bigcup_{1\leqslant i_1<i_2<\ldots<i_k\leqslant n-1} \{\{i_1^{i_1}, i_2^{i_2-i_1}, \ldots, i_k^{i_k-i_{k-1}}, n^{n-k}\}\}.$$

Proposition 2.5. [1]. *There is a one-to-one correspondence between the elements of the set $\Xi(n)$ and the elements of the set $C(n,+)$ of ordered partitions of a positive integer n.*

We prove this in several steps.

1. If an element $\xi \in \Xi$ has primary specification

$$[1^{\alpha(1)}, 2^{\alpha(2)}, \ldots, n^{\alpha(n)}],$$

then the following equality holds

$$\sum_{i=1}^{n} \alpha(i) = n.$$

Therefore, the exponents $\alpha(1), \alpha(2), \ldots, \alpha(n)$ of the first specification of an element ξ form some ordered partitions of a number n. Hence, we obtain reflection

$$\varphi : \xi \mapsto (\alpha(1), \alpha(2), \ldots, \alpha(n))$$

of the set $\Xi(n)$ in the set $C(n,+)$.

2. Injectiveness of reflection φ. Let

$$\xi_1 = (\xi_1(1), \xi_1(2), \ldots, \xi_1(n)), \;\; \xi_2 = (\xi_2(1), \xi_2(2), \ldots, \xi_2(n))$$

be two different elements of the set $\Xi(n)$ with primary specifications

$$[1^{\alpha_1(1)}, 2^{\alpha_1(2)}, \ldots, n^{\alpha_1(n)}], \ [1^{\alpha_2(1)}, 2^{\alpha_2(2)}, \ldots, n^{\alpha_2(n)}].$$

Let i be also the least index when the following inequality holds $\xi_1(i) \neq \xi_2(i)$. We assume that $\xi_1(i) < \xi_2(i)$. Then from the condition 2) of the definition 2.5 follows the inequality $\alpha_1(\xi_2(i)) < \alpha_2(\xi_2(i))$, i.e., different ordered partitions of the set $C(n, +)$ correspond to the elements ξ_1 and ξ_2.

3. We construct back reflection $\Xi(n)$ in $C(n, +)$ by the following algorithm. Let $p = (p(1), p(2), \ldots, p(s)) \in C(n, +)$.

p.1. *beginning*
p.2. $i := 1;\ \ p := p(i);\ \ j := 1$
p.3. $\xi(j) = \ldots = \xi(p) = p$
p.4. $j := p + 1;\ \ i := i + 1$
p.5. *if* $i \leq s$, *then* $p := p + p(i)$; *go to p.3.*
p.6. *end*

Since $1 \leq p(i)$, then after meeting p.4 and p.5 of this algorithm, the following inequality holds $j \leq p$, which together with the equalities of p.3. satisfy both conditions of Definition 2.5. □

Let $\xi \in \Xi(n)$ and r be a number of different components of an element ξ. The number $n - r$ is *a decrement* of an element ξ, and number $\varepsilon(\xi) = (-1)^{n-r}$ — its *sign*.

Remark 2.4. The bijection φ from the demonstration of the proposition 2.5 maintains the signs as the number of different components of an element ξ equals the number of non-zero components of the partition $\varphi(\xi)$.

From Propositions 2.3 and 2.5 it follows directly

Corollary 2.1. $|\Xi(n)| = 2^{n-1}$.

The set $\Xi(n)$ can also be constructed with the help of the following recurrent algorithm.

Proposition 2.6. [1]. *(i)* $\Xi(1) = \{(1)\}$.

(ii) If the set $\Xi(k)$ *is constructed already, then the elements of the set* $\Xi(k+1)$ *can be obtained by forming two elements of the set* $\Xi(k+1)$ *with the help of each element* $\xi = (\xi(1), \ldots, \xi(k))$ *of the set* $\Xi(k)$. *The first is ascribed to the* $(k+1)$ *place of a number* $k+1$, *and the second is formed with replacement of all the components equal to* k *by* $k+1$ *and is ascribed to the* $(k+1)$ *place of the component* $k+1$.

Proof. From the remark on Definition 2.5, $(k+1)$ place in each ordered multiset of the set $\Xi(k+1)$ is taken by a number $k+1$, therefore, ascribing the number $k+1$ to the elements of the set $\Xi(k)$ to the $(k+1)$ place, we obtain 2^{k-1} different elements of the set $\Xi(k+1)$. The set of these elements is denoted by $\Xi(k; k+1)$.

Replacement of a number k by a number $k+1$ in each multiset of the set $\Xi(k)$ as well as ascription of $k+1$ to the $(k+1)$ place does not violate conditions of the definition of the set $\Xi(n)$, for all the elements $\xi(i)$, less than k, satisfy these conditions, and a number $k+1$, which appeared in the j place, also takes all the consecutive places up to $(k+1)$ inclusive.

This procedure gives 2^{k-1} more different elements of the set $\Xi(k+1)$. Let us denote the set of these elements by $\Xi(k+1;k+1)$.

Multiplicity of occurrence of a number $k+1$ in each element of the set $\Xi(k;k+1)$ equals 1, and multiplicity of occurrence of the same number in each element of the set $\Xi(k+1;k+1)$ is more than 1. Therefore, all the elements of these two sets are different and belong to $\Xi(k+1)$–set.

Since $\Xi(k;k+1)\cup\Xi(k+1;k+1)\subseteq\Xi(k+1)$ and $|\Xi(k;k+1)\cup\Xi(k+1;k+1)|=2^k$, then from the corollary 2.1 follows the equality $\Xi(k;k+1)\cup\Xi(k+1;k+1)=\Xi(k+1)$. ■

Definition 2.6. [1]. *Elements $\xi_1,\xi_2\in\Xi(n)$ are called* ***amicable*** *if their bases satisfy the inequality*

$$[\xi_1]\cap[\xi_2]\neq\{n\}.$$

If not, these elements are called ***non-amicable****.*

The set of all the elements of the set $\Xi(n)$, which are non-amicable to the element α of this set, are denoted by $\Xi_\alpha(n)$.

We leave it for the reader to prove the following proposition.

Proposition 2.7. [1]. *There are exactly 3^{n-1} ordered pairs of non-amicable elements of the set $\Xi(n)$, i.e.,*

$$\sum_{\alpha\in\Xi(n)}|\Xi_\alpha(n)|=|\{(\alpha,\beta)\in\Xi(n)\times\Xi(n):[\alpha]\cap[\beta]=\{n\}\}|=3^{n-1}.$$

3. Definition of Triangular Matrices and Functions on Them

Due to the fact that a number of mathematical problems require introduction of triangular tables of numbers and some numerical functions over them, it is necessary to adjust the notion of matrix.

Matrix is any table of numbers from some number field.

To avoid confusion about the notions of rectangular and square matrices, we use the term "matrix" with the respective adjective, which characterizes its form. Thus, the notion of a triangular matrix is not confused with the notion of an upper or lower triangular matrix related only to rectangular and square matrices. Not to heap up redundant mathematical terms, however, in the theory of triangular matrices, we retain analogous notions of algebraic complement, row and column of a triangular matrix, its diagonal elements.

3.1. Definition of Triangular Matrices

Let K be some number field.

Definition 3.1. [6]. *A triangular table*

$$A=\begin{pmatrix} a_{11} & & & \\ a_{21} & a_{22} & & \\ \vdots & \vdots & \ddots & \\ a_{n1} & a_{n2} & \cdots & a_{nn}\end{pmatrix}_n \tag{3.1}$$

of numbers from a number field K is called **a triangular matrix**, *an element a_{11} is an upper element of this triangular matrix, and a number n is its order.*

Elements $a_{i1}, a_{i2}, \dots, a_{ii}$ form the i–th row of the triangular matrix (3.1), and elements $a_{jj}, a_{j+1,j}, \dots, a_{nj}$ form its j–th column. Elements $a_{11}, a_{22}, \dots, a_{nn}$ are *elements of the hypotenuse* of the triangular matrix; elements $a_{i,i-1},\ i = 2, 3, \dots, n$ are elements of the first subhypotenuse; elements $a_{i,i-2},\ i = 3, 4, \dots, n$ are elements of the second subhypotenuse etc.

Sometimes we shall use a contracted notation of the triangular matrix (3.1)

$$A = (a_{ij})_{1 \leqslant j \leqslant i \leqslant n}.$$

If in a triangular matrix, all elements, except elements of a hypotenuse, are zero, then this matrix is called *a triangular* 1*-matrix*. But if in a triangular matrix, only elements of a hypotenuse and elements of the first subhypotenuse are non-zero, then this triangular matrix is called *a triangular* 2*-matrix*. In the same way we write *a triangular* 3*-matrix* etc.

Example 3.1. In the following triangular matrix, some regularities are found.

$$B = \left(\frac{j}{i-j+1}\right)_{1 \leqslant j \leqslant i \leqslant n} = \begin{pmatrix} 1 & & & & & \\ \frac{1}{2} & 2 & & & & \\ \frac{1}{3} & \frac{2}{2} & 3 & & & \\ \frac{1}{4} & \frac{2}{3} & \frac{3}{2} & 4 & & \\ \vdots & \vdots & \vdots & \vdots & \ddots & \\ \frac{1}{n} & \frac{2}{n-1} & \frac{3}{n-2} & \frac{4}{n-3} & \vdots & n \end{pmatrix}_n . \tag{3.2}$$

The triangular matrix B is defined by the function of two variables

$$b_{ij} = \frac{j}{i-j+1},$$

where arguments i, j denote numbers of its row and column respectively.

Definition 3.2. [6]. *A triangular* 1*-matrix of the form*

$$I = (\delta_{ij})_{1 \leqslant j \leqslant i \leqslant n} = \begin{pmatrix} 1 & & & \\ 0 & 1 & & \\ \vdots & \vdots & \ddots & \\ 0 & 0 & \cdots & 1 \end{pmatrix}_n ,$$

where δ_{ij} is the Kronecker symbol, is an identity triangular matrix, and a matrix of the form

$$M = (M_i \cdot \delta_{ij})_{1 \leqslant j \leqslant i \leqslant n} = \begin{pmatrix} M_1 & & & \\ 0 & M_2 & & \\ \vdots & \vdots & \ddots & \\ 0 & 0 & \cdots & M_s \end{pmatrix}_n , \tag{3.3}$$

where $M_i,\ i = 1, \dots, s$ are some triangular matrices and rectangular zero tables are denoted by zeros, is a block triangular matrix.

Apart from the given above finite triangular matrices, infinite triangular matrices are also important

$$A = (a_{ij})_{1 \leqslant j \leqslant i < \infty}.$$

3.2. Operations with Triangular Matrices

We could define the sum of two triangular matrices of one order. It is similar to the corresponding operation with square matrices (lower triangular matrices).

It is known that linear transformations of vector spaces carry the main sense load of square matrices. Transformations of this kind require introduction of corresponding operations with square matrices. Since triangular matrices are usually defined by families of polynomials, then one of the functions triangular matrices serve is linear transformations of some linear space of polynomial vectors.

Definition 3.3. *A linear space of polynomial vectors is a linear space $\mathcal{P}_n$, the elements of which are* ***polynomial vectors*** *in the form of*

$$(f_0, f_1, \ldots, f_n),$$

where

$$f_i = a_{i0} + a_{i1}x + a_{i2}x^2 + \ldots + a_{ii}x^i,\ \ a_{ii} \neq 0,\ \ i = 1, 2, \ldots, n$$

are some polynomials of the i-th degree from variable x.

We need Aigner's idea of constructing connection coefficients between polynomial sequences (see [18], p. 110).

Let $\mathcal{P}_n$ be a linear space of polynomial vectors and $f = (f_0, f_1, \ldots, f_n)$ — its element. It is obvious that component of this vector form a basis in some linear space of polynomials P_n, the degree of which are not more than n. The vector of polynomials

$$e(x) = (1, x, x^2, \ldots, x^n) \in \mathcal{P}_n$$

is *a standard polynomial vector.*

Let us assume that besides the vector $f = (f_0, f_1, \ldots, f_n) \in \mathcal{P}_n$ one more polynomial vector $g = (g_0, g_1, \ldots, g_n)$ is given. We express components of the latter vector through basic polynomials $f_0, f_1, \ldots, f_n$. Thereby, we obtain the system of equalities

$$\left\{\begin{array}{rl} g_0 = & a_{00}f_0 \\ g_1 = & a_{10}f_0 + a_{11}f_1 \\ \cdots\cdots & \cdots\cdots\cdots\cdots\cdots\cdots\cdots\cdots \\ g_n = & a_{n0}f_0 + a_{n1}f_1 + a_{n2}f_2 + \ldots + a_{nn}f_n \end{array}\right., \tag{3.4}$$

or more concisely

$$g_i = \sum_{j=0}^{i} a_{ij}f_j,\ \ i = 0, 1, \ldots, n. \tag{3.5}$$

We shall write this system of equalities as a matrix

$$\begin{pmatrix} g_0 \\ g_1 \\ \vdots \\ g_n \end{pmatrix} \overset{def}{=} \begin{pmatrix} a_{00} & & & \\ a_{10} & a_{11} & & \\ \vdots & \ldots & \ddots & \\ a_{n0} & a_{n1} & \ldots & a_{nn} \end{pmatrix}_n \cdot \begin{pmatrix} f_0 \\ f_1 \\ \vdots \\ f_n \end{pmatrix}, \tag{3.6}$$

or

$$g = A \cdot f,$$

where

$$A = \begin{pmatrix} a_{00} & & & \\ a_{10} & a_{11} & & \\ \vdots & \dots & \ddots & \\ a_{n0} & a_{n1} & \dots & a_{nn} \end{pmatrix}_{n+1} . \tag{3.7}$$

From the equalities (3.4), (3.6) follows the rule of multiplication of the triangular matrix A by the polynomial vector f:

$$\begin{pmatrix} a_{00} & & & \\ a_{10} & a_{11} & & \\ \vdots & \dots & \ddots & \\ a_{n0} & a_{n1} & \dots & a_{nn} \end{pmatrix}_{n} \cdot \begin{pmatrix} f_0 \\ f_1 \\ \vdots \\ f_n \end{pmatrix} =$$

$$= \begin{pmatrix} a_{00}f_0 \\ a_{10}f_0 + a_{11}f_1 \\ \dots\dots\dots\dots\dots\dots\dots\dots \\ a_{n0}f_0 + a_{n1}f_1 + a_{n2}f_2 + \dots + a_{nn}f_n \end{pmatrix}.$$

Thus, the triangular matrix A can be interpreted as a linear operator A, which transforms the polynomial vector f of the space $\mathcal{P}_n$ into the polynomial vector g of this space.

It is obvious that the identity operator E is associated to the identity matrix

$$E = \begin{pmatrix} 1 & & & \\ 0 & 1 & & \\ \vdots & \dots & \ddots & \\ 0 & 0 & \dots & 1 \end{pmatrix}.$$

Using superposition of two operators, which are given by triangular matrices, it is possible to lay down a rule of multiplication of two triangular matrices:

$$C = AB = \left(\sum_{s=j}^{i} a_{is} b_{sj} \right)_{0 \leqslant j \leqslant i \leqslant n} . \tag{3.8}$$

3.3. Definition of Parafunctions of Triangular Matrices

In this section, we introduce the notion of functions of triangular matrices, which are called the paradeterminant and the parapermanent of triangular matrices. To begin with, we shall consider some subsidiary statements. To every element a_{ij} of the triangular matrix (3.1), we correspond $(i - j + 1)$ elements a_{ik}, $k \in \{j, ..., i\}$, which are called *derived elements* of a triangular matrix, generated by *a key element* a_{ij}. A key element of a triangular matrix is concurrently its derived element. The product of all derived elements generated by a key element a_{ij} is denoted by $\{a_{ij}\}$ and is called *a factorial product* of this key element, i.e.,

$$\{a_{ij}\} = \prod_{k=j}^{i} a_{ik}. \tag{3.9}$$

Let us show schematically elements of the matrix (3.1) by means of circles, key elements — filled circles, and derived elements — asterisks. Fig. 2.1. presents the triangular matrix of order 5, where a_{42} is a key element, and elements a_{42}, a_{43}, a_{44} are derived elements, which it generates.

$$\begin{pmatrix} \circ & & & & \\ \circ & \circ & & & \\ \circ & \circ & \circ & & \\ \circ & \bullet & * & * & \\ \circ & \circ & \circ & \circ & \circ \end{pmatrix}$$

Figure 2.1.

Definition 3.4. [6]. *A tuple of key elements of the matrix (3.1) is* ***a normal tuple*** *of this matrix, if they generate* ***a monotransversal****, i.e., a set of derived elements of cardinality n, no two of which belong to the same column of this matrix.*

For instance, in order to add the key element of the matrix schematically shown in Fig. 2.1. to a normal tuple of key elements, it is necessary to add two more key elements a_{11} and a_{55} to it.

Let $C(n,+)$ be the set of all ordered partitions of a positive integer n into positive integer summands. It turns out that there is a one-to-one correspondence between elements of this set and normal tuples of key elements of the matrix (3.1) of order n.

Let us consider some ordered r–partition $p = (p_1, ..., p_r)$. To each component p_s, $s \in \{1, ..., r\}$, of this partition, we correspond the key element a_{ij} of the matrix (3.1) with the help of the following algorithm :

p.1. beginning

p.2. $j := 1$; $s := 0$; $i := 0$

p.3. $s := s + 1$; $i := i + p_s$; key element$(s) := a_{ij}$

p.4. If $s < r$ then $j := j + p_s$; go to p.3

p.5. end.

Thereby we obtain a normal tuple of key elements generated by the partition p. It is also easy to establish a backward correspondence.

In general, there is a bijection between ordered r-partitions and normal tuple of key elements (Tarakanov V.Ye.):

$$(n_1, n_2, \ldots, n_r) \in C_r(n,+) \Leftrightarrow (a_{N_1,N_0+1}, a_{N_2,N_1+1}, \ldots, a_{N_r,N_r-1}), \tag{3.10}$$

$$N_0 = 0,\ N_s = \sum_{i=1}^{s} n_i,\ s = 1, 2, \ldots, r.$$

The algorithm given above describes one more geometric image of ordered partitions of a positive integer n into positive integer summands. Let us have a triangular matrix of order n. By the partition $p = (p_1, ..., p_r)$ of a positive integer n, we construct a normal tuple of elements of this matrix, which form its monotransversal.

To the first component p_1 of this partition, we correspond a row of elements of a triangular matrix, where there are exactly p_1 elements of this matrix. This row is the p_1–th row of the matrix. Then the first p_1 columns are ignored and a new triangular matrix of order $(n - p_1)$ is considered. To the second component p_2 of the partition, we correspond a row of a new matrix, which consists of p_2 elements, etc.

Example 3.2. Let us show a one-to-one correspondence between ordered partitions of the number 4 and normal tuples of key elements of a triangular matrix of order four according to the following schemes:

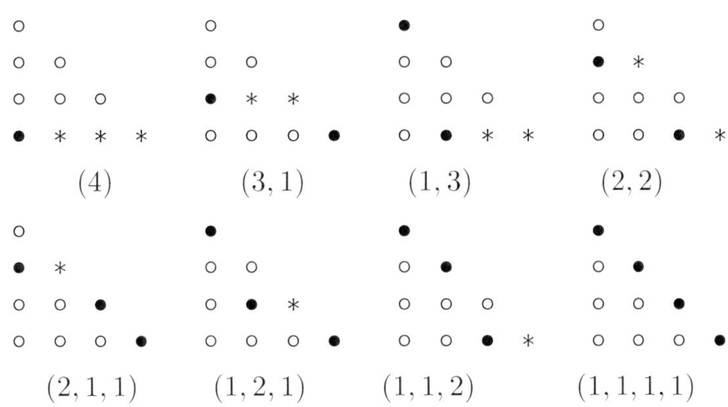

Figure 2.2.

To every normal tuple a of key elements, we correspond the sign $(-1)^{\varepsilon(a)}$, where $\varepsilon(a)$ is the sum of all the indices of the key elements of this tuple.

Definition 3.5. [6]. ***The paradeterminant*** *of the triangular matrix (3.1) is the number*

$$\mathrm{ddet}(A) = \left\langle \begin{array}{cccc} a_{11} & & & \\ a_{21} & a_{22} & & \\ \vdots & \vdots & \ddots & \\ a_{n1} & a_{n2} & \cdots & a_{nn} \end{array} \right\rangle =$$

$$= \sum_{(\alpha_1, \alpha_2, \ldots, \alpha_r) \in \mathbb{C}(n,+)} (-1)^{\varepsilon(a)} \prod_{s=1}^{r} \{a_{i(s),j(s)}\}, \tag{3.11}$$

where $a_{i(s),j(s)}$ is the key element corresponding to the s-th component of the partition $\alpha = (\alpha_1, \alpha_2, \ldots, \alpha_r)$, and the symbol $\varepsilon(a)$ is the sign of the normal tuple a of key elements.

In analogy to the notion of the permanent of a square matrix, we define the parapermanent of a triangular matrix.

Definition 3.6. [6]. ***The parapermanent*** *of the triangular matrix (3.1) is the number*

$$\operatorname{pper}(A) = \begin{bmatrix} a_{11} & & & \\ a_{21} & a_{22} & & \\ \vdots & \vdots & \ddots & \\ a_{n1} & a_{n2} & \cdots & a_{nn} \end{bmatrix} =$$

$$= \sum_{(\alpha_1,\alpha_2,\ldots,\alpha_r)\in\mathbb{C}(n,+)} \prod_{s=1}^{r} \{a_{i(s),j(s)}\}, \qquad (3.12)$$

where $a_{i(s),j(s)}$ is a key element corresponding to the s-th component of the partition $\alpha = (\alpha_1, \alpha_2, \ldots, \alpha_r)$.

Remark 3.1. The parapermanent of a triangular matrix can be defined as a sum of products of elements of all monotransversals of this matrix[5].

Remark 3.2. Sometimes it is convenient to denote the paradeterminant (3.11) and the parapermanent (3.12) by $\langle a_{ij}\rangle_{1\le j\le i\le n}$ and $[a_{ij}]_{1\le j\le i\le n}$.

Remark 3.3. Because of Proposition (2.3) on p. 190, the paradeterminant and the parapermanent of order n consists of 2^{n-1} summands.

In the sequel, where the paradeterminant and the parapermanent of a triangular matrix are presented simultaneously, we will use the term *parafunction of a triangular matrix.*

Example 3.3. Using Definition 3.5, find the value of the paradeterminant of the triangular matrix of order four (see the schemes on p. 199):

$$\left\langle \begin{matrix} a_{11} & & & \\ a_{21} & a_{22} & & \\ a_{31} & a_{32} & a_{33} & \\ a_{41} & a_{42} & a_{43} & a_{44} \end{matrix} \right\rangle =$$

$$= -a_{41}a_{42}a_{43}a_{44} + a_{31}a_{32}a_{33}a_{44} + a_{11}a_{42}a_{43}a_{44} +$$
$$+ a_{21}a_{22}a_{43}a_{44} - a_{21}a_{22}a_{33}a_{44} - a_{11}a_{32}a_{33}a_{44} -$$
$$- a_{11}a_{22}a_{43}a_{44} + a_{11}a_{22}a_{33}a_{44}.$$

□

We shall prove the theorem which could be the definition of the paradeterminant and the parapermanent of triangular matrices. This theorem, in principal, is based on the bijection (3.10).

Theorem 3.1. [5]. *If A is the triangular matrix (3.1), then the following equalities hold:*

$$\operatorname{ddet}(A) = \sum_{r=1}^{n} \sum_{p_1+\ldots+p_r=n} (-1)^{n-r} \prod_{s=1}^{r} \{a_{p_1+\ldots+p_s,\,p_1+\ldots+p_{s-1}+1}\}, \qquad (3.13)$$

[5]The permanent of a square matrix is jokingly called the determinant without signs. It can be defined as the sum of products of all elements of transversals of this matrix.

$$\text{pper(A)} = \sum_{r=1}^{n} \sum_{p_1+\ldots+p_r=n} \prod_{s=1}^{r} \{a_{p_1+\ldots+p_s, p_1+\ldots+p_{s-1}+1}\},$$

where the summation is over the set of natural solutions of the equality $p_1 + \ldots + p_r = n$.

Proof.

1) We shall prove that the result of the algorithm 3.3 on p. 198 is the tuple of key elements

$$a_{p_1,1}, a_{p_1+p_2,p_1+1}, \ldots, a_{p_1+\ldots+p_r,p_1+\ldots+p_{r-1}+1}.$$

For this purpose, we shall compute the final values of the indices i, j of the element a_{ij} in this algorithm. It is obvious they are the sums $i = p_1 + p_2 + \ldots,\ j = 1 + p_1 + \ldots$, moreover, because $s < r$, the last summand of the first sum is p_r, and of the second — p_{r-1}. It is also obvious that the obtained tuple of key elements of the triangular matrix (3.1) is the normal tuple of this matrix.

2) We shall prove that the sign $(-1)^{\varepsilon(a)}$ of the normal tuple of elements of the triangular matrix (3.1), in Definition 3.5, corresponding to the partition $p = (p_1, \ldots, p_r)$, coincides with the sign $(-1)^{n-r}$ of this ordered partition.

It is true that:

$$\sum_{s=1}^{r}((p_1 + \ldots + p_s) + (p_1 + \ldots + p_{s-1} + 1)) \equiv$$

$$\equiv (p_1 + \ldots + p_s + r) \equiv (n + r) \equiv n - r \pmod r.$$

■

We shall prove one more theorem, which could be the definition of parafunctions of triangular matrices. It is based on the notion of the set $\Xi(n)$.

Theorem 3.2. [1]. *If A is the triangular matrix (3.1), then the following equalities hold*

$$\text{ddet}(A) = \sum_{\xi\in\Xi(n)} (-1)^{n-r} \cdot a_{\xi(1),1}a_{\xi(2),2} \cdot \ldots \cdot a_{\xi(n),n},$$

$$\text{pper}(A) = \sum_{\xi\in\Xi(n)} a_{\xi(1),1}a_{\xi(2),2} \cdot \ldots \cdot a_{\xi(n),n},$$

where r is the number of elements in the basis of the multiset ξ or the number of elements belonging to this basis.

Proof. By the algorithm, cited in the demonstration of Proposition 2.5, the following equality holds

$$\prod_{s=1}^{r}\{a_{p_1+\ldots+p_s,\, p_1+\ldots+p_{s-1}+1}\} = a_{\xi(1),1}a_{\xi(2),2} \cdot \ldots \cdot a_{\xi(n),n}.$$

Therefore, the validity of this theorem follows directly from Theorem 3.1 and Remark 2.4.

■

3.4. Inverse Triangular Matrix

First of all, we shall note that an inverse triangular matrix, in principle, coincides with an inverse matrix of a lower triangular matrix. Therefore, we shall turn our attention only to its construction by means of paradeterminant of its corners.

A triangular matrix A^{-1} inverse to a triangular matrix A can be constructed with the help of the equality (3.6), which is determined by a linear operator of reflecting the vector of polynomials $f = (f_0, f_1, \ldots, f_n)$ on the vector of polynomials $g = (g_0, g_1, \ldots, g_n)$.

The elements of an inverse matrix can be obtained by solving simultaneous equations (3.4) under the system of unknown $(f_0, f_1, \ldots, f_n)$.

Theorem 3.3. *An triangular matrix A^{-1} inverse to the triangular matrix (3.7) is a matrix in the form of*

$$(b_{ij})_{0\leqslant j\leqslant i\leqslant n} = \left(\frac{(-1)^{i+j}}{a_{jj}} \left\langle \frac{a_{r+j+1,s+j}}{a_{r+j+1,s+j+1}} \right\rangle_{0\leqslant s\leqslant r\leqslant i-j-1} \right)_{0\leqslant j\leqslant i\leqslant n} . \tag{3.14}$$

Proof. We show that solution of the simultaneous equations (3.4) is

$$f_i = \sum_{j=0}^{i} \frac{(-1)^{i+j}}{a_{jj}} \cdot \left\langle \frac{a_{r+j+1,s+j}}{a_{r+j+1,s+j+1}} \right\rangle_{0\leqslant s\leqslant r\leqslant i-j-1} \cdot g_j, \; i = 0, 1, \ldots .$$

For $i = 0$, the simultaneous equations (3.4) have the following solution

$$f_0 = \sum_{j=0}^{0} \frac{(-1)^{0+j}}{a_{00}} \cdot \left\langle \frac{a_{r+j+1,s+j}}{a_{r+j+1,s+j+1}} \right\rangle_{0\leqslant s\leqslant r\leqslant -j-1} \cdot g_j = \frac{1}{a_{00}} \cdot g_0,$$

here we consider that

$$\left\langle \frac{a_{r+j+1,s+j}}{a_{r+j+1,s+j+1}} \right\rangle_{0\leqslant s\leqslant r\leqslant -1} = 1.$$

Let us assume that the simultaneous equations (3.4) for $i = k - 1$ have the following solution

$$f_{k-1} = \sum_{j=0}^{k-1} \frac{(-1)^{k-1+j}}{a_{jj}} \cdot \left\langle \frac{a_{r+j+1,s+j}}{a_{r+j+1,s+j+1}} \right\rangle_{0\leqslant s\leqslant r\leqslant k-j-2} \cdot g_j, \tag{3.15}$$

and prove thereby

$$f_k = \sum_{j=0}^{k} \frac{(-1)^{k+j}}{a_{jj}} \cdot \left\langle \frac{a_{r+j+1,s+j}}{a_{r+j+1,s+j+1}} \right\rangle_{0\leqslant s\leqslant r\leqslant k-j-1} \cdot g_j.$$

Let us find the polynomial f_k from the last equation of the system (3.4) for $i = k$:

$$f_k = -\sum_{i=0}^{k-1} \frac{a_{ki}}{a_{kk}} f_i + \frac{1}{a_{kk}} g_k.$$

Let us place in the last equality instead of polynomials $f_i,\ i = 0, 1, \ldots, k-1$ their known values (3.15):

$$f_k = -\sum_{i=0}^{k-1} \frac{a_{ki}}{a_{kk}} \sum_{j=0}^{i} \frac{(-1)^{i+j}}{a_{jj}} \left\langle \frac{a_{r+j+1,s+j}}{a_{r+j+1,s+j+1}} \right\rangle_{0 \leqslant s \leqslant r \leqslant i-j-1} \cdot g_j + \frac{1}{a_{kk}} g_k.$$

Let us change the order of summation:

$$f_k = \sum_{j=0}^{k-1} (-1)^{k+j} \frac{1}{a_{jj}} \times$$

$$\times \left(\sum_{i=j}^{k-1} (-1)^{i-k+1} \frac{a_{ki}}{a_{kk}} \left\langle \frac{a_{r+j+1,s+j}}{a_{r+j+1,s+j+1}} \right\rangle_{0 \leqslant s \leqslant r \leqslant i-j-1} \right) g_j + \frac{1}{a_{kk}} g_k.$$

But the expressions in the parentheses in the last equality is decomposition of the paradeterminant

$$\left\langle \frac{a_{r+j+1,s+j}}{a_{r+j+1,s+j+1}} \right\rangle_{0 \leqslant s \leqslant r \leqslant k-j-1}$$

by the elements of the last row; therefore, we have:

$$f_k = \sum_{j=0}^{k-1} (-1)^{k+j} \frac{1}{a_{jj}} \left\langle \frac{a_{r+j+1,s+j}}{a_{r+j+1,s+j+1}} \right\rangle_{0 \leqslant s \leqslant r \leqslant k-j-1} \cdot g_j + \frac{1}{a_{kk}} g_k =$$

$$= \sum_{j=0}^{k} (-1)^{k+j} \frac{1}{a_{jj}} \left\langle \frac{a_{r+j+1,s+j}}{a_{r+j+1,s+j+1}} \right\rangle_{0 \leqslant s \leqslant r \leqslant k-j-1} \cdot g_j.$$

■

Remark 3.4. *From Theorem 3.3 it follows that for an inverse triangular matrix to exist, it is necessary its all diagonal elements equal zero.*

Remark 3.5. *It is obvious all the diagonal elements of an inverse triangular matrix are not equal to zero.*

Example 3.4. A triangular matrix inverse to the matrix

$$A = \begin{pmatrix} a_{00} & & & \\ a_{10} & a_{11} & & \\ a_{20} & a_{21} & a_{22} & \\ a_{30} & a_{31} & a_{32} & a_{33} \end{pmatrix},$$

according to Theorem 3.3 is written as

$$A^{-1} =$$

$$
= \begin{pmatrix}
\frac{1}{a_{00}} & & & \\
-\frac{1}{a_{00}}\left\langle \frac{a_{10}}{a_{11}} \right\rangle & \frac{1}{a_{11}} & & \\
\frac{1}{a_{00}}\left\langle \begin{matrix} \frac{a_{10}}{a_{11}} & \\ \frac{a_{20}}{a_{21}} & \frac{a_{21}}{a_{22}} \end{matrix} \right\rangle & -\frac{1}{a_{11}}\left\langle \frac{a_{21}}{a_{22}} \right\rangle & \frac{1}{a_{22}} & \\
-\frac{1}{a_{00}}\left\langle \begin{matrix} \frac{a_{10}}{a_{11}} & & \\ \frac{a_{20}}{a_{21}} & \frac{a_{21}}{a_{22}} & \\ \frac{a_{30}}{a_{31}} & \frac{a_{31}}{a_{32}} & \frac{a_{32}}{a_{33}} \end{matrix} \right\rangle & \frac{1}{a_{11}}\left\langle \begin{matrix} \frac{a_{21}}{a_{22}} & \\ \frac{a_{31}}{a_{32}} & \frac{a_{32}}{a_{33}} \end{matrix} \right\rangle & -\frac{1}{a_{22}}\left\langle \frac{a_{32}}{a_{33}} \right\rangle & \frac{1}{a_{33}}
\end{pmatrix}.
$$

If we find the polynomial vector f from the simultaneous equations (3.6), then we obtain a new linear transformation $f = A^{-1}g$, which is a matrix inverse to the matrix A and transforms the polynomial vector g into the polynomial vector f.

Example 3.5. Let us find a triangular matrix inverse to the triangular matrix

$$
A = \begin{pmatrix}
a_1 & & & \\
a_1 & a_2 & & \\
\vdots & \dots & \dots & \\
a_1 & a_2 & \dots & a_n
\end{pmatrix}
$$

The paradeterminants of all corners $R_{i,j+1}(A')$, $2 \leqslant i - j - 1$ equal zero (see 4.5). The diagonal elements of the inverse matrix equal $\frac{1}{a_{ii}}$. Find the elements of the first subdiagonal.

$$
-\frac{1}{a_{j-1}}\left\langle \frac{a_{j-1}}{a_j} \right\rangle = -\frac{1}{a_j}.
$$

Thus, the inverse triangular matrix is written as:

$$
A^{-1} = \begin{pmatrix}
\frac{1}{a_1} & & & & \\
-\frac{1}{a_2} & \frac{1}{a_2} & & & \\
\vdots & \dots & \ddots & & \\
0 & 0 & \dots & \frac{1}{a_{n-1}} & \\
0 & 0 & \dots & -\frac{1}{a_n} & \frac{1}{a_n}
\end{pmatrix}.
$$

3.5. Paradeterminant Product of Triangular Matrices

In one of the preceding sections where operations with triangular matrices are covered, we have already considered the product of triangular matrices. But with this product of triangular matrices, the following equality does not hold

$$
\text{ddet}\,(AB) = \text{ddet}\,(A)\text{ddet}\,(B).
$$

The purpose of this section is to construct the product of triangular matrices, for which it does.

Definition 3.7. *The incomplete product of two paradeterminants* ddet (A) *and* ddet (B) *of triangular matrices* A *and* B *of order* n *is the expression* ddet $(A) \circ$ ddet (B), *which is* *defined by the equality*

$$\operatorname{ddet}(A) \circ \operatorname{ddet}(B) =$$

$$= \sum_{(\xi_i,\xi_j)\in\Xi(n)\times\Xi(n)} (-1)^{\varepsilon(\xi_i)+\varepsilon(\xi_j)} k(\xi_i,\xi_j) a_{\xi_i(1),1}\cdot\ldots\cdot a_{\xi_i(n),n} \times$$

$$\times b_{\xi_j(1),1}\cdot\ldots\cdot b_{\xi_j(n),n}, \tag{3.16}$$

where

$$k(\xi_i,\xi_j) = \begin{cases} 1, & [\xi_i]\cap[\xi_j] = \{n\}, \\ 0, & [\xi_i]\cap[\xi_j] \neq \{n\}, \end{cases} \tag{3.17}$$

and $\varepsilon(\xi_i), \varepsilon(\xi_j)$ *— signs (see p. 193) of elements* $\xi_i, \xi_j \in \Xi(n)$.

The equality (3.17) corresponds the number $k(\xi_i,\xi_j) = 1$ to the pair (ξ_1,ξ_j) of non-amicable elements ξ_i and ξ_j (see Definition 2.6 on p. 194), and the number $k(\xi_i,\xi_j) = 0$ to the pair of amicable elements.

In the same way, we can define the incomplete product of two parapermanents of triangular matrices:

Definition 3.8. *The incomplete product of two parapermanents* $\operatorname{pper}(A)$ *and* $\operatorname{pper}(B)$ *of triangular matrices* A *and* B *of order* n *is the expression* $\operatorname{pper}(A)\circ\operatorname{pper}(B)$, *which is defined by the equality*

$$\operatorname{pper}(A)\circ\operatorname{pper}(B) = \tag{3.18}$$

$$\sum\nolimits_{(\xi_i,\xi_j)\in\Xi(n)\times\Xi(n)} k(\xi_i,\xi_j) a_{\xi_i(1),1}\cdot\ldots\cdot a_{\xi_i(n),n}\cdot b_{\xi_j(1),1}\cdot\ldots\cdot b_{\xi_j(n),n} \quad .$$

Let us specify the equality (3.17). Number all elements of $\Xi(n)$-set in the order we obtain them by means of the recurrent algorithm, which is based on Proposition 2.6 (see p. 193). Thereby we obtain the sequence

$$\xi_1, \xi_2, \ldots, \xi_{2^{n-1}},$$

each term of which is some multiset of the basis $\{1,2,\ldots,n\}$. If all the values of the function defined by the equality (3.17) are written in the table, and one is replaced with a circle, and zero is replaced with an empty cell, then we obtain a fragment of a fractal figure[6] of the n-th generation.

For $n = 5$ we obtain a fragment of a fractal figure of the fifth generation, which is shown in the following table:

To construct an algorithm for finding an incomplete product of paradeterminants and an incomplete product of parapermanents, an important task is to describe pairs (i,j) of indices of function arguments

$$k(\xi_i,\xi_j),\ 1 \leqslant i \leqslant 2^{n-1},\ 2^{n-1} - i + 1 \leqslant j \leqslant 2^{n-1},$$

[6]Fractal is an infinitely self-similar geometric shape, each fragment of which is repeated when scale is reduced (see [19]). This notion was introduced by Benoit Mandelbrot in 1975. The birth of fractal geometry is associated with the publication of his monograph "The Fractal Geometry of Nature", 1977.

	ξ_1	ξ_2	ξ_3	ξ_4	ξ_5	ξ_6	ξ_7	ξ_8	ξ_9	ξ_{10}	ξ_{11}	ξ_{12}	ξ_{13}	ξ_{14}	ξ_{15}	ξ_{16}
ξ_1																•
ξ_2															•	•
ξ_3														•		•
ξ_4													•	•	•	•
ξ_5												•	•	•	•	•
ξ_6											•	•			•	•
ξ_7										•		•		•		•
ξ_8									•	•	•	•	•	•	•	•
ξ_9								•								•
ξ_{10}							•	•							•	•
ξ_{11}						•		•						•		•
ξ_{12}					•	•	•	•					•	•	•	•
ξ_{13}				•				•				•				•
ξ_{14}			•	•			•	•			•	•			•	•
ξ_{15}		•		•		•		•		•		•		•		•
ξ_{16}	•	•	•	•	•	•	•	•	•	•	•	•	•	•	•	•

to which in a fractal triangular of the n-th generation one can correspond 1.

Perepichka N.V. established dependence between the fragments of fractal figures and number triangles of zeros and ones, which he called *binary Pascal triangles.* A binary Pascal triangle can be obtained by replacing odd numbers with one and even numbers with zero in a classic Pascal triangle, i.e., with respective numbers of a classic Pascal triangle by module 2. Thus, we obtain an analogous recurrent algorithm for construction of a binary Pascal triangle by replacing in the expression

$$c_{ij} = c_{i-1,j-1} + c_{i-1,j}$$

the sum logical operation with the $\oplus$ "exclusive or" operation. Now it is easy to prove the equality:

$$k(\xi_i, \xi_j) = \binom{i-1}{i+j-(2^{n-1}+1)} mod\, 2, \tag{3.19}$$

where indices i, j satisfy the inequalities

$$1 \leqslant i \leqslant 2^{n-1},\ 2^{n-1} - i + 1 \leqslant j \leqslant 2^{n-1}.$$

Remark 3.6. *From the symmetry of a square table relative to its diagonal follows the truth of the equalities:*

$$\text{ddet}\,(A) \circ \text{ddet}\,(B) = \text{ddet}\,(B) \circ \text{ddet}\,(A)$$

$$\text{pper}(A) \circ \text{pper}(B) = \text{pper}(B) \circ \text{pper}(A)$$

Let $R_{ij}(A)$ and $R_{ij}(B)$ be the corners of matrices A and B. Let us denote the incomplete product of paradeterminants and parapermanents of these corners by d_{ij} and p_{ij} respectively, i.e.,

$$d_{ij} = \text{ddet}\,(R_{ij}(A)) \circ \text{ddet}\,(R_{ij}(B)) = \text{ddet}\,(R_{ij}(B)) \circ \text{ddet}\,(R_{ij}(A)),$$

$$p_{ij} = \text{pper}(R_{ij}(A)) \circ\ \text{pper}(R_{ij}(B)) = \text{pper}(R_{ij}(B)) \circ\ \text{pper}(R_{ij}(A)),$$

and we shall assume that

$$d_{i,i+1} = p_{i,i+1} = 1. \tag{3.20}$$

Definition 3.9. *The paradeterminant product of two triangular matrices A and B of order n is a triangular matrix $C = A \overset{d}{\circ} B$ of the same order, whose elements c_{ij} are defined by the equality*

$$c_{ij} = (-1)^{\delta_{ij}+1} \frac{d_{ij}}{d_{i,j+1}}, \tag{3.21}$$

here δ_{ij} is the Kronecker symbol, $1 \leqslant j \leqslant i \leqslant n$.

In the same way, we introduce the parapermanent product of two triangular matrices.

Definition 3.10. *The parapermanent product of two triangular matrices A and B of order n is a triangular matrix $C = A \overset{p}{\circ} B$ of the same order, whose elements c_{ij} are defined by the equality*

$$c_{ij} = \frac{p_{ij}}{p_{i,j+1}}. \tag{3.22}$$

The following equalities are obvious:

$$A \overset{d}{\circ} B = B \overset{d}{\circ} A, \tag{3.23}$$

$$(A \overset{d}{\circ} B) \overset{d}{\circ} C = A \overset{d}{\circ} (B \overset{d}{\circ} C). \tag{3.24}$$

Analogous equalities also hold for the parapermanent product of triangular matrices.

Theorem 3.4. *For triangular matrices A and B of the same order, the following equalities hold:*

$$\operatorname{ddet}(A \overset{d}{\circ} B) = \operatorname{ddet}(A)\operatorname{ddet}(B), \tag{3.25}$$

$$\operatorname{pper}(A \overset{p}{\circ} B) = \operatorname{pper}(A)\operatorname{pper}(B). \tag{3.26}$$

Proof. 1). First of all, we note that the factoral product of the element

$$c_{ij} = (-1)^{\delta_{ij}+1} \frac{d_{ij}}{d_{i,j+1}}$$

equals $(-1)^{i-j} d_{ij}$ and that in the paradeterminant

$$\operatorname{ddet}(A \overset{d}{\circ} B) = \operatorname{ddet}\left((-1)^{\delta_{ij}+1} \frac{d_{ij}}{d_{i,j+1}}\right)$$

the modules of all summands

$$d_{i(1),1} d_{i(2),i(1)+1} \cdot \ldots \cdot d_{i(r),i(r-1)+1} d_{n,i(r)+1} \tag{3.27}$$

are different. Let us place in these summands instead of incomplete paradeterminant products their values and we shall obtain the sum of different products of some summands of the paradeterminant of the matrix A by some summands of the paradeterminant of the matrix B.

2). We shall prove that the paradeterminant $\operatorname{ddet}(A \overset{d}{\circ} B)$ consists of $2^{2(n-1)}$ summands. To each ordered partition of a positive integer n into r components one can correspond the summand (3.27). The incomplete paradeterminant product d_{ij} consists of 3^{i-j} summands (see Proposition 2.7). Therefore, each summand in the form of (3.27), in its turn, consists of 3^{n-r} summands. But according to Proposition 2.2, there are exactly

$$\binom{n-1}{r-1}$$

r–partitions of a positive integer n. So,

$$3^{n-r}\binom{n-1}{r-1}$$

summands correspond to all r–partitions of a number n. Thus,

$$\sum_{r=1}^{n} 3^{n-r}\binom{n-1}{r-1} = (3+1)^{n-1} = 2^{2(n-1)}$$

summands correspond to all ordered partitions of a positive integer n. We shall obtain just enough different summands as a result of the product of the paradeterminants $\text{ddet}\,(A)$ and $\text{ddet}\,(B)$.

3) It follows from the equality (5.1) that the sign of each summand in the left-hand member of the equality (3.25) coincides with the sign of this summand in the right-hand member of this equality.

The second equality of this theorem is proved in the same way. ■

Proposition 3.1. *For any triangular matrix A the following equalities hold*

$$A \overset{d}{\circ} E = E \overset{d}{\circ} A = A,$$

$$A \overset{p}{\circ} E = E \overset{p}{\circ} A = A,$$

here E is an identity triangular matrix of the same order as a triangular matrix A.

Proof. The first equality of this proposition follows from obvious equalities

$$\text{ddet}\,(R_{ij}(A))_{i-j+1} \circ \text{ddet}\,(E)_{i-j+1} = \{a_{ij}\}$$

and the fact that the factorial product of the element

$$c_{ij} = (-1)^{\delta_{ij}+1}\frac{d_{ij}}{d_{i,j+1}}$$

equals

$$d_{ij} = \text{ddet}\,(R_{ij}(A))_{i-j+1} \circ \text{ddet}\,(E)_{i-j+1}\,.$$

The second equality is proved in the same way. ■

3.6. Scalar Product of Vector by Parafunction

Let us consider one more operation relating to parafunctions of triangular matrices. It occurs while studying partition polynomials, differentiating composed functions, inversing series etc.

Definition 3.11. [8]. *The scalar product of a vector* $(b_1, b_2, \ldots, b_n)$ *by the paradeterminant of the triangular matrix (3.1) is the number*

$$\begin{pmatrix} b_1 \\ b_2 \\ \vdots \\ b_n \end{pmatrix} \cdot \left\langle \begin{matrix} a_{11} & & & \\ a_{21} & a_{22} & & \\ \vdots & \vdots & \ddots & \\ a_{n1} & a_{n2} & \cdots & a_{nn} \end{matrix} \right\rangle_n \overset{def}{=} \tag{3.28}$$

$$= \sum_{r=1}^{n} b_r \cdot \sum_{p_1+\ldots+p_r=n} (-1)^{n-r} \prod_{s=1}^{r} \left\{a_{p_1+\ldots+p_s, p_1+\ldots+p_{s-1}+1}\right\}.$$

In the same way, we define the scalar product of a vector by the parapermanent of a triangular matrix.

Definition 3.12. *The scalar product of a vector* $(b_1, b_2, \ldots, b_n)$ *by the parapermanent of the triangular matrix (3.1) is the number*

$$(b_1, b_2, \ldots, b_n) \cdot \mathrm{pper}(A) \overset{def}{=} \tag{3.29}$$

$$= \sum_{r=1}^{n} b_r \cdot \sum_{p_1+\ldots+p_r=n} \prod_{s=1}^{r} \left\{a_{p_1+\ldots+p_s, p_1+\ldots+p_{s-1}+1}\right\}.$$

Example 3.6. Let us consider the product of a vector (b_1, b_2, b_3, b_4) by the paradeterminant of the triangular matrix (3.1) for $n = 4$.

$$\begin{pmatrix} b_1 \\ b_2 \\ b_3 \\ b_4 \end{pmatrix} \cdot \left\langle \begin{matrix} a_{11} & & & \\ a_{21} & a_{22} & & \\ a_{31} & a_{32} & a_{33} & \\ a_{41} & a_{42} & a_{43} & a_{44} \end{matrix} \right\rangle = \tag{3.30}$$

$$= -b_1\{a_{41}\} + b_2 \cdot (\{a_{31}\}\{a_{44}\} + \{a_{11}\}\{a_{42}\} + \{a_{21}\}\{a_{43}\}) -$$
$$b_3 \cdot (\{a_{11}\}\{a_{32}\}\{a_{44}\} + \{a_{11}\}\{a_{22}\}\{a_{43}\} + \{a_{21}\}\{a_{33}\}\{a_{44}\}) +$$
$$b_4\{a_{11}\}\{a_{22}\}\{a_{33}\}\{a_{44}\}.$$

Thus, when multiplying a vector by the paradeterminant of a triangular matrix, its r–th component is multiplied by the sum of all those paradeterminant summands corresponding to the partitions with r components. We also note that the product of a null vector or a vector, all the components of which equal one, by a parafunction of a triangular matrix is correspondingly equal to zero or parafunction of this triangular matrix.

Proposition 3.2. *Let* a, b *be two vectors of order* n*,* A *and* B *— triangular matrices of the same order, and* α *— some number of a number field* K*. Then the following equalities hold*

$$\alpha \cdot a \cdot \mathrm{ddet}\,(A) = (\alpha \cdot a) \cdot \mathrm{ddet}\,(A),$$

$$(a+b)\cdot \operatorname{ddet}(A) = a\cdot \operatorname{ddet}(A) + b\cdot \operatorname{ddet}(A),$$

$$a\cdot (\operatorname{ddet}(A) + \operatorname{ddet}(B)) = a\cdot \operatorname{ddet}(A) + a\cdot \operatorname{ddet}(B).$$

We leave for the reader *to prove* this proposition. □

Proposition 3.3. *For any triangular matrix (3.1) the following equality holds*

$$\begin{pmatrix} (-1)^{n-1} \\ (-1)^{n-2} \\ \vdots \\ (-1)^0 \end{pmatrix} \cdot \left\langle \begin{matrix} a_{11} & & & \\ a_{21} & a_{22} & & \\ \vdots & \vdots & \ddots & \\ a_{n1} & a_{n2} & \cdots & a_{nn} \end{matrix} \right\rangle_n = \left[\begin{matrix} a_{11} & & & \\ a_{21} & a_{22} & & \\ \vdots & \vdots & \ddots & \\ a_{n1} & a_{n2} & \cdots & a_{nn} \end{matrix} \right]. \tag{3.31}$$

Proof. According to the equality (3.28), we have

$$\begin{pmatrix} (-1)^{n-1} \\ (-1)^{n-2} \\ \vdots \\ (-1)^0 \end{pmatrix} \cdot \left\langle \begin{matrix} a_{11} & & & \\ a_{21} & a_{22} & & \\ \vdots & \vdots & \ddots & \\ a_{n1} & a_{n2} & \cdots & a_{nn} \end{matrix} \right\rangle_n =$$

$$= \sum_{r=1}^{n} (-1)^{n-r} \cdot \sum_{p_1+\ldots+p_r=n} (-1)^{n-r} \prod_{s=1}^{r} \left\{ a_{p_1+\ldots+p_s, p_1+\ldots+p_{s-1}+1} \right\},$$

from which, because of the second equality (3.13) (p. 200), follows the validity of this proposition. ■

In the same way we can prove the validity of the identity

$$\left\langle \begin{matrix} a_{11} & & & \\ a_{21} & a_{22} & & \\ \vdots & \vdots & \ddots & \\ a_{n1} & a_{n2} & \cdots & a_{nn} \end{matrix} \right\rangle_n = \begin{pmatrix} (-1)^{n-1} \\ (-1)^{n-2} \\ \vdots \\ (-1)^0 \end{pmatrix} \cdot \left[\begin{matrix} a_{11} & & & \\ a_{21} & a_{22} & & \\ \vdots & \vdots & \ddots & \\ a_{n1} & a_{n2} & \cdots & a_{nn} \end{matrix} \right]_n. \tag{3.32}$$

4. Properties of Parafunctions of Triangular Matrices

Unfortunately, so far there is no "physical definition" of a triangular matrix paradeterminant, which would make it possible to considerably simplify some demonstrations of this section. But it is interesting to know that despite the definition of triangular matrix parafunctions being substantially different from the definition of square matrix functions, there are a number of properties among properties of triangular matrix parafunctions, which are analogous to properties of determinants and permanents.

4.1. Algebraic Complements and Matrix Corners. Triangular Matrix Decomposition

Let us first consider some important notions relating to triangular matrices, which are analogs of the notions of a minor and an algebraic complement for square matrices.

Definition 4.1. [6]. *To each element a_{ij} of the given triangular matrix (3.1) we correspond a triangular matrix with this element in the bottom left corner, which we call* **a corner** *of the given triangular matrix and denote by* $R_{ij}(A)$.

It is obvious that the corner $R_{ij}(A)$ is a triangular matrix of order $(i - j + 1)$. The corner $R_{ij}(A)$ includes only those elements a_{rs} of the triangular matrix (3.1), the indices of which satisfy the relations $j \leqslant s \leqslant r \leqslant i$.

Below we shall consider that

$$\operatorname{ddet}(R_{01}(A)) = \operatorname{ddet}(R_{n,n+1}(A)) = \operatorname{pper}(R_{01}(A)) =$$

$$= \operatorname{pper}(R_{n,n+1}(A)) = 1. \tag{4.1}$$

Example 4.1. Let us have the triangular matrix:

$$A = \begin{pmatrix} a_{11} & & & & \\ a_{21} & a_{22} & & & \\ a_{31} & a_{32} & a_{33} & & \\ a_{41} & a_{42} & a_{43} & a_{44} & \\ a_{51} & a_{52} & a_{53} & a_{54} & a_{55} \end{pmatrix} \tag{4.2}$$

then the corner $R_{42}(A)$ is written as:

$$R_{42}(A) = \begin{pmatrix} a_{22} & & \\ a_{32} & a_{33} & \\ a_{42} & a_{43} & a_{44} \end{pmatrix}.$$

If the j-th column of the corner $R_{ij}(A)$ is replaced with respective elements of the k-th column $(k < j)$, and the rest of the elements of the corner are left unchanged, then this corner is denoted by $R_{i,\frac{j}{k}}(A)$. For instance, in the triangular matrix (4.2), we have:

$$R_{4\frac{2}{1}}(A) = \begin{pmatrix} a_{21} & & \\ a_{31} & a_{33} & \\ a_{41} & a_{43} & a_{44} \end{pmatrix}.$$

Below we shall consider that

$$\operatorname{ddet}(R_{n,\frac{n+1}{n}}(A)) = \operatorname{pper}(R_{n,\frac{n+1}{n}}(A)) = 0.$$

But if the i-th row of the corner R_{ij} of the triangular matrix A is replaced with respective elements of the k-th row $(k > i)$, and the rest of the elements of the corner are left unchanged, then this corner is denoted by $R_{\frac{i}{k},j}(A)$, or just by $R_{\frac{i}{k},j}$. Below we consider that

$$\operatorname{ddet}(R_{\frac{0}{i},1}) = \operatorname{pper}(R_{\frac{0}{i},1}) = 0. \tag{4.3}$$

Definition 4.2. [6]. *A rectangular table of elements of the triangular matrix (3.1) is inscribed in this matrix, if its one vertex coincides with an element a_{n1}, and its opposite one coincides with an element a_{ii}, $i \in \{1, \ldots, n\}$. This table is denoted by $T(i)$.*

If in the definition 4.2 we set $i = 1$, or $i = n$, then an inscribed rectangular table degenerates into the first column or into the n-th row of this triangular matrix respectively.

Example 4.2. In Fig. 3.1., the elements of the rectangular table $T(3)$, inscribed in a triangular matrix of the fourth order, are marked out with a rectangle

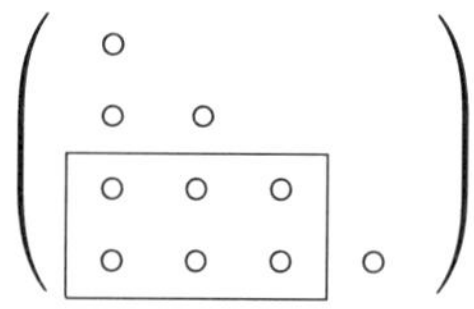

Figure 3.1.

When finding values of the paradeterminant and the parapermanent of triangular matrices, it is convenient to use *algebraic complements.*

Definition 4.3. [6]. ***Algebraic complements*** *D_{ij}, P_{ij} to a factorial product $\{a_{ij}\}$ of a key element a_{ij} of the matrix (3.1) are, respectively, numbers*

$$D_{ij} = (-1)^{i+j} \cdot \text{ddet}\,(R_{j-1,1}) \cdot \text{ddet}\,(R_{n,i+1}), \tag{4.4}$$

$$P_{ij} = \text{pper}(R_{j-1,1}) \cdot \text{pper}(R_{n,i+1}), \tag{4.5}$$

where $R_{j-1,1}$ and $R_{n,i+1}$ are corners of the triangular matrix (3.1).

Remark 4.1. To mark out corners, an algebraic complement to a factorial product of an element a_{ij} of the triangular matrix (3.1) it is convenient to use the following scheme, depicted in Fig.3.2.

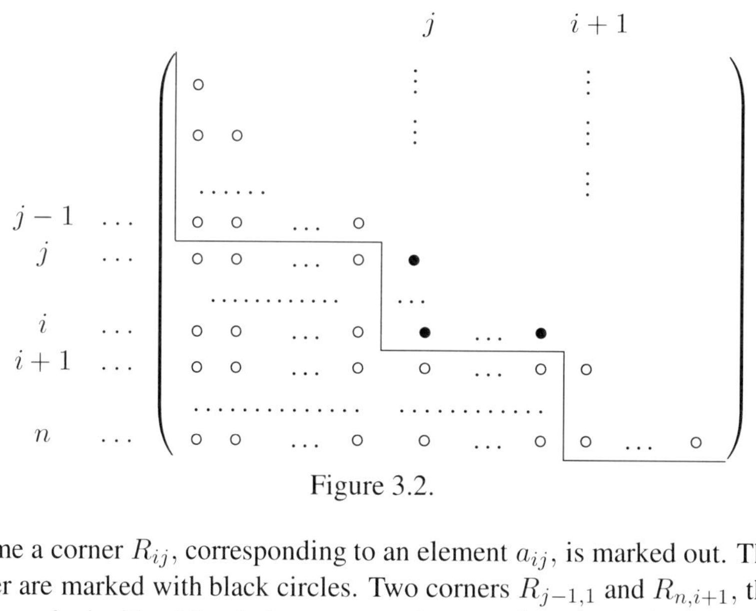

Figure 3.2.

In this scheme a corner R_{ij}, corresponding to an element a_{ij}, is marked out. The elements of this corner are marked with black circles. Two corners $R_{j-1,1}$ and $R_{n,i+1}$, the elements of which are marked with white circles, are constituents of an algebraic complement of the factorial product of an element a_{ij}.

Example 4.3. An algebraic complement D_{54} of a factorial product of an element a_{54} of a triangular matrix

$$(a_{ij})_{1\leqslant j\leqslant i\leqslant 7} = \begin{pmatrix} a_{11} & & & & & & \\ a_{21} & a_{22} & & & & & \\ a_{31} & a_{32} & a_{33} & & & & \\ a_{41} & a_{42} & a_{43} & a_{44} & & & \\ a_{51} & a_{52} & a_{53} & a_{54} & a_{55} & & \\ a_{61} & a_{62} & a_{63} & a_{64} & a_{65} & a_{66} & \\ a_{71} & a_{72} & a_{73} & a_{74} & a_{75} & a_{76} & a_{77} \end{pmatrix}$$

equals

$$D_{54} = (-1)^{5+4} \cdot \operatorname{ddet} \begin{pmatrix} a_{11} & & \\ a_{21} & a_{22} & \\ a_{31} & a_{32} & a_{33} \end{pmatrix} \cdot \operatorname{ddet} \begin{pmatrix} a_{66} & \\ a_{76} & a_{77} \end{pmatrix}$$

Now we shall study the properties of triangular matrix parafunctions.

Theorem 4.1. [6]. (Decomposition of a parafunction by elements of an inscribed rectangular table). *Let A be the triangular matrix (3.1), and $T(i)$ be some rectangular table of elements inscribed in it. Then the following equality holds:*

$$\operatorname{ddet}(A) = \sum_{s=1}^{i} \sum_{r=i}^{n} \{a_{rs}\}\, D_{rs}, \tag{4.6}$$

$$\operatorname{pper}(A) = \sum_{s=1}^{i} \sum_{r=i}^{n} \{a_{rs}\}\, P_{rs}, \tag{4.7}$$

where D_{rs} and P_{rs} are respectively algebraic complements to the factorial product of a key element a_{rs}, which belongs to $T(i)$.

Proof. Corners $R_{s-1,1}$ and $R_{n,r+1}$ consists of different elements of the matrix (3.1), which respectively have order $(s-1)-1+1 = s-1$ and $n-(r+1)+1 = n-r$. By Definitions 3.5 and 3.6 (p. 199), the paradeterminant and the parapermanent of these corners are respectively the sum 2^{s-2} and 2^{n-r-1} of different summands (here, according to the agreements (4.1), it is convenient to assume that $2^{-1} = 1$). Therefore the expressions $\{a_{rs}\}\, D_{rs}$, $\{a_{rs}\}\, P_{rs}$ for $n \in \{1, 2, \ldots\}$ consist of $2^{s-2} \cdot 2^{n-r-1}$ different summands. From Fig. 3.3. it is clear that a key element a_{rs} of the matrix (3.1) belongs to a normal tuple of key elements of this matrix only combined with normal tuples of key elements of its corners $R_{s-1,1}$, $R_{n,r+1}$. Note that each summand belonging to the right-hand member of the equalities (4.6), (4.7) consists of the product n of different elements of the matrix (3.1), for $(s-1)+(r-s+1)+(n-r) = n$, where $s-1$ and $n-r$ are orders of corners $R_{s-1,1}$ and $R_{n,r+1}$, and the product $\{a_{rs}\}$ consists of $r-s+1$ factors (see Fig. 3.3.).

The expressions in the right-hand member of the equalities (4.6), (4.7) consist of 2^{n-1} summands. Indeed,

$$\sum_{s=1}^{i} \sum_{r=i}^{n} 2^{s-2} \cdot 2^{n-r-1} =$$

$$= (1 + 2^0 + \ldots + 2^{i-2})(2^{n-i-1} + 2^{n-i-2} + \ldots + 2^0 + 1) = 2^{n-1}.$$

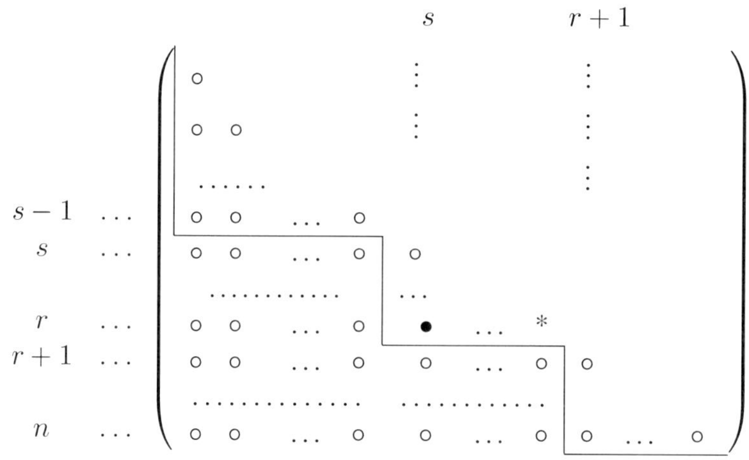

Figure 3.3.

Because all these summands are different, the theorem is proved. ■

Corollary 4.2. [6]. *If $i = 1$, then Theorem 4.1 gives decomposition of parafunctions by the elements of the fist column and the equalities (4.6), (4.7) are as follows:*

$$\operatorname{ddet}(A) = \sum_{r=1}^{n} \{a_{r1}\} D_{r1} = \sum_{r=1}^{n} (-1)^{r+1} \{a_{r1}\} \cdot \operatorname{ddet}(R_{n,r+1}), \tag{4.8}$$

$$\operatorname{pper}(A) = \sum_{r=1}^{n} \{a_{r1}\} P_{r1} = \sum_{r=1}^{n} \{a_{r1}\} \cdot \operatorname{pper}(R_{n,r+1}). \tag{4.9}$$

But if $i = n$, then we obtain decomposition of parafunctions by the elements of the last row:

$$\operatorname{ddet}(A) = \sum_{s=1}^{n} \{a_{ns}\} D_{ns} = \sum_{s=1}^{n} (-1)^{n+s} \{a_{ns}\} \cdot \operatorname{ddet}(R_{s-1,1}), \tag{4.10}$$

$$\operatorname{pper}(A) = \sum_{s=1}^{n} \{a_{ns}\} P_{ns} = \sum_{s=1}^{n} \{a_{ns}\} \cdot \operatorname{pper}(R_{s-1,1}). \tag{4.11}$$

Proof. For $i = 1$, an inscribed rectangle degenerates into the first column of the matrix (3.1). Thereby, considering the equalities (4.4), (4.5), because of the agreements (4.1), we obtain the equalities (4.8), (4.9).

The equalities (4.10), (4.11) are proved in the same way. ■

Corollary 4.3. *If all the elements of the i–th column ($i \in \{1, \ldots, n\}$) of the triangular matrix (3.1) are zeros, then*

$$\operatorname{ddet}(A) = \operatorname{pper}(A) = 0.$$

Proof. The validity of this corollary follows from the fact that all factorial products of elements of an inscribed rectangular table $T(i)$ equal zero. ■

Example 4.4. Let us find the value of the paradeterminant

$$Q_n = \begin{bmatrix} q_1 & & & & & \\ \frac{p_2}{q_2} & q_2 & & & & \\ 0 & \frac{p_3}{q_3} & q_3 & & & \\ \vdots & \cdots & \cdots & \ddots & & \\ 0 & 0 & 0 & \cdots & q_{n-1} & \\ 0 & 0 & 0 & \cdots & \frac{p_n}{q_n} & q_n \end{bmatrix}_n, \tag{4.12}$$

if $q_1 = 1,\ p_i + q_i = 1,\ i = 2, 3, \ldots$.

It is obvious that

$$Q_1 = [q_1] = 1,$$

$$Q_2 = \begin{bmatrix} 1 & \\ \frac{p_2}{q_2} & q_2 \end{bmatrix} = p_2 + q_2 = 1.$$

Let us decompose the parapermanent (4.12) by elements of the last row. By induction we have:

$$Q_n = q_n Q_{n-1} + p_n Q_{n-1} = q_n + p_n = 1.$$

4.2. Properties of Parafunctions of Some Classes of Triangular Matrices

Proposition 4.1. [6]. *If respective elements of all rows of an inscribed rectangular table* $T(i),\ i = 1, \ldots, n-1$ *of the triangular matrix (3.1) are equal, i.e., the following equalities hold*

$$a_{ij} = a_{i+1,j} = \ldots = a_{nj} = a_j,\ j = 1, \ldots, i, \tag{4.13}$$

then this equality also holds

$$\sum_{s=1}^{i} \{a_{is}\} D_{is} = -\sum_{s=1}^{i} \sum_{r=i+1}^{n} \{a_{rs}\} D_{rs}. \tag{4.14}$$

For the parapermanent of such a triangular matrix, the analogous equality holds:

$$\sum_{s=1}^{i} \{a_{is}\} P_{is} = \sum_{s=1}^{i} \sum_{r=i+1}^{n} \{a_{rs}\} P_{rs}. \tag{4.15}$$

Proof. We prove the equality (4.14). By Definition 4.3 of an algebraic complement (see the equality (4.4) on p.212), the expression $\{a_{is}\}D_{is}$, according to the equality (4.13), equals

$$\{a_{is}\} D_{is} = (-1)^{i+s} a_s a_{s+1} \ldots a_i \text{ddet}\,(R_{s-1,1}) \text{ddet}\,(R_{n,i+1}). \tag{4.16}$$

We shall consider the expression

$$\sum_{r=i+1}^{n} \{a_{rs}\} D_{rs}.$$

Since

$$\{a_{rs}\} = a_s \dots a_i \{a_{r,i+1}\}, \quad r = i+1, \dots, n,$$

then

$$\sum_{r=i+1}^{n} \{a_{rs}\} D_{rs} =$$

$$= \sum_{r=i+1}^{n} (-1)^{r+s} a_s \dots a_i \{a_{r,i+1}\} \text{ddet}\,(R_{s-1,1}) \text{ddet}\,(R_{n,r+1}) =$$

$$= a_s \dots a_i \text{ddet}\,(R_{s-1,1}) \sum_{r=i+1}^{n} (-1)^{r-s} \{a_{r,i+1}\} \text{ddet}\,(R_{n,r+1}) =$$

$$= a_s \dots a_i \text{ddet}\,(R_{s-1,1}) \sum_{r=i+1}^{n} (-1)^{r-s} \{a_{r,i+1}\} \text{ddet}\,(R_{n,r+1}).$$

But the sum

$$\sum_{r=i+1}^{n} (-1)^{r+i+1} \{a_{r,i+1}\} \text{ddet}\,(R_{n,r+1})$$

is the result of decomposition of the paradeterminant of a corner $R_{n,i+1}$ by elements of the first column, therefore, taking into consideration the equality (4.16), we have:

$$\sum_{r=i+1}^{n} \{a_{rs}\} D_{rs} =$$

$$= (-1)^{s-i-1+2i} a_s \dots a_i \text{ddet}\,(R_{s-1,1}) \text{ddet}\,(R_{n,i+1}) = -\{a_{is}\} D_{is}.$$

It is the last equality that proves the validity of the equality (4.14).

The equality (4.15) is proved in the same way as the equality proved above. ∎

Corollary 4.4. *If the respective elements of all rows of a rectangular table $T(i)$, $i = 1, \dots, n-1$, inscribed into the triangular matrix (3.1), are equal, i.e., the conditions (4.13) of Proposition 4.1 are true, then the paradeterminant of this triangular matrix equals zero, and for the parapermanent the following equality holds:*

$$\text{pper}(A) = 2 \cdot \text{pper}(R_{n,i+1}) \begin{bmatrix} a_{11} & & & & \\ a_{21} & a_{22} & & & \\ \vdots & \dots & \ddots & & \\ a_{i-1,1} & a_{i-1,2} & \cdots & a_{i-1,i-1} & \\ a_1 & a_2 & \cdots & a_{i-1} & a_i \end{bmatrix}. \tag{4.17}$$

Proof. Let us decompose the paradeterminant of the given triangular matrix by the element of the inscribed rectangular table $T(i)$:

$$\text{ddet}\,(A) = \sum_{s=1}^{i} \sum_{r=i}^{n} \{a_{rs}\} D_{rs} = \sum_{s=1}^{i} \left(\{a_{is}\} D_{is} + \sum_{r=i+1}^{n} \{a_{rs}\} D_{rs} \right).$$

The expression in the parenthesis of the last equality equals zero because of the equality (4.14). This is what proves the first part of the corollary.

Let us prove the second part of the corollary. Considering the equality (4.15) we obtain the following equalities in the same way:

$$\operatorname{pper}(A) = \sum_{s=1}^{i}\sum_{r=i}^{n}\{a_{rs}\}P_{rs} = \sum_{s=1}^{i}\left(\{a_{is}\}P_{is} + \sum_{r=i+1}^{n}\{a_{rs}\}P_{rs}\right) =$$

$$= 2\cdot \operatorname{pper}(R_{n,i+1})\sum_{s=1}^{i}\{a_{is}\}\operatorname{pper}(R_{s-1,1}).$$

Since the sum

$$\sum_{s=1}^{i}\{a_{is}\}\operatorname{pper}(R_{s-1,1})$$

is the result of the decomposition of the parapermanent of the corner

$$R_{i,1} = \begin{bmatrix} a_{11} & & & & \\ a_{21} & a_{22} & & & \\ \vdots & \dots & \ddots & & \\ a_{i-1,1} & a_{i-1,2} & \cdots & a_{i-1,i-1} & \\ a_1 & a_2 & \cdots & a_{i-1} & a_i \end{bmatrix}$$

by the elements of the last row, then this proves the second part of Corollary 4.4. ■

Corollary 4.5. *If all the elements of the first column of the triangular matrix (3.1) for* $(n > 1)$ *equal* a*, then its paradeterminant equals zero, and the parapermanent equals twice the product of this number by the parapermanent of the triangular matrix resulting from deleting its first column.*

Example 4.5. Let us find the values of the paradeterminants of triangular matrices

$$A = \left(\frac{i}{i-j+1}\right)_{1\leqslant j\leqslant i\leqslant n}, \quad B = (j-(j-1)\delta_{ij})_{1\leqslant j\leqslant i\leqslant n}.$$

The values of the first column elements of these triangular matrices equal

$$a_{i1} = \frac{i}{i-1+1} = 1, \; b_{i1} = 1, \; i = 1, 2, \dots, n.$$

Consequently, according to Corollary 4.5, the paradeterminants of these matrices equal 0 for $n > 1$. For $n = 1$ the paradeterminants of these triangular matrices equal 1.

Example 4.6. Let us find the value of the parapermanent of the triangular matrix

$$\begin{pmatrix} \frac{1}{2}a_1 & & & \\ \frac{1}{2}a_1 & \frac{1}{2}a_2 & & \\ \vdots & \dots & \ddots & \\ \frac{1}{2}a_1 & \frac{1}{2}a_2 & \dots & \frac{1}{2}a_n \end{pmatrix}_n = \left(\frac{1}{2}a_j\right)_{1\leqslant j\leqslant i\leqslant n}.$$

For this purpose we apply $(n-1)$ times the second part of Corollary 4.5 to the given triangular matrix. At that we obtain the following result:

$$\operatorname{pper}\left(\frac{1}{2}a_j\right)_{1\leqslant j\leqslant i\leqslant n}=\frac{1}{2}\prod_{i=1}^{n}a_i.$$

If in the given triangular matrix $a_j = j$, then, decomposing it by the elements of the last row, we obtain the identity for the factorial:

$$n!=\frac{n!}{2^{n-1}}+\sum_{s=2}^{n}\frac{(s-1)!s^{\overline{n-s+1}}}{2^{n-s+1}},$$

or

$$n!=\frac{2^{n-1}}{2^{n-1}-1}\cdot\sum_{s=2}^{n}\frac{(s-1)!s^{\overline{n-s+1}}}{2^{n-s+1}}.$$

Proposition 4.2. *Given the triangular matrix (3.1), then the following equalities hold:*

$$\sum_{r=j}^{n}\{a_{rj}\}\,D_{rj}=\operatorname{ddet}(R_{j-1,1})\operatorname{ddet}(R_{nj}),\tag{4.18}$$

$$\sum_{r=j}^{n}\{a_{rj}\}\,P_{rj}=\operatorname{pper}(R_{j-1,1})\operatorname{pper}(R_{nj}).\tag{4.19}$$

Proof.

$$\sum_{r=j}^{n}\{a_{rj}\}D_{rj}=\sum_{r=j}^{n}\{a_{rj}\}\cdot(-1)^{r+j}\operatorname{ddet}(R_{j-1,1})\operatorname{ddet}(R_{n,r+1})=$$

$$=\operatorname{ddet}(R_{j-1,1})\cdot\sum_{r=j}^{n}(-1)^{r+j}\{a_{rj}\}\operatorname{ddet}(R_{n,r+1}).$$

The last sum is the result of decomposing the paradeterminant $\operatorname{ddet}(R_{nj})$ by the elements of its first column, and so the equality (4.18) holds.

In the same way, the following equality is proved (4.19). ■

Proposition 4.3. *If all the elements of the j-th column $(1 < j < n)$ of the triangular matrix (3.1) equal a, then its paradeterminant equals the product of a number $(-a)$ by the paradeterminant of the triangular matrix B, which we obtain from the given one by deleting the j-th column and the $(j-1)$-th row.*

For the parapermanent of this triangular matrix, the following equality holds

$$\begin{bmatrix} a_{11} & & & & \\ \vdots & \ddots & & & \\ a_{j-1,1} & \cdots & a_{j-1,j-1} & & \\ a_{j1} & \cdots & a_{j,j-1} & a & \\ \vdots & \cdots & \cdots & \cdots & \ddots \\ a_{n1} & \cdots & a_{n,j-1} & a & \ldots\ a_{nn} \end{bmatrix}=$$

$$=a\cdot(2\operatorname{pper}(R_{j-1,1})\cdot\operatorname{pper}(R_{n,j+1})+\operatorname{pper}(B)).$$

Proof. We shall prove the first part of this proposition. Let us decompose the paradeterminant of the given triangular matrix by the elements of the inscribed rectangular table $T(j)$. This decomposition will include all the summands resulting from its decomposition by the elements of the j-th column. According to Proposition 4.2, all of them are included in the expression $\operatorname{ddet}(R_{j-1,1}) \cdot \operatorname{ddet}(R_{nj})$, the value of which equals zero (see Corollary 4.5 on p. 217), for $\operatorname{ddet}(R_{nj}) = 0$. But the expression includes $\operatorname{ddet}(R_{j-1,1}) \cdot \operatorname{ddet}(R_{nj})$ all the summands, the multiplier of which is at least one element of the $(j-1)$-th row, therefore when finding the paradeterminant of the triangular matrix (3.1) the $(j-1)$-th row may be deleted. The rest of summands of paradeterminant decomposition by the elements of an inscribed rectangular table has a factor a, therefore its factoring out equals deleting the j-th column of the triangular matrix (3.1). At that the sum of indices of a key element belonging to an inscribed rectangular table is reduced by one, and indices of the rest of key elements remain unchanged. Consequently, deletion of the j-th column changes evenness of the sum of indices of a key element of each summand and (-1) must be factored out.

We shall prove the second part of the proposition (4.3). Let us decompose the parapermanent of the given triangular matrix by the elements of the inscribed rectangular table $T(j)$:

$$\operatorname{pper}(A) = \sum_{s=1}^{j}\sum_{r=j}^{n} \{a_{rs}\}\, P_{rs} = \sum_{s=1}^{j-1}\sum_{r=j}^{n} \{a_{rs}\}\, P_{rs} + \sum_{r=j}^{n}\{a_{rj}\}P_{rj}. \tag{4.20}$$

The arguments, analogous to the arguments of the demonstration of the first part of this proposition, give us the following equality

$$\sum_{s=1}^{j-1}\sum_{r=j}^{n} \{a_{rs}\}\, P_{rs} = a \cdot \operatorname{pper}(B).$$

The second summand of the equality (4.20) is decomposition of the parapermanent of the given triangular matrix by the elements of the j–th column; therefore, according to Corollary (4.5), we have:

$$\sum_{r=j}^{n}\{a_{rj}\}P_{rj} = \operatorname{pper}(R_{j-1,1}) \cdot \operatorname{pper}(R_{nj}) =$$

$$= 2a \cdot \operatorname{pper}(R_{j-1,1})\operatorname{pper}(R_{n,j+1}).$$

■

Proposition 4.4. [6]. *If all the elements of the a_{ri}, i–th column of the triangular matrix (3.1) are the sum of some two elements, i.e., $a_{ri} = b_{ri} + c_{ri}$, where $r \in \{i, \dots, n\}$, then the following equality holds*

$$\operatorname{ddet}\begin{pmatrix} a_{11} & & & & & \\ a_{21} & a_{22} & & & & \\ \vdots & \cdots & \ddots & & & \\ a_{i1} & a_{i2} & \cdots & b_{ii}+c_{ii} & & \\ \vdots & \cdots & \cdots & \cdots & \ddots & \\ a_{n1} & a_{n2} & \cdots & b_{ni}+c_{ni} & \cdots & a_{nn} \end{pmatrix} =$$

$$\operatorname{ddet}\begin{pmatrix} a_{11} & & & & & \\ a_{21} & a_{22} & & & & \\ \vdots & \dots & \ddots & & & \\ a_{i1} & a_{i2} & \dots & b_{ii} & & \\ \vdots & \dots & \dots & \dots & \ddots & \\ a_{n1} & a_{n2} & \dots & b_{ni} & \dots & a_{nn} \end{pmatrix} +$$

$$+\operatorname{ddet}\begin{pmatrix} a_{11} & & & & & \\ a_{21} & a_{22} & & & & \\ \vdots & \dots & \ddots & & & \\ a_{i1} & a_{i2} & \dots & c_{ii} & & \\ \vdots & \dots & \dots & \dots & \ddots & \\ a_{n1} & a_{n2} & \dots & c_{ni} & \dots & a_{nn} \end{pmatrix}. \tag{4.21}$$

For parapermanents, an analogous equality holds.

Proof. By the definition of the paradeterminant, each if its summands is the product of n elements of the triangular matrix (3.1). And each of them includes one and only one element of the i-th column as a factor. Therefore, in each of 2^{n-1} summands of the paradeterminant of the left-hand member of this equality, there is a multiplier $(b_{ri} + c_{ri})$, $r \in \{i, \dots, n\}$. Removing the parenthesis in each summand and grouping respective summands, we obtain the right-hand member of the equality (4.4). ■

Remark 4.2. Using Proposition 4.4, by induction, we can prove the validity of an analogous proposition in case of k summands $b_{ri}^{(1)} + b_{ri}^{(2)} + \dots + b_{ri}^{(k)}$, $r \in \{i, \dots, n\}$.

Proposition 4.5. [6]. *If all the elements of the i–th* ($i \in \{1, \dots, n\}$) *column of the paradeterminant of the triangular matrix (3.1) satisfy the equalities* $a_{ri} = k \cdot b_{ri}$, $r \in \{i, \dots, n\}$, *then the following equality holds:*

$$\operatorname{ddet}\begin{pmatrix} a_{11} & & & & & \\ a_{21} & a_{22} & & & & \\ \vdots & \dots & \ddots & & & \\ a_{i1} & a_{i2} & \dots & k \cdot b_{ii} & & \\ \vdots & \dots & \dots & \dots & \ddots & \\ a_{n1} & a_{n2} & \dots & k \cdot b_{ni} & \dots & a_{nn} \end{pmatrix} =$$

$$k \cdot \operatorname{ddet}\begin{pmatrix} a_{11} & & & & & \\ a_{21} & a_{22} & & & & \\ \vdots & \dots & \ddots & & & \\ a_{i1} & a_{i2} & \dots & b_{ii} & & \\ \vdots & \dots & \dots & \dots & \ddots & \\ a_{n1} & a_{n2} & \dots & b_{ni} & \dots & a_{nn} \end{pmatrix}.$$

For parapermanents, an analogous proposition is true.

The demonstration of this proposition is analogous to the demonstration of Proposition 4.4. □

Proposition 4.6. [6]. *For the block-diagonal matrix (3.3) the following equalities hold:*

$$\operatorname{ddet}\begin{pmatrix} M_1 & & & \\ 0 & M_2 & & \\ \vdots & \dots & \ddots & \\ 0 & 0 & \dots & M_s \end{pmatrix} = \operatorname{ddet}(M_1)\cdot\ldots\cdot\operatorname{ddet}(M_s), \tag{4.22}$$

$$\operatorname{pper}\begin{pmatrix} M_1 & & & \\ 0 & M_2 & & \\ \vdots & \dots & \ddots & \\ 0 & 0 & \dots & M_s \end{pmatrix} = \operatorname{pper}(M_1)\cdot\ldots\cdot\operatorname{pper}(M_s), \tag{4.23}$$

where M_i, $1 \leqslant i \leqslant s$ are some triangular matrices.

Proof. We shall prove the equality (4.22). Let a_{i_s,i_s} be upper elements of matrices M_s. We decompose the paradeterminant of the matrix (3.3) by the elements of the table $T(i_2)$. Since the elements of this table, except the elements of the column i_2, equal zero, then according to Proposition 4.2 (equality (4.18), p. 218), we have the equality

$$\operatorname{ddet}\begin{pmatrix} M_1 & & & \\ 0 & M_2 & & \\ \vdots & \dots & \ddots & \\ 0 & 0 & \dots & M_s \end{pmatrix} = (\mathrm{M}_1)\cdot\operatorname{ddet}\begin{pmatrix} M_2 & & & \\ 0 & M_3 & & \\ \vdots & \dots & \ddots & \\ 0 & 0 & \dots & M_s \end{pmatrix}.$$

Applying Theorem 4.1 in sequence to the tables $T(i_3), \ldots, T(i_s)$, on the ground of Proposition 4.2, we obtain the equality (4.22).

The equality (4.23) is proved in the same way. ■

Proposition 4.7. [6]. *Let the elements of the matrix (3.1) be differentiable functions of the variable t, then the following equalities hold:*

$$\frac{d}{dt}(\operatorname{ddet}(A)) = \operatorname{ddet}\begin{pmatrix} a'_{11} & & & \\ a'_{21} & a_{22} & & \\ \vdots & \dots & \ddots & \\ a'_{n1} & a_{n2} & \dots & a_{nn} \end{pmatrix} + \tag{4.24}$$

$$+\operatorname{ddet}\begin{pmatrix} a_{11} & & & \\ a_{21} & a'_{22} & & \\ \vdots & \dots & \ddots & \\ a_{n1} & a'_{n2} & \dots & a_{nn} \end{pmatrix} + \ldots + \operatorname{ddet}\begin{pmatrix} a_{11} & & & \\ a_{21} & a_{22} & & \\ \vdots & \dots & \ddots & \\ a_{n1} & a_{n2} & \dots & a'_{nn} \end{pmatrix},$$

$$\tfrac{d}{dt}(\mathrm{pper}(A)) = \begin{bmatrix} a'_{11} & & & \\ a'_{21} & a_{22} & & \\ \vdots & \dots & \ddots & \\ a'_{n1} & a_{n2} & \dots & a_{nn} \end{bmatrix} + \tag{4.25}$$

$$+ \begin{bmatrix} a_{11} & & & \\ a_{21} & a'_{22} & & \\ \vdots & \dots & \ddots & \\ a_{n1} & a'_{n2} & \dots & a_{nn} \end{bmatrix} + \ldots + \begin{bmatrix} a_{11} & & & \\ a_{21} & a_{22} & & \\ \vdots & \dots & \ddots & \\ a_{n1} & a_{n2} & \dots & a'_{nn} \end{bmatrix}.$$

Proof. By the definition of a paradeterminant, it consists of 2^{n-1} summands, each of which is the product of n factors, taken by one from each column. Let $(-1)^{\varepsilon(a)} a_{i_1 1} a_{i_2 2} \cdot \ldots \cdot a_{i_n n}$ be one of the summands of this paradeterminant. It is known that

$$\frac{d}{dt}(a_{i_1 1} a_{i_2 2} \cdot \ldots \cdot a_{i_n n}) =$$

$$= a'_{i_1 1} a_{i_2 2} \cdot \ldots \cdot a_{i_n n} + a_{i_1 1} a'_{i_2 2} \cdot \ldots \cdot a_{i_n n} + \ldots + a_{i_1 1} a_{i_2 2} \cdot \ldots \cdot a'_{i_n n}. \tag{4.26}$$

We isolate the summands of paradeterminants in the right-hand member of the equality (4.24), respective to the summands $a_{i_1 1} a_{i_2 2} \cdot \ldots \cdot a_{i_n n}$ and find their sum, then we obtain the expression in the right-hand member of the equality (4.26). At that, according to the mentioned correspondence, the sign of the summand $a_{i_1 1} a_{i_2 2} \cdot \ldots \cdot a_{i_n n}$ coincides with the signs of the respective summands of the paradetrminants of the right-hand member of the equality (4.24). By summing all the summands $\frac{d}{dt}((-1)^{\varepsilon(a)} a_{i_1 1} a_{i_2 2} \cdot \ldots \cdot a_{i_n n})$ of the left-hand of the equality (4.24) and considering the equality (4.26), we obtain the equality (4.24). The equality (4.25) is proved similarly to the equality (4.24). ■

4.3. Transformations of Triangular Matrices that Fix Their Parafunctions

Proposition 4.8. *For any triangular matrix (3.1) and any number $k \in K$, where K is some number field, the following equalities hold:*

$$\mathrm{ddet}\begin{pmatrix} a_{11}+k & & & \\ a_{21}+k & a_{22} & & \\ \vdots & \vdots & \ddots & \\ a_{n1}+k & a_{n2} & \cdots & a_{nn} \end{pmatrix}_n = \mathrm{ddet}\begin{pmatrix} a_{11} & & & \\ a_{21} & a_{22} & & \\ \vdots & \vdots & \ddots & \\ a_{n1} & a_{n2} & \cdots & a_{nn} \end{pmatrix}_n \tag{4.27}$$

$$\begin{bmatrix} a_{11}-k & & & \\ a_{21}+k & a_{22} & & \\ \vdots & \vdots & \ddots & \\ a_{n1}+k & a_{n2} & \cdots & a_{nn} \end{bmatrix} = \begin{bmatrix} a_{11} & & & \\ a_{21} & a_{22} & & \\ \vdots & \vdots & \ddots & \\ a_{n1} & a_{n2} & \cdots & a_{nn} \end{bmatrix}. \tag{4.28}$$

Proof. We shall prove the second equality of this proposition, using Proposition 4.4 and Corollary 4.5.

$$\mathrm{pper}(A) =$$

$$\begin{bmatrix} a_{11}+k-k & & & \\ a_{21}+k-k & a_{22} & & \\ \vdots & \vdots & \ddots & \\ a_{n1}+k-k & a_{n2} & \cdots & a_{nn} \end{bmatrix} = \begin{bmatrix} a_{11}+k & & & \\ a_{21}+k & a_{22} & & \\ \vdots & \vdots & \ddots & \\ a_{n1}+k & a_{n2} & \cdots & a_{nn} \end{bmatrix} +$$

$$+ \begin{bmatrix} -k & & & \\ -k & a_{22} & & \\ \vdots & \vdots & \ddots & \\ -k & a_{n2} & \cdots & a_{nn} \end{bmatrix} = \begin{bmatrix} a_{11}+k & & & \\ a_{21}+k & a_{22} & & \\ \vdots & \vdots & \ddots & \\ a_{n1}+k & a_{n2} & \cdots & a_{nn} \end{bmatrix} +$$

$$+(-2k)\cdot \begin{bmatrix} a_{22} & & & \\ a_{32} & a_{33} & & \\ \vdots & \vdots & \ddots & \\ a_{n2} & a_{n3} & \cdots & a_{nn} \end{bmatrix} = \begin{bmatrix} a_{11}+k & & & \\ a_{21}+k & a_{22} & & \\ \vdots & \vdots & \ddots & \\ a_{n1}+k & a_{n2} & \cdots & a_{nn} \end{bmatrix} +$$

$$+ \begin{bmatrix} -2k & & & \\ 0 & a_{22} & & \\ \vdots & \vdots & \ddots & \\ 0 & a_{n2} & \cdots & a_{nn} \end{bmatrix} = \begin{bmatrix} a_{11}-k & & & \\ a_{21}+k & a_{22} & & \\ \vdots & \vdots & \ddots & \\ a_{n1}+k & a_{n2} & \cdots & a_{nn} \end{bmatrix}$$

The first equality is proved in the same way. ■

Proposition 4.9. *If in a triangular matrix A its upper element (see Definition 3.1) equals 0 and any element of the first column is multiplied by some number $k \neq 0$, and the respective element of the second column is divided by the same number, then the value of the paradeterminant and the parapermanent of the obtained triangular matrix A' remains unchanged.*

Proof. We shall prove the validity of this proposition for parapermanents. Let us compare the factorial products of the elements of the first column of triangular matrices A and A'. It is obvious that they are equal. Consequently, decompositions of the parapermanents of these triangular matrices by the elements of the first column coincide.

For paradeterminants the demonstration is analogous. ■

Remark 4.3. If in the triangular matrix (3.1), in its first column, besides the upper element, there are also zero elements, then they can be replaced with ones, but their respective elements of the second columns must be replaced with zeros. The validity of this remark is proved by the arguments analogous to the arguments of the demonstration of Proposition 4.9.

Theorem 4.6. *[9]. Given the triangular matrix*

$$A = (a_{rs})_{1 \le s \le r \le n},$$

then the following equality holds

$$ddet(A) = ddet(A'), \tag{4.29}$$

where

$$A' =$$

$$\begin{pmatrix} a_{11} \\ \vdots & \ddots \\ a_{j-1,1} & \dots & a_{j-1,j-1} \\ a_{j1} & \dots & a_{j,j-1} & a_{jj} \\ \vdots & \vdots & \vdots & \vdots & \ddots \\ a_{i-1,1} & \dots & a_{i-1,j-1} & a_{i-1,j} & \dots & a_{i-1,i-1} \\ 0 & \dots & 0 & x_{ij} & \dots & a_{i,i-1} & a_{ii} \\ a_{i+1,1} & \dots & a_{i+1,j-1} & a_{i+1,j} & \dots & a_{i+1,i-1} & a_{i+1,i} & a_{i+1,i+1} \\ \vdots & \vdots & \vdots & \vdots & \vdots & \vdots & \vdots & \vdots & \ddots \\ a_{n1} & \dots & a_{n,j-1} & a_{nj} & \dots & a_{n,i-1} & a_{ni} & a_{n,i+1} & \dots a_{nn} \end{pmatrix},$$

and

$$x_{ij} = a_{ij}\left(1 - \frac{\mathrm{ddet}\,(R_{\frac{j-1}{i},1})}{\mathrm{ddet}\,(R_{j-1,1})}\right). \tag{4.30}$$

Proof. To prove the equality (4.29) it is enough to compare an algebraic complement only of those factorial products of the elements belonging to the i-th row and columns starting from the first to the j-th.

In the left-hand member of the equality, we have

$$\big(\{a_{i1}\}(-1)^{i+1}\mathrm{ddet}\,(R_{n,i+1}) + \{a_{i2}\}(-1)^{i+2}\mathrm{ddet}\,(R_{11})\mathrm{ddet}\,(R_{n,i+1})$$

$$+ \ldots + \{a_{i,j-1}\}(-1)^{i+j-1}\mathrm{ddet}(R_{j-2,1})\mathrm{ddet}(R_{n,i+1})\big) +$$

$$+\{a_{ij}\}(-1)^{i+j}\mathrm{ddet}\,(R_{j-1,1})\mathrm{ddet}(R_{n,i+1}) =$$

$$= \{a_{ij}\}(-1)^{i+j-1}\mathrm{ddet}(R_{n,i+1})\,(a_{i,j-1}\mathrm{ddet}(R_{j-2,1}) -$$

$$-a_{i,j-2}a_{i,j-1}\mathrm{ddet}\,(R_{j-3,1}) + \ldots + (-1)^{j-2}a_{i1}\cdot\ldots\cdot a_{i,j-1}\big) +$$

$$+\{a_{ij}\}(-1)^{i+j}\mathrm{ddet}\,(R_{j-1,1})\mathrm{ddet}(R_{n,i+1}) =$$

$$= \{a_{ij}\}(-1)^{i+j-1}\mathrm{ddet}(R_{\frac{j-1}{i},1})\mathrm{ddet}(R_{n,i+1}) +$$

$$+\{a_{ij}\}(-1)^{i+j}\mathrm{ddet}(R_{j-1,1})\mathrm{ddet}(R_{n,i+1}) =$$

$$= \{a_{ij}\}(-1)^{i+j}\mathrm{ddet}(R_{n,i+1})\left(\mathrm{ddet}(R_{j-1,1}) - \mathrm{ddet}\left(R_{\frac{j-1}{i},1}\right)\right).$$

The sum of the respective elements of the right-hand member of the equality is obviously equal to

$$\{x_{ij}\}(-1)^{i+j}\mathrm{ddet}(R_{j-1,1})\mathrm{ddet}(R_{n,i+1}) =$$

$$= x_{ij}\{a_{i,j+1}\}(-1)^{i+j}\mathrm{ddet}(R_{j-1,1})\mathrm{ddet}(R_{n,i+1})$$

Comparing the obtained sums of the right-hand and left-hand members of the equality (4.29), we receive the equality (4.30.) ■ We shall give one useful corollary of Theorem 4.6, which allows deflating parafunctions by one.

Corollary 4.7. *For any triangular matrix (3.1), the following is true:*

$$\left\langle \begin{matrix} a_{11} & & & \\ a_{21} & a_{22} & & \\ \vdots & \cdots & \ddots & \\ a_{n1} & a_{n2} & \cdots & a_{nn} \end{matrix} \right\rangle_n = \left\langle \begin{matrix} (a_{11}-a_{21})\cdot a_{22} & & & \\ (a_{11}-a_{31})\cdot a_{32} & a_{33} & & \\ \vdots & \cdots & \ddots & \\ (a_{11}-a_{n1})\cdot a_{n2} & a_{n3} & \cdots & a_{nn} \end{matrix} \right\rangle_{n-1} \tag{4.31}$$

$$\left[\begin{matrix} a_{11} & & & \\ a_{21} & a_{22} & & \\ \vdots & \cdots & \ddots & \\ a_{n1} & a_{n2} & \cdots & a_{nn} \end{matrix} \right]_n = \left[\begin{matrix} (a_{11}+a_{21})\cdot a_{22} & & & \\ (a_{11}+a_{31})\cdot a_{32} & a_{33} & & \\ \vdots & \cdots & \ddots & \\ (a_{11}+a_{n1})\cdot a_{n2} & a_{n3} & \cdots & a_{nn} \end{matrix} \right]_{n-1} \tag{4.32}$$

Proof. Indeed, we shall apply Theorem 4.6 for $j = 2$, $i = 2, 3, \ldots, n$ to the paradeterminant and the parapermanent of the triangular matrix (3.1) respectively, the we get the equalities:

$$\left\langle \begin{matrix} a_{11} & & & \\ a_{21} & a_{22} & & \\ \vdots & \cdots & \ddots & \\ a_{n1} & a_{n2} & \cdots & a_{nn} \end{matrix} \right\rangle = \left\langle \begin{matrix} a_{11} & & & & \\ 0 & a_{22}\left(1-\frac{a_{21}}{a_{11}}\right) & & & \\ 0 & a_{32}\left(1-\frac{a_{31}}{a_{11}}\right) & a_{33} & & \\ \vdots & \cdots & \cdots & \ddots & \\ 0 & a_{n2}\left(1-\frac{a_{n1}}{a_{11}}\right) & a_{n3} & \cdots & a_{nn} \end{matrix} \right\rangle =$$

$$\left[\begin{matrix} a_{11} & & & \\ a_{21} & a_{22} & & \\ \vdots & \cdots & \ddots & \\ a_{n1} & a_{n2} & \cdots & a_{nn} \end{matrix} \right] = \left[\begin{matrix} a_{11} & & & & \\ 0 & a_{22}\left(1+\frac{a_{21}}{a_{11}}\right) & & & \\ 0 & a_{32}\left(1+\frac{a_{31}}{a_{11}}\right) & a_{33} & & \\ \vdots & \cdots & \cdots & \ddots & \\ 0 & a_{n2}\left(1+\frac{a_{n1}}{a_{11}}\right) & a_{n3} & \cdots & a_{nn} \end{matrix} \right]$$

from which evidently follows the proposition of the corollary. ∎

5. On Relation of Parafunctions with Other Functions of Matrices

5.1. On Relation of Paradeterminants to Parapermanents

Theorem 5.1. [6]. (The theorem on relation of the parapermanent to the paradeterminant). *If A is the triangular matrix (3.1), then the following equality holds*

$$\operatorname{pper}(a_{ij})_{1\leqslant j\leqslant i\leqslant n} = \operatorname{ddet}\left((-1)^{\delta_{ij}+1}a_{ij}\right)_{1\leqslant j\leqslant i\leqslant n}. \tag{5.1}$$

column. After some simplifications in the right-hand member, we obtain the equality

$$\sum_{i=1}^{k}(-1)^{i+1}\{a_{i1}\}\text{ddet}\,(R_{k,i+1}) = \sum_{i=1}^{k}(-1)^{i+1}b_{i1}\cdot B\begin{pmatrix} i & i+1\,, & \ldots & k \\ i & i+1\,, & \ldots & k \end{pmatrix},$$

which proves the fulfillment of the induction condition. ■

Remark 5.1. In the last equality, we assume that

$$\text{ddet}\,(R_{n,n+1}) = B\begin{pmatrix} n+1 & n \\ n+1 & n \end{pmatrix} = 1.$$

Corollary 5.4. *For any lower quasitriangular matrix (5.2), it is obvious that the following equality holds*

$$\begin{vmatrix} b_{11} & 1 & 0 & \ldots & 0 & 0 \\ b_{21} & b_{22} & 1 & \ldots & 0 & 0 \\ b_{31} & b_{32} & b_{33} & \ldots & 0 & 0 \\ \ldots & \ldots & \ldots & \ldots & \ldots & \ldots \\ b_{n-1,1} & b_{n-1,2} & b_{n-1,3} & \ldots & b_{n-1,n-1} & 1 \\ b_{n1} & b_{n2} & b_{n3} & \ldots & b_{n,n-1} & b_{nn} \end{vmatrix} =$$

$$= \text{ddet}\begin{pmatrix} b_{11} & & & & \\ \frac{b_{21}}{b_{22}} & b_{22} & & & \\ \frac{b_{31}}{b_{32}} & \frac{b_{32}}{b_{33}} & b_{33} & & \\ \vdots & \vdots & \vdots & \ddots & \\ \frac{b_{n1}}{b_{n2}} & \frac{b_{n2}}{b_{n3}} & \frac{b_{n3}}{b_{n4}} & \cdots & b_{nn} \end{pmatrix}. \tag{5.5}$$

Remark 5.2. Note that the equality (5.5) holds even when some elements of a lower quasi-triangular matrix equal 0, because when finding the value of the respective paradeterminant, zeros are canceled and uncertainty disappears.

Proposition 5.1. *The following equality holds*

$$\begin{vmatrix} a_{11} & 1 & 0 & \ldots & 0 & 0 \\ -a_{21} & a_{22} & 1 & \ldots & 0 & 0 \\ a_{31} & -a_{32} & a_{33} & \ldots & 0 & 0 \\ \vdots & \vdots & \vdots & \ldots & \vdots & \vdots \\ (-1)^{n-3}a_{n-2,1} & (-1)^{n-4}a_{n-2,2} & (-1)^{n-5}a_{n-2,3} & \cdots & 1 & 0 \\ (-1)^{n-2}a_{n-1,1} & (-1)^{n-3}a_{n-1,2} & (-1)^{n-4}a_{n-1,3} & \cdots & a_{n-1,n-1} & 1 \\ (-1)^{n-1}a_{n,1} & (-1)^{n-2}a_{n,2} & (-1)^{n-3}a_{n,3} & \cdots & -a_{n,n-1} & a_{n,n} \end{vmatrix} =$$

$$= \begin{vmatrix} a_{11} & -1 & 0 & \ldots & 0 & 0 \\ a_{21} & a_{22} & -1 & \ldots & 0 & 0 \\ a_{31} & a_{32} & a_{33} & \ldots & 0 & 0 \\ \vdots & \vdots & \vdots & \ldots & \vdots & \vdots \\ a_{n-2,1} & a_{n-2,2} & a_{n-2,3} & \cdots & -1 & 0 \\ a_{n-1,1} & a_{n-1,2} & a_{n-1,3} & \cdots & a_{n-1,n-1} & -1 \\ a_{n,1} & a_{n,2} & a_{n,3} & \cdots & a_{n,n-1} & a_{n,n} \end{vmatrix}. \tag{5.6}$$

Proof. For $n = 1$ and $n = 2$, it is obvious that the equality holds.

Let the equality (5.8) hold for $n = 1, 2, \ldots, k-1$. We shall prove its validity for $n = k$. Decompose the determinant of the left-hand member of the equality (5.8) by the elements of the last column into two determinants of order $(k - 1)$, multiply the second determinant and each element of its last row by (-1). Apply Proposition (5.8) for $n = k - 1$ to both determinants of order $(k - 1)$ and discover that the sum of two obtained summands results from decomposition of the determinant of the right-hand member of the equality (5.8) by the elements of the last column. ■

5.3. On Relation of Permanents to Determinants

Permanents are an important notion of linear algebra and can be applied in different areas of mathematics, especially in combinatorial analysis [20]. Despite the surface simplicity of permanents, the natural algorithm for finding their values, similar to the Gaussian algorithm for finding the values of determinants, has not been discovered yet. In this context, Polya in 1913 [21] defined the problem of finding a transformation, which could replace permanents with determinants, and proved himself that there is no way to ascribe the signs "+" and "−" simultaneously to the elements of a matrix of order n, $3 \leq n$ so that its permanent turns into the determinant. After fundamental generalization of Polya's result given by Marcus and Minc [22], the hope to find even linear transformations, which would transform the permanent of a matrix of order $n-$ $(3 \leqslant n)$ into the determinant of the same order, disappeared. In this regard, in [15] (p. 22-23), in principle, the following problem is posed:

From all matrices of order $n-$, we are to isolate a class of matrices, for which there is such a linear transformation on the set of their elements that the permanent of an initial matrix equals the determinant of a transformed matrix.

Definition 5.1. *Let*

$$A = (a_{ij})_{i,j=1,2,\ldots,n}$$

be a square matrix. ***Polya transformation*** *of this matrix is distribution of signs "+" and "−" before its elements, which transforms the permanent into the equal determinant. This transformation is denoted by $P(A)$.*

We shall isolate a matrix class, for which Polya transformation exists.

According to Theorem 5.3 (p. 226), for any triangular matrix (3.1) and lower quasitriangular matrix (5.2), the following identity holds

$$\operatorname{ddet}(A) = \det(B),$$

in which

$$b_{ij} = \{a_{ij}\} = \prod_{k=j}^{i} a_{ik}, \ 1 \leq j \leq i \leq n.$$

In this very point, there is Corollary 5.4, where the identity between the determinant and the paradeterminant of a triangular matrix is established.

In the first part of the proof of Theorem 5.3, the bijection between the summands of the paradeterminant of a triangular matrix A and the summands of the determinant of a quasi-triangular matrix B is established. In the second part of the theorem, the correspondence of the signs of these summands is proved. Since the permanents of square matrices differ from the determinants of these matrices only in signs, then from Corollary 5.4 of this theorem it follows that the equality holds

$$\text{per}(B) = \text{pper}(A), \tag{5.7}$$

where B and A are respectively the lower quasitriangular matrix (5.2) and the triangular matrix (3.1), and

$$a_{ij} = \frac{b_{ij}}{b_{i,j+1}}.$$

On p. 225 we proved the theorem on relation of the parapermanent to the paradeterminant and its corollary on relation of the paradeterminant to the parapermanent of a triangular matrix

$$\text{ddet}\,(a_{ij})_{1\le j\le i\le n} = \text{pper}((-1)^{\delta_{ij}+1}\cdot a_{ij})_{1\le j\le i\le n}.$$

From this identity, the equality follows

$$\text{pper}(a_{ij})_{1\le j\le i\le n} = \text{ddet}\,((-1)^{\delta_{ij}+1}\cdot a_{ij})_{1\le j\le i\le n}.$$

Thus, considering the equalities (5.7) and (5.1), we obtain the following relations:

$$\text{per}(b_{ij})_{i,j=1,2,\ldots,n} = \text{pper}\left(\frac{b_{ij}}{b_{i,j+1}}\right)_{1\leqslant j\leqslant i\leqslant n} =$$

$$= \text{ddet}\left((-1)^{\delta_{ij}+1}\frac{b_{ij}}{b_{i,j+1}}\right)_{1\leqslant j\leqslant i\leqslant n}$$

But since

$$\prod_{k=j}^{i}\frac{b_{ik}}{b_{i,k+1}} = b_{ij},$$

then the identity holds

$$\text{per}(b_{ij})_{i,j=1,2,\ldots,n} = \det\left((-1)^{i-j}b_{ij}\right)_{i,j=1,2,\ldots,n},$$

and so does the following

Theorem 5.5. [12]. *For the quasitriangular matrix (5.2) there is Polya transformation*

$$P(B) = (-1)^{i-j}b_{ij},\ i,j = 1,2,\ldots,n,$$

i.e., the following equality holds

$$\text{per}(B) = \det(P(B)).$$

It is interesting that if in the quasitriangular matrix (5.2) at least one non-zero element $b_{ij},\ i-j>-1$ is added, then there is no Polya transformation for such a matrix.

Indeed, let B^* be a square matrix obtained from the quasitriangular matrix (5.2) by adding a non-zero element $b_{ij},\ 2\leqslant j-i$. Then, due to the fact that the permanent of any matrix does not change after rearranging any of its columns or rows, we shall proceed to a new matrix, in which the $i-$th column changes places with the $j-$th column. At that we obtain the permanent of a matrix, in which the blocks can be isolated in sequence on its main diagonal:

$$[b_{11}],[b_{22}],\ldots,[b_{i-1,i-1}],[b_{ij}],[b_{i+1,i+1}],\ldots,[b_{j-2,j-2}],$$

$$\begin{bmatrix} b_{j-1,j-2} & b_{j-1,j-1} & b_{j-1,i} \\ b_{j,j-2} & b_{j,j-1} & b_{ji} \\ b_{j+1,j-2} & b_{j+1,j-1} & b_{j+1,i} \end{bmatrix},[b_{j+2,j+2}],\ldots,[b_{nn}].$$

Thus, in the initial permanent the elements of six transversals are constructed with the help of the elements of six transversals of an isolated block

$$\begin{bmatrix} b_{j-1,j-2} & b_{j-1,j-1} & b_{j-1,i} \\ b_{j,j-2} & b_{j,j-1} & b_{ji} \\ b_{j+1,j-2} & b_{j+1,j-1} & b_{j+1,i} \end{bmatrix}$$

and the elements of the main diagonal of this matrix. While three of them, which correspond to even permutations, must be positive, and other three – negative. But Polya proved [21] that one cannot place the signs in a matrix of order three in such a way[7]. This is what our proposition proves.

Thus, a quasitriangular matrix is maximum in terms of the number of non-zero elements of a matrix, for which Polya transformation does exist. Since a quasitriangular matrix consists of $(n^2+3n-2)/2$ elements, then the proved theorem agrees with Gibson's proposition [23] about the fact that if $(0,1)-$ matrix A of order n has a positive permanent and if the permanent A can be written in the form of the determinant, ascribing signs $\pm$ to the elements of a matrix A, then the number of elements in A is not more than $(n^2+3n-2)/2$.

Let us prove one more proposition, which allows us to find new Polya transformation for a lower quasitriangular matrix.

Proposition 5.2. *The following equality holds*

$$\begin{vmatrix} a_{11} & 1 & 0 & \cdots & 0 & 0 \\ -a_{21} & a_{22} & 1 & \cdots & 0 & 0 \\ a_{31} & -a_{32} & a_{33} & \cdots & 0 & 0 \\ \vdots & \vdots & \vdots & \cdots & \vdots & \vdots \\ (-1)^{n-3}a_{n-2,1} & (-1)^{n-4}a_{n-2,2} & (-1)^{n-5}a_{n-2,3} & \cdots & 1 & 0 \\ (-1)^{n-2}a_{n-1,1} & (-1)^{n-3}a_{n-1,2} & (-1)^{n-4}a_{n-1,3} & \cdots & a_{n-1,n-1} & 1 \\ (-1)^{n-1}a_{n,1} & (-1)^{n-2}a_{n,2} & (-1)^{n-3}a_{n,3} & \cdots & -a_{n,n-1} & a_{n,n} \end{vmatrix} =$$

[7]Let us assume that a transformation like this exists, then the number of minuses of each transversal, which corresponds to even permutation, must be even, and the total number of minuses is even. On the other hand, the number of minuses on each transversal, responsible for odd permutation, must be odd, and the total number of minuses is odd, which, actually, contradicts the preceding reasoning.

$$= \begin{vmatrix} a_{11} & -1 & 0 & \dots & 0 & 0 \\ a_{21} & a_{22} & -1 & \dots & 0 & 0 \\ a_{31} & a_{32} & a_{33} & \dots & 0 & 0 \\ \vdots & \vdots & \vdots & \dots & \vdots & \vdots \\ a_{n-2,1} & a_{n-2,2} & a_{n-2,3} & \dots & -1 & 0 \\ a_{n-1,1} & a_{n-1,2} & a_{n-1,3} & \dots & a_{n-1,n-1} & -1 \\ a_{n,1} & a_{n,2} & a_{n,3} & \dots & a_{n,n-1} & a_{n,n} \end{vmatrix}. \tag{5.8}$$

Proof. For $n = 1$ and $n = 2$, the equality is obviously true.

Let the equality (5.8) hold for $n = 1, 2, \dots, k-1$. We shall prove its validity for $n = k$. For this purpose, we decompose the determinant of the left-hand member of the equality (5.8) by the elements of the last column into two determinants of order $(k-1)$, multiply the second determinant and each element of its last row by (-1) and apply to both determinants of order $(k-1)$ Proposition (5.8) for $n = k-1$. But the sum of two obtained summands is the result of decomposition of the right-hand member of the equality (5.8) by the elements of the last column. ∎ Thus, the following also holds

Theorem 5.6. [12]. *For the quasitriangular matrix (5.2), there is Polya transformation*

$$P(B) = \begin{cases} b_{ij}, & 1 \leqslant j \leqslant i \leqslant n, \\ -1, & j - i = 1, \\ 0, & j - i \geqslant 2. \end{cases}$$

6. Some Theorems on Applying Triangular Matrix Calculus

Let us present a number of theorems on some application of triangular matrix calculus without proving them.

Mostly, it is difficult to solve a secular equation of a linear recurrence equation. In this case, it is useful to apply the following theorem, which is Stanley's theorem to the point, but formulated in terms of parapermanents (see [24], p. 301).

Theorem 6.1. [13]. *Suppose we are given two vectors*

$$a = (a_1, a_2, a_3, \dots, a_k),$$

$$b = (b_0 = 1, b_1, b_2, \dots, b_{k-1}).$$

For the sequence $\{u_n\}_{n=0}^{\infty}$, the following three equalities are equivalent:
1. Linear recurrence equality of the k–th order

$$u_n = a_1 u_{n-1} + a_2 u_{n-2} + a_3 u_{n-3} + \dots + a_k u_{n-k}, \quad n = k, k+1, k+2, \dots \tag{6.1}$$

with the initial conditions

$$u_0 = b_0 = 1, u_1 = b_1, u_2 = b_2, \dots, u_{k-1} = b_{k-1}. \tag{6.2}$$

2.

$$u_n = \operatorname{pper}(A_n) = \begin{bmatrix} a_1c_1 & & & & & & & \\ \frac{a_2}{a_1} & a_1c_2 & & & & & & \\ \cdots & \cdots & \ddots & & & & & \\ \frac{a_{k-1}}{a_{k-2}} & \frac{a_{k-2}}{a_{k-3}} & \cdots & a_1c_{k-1} & & & & \\ \frac{a_k}{a_{k-1}} & \frac{a_{k-1}}{a_{k-2}} & \cdots & \frac{a_2}{a_1} & a_1 & & & \\ 0 & \frac{a_k}{a_{k-1}} & \cdots & \frac{a_3}{a_2} & \frac{a_2}{a_1} & a_1 & & \\ \cdots & \cdots & \cdots & \cdots & \cdots & \cdots & \ddots & \\ 0 & 0 & \cdots & 0 & \frac{a_k}{a_{k-1}} & \cdots & \frac{a_2}{a_1} & a_1 \end{bmatrix}_n, \tag{6.3}$$

where the corrections c_i *are found from equalities*

$$c_i = b_i\left(\sum_{s=1}^{i} a_s b_{i-s}\right)^{-1} = \frac{b_i}{a_i + a_{i-1}b_1 + a_{i-2}b_2 + \cdots + a_2b_{i-2} + a_1b_{i-1}}, \tag{6.4}$$

$i = 1, \ldots, k-1;$

3.

$$1 + \sum_{i=1}^{\infty} u_i z^i =$$

$$= \frac{1 + b_1\left(1 - \frac{1}{c_1}\right)z^1 + b_2\left(1 - \frac{1}{c_2}\right)z^2 + \ldots + b_{k-1}\left(1 - \frac{1}{c_{k-1}}\right)z^{k-1}}{1 - a_1z - a_2z^2 - \ldots - a_kz^k}. \tag{6.5}$$

Theorem 6.2. *Suppose we are given the generating function*

$$f(z) = \frac{1 + d_1z + d_2z^2 + \ldots + d_{k-1}z^{k-1}}{1 - a_1z - a_2z^2 - \ldots - a_kz^k} \tag{6.6}$$

of the sequence $\{u_n\}_{n=1}^{\infty}$. *Then it satisfies the recurrence equation*

$$u_n = a_1u_{n-1} + a_2u_{n-2} + \ldots + a_ku_k$$

with the initial conditions

$$u_0 = 1,\ u_i = \begin{bmatrix} a_1 \mid d_1 & & & \\ \frac{a_2+d_2}{a_1} & a_1 & & \\ \cdots & \cdots & \ddots & \\ \frac{a_i+d_i}{a_{i-1}} & \frac{a_{i-1}}{a_{i-2}} & \cdots & a_1 \end{bmatrix}_i,\ i = 1, 2, \ldots, k-1. \tag{6.7}$$

Theorem 6.3. *Let* K *be some number field. For the triangular matrix*

$$F = \begin{pmatrix} \frac{a}{b}x_1 & & & & & & \\ \frac{a+r}{b+s}\frac{x_2}{x_1} & \frac{a}{b}x_1 & & & & & \\ \cdots & \cdots & \cdots & & & & \\ \frac{a+(n-1)r}{b+(n-1)s}\frac{x_n}{x_{n-1}} & \frac{a+(n-2)r}{b+(n-2)s}\frac{x_{n-1}}{x_{n-2}} & \cdots & \frac{a}{b}x_1 & & & \\ 0 & \frac{a+(n-1)r}{b+(n-1)s}\frac{x_n}{x_{n-1}} & \cdots & \frac{a+r}{b+s}\frac{x_2}{x_1} & \frac{a}{b}x_1 & & \\ \cdots & \cdots & \cdots & \cdots & \cdots & \cdots & \\ 0 & 0 & \cdots & \frac{a+(n-1)r}{b+(n-1)s}\frac{x_n}{x_{n-1}} & \cdots & \cdots & \frac{a}{b}x_1 \end{pmatrix}_m, \tag{6.8}$$

and the parameters a, b, r, s, the values of which belong to the number field K, the following identities hold:

$$\operatorname{ddet}(F) = \lozenge_{n,m} = \tag{6.9}$$

$$= \sum_{\lambda_1+\ldots+n\lambda_n=m} (-1)^{n-k} \frac{k!}{\lambda_1! \cdot \ldots \cdot \lambda_n!} \left(\frac{a^{1\{r\}}}{b^{1\{s\}}}\right)^{\lambda_1} \cdot \ldots \cdot \left(\frac{a^{n\{r\}}}{b^{n\{s\}}}\right)^{\lambda_n} x_1^{\lambda_1} \cdot \ldots \cdot x_n^{\lambda_n},$$

$$\operatorname{pper}(F) = \square_{n,m} = \tag{6.10}$$

$$= \sum_{\lambda_1+\ldots+n\lambda_n=m} \frac{k!}{\lambda_1! \cdot \ldots \cdot \lambda_n!} \left(\frac{a^{1\{r\}}}{b^{1\{s\}}}\right)^{\lambda_1} \cdot \ldots \cdot \left(\frac{a^{n\{r\}}}{b^{n\{s\}}}\right)^{\lambda_n} x_1^{\lambda_1} \cdot \ldots \cdot x_n^{\lambda_n}$$

If $A(z),\ B(z),\ X(z)$ are respective notations of formal power series with nonzero constant term

$$\sum_{i=0}^{\infty} a_i z^i,\ \sum_{i=0}^{\infty} b_i z^i,\ \sum_{i=0}^{\infty} x_i z^i,\ a_0 = b_0 = x_0 = 1$$

Then the following theorems are true:

Theorem 6.4. [14]. *If*

$$X(z) = \frac{A(z)}{B(z)}$$

then

$$x_i = \sum_{j=0}^{i-1} (-1)^j (a_{i-j} - b_{i-j}) \cdot \left\langle \frac{b_{s-r+1}}{b_{s-r}} \right\rangle_{1 \leqslant r \leqslant s \leqslant j},\ i = 1, 2, \ldots.$$

Here and below, we suppose that

$$\left\langle \frac{b_{s-r+1}}{b_{s-r}} \right\rangle_{1 \leqslant r \leqslant s \leqslant 0} = 1.$$

Theorem 6.5. [4] *If $X(z) = (A(z))^p$, here*

$$A(z) = 1 + \sum_{i=1}^{\infty} a_i z^i,$$

and p is some real number, then

$$x_n = (-1)^n \left\langle \frac{(i-j+1) \cdot p - (j-1)}{(i-j) \cdot p - j} \cdot \frac{a_{i-j+1}}{a_{i-j}} \right\rangle_{1 \leqslant j \leqslant i \leqslant n} =$$

$$= \left[(-1)^{\delta_{ij}} \frac{(i-j+1) \cdot p - (j-1)}{(i-j) \cdot p - j} \cdot \frac{a_{i-j+1}}{a_{i-j}} \right]_{1 \leqslant j \leqslant i \leqslant n} \tag{6.11}$$

Theorem 6.6. [14]. *Let f and f^{-1} be reciprocal functions, then if the following equality holds*

$$f(1 + a_1 z + a_2 z^2 + a_3 z^3 + \ldots) = 1 + b_1 z + b_2 z^2 + b_3 z^3 + \ldots, \tag{6.12}$$

where

$$b_i = \left\langle \tau_{sr} \frac{a_{s-r+1}}{a_{s-r}} \right\rangle_{1 \leqslant r \leqslant s \leqslant i}, \quad i = 1, 2, \ldots, \tag{6.13}$$

then the following equality also holds

$$f^{-1}(1 + a_1 z + a_2 z^2 + a_3 z^3 + \ldots) = 1 + x_1 z + x_2 z^2 + x_3 z^3 + \ldots, \tag{6.14}$$

where

$$x_i = \left\langle \tau_{s,s-r+1}^{-1} \frac{a_{s-r+1}}{a_{s-r}} \right\rangle_{1 \leqslant r \leqslant s \leqslant i}, \quad i = 1, 2, \ldots. \tag{6.15}$$

Remark 6.1. *According to Theorem 6.6, while constructing the triangular matrix*

$$\left\langle \tau_{s,s-r+1}^{-1} \frac{a_{s-r+1}}{a_{s-r}} \right\rangle_{1 \leqslant r \leqslant s \leqslant i}, \quad i = 1, 2, \ldots,$$

it is necessary to remember that the coefficients of its elements are inverse to the coefficients of the elements of the triangular matrix

$$\left\langle \tau_{sr} \frac{a_{s-r+1}}{a_{s-r}} \right\rangle_{1 \leqslant r \leqslant s \leqslant i}, \quad i = 1, 2, \ldots,$$

and written in each row in reverse order.

Theorem 6.7. [3]. (Theorem on composition of power series) *If the formal power series* $c(z) = \sum_{i=1}^{\infty} c_i z^i$ *is the result of the composition of the formal power series* $a(z) = \sum_{i=1}^{\infty} a_i z^i$ *and* $b(z) = \sum_{i=1}^{\infty} b_i z^i$, *then the following equalities hold:*

$$c_i = \begin{pmatrix} a_1 \\ a_2 \\ \vdots \\ a_i \end{pmatrix} \cdot \begin{bmatrix} b_1 & & & \\ \frac{b_2}{b_1} & b_1 & & \\ \vdots & \vdots & \ddots & \\ \frac{b_i}{b_{i-1}} & \frac{b_{i-1}}{b_{i-2}} & \cdots & b_1 \end{bmatrix}_i .$$

Theorem 6.8. [3]. (Theorem on inversion of a series) *Suppose we are given two formal power series*

$$a(x) = y = x + a_2 x^2 + a_3 x^3 + a_4 x^4 + \ldots$$

and

$$b(y) = x = y + b_2 y^2 + b_3 y^3 + b_4 y^4 + \ldots,$$

then the following equalities are true

$$b_n = \frac{1}{n} \cdot (-1)^{n-1} \left\langle \frac{(i-j+1)n + (j-1)}{(i-j)n + j} \cdot \frac{a_{i-j+2}}{a_{i-j+1}} \right\rangle_{1 \leqslant j \leqslant i \leqslant n-1} =$$

$$= \frac{1}{n} \left[(-1)^{\delta_{ij}} \cdot \frac{(i-j+1)n + (j-1)}{(i-j)n + j} \cdot \frac{a_{i-j+2}}{a_{i-j+1}} \right]_{1 \leqslant j \leqslant i \leqslant n-1}, \quad a_1 = 1. \tag{6.16}$$

7. Conclusion

This short introduction to the Theory of Triangular Matrices (Tables) includes only its main notions and theorems, which make it possible to become familiar with this theory skin-deep. For a more detailed familiarization with it, the reader should consider the monograph [25], where the main results obtained by the author and his students before 2010 are systemized.

Any mathematical theory is of no importance if it is not connected with other areas of mathematics and does not solve applied problems. In this introduction, Chapter 7 covers these aspects where one can find a number of theorems without their proving. It should be mentioned that with the help of parafunctions of triangular matrices, the principles of recurrence fraction theory are established. This theory is the most natural n-dimensional generalization of chain fractions.

References

[1] R.A. Zatorsky, *Some Methods and Problems of Combinatorial Analysis* (Specialized Course in Mathematics) Ivano-Frankivsk: Lik, 2006. (in Russian).

[2] R.A.Zatorsky, On the Number of Combinations on Multisets, *Bulletin of KNU, Series: Physics and Mathematics*, 3 (2000) 42–47 (in Russian).

[3] R.A.Zatorsky, Operations with Formal Power Series with Zero Constant Term, *Bulletin of KNU, Series: Physics and Mathematics* 1 (2010) 7–10 (in Russian).

[4] R.A. Zatorsky, Parafunctions and Combinatorial Identities, Herald of Chernivtsi University, Collection of Scientific Papers, *Mathematics* 336-337, pp. 79–84, Chernivtsi: ChNU, 2007. (in Russian).

[5] A.H. Haniushkin, R.A. Zatorsky, I.I. Lishchynsky, On Paradeterminants and Parapermanents, *Bulletin of KNU, Series: Physics and Mathematics* 1 (2005) 35–41 (in Russian).

[6] R.A. Zatorsky, On Paradeterminants and Parapermanents of Triangular Matrices, *Mathematical Studies* 17(1) (2002) 3–17 (in Russian).

[7] R.A. Zatorsky, Theory of paradeterminants and its applications, *Algebra and Diskrete Mathematics* 1 (2007) 109–138.

[8] R.A. Zatorsky, Scalar Product of the Vector by the Paradeterminant of a Triangular Matrix and Its Applications, *PreCarpathian Bulletin of ShSS* 1 (2008) 22-31.

[9] R.A. Zatorsky, O.R. Maliarchuk, Reduction of Paradeterminants of Triangular Matrices to the k-diagonal Form, *Mathematical Studies* 34(1) (2010) 20–29 (in Ukrainian).

[10] R.A. Zatorsky, I.I. Lishchynsky, On the Relation of Determinants to Paradeterminants, *Mathematical Studies* 25(1) (2006) 97–102 (in Russian).

[11] R.A. Zatorsky, Determinants of Triangular Matrices and Trajectories on the Ferrer Diagrams, *Mathematical Notes* 72(6) (2002) 834–852 (in Russian).

[12] V.Ye.Tarakanov, R.A. Zatorsky, On Relation of Determinants to Permanents, *Mathematical Notes* 85(2) (2009) 292-299 (in Russian).

[13] R.A. Zatorsky, I.I. Lishchynsky, Application of Parapermanents to Linear Recurrence Equations, *Herald of NTUU "KPI"* 5 (2008) 122–128 (in Russian).

[14] R.A. Zatorsky, Operations with Formal Power Series with Nonzero Constant Term, *Bulletin of KNU, Series: Physics and Mathematics* 1 (2008) 36–39 (in Russian).

[15] H. Minc, *Permanents*, Henryk Minc Cambridge University Press, 1984.

[16] R. Bellman, *Introduction to matrix analysis* (2nd ed.), Society for Industrial and Applied Mathematics Philadelphia, 1997.

[17] A.G. Kurosh, *Higher Algebra*, Mir, 1975.

[18] M. Aigner, *Combinatorial Theory*, New York: Springer-Verlag, 1997.

[19] J. Feder, *Fractals*, Plenum Press, NY, 1988.

[20] V.Ye.Tarakanov, Combinatorial Problems and $(0, 1)$–Matrices, in *Problems of Science and Technological Progress*, Nauka, Moscow, 1985 (in Russian).

[21] G. Polya, Uber eine Aufgabe der Wahrscheinlichkeitsrechnung betreffend die Irrfahrt im Strassennetz, *Mathematische Annalen* 84 (1921) 149–160.

[22] M. Marcus, H.Minc, On the relation between the determinant and the permanent, *Illinois J. Math.* 5 (1961) 376-381.

[23] P.M. Gibson, Conversion of the permanent into the determinant, *Proc. Amer. Math. Soc.* 27 (1971) 471–476.

[24] R. Stanley, *Enumerative Combinatorics*, second edition. Cambridge University Press, 1999.

[25] R.A. Zatorsky, *Calculus of Triangular Matrices and Its Applications.* Ivano-Frankivsk, Simyk, 2010. (in Ukrainian).

In: Advances in Linear Algebra Research
Editor: Ivan Kyrchei, pp. 239-285

ISBN: 978-1-63463-565-3

Chapter 7

RECENT DEVELOPMENTS IN ITERATIVE ALGORITHMS FOR SOLVING LINEAR MATRIX EQUATIONS

Masoud Hajarian*
Department of Mathematics, Faculty of Mathematical Sciences,
Shahid Beheshti University, General Campus, Evin, Tehran, Iran

Abstract

The aim of this chapter is to present the latest developments in iterative methods for solving linear matrix equations. The iterative methods are obtained by extending the methods presented to solve the linear system $Ax = b$. Numerical examples are investigated to confirm the efficiency of the methods.

Keywords: Linear matrix equation; Sylvester matrix equation; Iterative method

AMS Subject Classification: 15A24; 93E24; 65F10; 65F30

1. Introduction

It is well-known that the linear matrix equations such as (coupled) Lyapunov, (coupled) Sylvester, and Sylvester-transpose matrix equations are important equations which play a fundamental role in the various fields of engineering theory, particularly in theories and applications of stability and control [31–33,71]. The Sylvester and Lyapunov matrix equations arise in stability analysis of linear systems [53], model reduction [1] and in the solution of the algebraic Riccati matrix equation [54]. The coupled Lyapunov matrix equations

$$A_i^T X_i + X_i A_i + Q_i + \sum_{j=1}^{N} p_{ij} X_j = 0, \quad Q_i > 0, \quad i \in \{1, 2, ..., N\}, \tag{1}$$

*E-mail addresses: m_hajarian@sbu.ac.ir; mhajarian@aut.ac.ir

and

$$X_i = A_i^T(\sum_{j=1}^{N} p_{ij}X_j)A_i + Q_i, \ \ Q_i > 0, \ \ i \in \{1, 2, ..., N\}, \tag{2}$$

with unknown matrices X_i are often encountered in stability analysis of linear jump systems with Markovian transitions [3,55]. In the analysis and design problems associated with the linear system

$$\dot{x}(t) = Ax(t) + Bu(t),$$

the generalized Sylvester matrix equation

$$AX - XF = BY + R, \tag{3}$$

is encountered [70]. Also it can be shown that certain control problems, such as pole/eigenstructure assignment and observer design of the second-order linear systems

$$M\ddot{x}(t) + D\dot{x}(t) + Kx(t) = Bu(t), \tag{4}$$

are closely related with the second-order Sylvester matrix equation [5,32,45–47]

$$MXF^2 + DXF + KX = BY. \tag{5}$$

The above applications have motivated both mathematicians and engineers to construct a large number of methods catering to solving linear matrix equations [8–12,43,48–52,62–68,72–79].

In this chapter, we survey some of the latest developments in iterative algorithms for solving several linear matrix equations. The outline of this chapter is as follows. In Section 2, we propose four iterative algorithms based on the conjugate gradient method on the normal equations (CGNE) for finding the generalized centro-symmetric and generalized bisymmetric solutions of general matrix equations. The gradient based iterative algorithms are proposed for solving matrix equations over the generalized centro-symmetric and generalized bisymmetric matrices in Section 3. In Section 4, the least-squares QR-factorization (LSQR) algorithms are extended for solving the generalized Sylvester-transpose and periodic Sylvester matrix equations. The matrix form of bi-conjugate gradients (Bi-CG) and bi-conjugate residual (Bi-CR) algorithms are given to solve the generalized Sylvester-transpose matrix equation in Section 5. In Section 6, the conjugate gradients squared (CGS) method is developed to obtain an algorithm for solving the general coupled matrix equations. We propose the extended bi-conjugate gradient stabilized (Bi-CGSTAB) method for finding the solutions of coupled Sylvester matrix equations in Section 7. In Section 8, the matrix form of quasi-minimal residual variant of the Bi-CGSTAB algorithm (QMR-CGSTAB) is proposed to find the solutions of the general coupled and periodic coupled matrix equations. Numerical results are reported in Section 9. Finally Section 10 ends this chapter with a brief conclusion.

In this work, we will use the following notations. Let $\mathbb{R}^{m\times n}$ and $\mathbb{SOR}^{n\times n}$ denote the set of $m \times n$ real and symmetric orthogonal matrices, respectively. The symbols A^T and $\mathrm{tr}(A)$ represent the transpose and the trace of a matrix A, respectively. The unit matrix of order n is denoted by I_n. We also write it as I, when the dimensions of this matrix is clear. For an $m \times n$ matrix A, the so-called vectorization operator $\mathrm{vec}(A)$ is defined by the following

$$\mathrm{vec}(A) = (\, a_1^T \;\; a_2^T \;\; \dots \;\; a_n^T \,)^T,$$

where a_k is the k-th column of A. The notation $A \otimes B$ stands for the Kronecker product of matrices A and B. A well-known property of Kronecker product is, for matrices A, B and X with appropriate dimension

$$\mathrm{vec}(AXB) = (B^T \otimes A)\mathrm{vec}(X).$$

The inner product $\langle .,. \rangle$ in $\mathbb{R}^{m\times n}$ is defined as follows.

$$\langle A, B \rangle = \mathrm{tr}(B^T A) \quad \text{for} \quad A, B \in \mathbb{R}^{m\times n}.$$

The induced matrix norm is $||A|| = \sqrt{\langle A, A \rangle} = \sqrt{\mathrm{tr}(A^T A)}$, which is the Frobenius norm. The generalized centro-symmetric and generalized bisymmetric matrices have wide applications in many fields [6]. These matrices can be defined as follows.

Definition 1. *Let $P \in \mathbb{R}^{n\times n}$ and $Q \in \mathbb{R}^{n\times n}$ be two symmetric orthogonal matrices, i.e., $P = P^T = P^{-1}$ and $Q = Q^T = Q^{-1}$. If $A = PAQ$ ($A = A^T = PAQ$) then A is called a generalized centro-symmetric (generalized bisymmetric) matrix with respect to (P, Q). The symbol $\mathbb{SR}^{n\times n}_{P,Q}$ ($\mathbb{BSR}^{n\times n}_{P,Q}$) denotes the set of order n generalized centro-symmetric (generalized bisymmetric) matrices with respect to (P, Q) where $P, Q \in \mathbb{SOR}^{n\times n}$. It is obvious that every $n \times n$ (symmetric) matrix is also a generalized centro-symmetric (generalized bisymmetric) with respect to (I_n, I_n).*

2. The Matrix Form of CGNE Algorithm

In this section, first we briefly review the CGNE algorithm. Second the matrix form of CGNE algorithm is proposed for solving various linear matrix equations. The CGNE algorithm is known as one of the most efficient methods for finding the solution of non-symmetric linear equations. The CGNE algorithm for solving $Ax = b$ can be summarized as following [2, 7].

Algorithm 1. ***(CGNE algorithm)***
$x(1)$ *initial vector,* $\quad r(1) = b - Ax(1), \quad p(1) = A^T r(1),$
For $k = 2, 3, \dots$
$\alpha(k-1) = \frac{\langle r(k-1), r(k-1)\rangle}{\langle p(k-1), p(k-1)\rangle},$
$x(k) = x(k-1) + \alpha(k-1)p(k-1), \quad r(k) = r(k-1) - \alpha(k-1)Ap(k-1),$
$\beta(k) = \frac{\langle r(k), r(k)\rangle}{\langle r(k-1), r(k-1)\rangle},$
$p(k) = A^T r(k) + \beta(k)p(k-1).$

The CGNE algorithm is mathematically equivalent to the LSQR which possesses more favorable numerical properties, for more details about CGNE see [2]. In recent years, various methods based on the CGNE algorithm were proposed for solving linear matrix equations [16–21, 23, 24, 57]. In this section, four extended CGNE algorithms are presented for solving general matrix equations. First we consider the following problems.

Problem 1. *For given matrices $A_{ij} \in \mathbb{R}^{r_i \times n_j}$, $B_{ij} \in \mathbb{R}^{n_j \times s_i}$, $M_i \in \mathbb{R}^{r_i \times s_i}$ and the symmetric orthogonal matrices $R_j, S_j \in \mathbb{SOR}^{n_j \times n_j}$, find the generalized centro-symmetric matrix group $(X_1, X_2, ..., X_p)$ with $X_j \in \mathbb{SR}^{n_j \times n_j}_{R_j,S_j}$, $j = 1, 2, ..., p$, such that*

$$\sum_{j=1}^{p} A_{ij} X_j B_{ij} = M_i, \qquad i = 1, 2, ..., p. \tag{6}$$

Problem 2. *For given matrices $A_{ij} \in \mathbb{R}^{r_i \times n_j}$, $B_{ij} \in \mathbb{R}^{n_j \times s_i}$, $M_i \in \mathbb{R}^{r_i \times s_i}$ and the symmetric orthogonal matrices $R_j, S_j \in \mathbb{SOR}^{n_j \times n_j}$, find the generalized bisymmetric matrix group $(X_1, X_2, ..., X_p)$ with $X_j \in \mathbb{BSR}^{n_j \times n_j}_{R_j,S_j}$, $j = 1, 2, ..., p$, such that (6) holds.*

Problem 3. *For given linear operators $\mathcal{A}_1, \mathcal{A}_2, ..., \mathcal{A}_l$ from $\mathbb{R}^{n \times n}$ onto $\mathbb{R}^{p \times q}$, the symmetric orthogonal matrix $P, Q \in \mathbb{SOR}^{n \times n}$ and $C_1, C_2, ..., C_r \in \mathbb{R}^{p \times q}$, find the generalized centro-symmetric matrix $X \in \mathbb{SR}^{n \times n}_{P,Q}$ such that*

$$\begin{cases} \mathcal{A}_1(X) = C_1, \\ \mathcal{A}_2(X) = C_2, \\ \vdots \quad \quad \vdots \ \vdots \\ \mathcal{A}_l(X) = C_l. \end{cases} \tag{7}$$

Problem 4. *For given linear operators $\mathcal{A}_1, \mathcal{A}_2, ..., \mathcal{A}_l$ from $\mathbb{R}^{n \times n}$ onto $\mathbb{R}^{p \times q}$, the symmetric orthogonal matrix $P, Q \in \mathbb{SOR}^{n \times n}$ and $C_1, C_2, ..., C_r \in \mathbb{R}^{p \times q}$, find the generalized bisymmetric matrix $X \in \mathbb{BSR}^{n \times n}_{P,Q}$ such that (7) holds.*

Obviously, the matrix equations (6) and (7) include several linear matrix equations such as (1)-(5). By similar ways to those in [14, 15, 41], we can extend the CGNE algorithm to obtain the following iterative algorithms for solving Problems 1-4.

Algorithm 2. ***(To solve Problem 1)***

Step 2.1. *Choose arbitrary initial matrices $X_j(1) \in \mathbb{SR}^{n_j \times n_j}_{R_j,S_j}$ for $j = 1, 2, ..., p$;*

Step 2.2. *For $i = 1, 2, ..., p$, calculate*

$$R(1) = \text{diag}\Big(M_1 - \sum_{t=1}^{p} A_{1t} X_t(1) B_{1t}, M_2 - \sum_{t=1}^{p} A_{2t} X_t(1) B_{2t}, ..., M_p - \sum_{t=1}^{p} A_{pt} X_t(1) B_{pt}\Big),$$

$$P_i(1) = \frac{1}{2}\Big\{\sum_{s=1}^{p}\Big[A_{si}^T\Big(M_s - \sum_{t=1}^{p} A_{st} X_t(1) B_{st}\Big)B_{si}^T\Big] + \sum_{s=1}^{p}\Big[R_i A_{si}^T\Big(M_s - \sum_{t=1}^{p} A_{st} X_t(1) B_{st}\Big)B_{si}^T S_i\Big]\Big\},$$

$k := 1;$

Step 2.3. *If $||R(k)|| = 0$ or $||R(k)|| \neq 0$, $||P_m(k)|| = 0$ for all $m = 1, 2, .., p$, then stop; else $k := k + 1$;*

Step 2.4. *Calculate*

$$\begin{aligned}
X_i(k) =& X_i(k-1) + \frac{||R(k-1)||_F^2}{\sum_{t=1}^{p}||P_t(k-1)||_F^2} P_i(k-1), \quad i = 1, 2, ..., p, \\
R(k) =& \mathrm{diag}\Big(M_1 - \sum_{t=1}^{p} A_{1t}X_t(k)B_{1t},\ M_2 - \sum_{t=1}^{p} A_{2t}X_t(k)B_{2t}, ...,\ M_p - \sum_{t=1}^{p} A_{pt}X_t(k)B_{pt}\Big) \\
=& R(k-1) - \frac{||R(k-1)||_F^2}{\sum_{t=1}^{p}||P_t(k-1)||_F^2} \\
&\times \mathrm{diag}\Big(\sum_{t=1}^{p} A_{1t}P_t(k-1)B_{1t},\ \sum_{t=1}^{p} A_{2t}P_t(k-1)B_{2t}, ...,\ \sum_{t=1}^{p} A_{pt}P_t(k-1)B_{pt}\Big), \\
P_i(k) =& \frac{1}{2}\Big[\sum_{s=1}^{p} A_{si}^T\Big(M_s - \sum_{t=1}^{p} A_{st}X_t(k)B_{st}\Big)B_{si}^T + \sum_{s=1}^{p} R_iA_{si}^T\Big(M_s - \sum_{t=1}^{p} A_{st}X_t(k)B_{st}\Big)B_{si}^TS_i\Big] \\
&+ \frac{||R(k)||_F^2}{||R(k-1)||_F^2} P_i(k-1), \quad i = 1, 2, ..., p;
\end{aligned}$$

Step 2.5. *Go to Step 2.3.*

Algorithm 3. ***(To solve Problem 2)***

Step 3.1. *Choose arbitrary initial matrices* $X_j(1) \in \mathbb{BSR}_{R_j,S_j}^{n_j \times n_j}$ *for* $j = 1, 2, ..., p$;

Step 3.2. *For* $i = 1, 2, ..., p$, *calculate*

$$\begin{aligned}
R(1) =& \mathrm{diag}\Big(M_1 - \sum_{t=1}^{p} A_{1t}X_t(1)B_{1t}, M_2 - \sum_{t=1}^{p} A_{2t}X_t(1)B_{2t}, ..., M_p - \sum_{t=1}^{p} A_{pt}X_t(1)B_{pt}\Big), \\
P_i(1) =& \frac{1}{4}\Big\{\sum_{s=1}^{p}\Big[A_{si}^T\Big(M_s - \sum_{t=1}^{p} A_{st}X_t(1)B_{st}\Big)B_{si}^T\Big] + \sum_{s=1}^{p}\Big[B_{si}\Big(M_s - \sum_{t=1}^{p} A_{st}X_t(1)B_{st}\Big)^T A_{si}\Big] \\
&+ \sum_{s=1}^{p}\Big[R_iA_{si}^T\Big(M_s - \sum_{t=1}^{p} A_{st}X_t(1)B_{st}\Big)B_{si}^TS_i\Big] + \sum_{s=1}^{p}\Big[R_iB_{si}\Big(M_s - \sum_{t=1}^{p} A_{st}X_t(1)B_{st}\Big)^T A_{si}S_i\Big]\Big\}, \\
k :=& 1;
\end{aligned}$$

Step 3.3. *If* $||R(k)|| = 0$ *or* $||R(k)|| \neq 0$, $||P_m(k)|| = 0$ *for all* $m = 1, 2, .., p$, *then stop; else* $k := k + 1$;

Step 3.4. *Calculate*

$$\begin{aligned}
X_i(k) =& X_i(k-1) + \frac{||R(k-1)||_F^2}{\sum_{t=1}^{p}||P_t(k-1)||_F^2} P_i(k-1), \quad i = 1, 2, ..., p, \\
R(k) =& \mathrm{diag}\Big(M_1 - \sum_{t=1}^{p} A_{1t}X_t(k)B_{1t},\ M_2 - \sum_{t=1}^{p} A_{2t}X_t(k)B_{2t}, ...,\ M_p - \sum_{t=1}^{p} A_{pt}X_t(k)B_{pt}\Big) \\
=& R(k-1) - \frac{||R(k-1)||_F^2}{\sum_{t=1}^{p}||P_t(k-1)||_F^2} \\
&\times \mathrm{diag}\Big(\sum_{t=1}^{p} A_{1t}P_t(k-1)B_{1t},\ \sum_{t=1}^{p} A_{2t}P_t(k-1)B_{2t}, ...,\ \sum_{t=1}^{p} A_{pt}P_t(k-1)B_{pt}\Big), \\
P_i(k) =& \frac{1}{4}\Big[\sum_{s=1}^{p} A_{si}^T\Big(M_s - \sum_{t=1}^{p} A_{st}X_t(k)B_{st}\Big)B_{si}^T + \sum_{s=1}^{p} B_{si}\Big(M_s - \sum_{t=1}^{p} A_{st}X_t(k)B_{st}\Big)^T A_{si} \\
&+ \sum_{s=1}^{p} R_iA_{si}^T\Big(M_s - \sum_{t=1}^{p} A_{st}X_t(k)B_{st}\Big)B_{si}^TS_i + \sum_{s=1}^{p} R_iB_{si}\Big(M_s - \sum_{t=1}^{p} A_{st}X_t(k)B_{st}\Big)^T A_{si}S_i\Big] \\
&+ \frac{||R(k)||_F^2}{||R(k-1)||_F^2} P_i(k-1), \quad i = 1, 2, ..., p;
\end{aligned}$$

Step 3.5. *Go to Step 3.3.*

Algorithm 4. ***(To solve Problem 3)***

Step 4.1. *Choose an arbitrary initial matrix* $X(1) \in \mathbb{SR}^{n\times n}_{P,Q}$*;*

Step 4.2. *Compute*

$$\begin{aligned}
R_j(1) =& C_j - \mathcal{A}_j(X(1)), \quad \text{for} \quad j = 1, 2, ..., l, \\
Q(1) =& \frac{1}{2}\sum_{r=1}^{l} \Big[\mathcal{A}_r^T(R_r(1)) + P\mathcal{A}_r^T(R_r(1))Q\Big], \\
k :=& 1;
\end{aligned}$$

Step 4.3. *If* $\sum_{r=1}^{l} ||R_r(k)|| = 0$ *or* $\sum_{r=1}^{l} ||R_r(k)|| \neq 0$*,* $||Q(k)|| = 0$*, then stop; else* $k = k + 1$*;*

Step 4.4. *Compute*

$$\begin{aligned}
X(k) =& X(k-1) + \frac{\sum_{r=1}^{l} ||R_r(k-1)||^2}{||Q(k-1)||^2} Q(k-1); \\
R_j(k) =& C_j - \mathcal{A}_j(X(k)) \\
=& R_j(k-1) - \frac{\sum_{r=1}^{l} ||R_r(k-1)||^2}{||Q(k-1)||^2} \mathcal{A}_j(Q(k-1)), \quad \text{for} \quad j = 1, 2, ..., l; \\
Q(k) =& \frac{1}{2}\sum_{r=1}^{l} \Big[\mathcal{A}_r^T(R_r(k)) + P\mathcal{A}_r^T(R_r(k))Q\Big] \\
&+ \frac{\sum_{r=1}^{l} ||R_r(k)||^2}{\sum_{r=1}^{l} ||R_r(k-1)||^2} Q(k-1);
\end{aligned}$$

Step 4.5. *Go to step 4.3.*

Algorithm 5. ***(To solve Problem 4)***

Step 5.1. *Choose an arbitrary initial matrix* $X(1) \in \mathbb{BSR}^{n\times n}_{P,Q}$*;*

Step 5.2. *Compute*

$$\begin{aligned}
R_j(1) =& C_j - \mathcal{A}_j(X(1)), \quad \text{for} \quad j = 1, 2, ..., l; \\
Q(1) =& \frac{1}{4}\sum_{r=1}^{l} \Big\{\mathcal{A}_r^T(R_r(1)) + \Big[\mathcal{A}_r^T(R_r(1))\Big]^T + P\Big[\mathcal{A}_r^T(R_r(1)) + \Big[\mathcal{A}_r^T(R_r(1))\Big]^T\Big]Q\Big\}; \\
k :=& 1;
\end{aligned}$$

Step 5.3. *If* $\sum_{r=1}^{l} ||R_r(k)|| = 0$ *or* $\sum_{r=1}^{l} ||R_r(k)|| \neq 0$*,* $||Q(k)|| = 0$*, then stop; else* $k = k + 1$*;*

Step 5.4. *Compute*

$$\begin{aligned}
X(k) =& X(k-1) + \frac{\sum_{r=1}^{l} ||R_r(k-1)||^2}{||Q(k-1)||^2} Q(k-1);\\
R_j(k) =& C_j - \mathcal{A}_j(X(k))\\
=& R_j(k-1) - \frac{\sum_{r=1}^{l} ||R_r(k-1)||^2}{||Q(k-1)||^2} \mathcal{A}_j(Q(k-1)), \text{ for } j = 1,2,...,l;\\
Q(k) =& \frac{1}{4}\sum_{r=1}^{l} \Big\{ \mathcal{A}_r^T(R_r(k)) + \big[\mathcal{A}_r^T(R_r(k))\big]^T + P\Big[\mathcal{A}_r^T(R_r(k)) + \big[\mathcal{A}_r^T(R_r(k))\big]^T\Big]Q\Big\}\\
&+ \frac{\sum_{r=1}^{l} ||R_r(k)||^2}{\sum_{r=1}^{l} ||R_r(k-1)||^2} Q(k-1);
\end{aligned}$$

Step 5.5. *Go to step 5.3.*

By similar proofs to those in [14, 15, 41], it can be proven that Algorithms 2-5 converge within a finite number of iterations in the absence of roundoff errors. Also by using Algorithms 2-5, the solvability of Problems 1-4 can be determined in the absence of roundoff errors respectively, for more details see [14, 15, 41].

Recently in [42], the CGNE algorithm was also extended to the periodic discrete-time generalized coupled Sylvester equations

$$\begin{cases} A_k X_k B_k + C_k Y_k D_k = M_k, \\ E_k X_{k+1} F_k + G_k Y_k H_k = N_k, \end{cases} \tag{8}$$

for $k = 1, 2, ...,$ where the coefficient matrices $A_k \in \mathbb{R}^{p\times n}$, $C_k \in \mathbb{R}^{p\times m}$, $B_k \in \mathbb{R}^{n\times q}$, $D_k \in \mathbb{R}^{m\times q}$, $E_k \in \mathbb{R}^{r\times n}$, $G_k \in \mathbb{R}^{r\times m}$, $F_k \in \mathbb{R}^{n\times s}$, $H_k \in \mathbb{R}^{m\times s}$, $M_k \in \mathbb{R}^{p\times q}$, $N_k \in \mathbb{R}^{r\times s}$, and the solutions $X_k \in \mathbb{R}^{n\times n}$, $Y_k \in \mathbb{R}^{m\times m}$ are periodic with period ϕ, i.e., $A_{k+\phi} = A_k$, $B_{k+\phi} = B_k$, $C_{k+\phi} = C_k$, $D_{k+\phi} = D_k$, $E_{k+\phi} = E_k$, $F_{k+\phi} = F_k$, $G_{k+\phi} = G_k$, $H_{k+\phi} = H_k$, $M_{k+\phi} = M_k$, $N_{k+\phi} = N_k$, $X_{k+\phi} = X_k$ and $Y_{k+\phi} = Y_k$. This class of periodic matrix equations contains various linear discrete-time periodic matrix equations such as the periodic discrete-time coupled Sylvester matrix equations. The extended CGNE algorithm to solve (8) can be summarized as follows.

Algorithm 6. ***(To solve (8))***

Step 6.1. *Choose the initial matrices* $X_k(1) \in \mathbb{R}^{n\times n}$ *and* $Y_k(1) \in \mathbb{R}^{m\times m}$ *for* $k = 1, 2, ..., \phi$;
Set $X_{\phi+1}(1) = X_1(1)$;
For $k = 1, 2, ..., \phi$ *compute*

$$R_{1,k}(1) = M_k - A_k X_k(1) B_k - C_k Y_k(1) D_k,$$

$$R_{2,k}(1) = N_k - E_k X_{k+1}(1) F_k - G_k Y_k(1) H_k;$$

Set $R_{2,0}(1) = R_{2,\phi}(1)$, $E_0 = E_\phi$ *and* $F_0 = F_\phi$;

For $k = 1, 2, ..., \phi$ *compute*

$$S_k(1) = A_k^H R_{1,k}(1) B_k^H + E_{k-1}^H R_{2,k-1}(1) F_{k-1}^H,$$

$$T_k(1) = C_k^H R_{1,k}(1) D_k^H + G_k^H R_{2,k}(1) H_k^H;$$

Set $S_{\phi+1}(1) = S_1(1)$ *and* $i := 1$*;*
Step 6.2. *If* $||R_{1,k}(i)|| = 0$ *and* $||R_{2,k}(i)|| = 0$ *for* $k = 1, 2, ..., \phi$*, then stop, else,* $i := i+1$*;*
For $k = 1, 2, ..., \phi$ *compute*

$$X_k(i) = X_k(i-1) + \frac{\sum_{j=1}^{\phi}[||R_{1,j}(i-1)||^2 + ||R_{2,j}(i-1)||^2]}{\sum_{j=1}^{\phi}[||S_j(i-1)||^2 + ||T_j(i-1)||^2]} S_k(i-1),$$

set $X_{\phi+1}(i) = X_1(i)$,

$$Y_k(i) = Y_k(i-1) + \frac{\sum_{j=1}^{\phi}[||R_{1,j}(i-1)||^2 + ||R_{2,j}(i-1)||^2]}{\sum_{j=1}^{\phi}[||S_j(i-1)||^2 + ||T_j(i-1)||^2]} T_k(i-1),$$

$$R_{1,k}(i) = M_k - A_k X_k(i) B_k - C_k Y_k(i) D_k$$

$$= R_{1,k}(i-1) - \frac{\sum_{j=1}^{\phi}[||R_{1,j}(i-1)||^2 + ||R_{2,j}(i-1)||^2]}{\sum_{j=1}^{\phi}[||S_j(i-1)||^2 + ||T_j(i-1)||^2]} [A_k S_k(i-1) B_k + C_k T_k(i-1) D_k],$$

$$R_{2,i}(i) = N_k - E_k X_{k+1}(i) F_k - G_k Y_k(i) H_k$$

$$= R_{2,k}(i-1) - \frac{\sum_{j=1}^{\phi}[||R_{1,j}(i-1)||^2 + ||R_{2,j}(i-1)||^2]}{\sum_{j=1}^{\phi}[||S_j(i-1)||^2 + ||T_j(i-1)||^2]} [E_k S_{k+1}(i-1) F_k + G_k T_k(i-1) H_k];$$

Set $R_{2,0}(i) = R_{2,\phi}(i)$ *and for* $k = 1, 2, ..., \phi$ *compute*

$$S_k(i) = A_k^H R_{1,k}(i) B_k^H + E_{k-1}^H R_{2,k-1}(i) F_{k-1}^H$$

$$+ \frac{\sum_{j=1}^{\phi}[||R_{1,j}(i)||^2 + ||R_{2,j}(i)||^2]}{\sum_{j=1}^{\phi}[||R_{1,j}(i-1)||^2 + ||R_{2,j}(i-1)||^2]} S_k(i-1),$$

$$T_k(i) = C_k^H R_{1,k}(i) D_k^H + G_k^H R_{2,k}(i) H_k^H$$

$$+ \frac{\sum_{j=1}^{\phi}[||R_{1,j}(i)||^2 + ||R_{2,j}(i)||^2]}{\sum_{j=1}^{\phi}[||R_{1,j}(i-1)||^2 + ||R_{2,j}(i-1)||^2]} T_k(i-1);$$

Set $S_{\phi+1}(i) = S_1(i)$ *and and go to Step 2.*

In [42], it was shown that Algorithm 6 converges within a finite number of iterations in the absence of roundoff errors.

3. The Gradient Iterative Algorithms

One of the famous method for solving the linear system $Ax = b$ is gradient based iterative algorithm [25–27] as follows.

$$x^{(k+1)} = x^{(k)} + \mu A^T(b - Ax^{(k)}), \quad 0 < \mu \leq \frac{2}{||A||^2}. \tag{9}$$

Ding and Chen [26–28] introduced gradient based iterative algorithms and least squares based iterative algorithm for solving (coupled) matrix equations. In [22, 25, 35, 36, 44, 75, 78], several gradient based iterative algorithms were proposed for solving linear matrix equations. By applying Kronecker product and vectorization operator, we can generalize the gradient based iterative algorithm (9) to solve Problems 1 and 2.

Algorithm 7. ***(To solve Problem 1)***

Step 7.1. *Choose arbitrary initial matrices* $X_j(1) \in \mathbb{SR}^{n_j \times n_j}_{R_j,S_j}$ *for* $j = 1, 2, ..., p$ *and a parameter*

$$0 < \mu < \frac{2}{\sum_{i=1}^{p}\sum_{j=1}^{p} ||A_{ij}B_{ij}||^2}. \tag{10}$$

Compute

$$R_s(1) = M_s - \sum_{i=1}^{p} A_{si}X_i(1)B_{si} \qquad \text{for} \qquad s = 1, 2, ..., p;$$

Step 7.2. *For* $t = 1, 2, ...,$ *compute*

$$\begin{aligned} X_s(t+1) =& X_s(t) + \frac{\mu}{2}\Big[\sum_{i=1}^{p} A_{is}^T R_i(t) B_{is}^T + \sum_{i=1}^{p} P_s A_{is}^T R_i(t) B_{is}^T P_s\Big] \ \ for \ \ s = 1, 2, ..., p, \\ R_s(t+1) =& M_s - \sum_{i=1}^{p} A_{si}X_i(t+1)B_{si} \ \ for \ \ s = 1, 2, ..., p. \end{aligned}$$

Algorithm 8. ***(To solve Problem 2)***

Step 8.1. *Choose arbitrary initial matrices* $X_j(1) \in \mathbb{BSR}^{n_j \times n_j}_{R_j,S_j}$ *for* $j = 1, 2, ..., p$ *and a parameter*

$$0 < \mu < \frac{2}{\sum_{i=1}^{p}\sum_{j=1}^{p} ||A_{ij}B_{ij}||^2}. \tag{11}$$

Compute

$$R_s(1) = M_s - \sum_{i=1}^{p} A_{si}X_i(1)B_{si} \qquad for \qquad s = 1, 2, ..., p;$$

Step 8.2. *For* $t = 1, 2, ..., $ *compute*

$$X_s(t+1) = X_s(t) + \frac{\mu}{4}\Big[\sum_{i=1}^{p} A_{is}^T R_i(t) B_{is}^T + \sum_{i=1}^{p} P_s A_{is}^T R_i(t) B_{is}^T P_s + \sum_{i=1}^{p} B_{is} R_i(t)^T A_{is} + \sum_{i=1}^{p} P_s B_{is} R_i(t)^T A_{is} P_s\Big] \quad for \quad s = 1, 2, ..., p,$$

$$R_s(t+1) = M_s - \sum_{i=1}^{p} A_{si} X_i(t+1) B_{si} \quad for \quad s = 1, 2, ..., p.$$

Similar to the proofs in [22,35,36,44], we can show that Algorithms 7 and 8 converge.

4. The Matrix Form of LSQR Algorithms

Based on the Golub-Kahan bidiagonalization process, two different forms of the LSQR method are presented for computing the solution x of the linear systems $Ax = b$ and unconstrained least-squares problem $\min_x ||Ax - b||$. The LSQR algorithms can be summarized as follows [56].

Algorithm 9. ***(The variant 1 of LSQR method)***
$\tau(0) := 1;\ \xi(0) := -1;\ \omega(0) := 0;\ w(0) := 0;\ z(0) := 0;$
$\beta(1)u(1) = b;\ \ \alpha(1)v(1) = A^T u(1);$
For $i = 1, 2, ...,$ *until convergence, do:*
$\xi(i) = -\xi(i-1)\beta(i)/\alpha(i);$
$z(i) = z(i-1) + \xi(i)v(i);$
$w(i) = (\tau(i-1) - \beta(i)w(i-1))/\alpha(i);$
$\omega(i) = \omega(i-1) + w(i)v(i);$
$\beta(i+1)u(i+1) = Av(i) - \alpha(i)u(i);$
$\tau(i) = -\tau(i-1)\alpha(i)/\beta(i+1);$
$\alpha(i+1)v(i+1) = A^T u(i+1) - \beta(i+1)v(i);$
$\gamma(i) = \beta(i+1)\xi(i)/(\beta(i+1)w(i) - \tau(i));$
$x(i) = z(i) - \gamma(i)\omega(i).$

Algorithm 10. ***(The variant 2 of LSQR method)***
$\theta(1)v(1) = A^T b;\ \rho(1)p(1) = Av(1);$
$\omega(1) = v(1)/\rho(1);\ \xi(1) = \theta(1)/\rho(1);\ x(1) = \xi(1)\omega(1);$
For $i = 1, 2, ...,$ *until convergence, do:*
$\theta(i+1)v(i+1) = A^T p(i) - \rho(i)v(i);$
$\rho(i+1)p(i+1) = Av(i+1) - \theta(i+1)p(i);$
$\omega(i+1) = (v(i+1) - \theta(i+1)\omega(i))/\rho(i+1);$
$\xi(i+1) = -\xi(i)\theta(i+1)/\rho(i+1);$
$x(i+1) = x(i) + \xi(i+1)\omega(i+1).$

In this section, first we obtain the matrix form of the above LSQR methods for solving the generalized Sylvester-transpose matrix equation

$$\sum_{k=1}^{p}(A_kXB_k + C_kX^TD_k) + \sum_{j=1}^{q}(E_jYF_j + G_jY^TH_j) = M, \tag{12}$$

and the minimum Frobenius norm residual problem

$$||M - \sum_{k=1}^{p}(A_kXB_k + C_kX^TD_k) - \sum_{j=1}^{q}(E_jYF_j + G_jY^TH_j)||. \tag{13}$$

where $A_k, C_k \in \mathbb{R}^{s\times n}$, $B_k, D_k \in \mathbb{R}^{n\times t}$, $E_j, G_j \in \mathbb{R}^{s\times m}$, $F_j, G_j \in \mathbb{R}^{m\times t}$ and $M \in \mathbb{R}^{s\times t}$ are known matrices and $X \in \mathbb{R}^{n\times n}$, $Y \in \mathbb{R}^{m\times m}$ are the matrices to be determined. Second we present the matrix form of the LSQR methods for finding solutions of the periodic Sylvester matrix equation

$$\widetilde{A}_l\widetilde{X}_l\widetilde{B}_l + C_l\widetilde{X}_{l+1}\widetilde{D}_l = \widetilde{E}_l, \tag{14}$$

for $l = 1, 2, ...$, where the coefficient matrices $\widetilde{A}_l, \widetilde{C}_l \in \mathsf{R}^{p_1\times n}$, $\widetilde{B}_l, \widetilde{D}_l \in \mathbb{R}^{n\times q_1}$, $\widetilde{E}_l \in \mathbb{R}^{p_1\times q_1}$ and the solutions $\widetilde{X}_l \in \mathbb{R}^{n\times n}$ are periodic with period λ, i.e., $\widetilde{A}_{l+\lambda} = \widetilde{A}_l$, $\widetilde{B}_{l+\lambda} = \widetilde{B}_l$, $\widetilde{C}_{l+\lambda} = \widetilde{C}_l$, $\widetilde{D}_{l+\lambda} = \widetilde{D}_l$, $\widetilde{E}_{l+\lambda} = \widetilde{E}_l$ and $\widetilde{X}_{l+\lambda} = \widetilde{X}_l$.

By using the Kronecker product, we can transform the generalized Sylvester-transpose matrix equation (12) into the following linear system

$$\underbrace{\left(\sum_{k=1}^{p}(B_k^T \otimes A_k + (D_k^T \otimes C_k)P) \quad \sum_{j=1}^{q}(F_j^T \otimes E_j + (H_j^T \otimes G_j)Q) \right)}_{A}$$

$$\times \underbrace{\begin{pmatrix} \mathrm{vec}(X) \\ \mathrm{vec}(Y) \end{pmatrix}}_{x} = \underbrace{\left(\mathrm{vec}(M) \right)}_{b}, \tag{15}$$

where $P \in \mathbb{R}^{st\times n^2}$ and $Q \in \mathbb{R}^{st\times m^2}$ are unitary matrices [80]. It is obvious that the size of the above linear systems is large. The LSQR methods will consume more computer time and memory space once the size of the linear systems is large. To overcome the complications, we directly develop the LSQR methods for solving the generalized Sylvester-transpose matrix equation (12) and the minimum Frobenius norm residual problem (13). By considering the linear systems (15), we rewrite the vectors $u(i)$, $v(i)$, $p(i)$, $z(i)$, $\omega(i)$, $Av(i)$, $A^Tp(i)$ and $A^Tu(i)$ of Algorithms 9 and 10 in the matrix forms. For Algorithm 9, we can write

$$\beta(1)u(1) = b \rightarrow \beta(1)u(1) = \mathrm{vec}(M), \tag{16}$$

$$\alpha(1)v(1) = A^Tu(1) \rightarrow \alpha(1)v(1)$$

$$= \left(\sum_{k=1}^{p}(B_k^T \otimes A_k + (D_k^T \otimes C_k)P) \quad \sum_{j=1}^{q}(F_j^T \otimes E_j + (H_j^T \otimes G_j)Q) \right)^T u(1)$$

$$= \begin{pmatrix} \sum_{k=1}^{p}(B_k \otimes A_k^T + P(D_k \otimes C_k^T)) \\ \sum_{j=1}^{q}(F_j \otimes E_j^T + Q(H_j \otimes G_j^T)) \end{pmatrix} u(1), \tag{17}$$

$$\beta(i+1)u(i+1) = Av(i) - \alpha(i)u(i) \rightarrow \beta(i+1)u(i+1)$$

$$= \left(\sum_{k=1}^{p}(B_k^T \otimes A_k + (D_k^T \otimes C_k)P) \quad \sum_{j=1}^{q}(F_j^T \otimes E_j + (H_j^T \otimes G_j)Q) \right) v(i) - \alpha(i)u(i), \tag{18}$$

$$\alpha(i+1)v(i+1) = A^T u(i+1) - \beta(i+1)v(i) \rightarrow \alpha(i+1)v(i+1)$$

$$= \begin{pmatrix} \sum_{k=1}^{p}(B_k \otimes A_k^T + P(D_k \otimes C_k^T)) \\ \sum_{j=1}^{q}(F_j \otimes E_j^T + Q(H_j \otimes G_j^T)) \end{pmatrix} u(i+1) - \beta(i+1)v(i). \tag{19}$$

By considering (68)-(71), we define

$$u(i) = \mathrm{vec}(U(i)), \quad v(i) = \begin{pmatrix} \mathrm{vec}(V_1(i)) \\ \mathrm{vec}(V_2(i)) \end{pmatrix}, \tag{20}$$

where $U(i)\mathbb{R}^{s\times t}$, $V_1(i)\mathbb{R}^{n\times n}$ and $V_2(i)\mathbb{R}^{m\times m}$ for $i = 0, 1, ...$ This implies that

$$\beta(1)U(1) = M, \tag{21}$$

$$\alpha(1)V_1(1) = \sum_{k=1}^{p}(A_k^T U(1)B_k^T + D_k U(1)^T C_k), \quad \alpha(1)V_2(1) = \sum_{j=1}^{q}(E_j^T U(1)F_j^T + H_j U(1)^T G_j), \tag{22}$$

$$\beta(i+1)U(i+1) = \sum_{k=1}^{p}(A_k V_1(i)B_k + C_k V_1(i)^T D_k) + \sum_{j=1}^{q}(E_j V_2(i)F_j + G_j V_2(i)^T H_j) - \alpha(i)U(i), \tag{23}$$

$$\alpha(i+1)V_1(i+1) = \sum_{k=1}^{p}(A_k^T U(i+1)B_k^T + D_k U(i+1)^T C_k) - \beta(i+1)V_1(i), \tag{24}$$

$$\alpha(i+1)V_2(i+1) = \sum_{j=1}^{q}(E_j^T U(i+1)F_j^T + H_j U(i+1)^T G_j) - \beta(i+1)V_2(i). \tag{25}$$

By applying (20)-(25), we can develop Algorithm 9 for solving (12) and (13) as follows.

Algorithm 11. ***(The variant 1 of matrix LSQR method to solve (12) and (13))***

$\tau(0) := 1;$ $\xi(0) := -1;$ $\Omega_1(0) := 0;$ $\Omega_2(0) := 0;$ $w(0) := 0;$ $Z_1(0) := 0;$ $Z_2(0) := 0;$ $X(0) := 0;$ $Y(0) := 0;$

$\beta(1) = ||M||;$ $U(1) = M/\beta(1);$

$\alpha(1) = (||\sum_{k=1}^{p}(A_k^T U(1)B_k^T + D_k U(1)^T C_k)||^2 + ||\sum_{j=1}^{q}(E_j^T U(1)F_j^T + H_j U(1)^T G_j)||^2)^{1/2};$

$\alpha(1)V_1(1) = \sum_{k=1}^{p}(A_k^T U(1)B_k^T + D_k U(1)^T C_k),$ $\alpha(1)V_2(1) = \sum_{j=1}^{q}(E_j^T U(1)F_j^T + H_j U(1)^T G_j);$

For $i = 1, 2, ...,$ *until convergence, do:*

$\xi(i) = -\xi(i-1)\beta(i)/\alpha(i);$

$Z_1(i) = Z_1(i-1) + \xi(i)V_1(i);$

$Z_2(i) = Z_2(i-1) + \xi(i)V_2(i);$

$w(i) = (\tau(i-1) - \beta(i)w(i-1))/\alpha(i);$

$\Omega_1(i) = \Omega_1(i-1) + w(i)V_1(i);$

$\Omega_2(i) = \Omega_2(i-1) + w(i)V_2(i);$

$\beta(i+1) = ||\sum_{k=1}^{p}(A_kV_1(i)B_k + C_kV_1(i)^TD_k) + \sum_{j=1}^{q}(E_jV_2(i)F_j + G_jV_2(i)^TH_j) - \alpha(i)U(i)||;$

$\beta(i+1)U(i+1) = \sum_{k=1}^{p}(A_kV_1(i)B_k + C_kV_1(i)^TD_k) + \sum_{j=1}^{q}(E_jV_2(i)F_j + G_jV_2(i)^TH_j) - \alpha(i)U(i);$

$\tau(i) = -\tau(i-1)\alpha(i)/\beta(i+1);$

$\alpha(i+1) = (||\sum_{k=1}^{p}(A_k^TU(i+1)B_k^T + D_kU(i+1)^TC_k) - \beta(i+1)V_1(i)||^2 + ||\sum_{j=1}^{q}(E_j^TU(i+1)F_j^T + H_jU(i+1)^TG_j) - \beta(i+1)V_2(i)||^2)^{1/2};$

$\alpha(i+1)V_1(i+1) = \sum_{k=1}^{p}(A_k^TU(i+1)B_k^T + D_kU(i+1)^TC_k) - \beta(i+1)V_1(i);$

$\alpha(i+1)V_2(i+1) = \sum_{j=1}^{q}(E_j^TU(i+1)F_j^T + H_jU(i+1)^TG_j) - \beta(i+1)V_2(i);$

$\gamma(i) = \beta(i+1)\xi(i)/(\beta(i+1)w(i) - \tau(i));$

$X(i) = Z_1(i) - \gamma(i)\Omega_1(i);$

$Y(i) = Z_2(i) - \gamma(i)\Omega_2(i).$

Similarly for Algorithm 10, we can obtain

$$\theta(1)v(1) = A^Tb \to \theta(1)v(1) = \begin{pmatrix} \sum_{k=1}^{p}(B_k \otimes A_k^T + P(D_k \otimes C_k^T)) \\ \sum_{j=1}^{q}(F_j \otimes E_j^T + Q(H_j \otimes G_j^T)) \end{pmatrix} \mathrm{vec}(M), \quad (26)$$

$$\rho(1)p(1) = Av(1) \to \rho(1)p(1)$$

$$= \left(\sum_{k=1}^{p}(B_k^T \otimes A_k + (D_k^T \otimes C_k)P) \quad \sum_{j=1}^{q}(F_j^T \otimes E_j + (H_j^T \otimes G_j)Q) \right) v(1), \quad (27)$$

$$\theta(i+1)v(i+1) = A^Tp(i) - \rho(i)v(i) \to \theta(i+1)v(i+1)$$

$$= \begin{pmatrix} \sum_{k=1}^{p}(B_k \otimes A_k^T + P(D_k \otimes C_k^T)) \\ \sum_{j=1}^{q}(F_j \otimes E_j^T + Q(H_j \otimes G_j^T)) \end{pmatrix} p(i) - \rho(i)v(i) \quad (28)$$

$$\rho(i+1)p(i+1) = Av(i+1) - \theta(i+1)p(i) \to \rho(i+1)p(i+1)$$

$$= \left(\sum_{k=1}^{p}(B_k^T \otimes A_k + (D_k^T \otimes C_k)P) \quad \sum_{j=1}^{q}(F_j^T \otimes E_j + (H_j^T \otimes G_j)Q) \right) v(i+1) - \theta(i+1)p(i). \quad (29)$$

From (26)-(29), we define

$$P(i) = \text{vec}(P(i)), \quad v(i) = \begin{pmatrix} \text{vec}(V_1(i)) \\ \text{vec}(V_2(i)) \end{pmatrix}, \tag{30}$$

where $P(i)\mathbb{R}^{s\times t}$, $V_1(i)\mathbb{R}^{n\times n}$ and $V_2(i)\mathbb{R}^{m\times m}$ for $i = 1, 2, \ldots$. We can get

$$\theta(1)V_1(1) = \sum_{k=1}^{p}(A_k^T M B_k^T + D_k M^T C_k), \quad \theta(1)V_2(1) = \sum_{j=1}^{q}(E_j^T M F_j^T + H_j M^T G_j), \tag{31}$$

$$\rho(1)P(1) = \sum_{k=1}^{p}(A_k V_1(1) B_k + C_k V_1(1)^T D_k) + \sum_{j=1}^{q}(E_j V_2(1) F_j + G_j V_2(1)^T H_j), \tag{32}$$

$$\theta(i+1)V_1(i+1) = \sum_{k=1}^{p}(A_k^T P_i B_k^T + D_k P_i^T C_k) - \rho(i)V_1(i+1), \tag{33}$$

$$\theta(i+1)V_2(i+1) = \sum_{j=1}^{q}(E_j^T P_i F_j^T + H_j P_i^T G_j) - \rho(i)V_2(i+1), \tag{34}$$

$$\rho(i+1)P(i+1) = \sum_{k=1}^{p}(A_k V_1(i+1) B_k + C_k V_1(i+1)^T D_k) + \sum_{j=1}^{q}(E_j V_2(i+1) F_j + G_j V_2(i+1)^T H_j) - \theta(i+1)P(i). \tag{35}$$

By considering (31)-(35), we can extend Algorithm 10 for solving (12) and (13) as follows.

Algorithm 12. ***(The variant 2 of matrix LSQR method to solve (12) and (13))***

$\theta(1) = (||\sum_{k=1}^{p}(A_k^T M B_k^T + D_k M^T C_k)||^2 + ||\sum_{j=1}^{q}(E_j^T M F_j^T + H_j M^T G_j)||^2)^{1/2}$;

$\theta(1)V_1(1) = \sum_{k=1}^{p}(A_k^T M B_k^T + D_k M^T C_k)$, $\theta(1)V_2(1) = \sum_{j=1}^{q}(E_j^T M F_j^T + H_j M^T G_j)$;

$\rho(1) = ||\sum_{k=1}^{p}(A_k V_1(1) B_k + C_k V_1(1)^T D_k) + \sum_{j=1}^{q}(E_j V_2(1) F_j + G_j V_2(1)^T H_j)||$;

$\rho(1)P(1) = \sum_{k=1}^{p}(A_k V_1(1) B_k + C_k V_1(1)^T D_k) + \sum_{j=1}^{q}(E_j V_2(1) F_j + G_j V_2(1)^T H_j)$;

$\Omega_1(1) = V_1(1)/\rho(1)$; $\Omega_2(1) = V_2(1)/\rho(1)$; $\xi(1) = \theta(1)/\rho(1)$; $X(1) = \xi(1)\Omega_1(1)$; $Y(1) = \xi(1)\Omega_2(1)$;

For $i = 1, 2, \ldots$, *until convergence, do:*

$\theta(i+1) = (||\sum_{k=1}^{p}(A_k^T P_i B_k^T + D_k P_i^T C_k) - \rho(i)V_1(i+1)||^2 + ||\sum_{j=1}^{q}(E_j^T P_i F_j^T + H_j P_i^T G_j) - \rho(i)V_2(i+1)||^2)^{1/2}$;

$\theta(i+1)V_1(i+1) = \sum_{k=1}^{p}(A_k^T P_i B_k^T + D_k P_i^T C_k) - \rho(i)V_1(i+1)$;

$\theta(i+1)V_2(i+1) = \sum_{j=1}^{q}(E_j^T P_i F_j^T + H_j P_i^T G_j) - \rho(i)V_2(i+1);$

$\rho(i+1) = ||\sum_{k=1}^{p}(A_k V_1(i+1)B_k + C_k V_1(i+1)^T D_k) + \sum_{j=1}^{q}(E_j V_2(i+1)F_j + G_j V_2(i+1)^T H_j) - \theta(i+1)P(i)||;$

$\rho(i+1)P(i+1) = \sum_{k=1}^{p}(A_k V_1(i+1)B_k + C_k V_1(i+1)^T D_k) + \sum_{j=1}^{q}(E_j V_2(i+1)F_j + G_j V_2(i+1)^T H_j) - \theta(i+1)P(i);$

$\Omega_1(i+1) = (V_1(i+1) - \theta(i+1)\Omega_1(i))/\rho(i+1);$ $\Omega_2(i+1) = (V_2(i+1) - \theta(i+1)\Omega_2(i))/\rho(i+1);$

$\xi(i+1) = -\xi(i)\theta(i+1)/\rho(i+1);$

$X(i+1) = X(i) + \xi(i+1)\Omega_1(i+1);$

$Y(i+1) = Y(i) + \xi(i+1)\Omega_2(i+1).$

The stopping criteria on the above algorithms can be used as

$$||M - \sum_{k=1}^{p}(A_k X(i)B_k + C_k X(i)^T D_k) - \sum_{j=1}^{q}(E_j Y(i)F_j + G_j Y(i)^T H_j)|| \leq \varepsilon,$$

where $\varepsilon > 0$ is a small tolerance.

We can easily show that the periodic Sylvester matrix equation (14) is equivalent to the following Sylvester matrix equation

$$\mathcal{A}\mathcal{X}\mathcal{B} + \mathcal{C}\mathcal{X}\mathcal{D} = \mathcal{E}, \tag{36}$$

where

$$\mathcal{A} = \begin{pmatrix} 0 & \cdots & 0 & \widetilde{A}_1 \\ \widetilde{A}_2 & & & 0 \\ & \ddots & & \vdots \\ 0 & & \widetilde{A}_\lambda & 0 \end{pmatrix}, \quad \mathcal{B} = \begin{pmatrix} 0 & \widetilde{B}_2 & & 0 \\ \vdots & & \ddots & \\ 0 & & & \widetilde{B}_\lambda \\ \widetilde{B}_1 & 0 & \cdots & 0 \end{pmatrix}, \quad \mathcal{C} = \mathrm{diag}\left(\widetilde{C}_1, \widetilde{C}_2, ..., \widetilde{C}_\lambda\right),$$

$$\mathcal{D} = \mathrm{diag}\left(\widetilde{D}_1, \widetilde{D}_2, ..., \widetilde{D}_\lambda\right), \quad \mathcal{E} = \mathrm{diag}\left(\widetilde{E}_1, \widetilde{E}_2, ..., \widetilde{E}_\lambda\right), \quad \mathcal{X} = \mathrm{diag}\left(\widetilde{X}_2, \widetilde{X}_3, ..., \widetilde{X}_\lambda, \widetilde{X}_1\right).$$

By Algorithms 11 and 12, we can compute the solutions of the Sylvester matrix equation (36). But obviously the size of (36) is also large. By considering (14), (36) and applying Algorithms 11 and 12, we propose the following algorithms for solving the periodic Sylvester matrix equation (14).

Algorithm 13. ***(The variant 1 of matrix LSQR method to solve (14))***

$\tau(0) := 1;$ $\xi(0) := -1;$ $w(0) := 0;$ $\Omega_l(0) := 0;$ $Z_l(0) := 0;$ $\widetilde{X}_l(0) :=$

0, *for* $l = 1, 2, ..., \lambda$;

$\beta(1) = (\sum_l^{\lambda} ||\widetilde{E}_l||^2)^{1/2}$; $U_l(1) = \widetilde{E}_l/\beta(1)$ *for* $l = 1, 2, ..., \lambda$;

Set $\widetilde{C}_0 = \widetilde{C}_\lambda$, $\widetilde{D}_0 = \widetilde{D}_\lambda$, $U_0(1) = U_\lambda(1)$;

$\alpha(1) = (\sum_{l=1}^{\lambda} ||\widetilde{A}_l^T U_l(1)\widetilde{B}_l^T + \widetilde{C}_{l-1}^T U_{l-1}(1)\widetilde{D}_{l-1}^T||^2)^{1/2}$;

$\alpha(1)V_l(1) = \widetilde{A}_l^T U_l(1)\widetilde{B}_l^T + \widetilde{C}_{l-1}^T U_{l-1}(1)\widetilde{D}_{l-1}^T$, *for* $l = 1, 2, ..., \lambda$;

Set $V_{\lambda+1}(1) = V_1(1)$;

For $i = 1, 2, ...,$ *until convergence, do:*

$\xi(i) = -\xi(i-1)\beta(i)/\alpha(i)$;

$Z_l(i) = Z_l(i-1) + \xi(i)V_l(i)$, *for* $l = 1, 2, ..., \lambda$;

$w(i) = (\tau(i-1) - \beta(i)w(i-1))/\alpha(i)$;

$\Omega_l(i) = \Omega_l(i-1) + w(i)V_l(i)$, *for* $l = 1, 2, ..., \lambda$;

$\beta(i+1) = (\sum_{l=1}^{\lambda} ||\widetilde{A}_l V_l(i)\widetilde{B}_l + \widetilde{C}_l V_{l+1}(i)\widetilde{D}_l - \alpha(i)U_l(i)||^2)^{1/2}$;

$\beta(i+1)U_l(i+1) = \widetilde{A}_l V_l(i)\widetilde{B}_l + \widetilde{C}_l V_{l+1}(i)\widetilde{D}_l - \alpha(i)U_l(i)$, *for* $l = 1, 2, ..., \lambda$;

$\tau(i) = -\tau(i-1)\alpha(i)/\beta(i+1)$;

Set $U_0(i+1) = U_l(i+1)$;

$\alpha(i+1) = (\sum_{l=1}^{\lambda} ||\widetilde{A}_l^T U_l(i+1)\widetilde{B}_l^T + \widetilde{C}_{l-1}^T U_{l-1}(i+1)\widetilde{D}_{l-1}^T - \beta(i+1)V_l(i)||^2)^{1/2}$;

$\alpha(i+1)V_l(i+1) = \widetilde{A}_l^T U_l(i+1)\widetilde{B}_l^T + \widetilde{C}_{l-1}^T U_{l-1}(i+1)\widetilde{D}_{l-1}^T - \beta(i+1)V_l(i)$, *for* $l = 1, 2, ..., \lambda$;
Set $V_{\lambda+1}(i+1) = V_1(i+1)$;

$\gamma(i) = \beta(i+1)\xi(i)/(\beta(i+1)w(i) - \tau(i))$;

$\widetilde{X}_l(i) = Z_l(i) - \gamma(i)\Omega_l(i)$, *for* $l = 1, 2, ..., \lambda$;

Set $\widetilde{X}_{\lambda+1}(i) = \widetilde{X}_1(i)$.

Algorithm 14. ***(The variant 2 of matrix LSQR method to solve (14))***

Set $\widetilde{C}_0 = \widetilde{C}_\lambda$, $\widetilde{D}_0 = \widetilde{D}_\lambda$, $\widetilde{E}_0 = \widetilde{E}_\lambda$;

$\theta(1) = (\sum_{l=1}^{\lambda} ||\widetilde{A}_l^T \widetilde{E}_l \widetilde{B}_l^T + \widetilde{C}_{l-1}^T \widetilde{E}_{l-1} \widetilde{D}_{l-1}^T||^2)^{1/2}$;

$\theta(1)V_l(1) = \widetilde{A}_l^T \widetilde{E}_l \widetilde{B}_l^T + \widetilde{C}_{l-1}^T \widetilde{E}_{l-1} \widetilde{D}_{l-1}^T$, *for* $l = 1, 2, ..., \lambda$;

$\rho(1) = (\sum_{l=1}^{\lambda} ||\widetilde{A}_l V_l(i) \widetilde{B}_l + \widetilde{C}_l V_{l+1}(i) \widetilde{D}_l||^2)^{1/2}$;

$\rho(1)P_l(1) = \widetilde{A}_l V_l(1) \widetilde{B}_l + \widetilde{C}_l V_{l+1}(1) \widetilde{D}_l$, *for* $l = 1, 2, ..., \lambda$;

$\Omega_l(1) = V_l(1)/\rho(1)$; $\xi(1) = \theta(1)/\rho(1)$; $X_l(1) = \xi(1)\Omega_l(1)$, *for* $l = 1, 2, ..., \lambda$;

For $i = 1, 2, ...,$ *until convergence, do:*

Set $P_0(i) = P_\lambda(i)$;

$\theta(i+1) = (\sum_{l=1}^{\lambda} ||\widetilde{A}_l^T P_l(i) \widetilde{B}_l^T + \widetilde{C}_{l-1}^T P_{l-1}(i) \widetilde{D}_{l-1}^T - \rho(i)V_l(i+1)||^2)^{1/2}$;

$\theta(i+1)V_l(i+1) = \widetilde{A}_l^T P_l(i) \widetilde{B}_l^T + \widetilde{C}_{l-1}^T P_{l-1}(i) \widetilde{D}_{l-1}^T - \rho(i)V_l(i+1)$, *for* $l = 1, 2, ..., \lambda$;

Set $V_{\lambda+1}(i+1) = V_1(i+1)$;

$\rho(i+1) = (\sum_{l=1}^{\lambda} ||\widetilde{A}_l V_l(i+1) \widetilde{B}_l + \widetilde{C}_l V_{l+1}(i+1) \widetilde{D}_l - \theta(i+1)P_l(i)||^2)^{1/2}$;

$\rho(i+1)P_l(i+1) = \widetilde{A}_l V_l(i+1) \widetilde{B}_l + \widetilde{C}_l V_{l+1}(i+1) \widetilde{D}_l - \theta(i+1)P_l(i)$, *for* $l = 1, 2, ..., \lambda$;

$\Omega_l(i+1) = (V_l(i+1) - \theta(i+1)\Omega_l(i))/\rho(i+1)$, *for* $l = 1, 2, ..., \lambda$;

$\xi(i+1) = -\xi(i)\theta(i+1)/\rho(i+1)$;

$\widetilde{X}_l(i+1) = \widetilde{X}_l(i) + \xi(i+1)\Omega_l(i+1)$, *for* $l = 1, 2, ..., \lambda$;

Set $\widetilde{X}_{\lambda+1}(i+1) = \widetilde{X}_1(i+1)$.

The stopping criteria on Algorithms 13 and 14 can be used as

$$\sqrt{\sum_{l=1}^{\lambda} ||\widetilde{E}_l - \widetilde{A}_l \widetilde{X}_l(i) \widetilde{B}_l - \widetilde{C}_l \widetilde{X}_{l+1}(i) \widetilde{D}_l||^2} \leq \varepsilon,$$

where $\varepsilon > 0$ is a small tolerance.

5. The Matrix Form of Bi-CG and Bi-CR Algorithms

To solve nonsymmetric linear systems $Ax = b$ where A is an $m \times m$ real nonsymmetric matrix and b is an m-vector, the Bi-CG and Bi-CR methods have been proposed as an extension of CG and CR, respectively.
First we present the Bi-CG algorithm and then based on the Bi-CG derivation, we present

the Bi-CR algorithm. There are several ways to derive the algorithm of Bi-CG. Here we give the details of one of the simplest derivations [61]. By using $Ax = b$ and a dual linear system $A^T x^* = b^*$, we obtain the following $2m \times 2m$ symmetric linear system

$$\begin{pmatrix} O & A \\ A^T & O \end{pmatrix} \begin{pmatrix} x^* \\ x \end{pmatrix} = \begin{pmatrix} b \\ b^* \end{pmatrix}, \quad \text{or} \quad \hat{A}\hat{x} = \hat{b}. \tag{37}$$

Now for solving (37), we apply the CG algorithm with the following preconditioner:

$$P = \begin{pmatrix} O & I \\ I & O \end{pmatrix}. \tag{38}$$

Hence the resulting algorithm at the n-th iteration step can be written as:

$$\hat{p}_n = P^{-1}\hat{r}_n + \beta_{n-1}\hat{p}_{n-1},$$

$$\alpha_n = \frac{(P^{-1}\hat{r}_n, \hat{r}_n)}{(\hat{p}_n, \hat{A}\hat{p}_n)},$$

$$\hat{x}_{n+1} = \hat{x}_n + \alpha_n \hat{p}_n,$$

$$\hat{r}_{n+1} = \hat{r}_n - \alpha_n \hat{A}\hat{p}_n,$$

$$\beta_n = \frac{(P^{-1}\hat{r}_{n+1}, \hat{r}_{n+1})}{(P^{-1}\hat{r}_n, \hat{r}_n)}.$$

Substituting P^{-1} of (38) and the vectors

$$\hat{x}_n := \begin{pmatrix} x_n^* \\ x_n \end{pmatrix} \qquad \hat{r}_n := \begin{pmatrix} r_n^* \\ r_n \end{pmatrix} \qquad \hat{p}_n := \begin{pmatrix} p_n^* \\ p_n \end{pmatrix}, \tag{39}$$

into the previous recurrences, we obtain the following Bi-CG algorithm [58].

Algorithm 15. ***(Bi-CG algorithm)***

x_0 is an initial guess, $r_0 = b - Ax_0$,
choose r_0^* (for example $r_0^* = r_0$),
set $p_{-1}^* = p_{-1} = 0$, $\beta_{-1} = 0$,
for $n = 0, 1, \ldots,$ *until convergence, do:*
$p_n = r_n + \beta_{n-1} p_{n-1}$,
$p_n^* = r_n^* + \beta_{n-1} p_{n-1}^*$,
$s_n = Ap_n$,
$s_n^* = A^T p_n^*$,
$\alpha_n = \frac{\langle r_n^*, r_n \rangle}{\langle p_n^*, s_n \rangle}$,
$x_{n+1} = x_n + \alpha_n p_n$,
$r_{n+1} = r_n - \alpha_n s_n$,
$r_{n+1}^* = r_n^* - \alpha_n s_n^*$,
$\beta_n = \frac{\langle r_{n+1}^*, r_{n+1} \rangle}{\langle r_n^*, r_n \rangle}$.

By using the preconditioned CR method with the preconditioner (38) to symmetric linear system (37), we have

$$\hat{p}_n = P^{-1}\hat{r}_n + \beta_{n-1}\hat{p}_{n-1},$$

$$\alpha_n = \frac{(P^{-1}\hat{r}_n, \hat{A}P^{-1}\hat{r}_n)}{(P^{-1}\hat{A}\hat{p}_n, \hat{A}\hat{p}_n)},$$

$$\hat{x}_{n+1} = \hat{x}_n + \alpha_n\hat{p}_n,$$

$$\hat{r}_{n+1} = \hat{r}_n - \alpha_n\hat{A}\hat{p}_n,$$

$$\beta_n = \frac{P^{-1}\hat{r}_{n+1}, \hat{A}P^{-1}\hat{r}_{n+1}}{(P^{-1}\hat{r}_n, \hat{A}P^{-1}\hat{r}_n)}.$$

Substituting P^{-1} of (38) and vectors (39) into the previous recurrences, we obtain the following Bi-CR algorithm [58].

Algorithm 16. ***(Bi-CR algorithm)***
x_0 is an initial guess, $r_0 = b - Ax_0$,
choose r_0^* (for example $r_0^* = r_0$),
set $p_{-1}^* = p_{-1} = 0$, $\beta_{-1} = 0$,
for $n = 0, 1, ...,$ until convergence, do:
$p_n = r_n + \beta_{n-1}p_{n-1}$,
$p_n^* = r_n^* + \beta_{n-1}p_{n-1}^*$,
$s_n = Ap_n$,
$s_n^* = A^T p_n^*$,
$t_n = Ar_n$,
$\alpha_n = \frac{\langle r_n^*, t_n\rangle}{\langle s_n^*, s_n\rangle}$,
$x_{n+1} = x_n + \alpha_n p_n$,
$r_{n+1} = r_n - \alpha_n s_n$,
$r_{n+1}^* = r_n^* - \alpha_n s_n^*$,
$t_{n+1} = Ar_{n+1}$,
$\beta_n = \frac{\langle r_{n+1}^*, t_{n+1}\rangle}{\langle r_n^*, t_n\rangle}$.

In [37], Algorithms 15 and 16 were developed to solve the generalized Sylvester-transpose matrix equation

$$\sum_{i=1}^{p}(A_i X B_i + C_i X^T D_i) = E, \tag{40}$$

where $A_i, B_i, C_i, D_i, E \in \mathbb{R}^{m\times m}$ $(i = 1, 2, ..., p)$ and $X \in \mathbb{R}^{m\times m}$. By using Kronecker product and vectorization operator, the generalized Sylvester-transpose matrix equation (40) can be transformed into the following nonsymmetric linear systems

$$\underbrace{\left(\ \sum_{i=1}^{p}(B_i^T \otimes A_i + (D_i^T \otimes C_i)P)\ \right)}_{A}\underbrace{\mathrm{vec}(X)}_{x} = \underbrace{\mathrm{vec}(E)}_{b}, \tag{41}$$

where $A \in \mathbb{R}^{m^2\times m^2}$, $x, b \in \mathbb{R}^{m^2}$ and $P \in \mathbb{R}^{m^2\times m^2}$ is a unitary matrix [80]. It is obvious that the size of the above system is large. However, iterative methods will consume more computer time and memory space once the size of the system is large. To overcome this complication, Algorithms 15 and 16 are directly extend to solve (40). By considering the linear systems (41) and using the vectorization operator, we rewrite vectors r_n, r_n^*, p_n, p_n^*, s_n, s_n^*, t_n and x_n of Algorithms 15 and 16 into the matrix forms. We can write

$$r_0 = b - Ax_0 \rightarrow r_0 = \mathrm{vec}(E)$$

$$- \left(\sum_{i=1}^{p}(B_i^T \otimes A_i + (D_i^T \otimes C_i)P)\right) x_0, \tag{42}$$

$$s_n = Ap_n \rightarrow s_n$$

$$= \left(\sum_{i=1}^{p}(B_i^T \otimes A_i + (D_i^T \otimes C_i)P)\right) p_n, \tag{43}$$

$$s_n^* = A^T p_n^* \rightarrow s_n^*$$

$$= \left(\sum_{i=1}^{p}(B_i^T \otimes A_i + (D_i^T \otimes C_i)P)\right)^T p_n^*$$

$$= \left(\sum_{i=1}^{p}(B_i \otimes A_i^T + P(D_i \otimes C_i^T))\right) p_n^*, \tag{44}$$

$$t_n = Ar_n \rightarrow t_n$$

$$= \left(\sum_{i=1}^{p}(B_i^T \otimes A_i + (D_i^T \otimes C_i)P)\right) r_n. \tag{45}$$

By considering (42)-(45), we define:

$$x_n = \mathrm{vec}(X_n), \quad s_n = \mathrm{vec}(S_n), \tag{46}$$

$$p_n = \mathrm{vec}(P_n), \quad r_n = \mathrm{vec}(R_n), \tag{47}$$

$$s_n^* = \mathrm{vec}(S_n^*), \quad p_n^* = \mathrm{vec}(P_n^*), \tag{48}$$

$$r_n^* = \mathrm{vec}(R_n^*), \quad \text{and} \quad t_n = \mathrm{vec}(T_n), \tag{49}$$

where $X_n, S_n, S_n^*, R_n, R_n^*, P_n, P_n^*, T_n \in \mathbb{R}^{m\times m}$. From (46)-(49), we can get

$$R_0 = E - \sum_{i=1}^{p}(A_i X_0 B_i + C_i X_0^T D_i), \tag{50}$$

$$S_n = \sum_{i=1}^{p}(A_i P_n B_i + C_i P_n^T D_i), \tag{51}$$

$$S_n^* = \sum_{i=1}^{p}(A_i^T P_n^* B_i^T + D_i P_n^{*T} C_i), \tag{52}$$

$$T_n = \sum_{i=1}^{p}(A_i R_n B_i + C_i R_n^T D_i), \tag{53}$$

$$P_n = R_n + \beta_{n-1} P_{n-1}, \qquad P_n^* = R_n^* + \beta_{n-1} P_{n-1}^* \tag{54}$$

$$X_{n+1} = X_n + \alpha_n P_n, \tag{55}$$

$$R_{n+1} = R_n - \alpha_n S_n, \qquad R_{n+1}^* = R_n^* - \alpha_n S_n^*. \tag{56}$$

For Algorithms 15 and 16, parameters α_n and β_n can be determined respectively as

$$\alpha_n = \frac{\langle r_n^*, r_n \rangle}{\langle p_n^*, s_n \rangle} = \frac{\langle \text{vec}(R_n^*), \text{vec}(R_n) \rangle}{\langle \text{vec}(P_n^*), \text{vec}(S_n) \rangle} = \frac{\langle R_n^*, R_n \rangle}{\langle P_n^*, S_n \rangle}, \tag{57}$$

$$\beta_n = \frac{\langle r_{n+1}^*, r_{n+1} \rangle}{\langle r_n^*, r_n \rangle} = \frac{\langle \text{vec}(R_{n+1}^*), \text{vec}(R_{n+1}) \rangle}{\langle \text{vec}(R_n^*), \text{vec}(R_n) \rangle}$$
$$= \frac{\langle R_{n+1}^*, R_{n+1} \rangle}{\langle R_n^*, R_n \rangle}, \tag{58}$$

and

$$\alpha_n = \frac{\langle r_n^*, t_n \rangle}{\langle s_n^*, s_n \rangle} = \frac{\langle \text{vec}(R_n^*), \text{vec}(T_n) \rangle}{\langle \text{vec}(S_n^*), \text{vec}(S_n) \rangle} = \frac{\langle R_n^*, T_n \rangle}{\langle S_n^*, S_n \rangle}, \tag{59}$$

$$\beta_n = \frac{\langle r_{n+1}^*, t_{n+1} \rangle}{\langle r_n^*, t_n \rangle} = \frac{\langle \text{vec}(R_{n+1}^*), \text{vec}(T_{n+1}) \rangle}{\langle \text{vec}(R_n^*), \text{vec}(T_n) \rangle}$$
$$= \frac{\langle R_{n+1}^*, T_{n+1} \rangle}{\langle R_n^*, T_n \rangle}. \tag{60}$$

Here by applying (46)-(60), we present the matrix form of Algorithms 15 and 16 for finding the solution of (40).

Algorithm 17. ***(Matrix form of Bi-CG algorithm to solve (40))***

$X_0 \in \mathbb{R}^{m \times m}$ *is an initial guess and*
$R_0 = E - \sum_{i=1}^{p} (A_i X_0 B_i + C_i X_0^T D_i)$,
choose R_0^* *(for example* $R_0^* = R_0$*),*
set $P_{-1}^* = P_{-1} = 0$, $\beta_{-1} = 0$,
for $n = 0, 1, \ldots$, *until convergence, do:*
$P_n = R_n + \beta_{n-1} P_{n-1}$,
$P_n^* = R_n^* + \beta_{n-1} P_{n-1}^*$,
$S_n = \sum_{i=1}^{p} (A_i P_n B_i + C_i P_n^T D_i)$,
$S_n^* = \sum_{i=1}^{p} (A_i^T P_n^* B_i^T + D_i P_n^{*T} C_i)$,
$\alpha_n = \frac{\langle R_n^*, R_n \rangle}{\langle P_n^*, S_n \rangle}$,
$X_{n+1} = X_n + \alpha_n P_n$,
$R_{n+1} = R_n - \alpha_n S_n$,
$R_{n+1}^* = R_n^* - \alpha_n S_n^*$,
$\beta_n = \frac{\langle R_{n+1}^*, R_{n+1} \rangle}{\langle R_n^*, R_n \rangle}$.

Algorithm 18. ***(Matrix form of Bi-CR algorithm to solve (40))***

$X_0 \in \mathbb{R}^{m \times m}$ *is an initial guess and*
$R_0 = E - \sum_{i=1}^{p} (A_i X_0 B_i + C_i X_0^T D_i)$,
choose R_0^* *(for example* $R_0^* = R_0$*),*
set $P_{-1}^* = P_{-1} = 0$, $\beta_{-1} = 0$,
for $n = 0, 1, \ldots$, *until convergence, do:*
$P_n = R_n + \beta_{n-1} P_{n-1}$,
$P_n^* = R_n^* + \beta_{n-1} P_{n-1}^*$,

$S_n = \sum_{i=1}^{p}(A_iP_nB_i + C_iP_n^TD_i),$
$S_n^* = \sum_{i=1}^{p}(A_i^TP_n^*B_i^T + D_iP_n^{*T}C_i),$
$T_n = \sum_{i=1}^{p}(A_iR_nB_i + C_iR_n^TD_i),$
$\alpha_n = \frac{\langle R_n^*,T_n\rangle}{\langle S_n^*,S_n\rangle},$
$X_{n+1} = X_n + \alpha_nP_n,$
$R_{n+1} = R_n - \alpha_nS_n,$
$R_{n+1}^* = R_n^* - \alpha_nS_n^*,$
$T_{n+1} = \sum_{i=1}^{p}(A_iR_{n+1}B_i + C_iR_{n+1}^TD_i),$
$\beta_n = \frac{\langle R_{n+1}^*,T_{n+1}\rangle}{\langle R_n^*,T_n\rangle}.$

6. The Matrix Form of CGS Algorithm

The CGS method is based on the Bi-CG method and not the original CG algorithm for solving $Ax = b$. One major drawback of Bi-CG method is that it requires a multiplication with the transpose of A. A number of hybrid Bi-CG methods such as CGS have been presented to improve the convergence of Bi-CG and to avoid multiplication by A^T. The CGS method avoids using A^T and accelerates the convergence by squaring the Bi-CG polynomials [59]. The CGS algorithm can be summarized as follows [34].

Algorithm 19. ***(CGS Algorithm)***
Choose $x_0 \in \mathbb{R}^m$;
Set $p_0 = u_0 = r_0 = b - Ax_0$, $v_0 = Ap_0$;
Choose $\widetilde{r}_0$ *such that* $\rho_0 = \langle r_0, \widetilde{r}_0\rangle \neq 0$ *(for example* $\widetilde{r}_0 = r_0$*);*
For $n = 1, 2, ...$ *until* $(||r_{n-1}||/||b||) \leq \varepsilon$, *do:*
Set $\sigma_{n-1} = \langle v_{n-1}, \widetilde{r}_0\rangle$, $\alpha_{n-1} = \rho_{n-1}/\sigma_{n-1}$;
$q_n = u_{n-1} - \alpha_{n-1}v_{n-1}$;
Set $x_n = x_{n-1} + \alpha_{n-1}(u_{n-1} + q_n)$;
$r_n = r_{n-1} - \alpha_{n-1}A(u_{n-1} + q_n)$;
Set $\rho_n = \langle r_n, \widetilde{r}_0\rangle$, $\beta_n = \rho_n/\rho_{n-1}$;
$u_n = r_n + \beta_nq_n$;
$p_n = u_n + \beta_n(q_n + \beta_np_{n-1})$; $v_n = Mp_n$.

In the above algorithm, the stopping tolerance ε is a small positive number. In exact arithmetic, Algorithm 19 terminates after a finite number, say n^*, of iterations. In [38], Algorithm 19 was generalized to solve the general coupled matrix equations

$$\sum_{j=1}^{p} A_{ij}X_jB_{ij} = C_i, \quad i = 1, 2, ..., p, \tag{61}$$

where $A_{ij}, X_j, B_{ij} \in \mathbb{R}^{m\times m}$ for $i, j = 1, 2, ..., p$.

By means of Kronecker product and vectorization operator, the general coupled matrix equations (61) can be transformed into the following linear systems:

$$\underbrace{\begin{pmatrix} B_{11}^T \otimes A_{11} & B_{12}^T \otimes A_{12} & \dots & B_{1p}^T \otimes A_{1p} \\ B_{21}^T \otimes A_{21} & B_{22}^T \otimes A_{22} & \dots & B_{2p}^T \otimes A_{2p} \\ \vdots & \vdots & \vdots & \vdots \\ B_{p1}^T \otimes A_{p1} & B_{p2}^T \otimes A_{p2} & \dots & B_{pp}^T \otimes A_{pp} \end{pmatrix}}_{A} \underbrace{\begin{pmatrix} \mathrm{vec}(X_1) \\ \mathrm{vec}(X_2) \\ \vdots \\ \mathrm{vec}(X_p) \end{pmatrix}}_{x} = \underbrace{\begin{pmatrix} \mathrm{vec}(C_1) \\ \mathrm{vec}(C_2) \\ \vdots \\ \mathrm{vec}(C_p) \end{pmatrix}}_{b}. \tag{62}$$

The dimension of the associate matrix A is high when m is large. In order to overcome this problem, we obtain the matrix form of Algorithm 19 for solving (62). Applying Algorithm 19 for (62) gives us:

$$p_0 = u_0 = r_0 = b - Ax_0$$

$$= b - \begin{pmatrix} B_{11}^T \otimes A_{11} & B_{12}^T \otimes A_{12} & \dots & B_{1p}^T \otimes A_{1p} \\ B_{21}^T \otimes A_{21} & B_{22}^T \otimes A_{22} & \dots & B_{2p}^T \otimes A_{2p} \\ \vdots & \vdots & \vdots & \vdots \\ B_{p1}^T \otimes A_{p1} & B_{p2}^T \otimes A_{p2} & \dots & B_{pp}^T \otimes A_{pp} \end{pmatrix} x_0, \tag{63}$$

$$v_0 = Ap_0 = \begin{pmatrix} B_{11}^T \otimes A_{11} & B_{12}^T \otimes A_{12} & \dots & B_{1p}^T \otimes A_{1p} \\ B_{21}^T \otimes A_{21} & B_{22}^T \otimes A_{22} & \dots & B_{2p}^T \otimes A_{2p} \\ \vdots & \vdots & \vdots & \vdots \\ B_{p1}^T \otimes A_{p1} & B_{p2}^T \otimes A_{p2} & \dots & B_{pp}^T \otimes A_{pp} \end{pmatrix} p_0, \tag{64}$$

$$r_n = r_{n-1} - \alpha_{n-1} A(u_{n-1} + q_n) = r_{n-1} - \alpha_{n-1}$$

$$\times \begin{pmatrix} B_{11}^T \otimes A_{11} & B_{12}^T \otimes A_{12} & \dots & B_{1p}^T \otimes A_{1p} \\ B_{21}^T \otimes A_{21} & B_{22}^T \otimes A_{22} & \dots & B_{2p}^T \otimes A_{2p} \\ \vdots & \vdots & \vdots & \vdots \\ B_{p1}^T \otimes A_{p1} & B_{p2}^T \otimes A_{p2} & \dots & B_{pp}^T \otimes A_{pp} \end{pmatrix} (u_{n-1} + q_n). \tag{65}$$

Noting that (63)-(65), the following vectors are defined

$$b = \begin{pmatrix} \mathrm{vec}(C_1) \\ \mathrm{vec}(C_2) \\ \vdots \\ \mathrm{vec}(C_p) \end{pmatrix}, \; x_n = \begin{pmatrix} \mathrm{vec}(X_{1,n}) \\ \mathrm{vec}(X_{2,n}) \\ \vdots \\ \mathrm{vec}(X_{p,n}) \end{pmatrix}, \; u_n = \begin{pmatrix} \mathrm{vec}(U_{1,n}) \\ \mathrm{vec}(U_{2,n}) \\ \vdots \\ \mathrm{vec}(U_{p,n}) \end{pmatrix}, \tag{66}$$

$$v_n = \begin{pmatrix} \mathrm{vec}(V_{1,n}) \\ \mathrm{vec}(V_{2,n}) \\ \vdots \\ \mathrm{vec}(V_{p,n}) \end{pmatrix}, \; p_n = \begin{pmatrix} \mathrm{vec}(P_{1,n}) \\ \mathrm{vec}(P_{2,n}) \\ \vdots \\ \mathrm{vec}(P_{p,n}) \end{pmatrix}, \; q_n = \begin{pmatrix} \mathrm{vec}(Q_{1,n}) \\ \mathrm{vec}(Q_{2,n}) \\ \vdots \\ \mathrm{vec}(Q_{p,n}) \end{pmatrix}, \tag{67}$$

$$r_n = \begin{pmatrix} \mathrm{vec}(R_{1,n}) \\ \mathrm{vec}(R_{2,n}) \\ \vdots \\ \mathrm{vec}(R_{p,n}) \end{pmatrix}, \quad \text{and } \widetilde{r}_0 = \begin{pmatrix} \mathrm{vec}(\widetilde{R}_{1,0}) \\ \mathrm{vec}(\widetilde{R}_{2,0}) \\ \vdots \\ \mathrm{vec}(\widetilde{R}_{p,0}) \end{pmatrix}, \tag{68}$$

where $C_i, X_{i,n}, P_{i,n}, Q_{i,n}, U_{i,n}, V_{i,n}, R_{i,n}, \widetilde{R}_{i,0} \in \mathbb{R}^{m\times m}$ for $i = 1, 2, .., p$ and $n = 0, 1, 2, \ldots$. Substituting (66)-(68) into (63)-(65) gives us

$$\begin{pmatrix} \mathrm{vec}(P_{1,0}) \\ \mathrm{vec}(P_{2,0}) \\ \vdots \\ \mathrm{vec}(P_{p,0}) \end{pmatrix} = \begin{pmatrix} \mathrm{vec}(U_{1,0}) \\ \mathrm{vec}(U_{2,0}) \\ \vdots \\ \mathrm{vec}(U_{p,0}) \end{pmatrix} = \begin{pmatrix} \mathrm{vec}(R_{1,0}) \\ \mathrm{vec}(R_{2,0}) \\ \vdots \\ \mathrm{vec}(R_{p,0}) \end{pmatrix} = \begin{pmatrix} \mathrm{vec}(C_1 - \sum_{j=1}^{p} A_{1j}X_{j,0}B_{1j}) \\ \mathrm{vec}(C_2 - \sum_{j=1}^{p} A_{2j}X_{j,0}B_{2j}) \\ \vdots \\ \mathrm{vec}(C_p - \sum_{j=1}^{p} A_{pj}X_{j,0}B_{pj}) \end{pmatrix}, \tag{69}$$

$$\begin{pmatrix} \mathrm{vec}(V_{1,n}) \\ \mathrm{vec}(V_{2,n}) \\ \vdots \\ \mathrm{vec}(V_{p,n}) \end{pmatrix} = \begin{pmatrix} \mathrm{vec}(\sum_{j=1}^{p} A_{1j}P_{j,n}B_{1j}) \\ \mathrm{vec}(\sum_{j=1}^{p} A_{2j}P_{j,n}B_{2j}) \\ \vdots \\ \mathrm{vec}(\sum_{j=1}^{p} A_{pj}P_{j,n}B_{pj}) \end{pmatrix}, \tag{70}$$

$$\begin{pmatrix} \mathrm{vec}(R_{1,n}) \\ \mathrm{vec}(R_{2,n}) \\ \vdots \\ \mathrm{vec}(R_{p,n}) \end{pmatrix} = \begin{pmatrix} \mathrm{vec}(R_{1,n-1}) \\ \mathrm{vec}(R_{2,n-1}) \\ \vdots \\ \mathrm{vec}(R_{p,n-1}) \end{pmatrix} - \alpha_{n-1} \begin{pmatrix} \mathrm{vec}(\sum_{j=1}^{p} A_{1j}(U_{j,n-1}+Q_{j,n})B_{1j}) \\ \mathrm{vec}(\sum_{j=1}^{p} A_{2j}(U_{j,n-1}+Q_{j,n})B_{2j}) \\ \vdots \\ \mathrm{vec}(\sum_{j=1}^{p} A_{pj}(U_{j,n-1}+Q_{j,n})B_{pj}) \end{pmatrix}. \tag{71}$$

Also by using (66) and (68), the parameters σ_{n-1} and ρ_n can be written as

$$\begin{aligned} \sigma_{n-1} =& \langle v_{n-1}, \widetilde{r}_0\rangle = \left\langle \begin{pmatrix} \mathrm{vec}(V_{1,n-1}) \\ \mathrm{vec}(V_{2,n-1}) \\ \vdots \\ \mathrm{vec}(V_{p,n-1}) \end{pmatrix}, \begin{pmatrix} \mathrm{vec}(\widetilde{R}_{1,0}) \\ \mathrm{vec}(\widetilde{R}_{2,0}) \\ \vdots \\ \mathrm{vec}(\widetilde{R}_{p,0}) \end{pmatrix} \right\rangle \\ =& \sum_{i=1}^{p} \left\langle \mathrm{vec}(V_{i,n-1}), \mathrm{vec}(\widetilde{R}_{i,0}) \right\rangle = \sum_{i=1}^{p} \left\langle V_{i,n-1}, \widetilde{R}_{i,0} \right\rangle, \end{aligned}$$

and

$$\rho_n = \langle r_n, \widetilde{r}_0\rangle = \left\langle \begin{pmatrix} \mathrm{vec}(R_{1,n}) \\ \mathrm{vec}(R_{2,n}) \\ \vdots \\ \mathrm{vec}(R_{p,n}) \end{pmatrix}, \begin{pmatrix} \mathrm{vec}(\widetilde{R}_{1,0}) \\ \mathrm{vec}(\widetilde{R}_{2,0}) \\ \vdots \\ \mathrm{vec}(\widetilde{R}_{p,0}) \end{pmatrix} \right\rangle = \sum_{i=1}^{p} \left\langle R_{i,n}, \widetilde{R}_{i,0} \right\rangle.$$

From the above discussion, the following matrix algorithm can be constructed for solving the general coupled matrix equations (61).

Algorithm 20. ***(Matrix form of CGS Algorithm for (61))***
Choose the initial matrices $X_{i,0} \in \mathbb{R}^{m\times m}$, *for* $i = 1, 2, ..., p$;
Set $P_{i,0} = U_{i,0} = R_{i,0} = C_i - \sum_{j=1}^{p} A_{ij}X_{j,0}B_{ij}$, $V_{i,0} = \sum_{j=1}^{p} A_{ij}P_{j,0}B_{ij}$, *for* $i = 1, 2, ..., p$;
Choose $\widetilde{R}_{i,0}$ *such that* $\rho_0 = \sum_{i=1}^{p}\langle R_{i,0}, \widetilde{R}_{i,0}\rangle \neq 0$, *for* $i = 1, 2, ..., p$ *(for example* $\widetilde{R}_{i,0} = R_{i,0}$ *for* $i = 1, 2, ..., p$*);*
For $n = 1, 2, \ldots$ *until* $\sqrt{\sum_{i=1}^{p} ||R_{i,n-1}||^2} \leq \varepsilon$, *do:*
Set $\sigma_{n-1} = \sum_{i=1}^{p}\langle V_{i,n-1}, \widetilde{R}_{i,0}\rangle$, $\alpha_{n-1} = \rho_{n-1}/\sigma_{n-1}$;

$Q_{i,n} = U_{i,n} - \alpha_{n-1} V_{i,n}$, *for* $i = 1, 2, ..., p$;
Set $X_{i,n} = X_{i,n-1} + \alpha_{n-1}(U_{i,n-1} + Q_{i,n})$, *for* $i = 1, 2, ..., p$;
$R_{i,n} = R_{i,n-1} - \alpha_{n-1} \sum_{j=1}^{p} A_{ij}(U_{j,n-1} + Q_{j,n})B_{ij}$, *for* $i = 1, 2, ..., p$;
Set $\rho_n = \sum_{i=1}^{p} \left\langle R_{i,n}, \widetilde{R}_{i,0} \right\rangle$, $\beta_n = \rho_n / \rho_{n-1}$;
$U_{i,n} = R_{i,n} + \beta_n Q_{i,n}$, *for* $i = 1, 2, ..., p$;
$P_{i,n} = U_{i,n} + \beta_n (Q_{i,n} + \beta_n Q_{i,n-1})$;
$V_{i,n} = \sum_{j=1}^{p} A_{ij} P_{j,n} B_{ij}$, *for* $i = 1, 2, ..., p$.

From Algorithm 20 is the matrix form of the original CGS algorithm, therefore in general Algorithm 20 has the same properties as Algorithm 19. For example, in exact arithmetic, Algorithm 20 terminates after a finite number of iterations.

7. The Matrix Form of Bi-CGSTAB Algorithm

One major drawback of Bi-CG algorithm is that it requires a multiplication with the matrix A^T. In many areas, the transpose matrix is not readily available or the multiplication is difficult to perform. To avoid calculating the matrix A^T and to improve the convergence rate in Bi-CG, many efforts have been devoted to investigating more efficient methods from restructuring Bi-CG algorithm [69]. In [60], van der Vorst introduced one of the most successful improvement of Bi-CG, known as Bi-CGSTAB. In Bi-CGSTAB, the accelerating polynomial is defined by using two-term recurrence relations to design the residual polynomial of Bi-CGSTAB. The Bi-CGSTAB is characterized by residuals as:

$$r_n = \tau_n(A)\phi_n(A)r_0 = (I - \omega_1 A)(I - \omega_2 A)...(I - \omega_n A)\phi_n(A)r_0,$$

where $r_0 = b - Ax_0$, $\phi_n \in \mathcal{P}_n$, $\phi_n(0) = 1$, and the ω_i's are chosen to locally minimize the residual by a steepest descent method, for more details see [29, 60]. By computing the above residual, a more smoothly converging algorithm is obtained. The Bi-CGSTAB algorithm can be summarized in as follows.

Algorithm 21. ***(The Bi-CGSTAB algorithm)***
Choose x_0 *and compute* $r_0 = b - Ax_0$;

Pick an arbitrary vector $\widetilde{r}_0$ *(for example* $\widetilde{r}_0 = r_0$*)*;

Set $v_0 = p_0 = 0$; $\rho_0 = \alpha_1 = \omega_0 = 1$;

For $n = 1, 2, ...,$ *until convergence*

$\rho_n = \langle r_{n-1}, \widetilde{r}_0 \rangle$; $\beta_n = (\frac{\rho_n}{\rho_{n-1}})(\frac{\alpha_n}{\omega_{n-1}})$;

$p_n = r_{n-1} + \beta_n (p_{n-1} - \omega_{n-1} v_{n-1})$;

$v_n = Ap_n$;

$\sigma_n = \langle v_n, \widetilde{r}_0 \rangle$; $\alpha_n = \frac{\rho_n}{\sigma_n}$;

$s_n = r_{n-1} - \alpha_n v_n$; $t_n = As_n$;

$\omega_n = \frac{\langle s_n, t_n \rangle}{\langle t_n, t_n \rangle}$;

$r_n = s_n - \omega_n t_n;$

$x_n = x_{n-1} + \alpha_n p_n + \omega_n s_n.$

As explained in [60], in exact arithmetic, the Bi-CGSTAB algorithm terminates with true solution after $n \leq m$ steps. In [40], the Bi-CGSTAB algorithm was developed for solving the coupled Sylvester matrix equations

$$\begin{cases} \sum_{i=1}^{k}(A_i X B_i + C_i Y D_i) = M, \\ \sum_{i=1}^{k}(E_i X F_i + G_i Y H_i) = N, \end{cases} \tag{72}$$

where $A_i, B_i, C_i, D_i, E_i, F_i, G_i, H_i, M, N, X, Y \in \mathbb{R}^{m\times m}$. By means Kronecker product and vectorization operator, we can transform the coupled Sylvester matrix equations (72) into the linear systems of equations $Ax = b$ with the following parameters:

$$A = \begin{pmatrix} \sum_{i=1}^{k} B_i^T \otimes A_i & \sum_{i=1}^{k} D_i^T \otimes C_i \\ \sum_{i=1}^{k} F_i^T \otimes E_i & \sum_{i=1}^{k} H_i^T \otimes G_i \end{pmatrix} \in \mathbb{R}^{2m^2\times 2m^2}, \quad x = \begin{pmatrix} \text{vec}(X) \\ \text{vec}(Y) \end{pmatrix} \in \mathbb{R}^{2m^2},$$
$$b = \begin{pmatrix} \text{vec}(M) \\ \text{vec}(N) \end{pmatrix} \in \mathbb{R}^{2m^2}. \tag{73}$$

Obviously the dimension of the above associate matrix A is high when m. In order to overcome this problem, the Bi-CGSTAB algorithm is directly developed for solving the coupled Sylvester matrix equations (72). By applying the Bi-CGSTAB algorithm for $Ax = b$ with parameters (73), we can write

$$r_0 = b - Ax_0 = \begin{pmatrix} \text{vec}(M) \\ \text{vec}(N) \end{pmatrix} - \begin{pmatrix} \sum_{i=1}^{k} B_i^T \otimes A_i & \sum_{i=1}^{k} D_i^T \otimes C_i \\ \sum_{i=1}^{k} F_i^T \otimes E_i & \sum_{i=1}^{k} H_i^T \otimes G_i \end{pmatrix} x_0, \tag{74}$$

$$v_n = Ap_n = \begin{pmatrix} \sum_{i=1}^{k} B_i^T \otimes A_i & \sum_{i=1}^{k} D_i^T \otimes C_i \\ \sum_{i=1}^{k} F_i^T \otimes E_i & \sum_{i=1}^{k} H_i^T \otimes G_i \end{pmatrix} p_n, \tag{75}$$

$$t_n = As_n = \begin{pmatrix} \sum_{i=1}^{k} B_i^T \otimes A_i & \sum_{i=1}^{k} D_i^T \otimes C_i \\ \sum_{i=1}^{k} F_i^T \otimes E_i & \sum_{i=1}^{k} H_i^T \otimes G_i \end{pmatrix} s_n. \tag{76}$$

By considering the above equations, we define

$$x_n = \begin{pmatrix} \text{vec}(X_n) \\ \text{vec}(Y_n) \end{pmatrix}, \quad p_n = \begin{pmatrix} \text{vec}(P_{1,n}) \\ \text{vec}(P_{2,n}) \end{pmatrix} \quad s_n = \begin{pmatrix} \text{vec}(S_{1,n}) \\ \text{vec}(S_{2,n}) \end{pmatrix}, \tag{77}$$

$$\widetilde{r}_0 = \begin{pmatrix} \text{vec}(\widetilde{R}_{1,0}) \\ \text{vec}(\widetilde{R}_{2,0}) \end{pmatrix}, \quad r_n = \begin{pmatrix} \text{vec}(R_{1,n}) \\ \text{vec}(R_{2,n}) \end{pmatrix}, \quad t_n = \begin{pmatrix} \text{vec}(T_{1,n}) \\ \text{vec}(T_{2,n}) \end{pmatrix} \quad \text{and} \quad v_n = \begin{pmatrix} \text{vec}(V_{1,n}) \\ \text{vec}(V_{2,n}) \end{pmatrix}, \tag{78}$$

where $X_n, Y_n, P_{i,n}, S_{i,n}, T_{i,n}, P_{i,n}, R_{i,n}, \widetilde{R}_{i,0} \in \mathbb{R}^{m\times m}$ for $i = 1, 2$ and $n = 0, 1, 2, \ldots$. By using these definitions, the scalars ρ_n, σ_n and ω_n can be written as

$$\rho_n = \langle r_{n-1}, \widetilde{r}_0 \rangle = \left\langle \begin{pmatrix} \text{vec}(R_{1,n-1}) \\ \text{vec}(R_{2,n-1}) \end{pmatrix}, \begin{pmatrix} \text{vec}(\widetilde{R}_{1,0}) \\ \text{vec}(\widetilde{R}_{2,0}) \end{pmatrix} \right\rangle$$

$$= \left\langle \text{vec}(R_{1,n-1}), \text{vec}(\widetilde{R}_{1,0}) \right\rangle + \left\langle \text{vec}(R_{2,n-1}), \text{vec}(\widetilde{R}_{2,0}) \right\rangle = \left\langle R_{1,n-1}, \widetilde{R}_{1,0} \right\rangle + \left\langle R_{2,n-1}, \widetilde{R}_{2,0} \right\rangle, \tag{79}$$

$$\sigma_n = \langle v_n, \widetilde{r}_0 \rangle = \left\langle V_{1,n}, \widetilde{R}_{1,0} \right\rangle + \left\langle V_{2,n}, \widetilde{R}_{2,0} \right\rangle, \tag{80}$$

$$\omega_n = \frac{\langle s_n, t_n \rangle}{\langle t_n, t_n \rangle} = \frac{\left\langle S_{1,n}, T_{1,n} \right\rangle + \left\langle S_{2,n}, T_{2,n} \right\rangle}{\left\langle T_{1,n}, T_{1,n} \right\rangle + \left\langle T_{2,n}, T_{2,n} \right\rangle} = \frac{\left\langle S_{1,n}, T_{1,n} \right\rangle + \left\langle S_{2,n}, T_{2,n} \right\rangle}{||T_{1,n}||^2 + ||T_{2,n}||^2}. \tag{81}$$

From the discussion above, we obtain the matrix form of the Bi-CGSTAB algorithm for solving the coupled Sylvester matrix equations (72).

Algorithm 22. ***(Matrix form of the Bi-CGSTAB algorithm for solving (72))***
Choose initial matrices $X_0, Y_0 \in \mathbb{R}^{m\times m}$;

Compute $R_{1,0} = M - \sum_{i=1}^{k}(A_iX_0B_i + C_iY_0D_i)$ *and* $R_{2,0} = N - \sum_{i=1}^{k}(E_iX_0F_i + G_iY_0H_i)$;

Pick arbitrary matrices $\widetilde{R}_{1,0}, \widetilde{R}_{2,0} \in \mathbb{R}^{n\times n}$ *(for example* $\widetilde{R}_{1,0} = R_{1,0}$ *and* $\widetilde{R}_{2,0} = R_{2,0}$*);*

Set $V_{1,0} = V_{2,0} = P_{1,0} = P_{2,0} = 0$; $\rho_0 = \alpha_1 = \omega_0 = 1$;

For $n = 1, 2, \ldots$, *until convergence*

$\rho_n = \langle R_{1,n-1}, \widetilde{R}_{1,0}\rangle + \langle R_{2,n-1}, \widetilde{R}_{2,0}\rangle$; $\beta_n = (\frac{\rho_n}{\rho_{n-1}})(\frac{\alpha_n}{\omega_{n-1}})$;

$P_{1,n} = R_{1,n-1} + \beta_n(P_{1,n-1} - \omega_{n-1}V_{1,n-1})$;

$P_{2,n} = R_{2,n-1} + \beta_n(P_{2,n-1} - \omega_{n-1}V_{2,n-1})$;

$V_{1,n} = \sum_{i=1}^{k}(A_iP_{1,n}B_i + C_iP_{2,n}D_i)$;

$V_{2,n} = \sum_{i=1}^{k}(E_iP_{1,n}F_i + G_iP_{2,n}H_i)$;

$\sigma_n = \langle V_{1,n}, \widetilde{R}_{1,0}\rangle + \langle V_{2,n}, \widetilde{R}_{2,0}\rangle$; $\alpha_n = \frac{\rho_n}{\sigma_n}$;

$S_{1,n} = R_{1,n-1} - \alpha_nV_{1,n}$; $S_{2,n} = R_{2,n-1} - \alpha_nV_{2,n}$;

$T_{1,n} = \sum_{i=1}^{k}(A_iS_{1,n}B_i + C_iS_{2,n}D_i)$;

$T_{2,n} = \sum_{i=1}^{k}(E_iS_{1,n}F_i + G_iS_{2,n}H_i)$;

$\omega_n = \frac{\langle S_{1,n}, T_{1,n}\rangle + \langle S_{2,n}, T_{2,n}\rangle}{||T_{1,n}||^2 + ||T_{2,n}||^2}$;

$R_{1,n} = S_{1,n} - \omega_nT_{1,n}$; $R_{2,n} = S_{2,n} - \omega_nT_{2,n}$;

$X_n = X_{n-1} + \alpha_nP_{1,n} + \omega_nS_{1,n}$; $Y_n = Y_{n-1} + \alpha_nP_{2,n} + \omega_nS_{2,n}$.

The stopping criteria on the matrix form of the Bi-CGSTAB algorithm can be used as

$$\sqrt{||M - \sum_{i=1}^{k}(A_iX_nB_i + C_iY_nD_i)||^2 + ||N - \sum_{i=1}^{k}(E_iX_nF_i + G_iY_nH_i)||^2} \leq \varepsilon,$$

or

$$\sqrt{||X_n - X_{n-1}||^2 + ||Y_n - Y_{n-1}||^2} \leq \varepsilon,$$

where $\varepsilon > 0$ is a small tolerance.

8. The Matrix Form of QMRCGSTAB Algorithm

The QMRCGSTAB algorithm is a biorthogonalisation method for solving nonsymmetric linear systems $Ax = b$ which improves the irregular behaviour of Bi-CGSTAB algorithm. This method is based on the quasi-minimisation of the residual applying the standard Givens rotations that lead to iterations with short term recurrences [4]. The pseudocode for the QMRCGSTAB algorithm can be summarized as follows.

Algorithm 23. ***(QMRCGSTAB algorithm)***

Choose the initial guess $x_0 \in \mathbb{R}^n$, $r_0 = b - Ax_0$,

choose r_0^* *(for example* $r_0^* = r_0$*),*

$p_0 = v_0 = u_0 = 0,\ \rho_0 = \alpha_0 = \omega_0 = 1;\ \tau = ||r_0||,\ \theta_0 = 0,\ \eta_0 = 0,$

for $k = 1, 2, \ldots$ *do*

$\rho_k = \langle r_0^*, r_{k-1}\rangle;\ \beta_k = (\rho_k \alpha_{k-1})/(\rho_{k-1}\omega_{k-1}),$

$p_k = r_{k-1} + \beta_k(p_{k-1} - \omega_{k-1}v_{k-1}),$

$v_k = Ap_k,$

$\alpha_k = \rho_k/\langle r_0^*, v_k\rangle,$

$s_k = r_{k-1} - \alpha_k v_k,$

$\widetilde{\theta}_k = ||s_k||/\tau;\ c = 1/\sqrt{1 + \widetilde{\theta}_k^2};\ \widetilde{\tau} = \tau\widetilde{\theta}_k c,$

$\widetilde{\eta}_k = c^2\alpha_k;\ \widetilde{u}_k = p_k + \frac{\theta_{k-1}^2\eta_{k-1}}{\alpha_k}u_{k-1},$

$\widetilde{x}_k = x_{k-1} + \widetilde{\eta}_k\widetilde{u}_k,$

$t_k = As_k;\ \omega_k = \langle s_k, t_k\rangle/\langle t_k, t_k\rangle,$

$r_k = s_k - \omega_k t_k;\ \theta_k = ||r_k||/\widetilde{\tau};\ c = 1/\sqrt{1 + \widetilde{\theta}_k^2};\ \tau = \widetilde{\tau}\theta_k c,$

$\eta_k = c^2\omega_k;\ u_k = s_k + \frac{\widetilde{\theta}_k^2\widetilde{\eta}_k}{\omega_k}\widetilde{u}_k,$

$x_k = \widetilde{x}_k + \eta_k u_k,$

if x_k *is accurate enough, then quit,*

end.

In exact arithmetic, the QMRCGSTAB algorithm terminates after a finite number, say k^*, of iterations (for more details about QMRCGSTAB algorithm see [4]). In [39], the QMRCGSTAB algorithm was extended to find the solution of the general coupled matrix equations

$$\sum_{j=1}^{l}(A_{i,1,j}X_1B_{i,1,j} + A_{i,2,j}X_2B_{i,2,j} + ... + A_{i,l,j}X_lB_{i,l,j}) = C_i \quad \text{for} \quad i = 1, 2, .., l, \tag{82}$$

where $A_{i,k,j}, B_{i,k,j}, X_i \in \mathbb{R}^{n\times n}$ for $i, j, k = 1, 2, .., l$. Also the extended QMRCGSTAB algorithm was proposed to solve the periodic coupled matrix equations

$$\begin{cases} \mathcal{A}_{1,t}\mathcal{X}_t\mathcal{B}_{1,t} + \mathcal{C}_{1,t}\mathcal{X}_{t+1}\mathcal{D}_{1,t} + \mathcal{E}_{1,t}\mathcal{Y}_t\mathcal{F}_{1,t} = \mathcal{G}_{1,t}, \\ \mathcal{A}_{2,t}\mathcal{X}_t\mathcal{B}_{2,t} + \mathcal{C}_{2,t}\mathcal{X}_{t+1}\mathcal{D}_{2,t} + \mathcal{E}_{2,t}\mathcal{Y}_t\mathcal{F}_{2,t} = \mathcal{G}_{2,t}, \end{cases} \quad \text{for} \quad t = 1, 2, .., \phi, \tag{83}$$

where the coefficient matrices $\mathcal{A}_{j,t}, \mathcal{B}_{j,t}, \mathcal{C}_{j,t}, \mathcal{D}_{j,t}, \mathcal{E}_{j,t}, \mathcal{F}_{j,t}, \mathcal{G}_{j,t} \in \mathcal{R}^{n\times n}$ and the solutions $\mathcal{X}_t, \mathcal{Y}_t \in \mathcal{R}^{n\times n}$ are ϕ-periodic for $j = 1, 2$.

To solve (82) by the QMRCGSTAB algorithm, we need to transform (82) into linear system $Ax = b$. By using the Kronecker product and vectorization operator, we can transform (82) into the following system:

$$\underbrace{\begin{pmatrix} \sum_{j=1}^{l}(B_{1,1,j}^T \otimes A_{1,1,j}) & \sum_{j=1}^{l}(B_{1,2,j}^T \otimes A_{1,2,j}) & \cdots & \sum_{j=1}^{l}(B_{1,l,j}^T \otimes A_{1,l,j}) \\ \sum_{j=1}^{l}(B_{2,1,j}^T \otimes A_{2,1,j}) & \sum_{j=1}^{l}(B_{2,2,j}^T \otimes A_{2,2,j}) & \cdots & \sum_{j=1}^{l}(B_{2,l,j}^T \otimes A_{2,l,j}) \\ \vdots & \vdots & \vdots & \vdots \\ \sum_{j=1}^{l}(B_{l,1,j}^T \otimes A_{l,1,j}) & \sum_{j=1}^{l}(B_{l,2,j}^T \otimes A_{l,2,j}) & \cdots & \sum_{j=1}^{l}(B_{l,l,j}^T \otimes A_{l,l,j}) \end{pmatrix}}_{A}$$

$$\times \underbrace{\begin{pmatrix} \text{vec}(X_1) \\ \text{vec}(X_2) \\ \vdots \\ \text{vec}(X_l) \end{pmatrix}}_{x} = \underbrace{\begin{pmatrix} \text{vec}(C_1) \\ \text{vec}(C_2) \\ \vdots \\ \text{vec}(C_l) \end{pmatrix}}_{b}, \tag{84}$$

where $A \in \mathbb{R}^{n^2l\times n^2l}$ and $x, b \in \mathbb{R}^{n^2l}$. Obviously the dimension of the associate matrix A is high when n or l is large. Iterative methods like the QMRCGSTAB algorithm take much computer time and memory for solving the large systems (84). Also the solutions obtained by iterative methods are not accurate enough. To overcome these complications, we directly extend the QMRCGSTAB algorithm for solving (82). For this purpose, we substitute the parameters of linear systems (84) into the QMRCGSTAB algorithm. We have

$$r_0 = b - Ax_0 \rightarrow$$

$$r_0 = \begin{pmatrix} \text{vec}(C_1) \\ \text{vec}(C_2) \\ \vdots \\ \text{vec}(C_l) \end{pmatrix} - \begin{pmatrix} \sum_{j=1}^{l}(B_{1,1,j}^T \otimes A_{1,1,j}) & \sum_{j=1}^{l}(B_{1,2,j}^T \otimes A_{1,2,j}) & \cdots & \sum_{j=1}^{l}(B_{1,l,j}^T \otimes A_{1,l,j}) \\ \sum_{j=1}^{l}(B_{2,1,j}^T \otimes A_{2,1,j}) & \sum_{j=1}^{l}(B_{2,2,j}^T \otimes A_{2,2,j}) & \cdots & \sum_{j=1}^{l}(B_{2,l,j}^T \otimes A_{2,l,j}) \\ \vdots & \vdots & \vdots & \vdots \\ \sum_{j=1}^{l}(B_{l,1,j}^T \otimes A_{l,1,j}) & \sum_{j=1}^{l}(B_{l,2,j}^T \otimes A_{l,2,j}) & \cdots & \sum_{j=1}^{l}(B_{l,l,j}^T \otimes A_{l,l,j}) \end{pmatrix} x_0, \tag{85}$$

$$v_k = Ap_k \rightarrow v_k = \begin{pmatrix} \sum_{j=1}^{l}(B_{1,1,j}^T \otimes A_{1,1,j}) & \sum_{j=1}^{l}(B_{1,2,j}^T \otimes A_{1,2,j}) & \cdots & \sum_{j=1}^{l}(B_{1,l,j}^T \otimes A_{1,l,j}) \\ \sum_{j=1}^{l}(B_{2,1,j}^T \otimes A_{2,1,j}) & \sum_{j=1}^{l}(B_{2,2,j}^T \otimes A_{2,2,j}) & \cdots & \sum_{j=1}^{l}(B_{2,l,j}^T \otimes A_{2,l,j}) \\ \vdots & \vdots & \vdots & \vdots \\ \sum_{j=1}^{l}(B_{l,1,j}^T \otimes A_{l,1,j}) & \sum_{j=1}^{l}(B_{l,2,j}^T \otimes A_{l,2,j}) & \cdots & \sum_{j=1}^{l}(B_{l,l,j}^T \otimes A_{l,l,j}) \end{pmatrix} p_k, \tag{86}$$

and

$$t_k = As_k \rightarrow t_k = \begin{pmatrix} \sum_{j=1}^{l}(B_{1,1,j}^T \otimes A_{1,1,j}) & \sum_{j=1}^{l}(B_{1,2,j}^T \otimes A_{1,2,j}) & \cdots & \sum_{j=1}^{l}(B_{1,l,j}^T \otimes A_{1,l,j}) \\ \sum_{j=1}^{l}(B_{2,1,j}^T \otimes A_{2,1,j}) & \sum_{j=1}^{l}(B_{2,2,j}^T \otimes A_{2,2,j}) & \cdots & \sum_{j=1}^{l}(B_{2,l,j}^T \otimes A_{2,l,j}) \\ \vdots & \vdots & \vdots & \vdots \\ \sum_{j=1}^{l}(B_{l,1,j}^T \otimes A_{l,1,j}) & \sum_{j=1}^{l}(B_{l,2,j}^T \otimes A_{l,2,j}) & \cdots & \sum_{j=1}^{l}(B_{l,l,j}^T \otimes A_{l,l,j}) \end{pmatrix} s_k. \tag{87}$$

Now from the above equations and the QMRCGSTAB algorithm, we define:

$$r_k = \begin{pmatrix} \text{vec}(R_1(k)) \\ \text{vec}(R_2(k)) \\ \vdots \\ \text{vec}(R_l(k)) \end{pmatrix}, \quad v_k = \begin{pmatrix} \text{vec}(V_1(k)) \\ \text{vec}(V_2(k)) \\ \vdots \\ \text{vec}(V_l(k)) \end{pmatrix}, \quad s_k = \begin{pmatrix} \text{vec}(S_1(k)) \\ \text{vec}(S_2(k)) \\ \vdots \\ \text{vec}(S_l(k)) \end{pmatrix}, \quad p_k = \begin{pmatrix} \text{vec}(P_1(k)) \\ \text{vec}(P_2(k)) \\ \vdots \\ \text{vec}(P_l(k)) \end{pmatrix}, \tag{88}$$

$$r_k^* = \begin{pmatrix} \text{vec}(R_1^*(k)) \\ \text{vec}(R_2^*(k)) \\ \vdots \\ \text{vec}(R_l^*(k)) \end{pmatrix}, \quad t_k = \begin{pmatrix} \text{vec}(T_1(k)) \\ \text{vec}(T_2(k)) \\ \vdots \\ \text{vec}(T_l(k)) \end{pmatrix}, \quad x_k = \begin{pmatrix} \text{vec}(X_1(k)) \\ \text{vec}(X_2(k)) \\ \vdots \\ \text{vec}(X_l(k)) \end{pmatrix}, \quad u_k = \begin{pmatrix} \text{vec}(U_1(k)) \\ \text{vec}(U_2(k)) \\ \vdots \\ \text{vec}(U_l(k)) \end{pmatrix}, \tag{89}$$

$$\widetilde{x}_k = \begin{pmatrix} \text{vec}(\widetilde{X}_1(k)) \\ \text{vec}(\widetilde{X}_2(k)) \\ \vdots \\ \text{vec}(\widetilde{X}_l(k)) \end{pmatrix}, \quad \widetilde{u}_k = \begin{pmatrix} \text{vec}(\widetilde{U}_1(k)) \\ \text{vec}(\widetilde{U}_2(k)) \\ \vdots \\ \text{vec}(\widetilde{U}_l(k)) \end{pmatrix}, \tag{90}$$

where $R_i(k), V_i(k), S_i(k), P_i(k), R_i^*(k), T_i(k), X_i(k), U_i(k), \widetilde{X}_i(k), \widetilde{U}_i(k) \in \mathbb{R}^{n\times n}$ for $i = 1, 2, ..l$ and $k = 0, 1, 2, ...$ By substituting the definitions in (86)-(87), we can get

$$\begin{pmatrix} \text{vec}(R_1(0)) \\ \text{vec}(R_2(0)) \\ \vdots \\ \text{vec}(R_l(0)) \end{pmatrix} = \begin{pmatrix} \text{vec}(C_1 - \sum_{j=1}^{l}(A_{1,1,j}X_1(0)B_{1,1,j} + A_{1,2,j}X_2(0)B_{1,2,j} + ... + A_{1,l,j}X_l(0)B_{1,l,j})) \\ \text{vec}(C_2 - \sum_{j=1}^{l}(A_{2,1,j}X_1(0)B_{2,1,j} + A_{2,2,j}X_2(0)B_{2,2,j} + ... + A_{2,l,j}X_l(0)B_{2,l,j})) \\ \vdots \\ \text{vec}(C_l - \sum_{j=1}^{l}(A_{l,1,j}X_1(0)B_{l,1,j} + A_{l,2,j}X_2(0)B_{l,2,j} + ... + A_{l,l,j}X_l(0)B_{l,l,j})) \end{pmatrix}, \tag{91}$$

$$\begin{pmatrix} \text{vec}(V_1(k)) \\ \text{vec}(V_2(k)) \\ \vdots \\ \text{vec}(V_l(k)) \end{pmatrix} = \begin{pmatrix} \text{vec}(\sum_{j=1}^{l}(A_{1,1,j}P_1(k)B_{1,1,j} + A_{1,2,j}P_2(k)B_{1,2,j} + ... + A_{1,l,j}P_l(k)B_{1,l,j})) \\ \text{vec}(\sum_{j=1}^{l}(A_{2,1,j}P_1(k)B_{2,1,j} + A_{2,2,j}P_2(k)B_{2,2,j} + ... + A_{2,l,j}P_l(k)B_{2,l,j})) \\ \vdots \\ \text{vec}(\sum_{j=1}^{l}(A_{l,1,j}P_1(k)B_{l,1,j} + A_{l,2,j}P_2(k)B_{l,2,j} + ... + A_{l,l,j}P_l(k)B_{l,l,j})) \end{pmatrix}, \tag{92}$$

and

$$\begin{pmatrix} \mathrm{vec}(T_1(k)) \\ \mathrm{vec}(T_2(k)) \\ \vdots \\ \mathrm{vec}(T_l(k)) \end{pmatrix} = \begin{pmatrix} \mathrm{vec}(\sum_{j=1}^{l}(A_{1,1,j}S_1(k)B_{1,1,j} + A_{1,2,j}S_2(k)B_{1,2,j} + ... + A_{1,l,j}S_l(k)B_{1,l,j})) \\ \mathrm{vec}(\sum_{j=1}^{l}(A_{2,1,j}S_1(k)B_{2,1,j} + A_{2,2,j}S_2(k)B_{2,2,j} + ... + A_{2,l,j}S_l(k)B_{2,l,j})) \\ \vdots \\ \mathrm{vec}(\sum_{j=1}^{l}(A_{l,1,j}S_1(k)B_{l,1,j} + A_{l,2,j}S_2(k)B_{l,2,j} + ... + A_{l,l,j}S_l(k)B_{l,l,j})) \end{pmatrix}. \quad (93)$$

Also we have

$$\rho_k = \langle r_0^*, r_{k-1}\rangle = \left\langle \begin{pmatrix} \mathrm{vec}(R_1^*(0)) \\ \mathrm{vec}(R_2^*(0)) \\ \vdots \\ \mathrm{vec}(R_l^*(0)) \end{pmatrix}, \begin{pmatrix} \mathrm{vec}(R_1(k-1)) \\ \mathrm{vec}(R_2(k-1)) \\ \vdots \\ \mathrm{vec}(R_l(k-1)) \end{pmatrix} \right\rangle = \sum_{i=1}^{l}\langle R_i^*(0), R_i(k-1)\rangle, \quad (94)$$

$$\alpha_k = \rho_k/\langle r_0^*, v_k\rangle = \rho_k/\sum_{i=1}^{l}\langle R_i^*(0), V_i(k)\rangle, \quad (95)$$

and

$$\omega_k = \langle s_k, t_k\rangle/\langle t_k, t_k\rangle = \sum_{i=1}^{l}\langle S_i(k), T_i(k)\rangle/\sum_{i=1}^{l}\langle T_i(k), T_i(k)\rangle. \quad (96)$$

From the discussion above, the matrix form of QMRCGSTAB algorithm for solving (82) can be given as follows.

Algorithm 24. *(The matrix form of QMRCGSTAB algorithm to solve (82))*

Choose the initial guesses $X_i(0) \in \mathbb{R}^{n\times n}$ for $i = 1, 2, .., l$,

$R_i(0) = C_i - \sum_{j=1}^{l}(A_{i,1,j}X_1(0)B_{i,1,j} + A_{i,2,j}X_2(0)B_{i,2,j} + ... + A_{i,l,j}X_l(0)B_{i,l,j})$, for $i = 1, 2, .., l$
choose $R_i^*(0)$ (for example $R_i^*(0) = R_i(0)$), for $i = 1, 2, .., l$

$P_i(0) = V_i(0) = U_i(0) = 0$, for $i = 1, 2, .., l$ $\rho_0 = \alpha_0 = \omega_0 = 1$; $\tau = (\sum_{i=1}^{l}||R_i(0)||^2)^{1/2}$, $\theta_0 = 0$, $\eta_0 = 0$,
for $k = 1, 2, ...$ do

$\rho_k = \sum_{i=1}^{l}\langle R_i^*(0), R_i(k-1)\rangle$; $\beta_k = (\rho_k\alpha_{k-1})/(\rho_{k-1}\omega_{k-1})$,

$P_i(k) = R_i(k-1) + \beta_k(P_i(k-1) - \omega_{k-1}V_i(k-1))$, for $i = 1, 2, .., l$,

$V_i(k) = \sum_{j=1}^{l}(A_{i,1,j}P_1(k)B_{i,1,j} + A_{i,2,j}P_2(k)B_{i,2,j} + ... + A_{i,l,j}P_l(k)B_{i,l,j})$, for $i = 1, 2, .., l$,

$\alpha_k = \rho_k/\sum_{i=1}^{l}\langle R_i^*(0), V_i(k)\rangle$,

$S_i(k) = R_i(k-1) - \alpha_k V_i(k)$, for $i = 1, 2, .., l$,

$\widetilde{\theta}_k = (\sum_{i=1}^{l}||S_i(k)||^2)^{1/2}/\tau$; $c = 1/\sqrt{1+\widetilde{\theta}_k^2}$; $\widetilde{\tau} = \tau\widetilde{\theta}_k c$,

$\widetilde{\eta}_k = c^2\alpha_k$; $\widetilde{U}_i(k) = P_i(k) + \frac{\theta_{k-1}^2\eta_{k-1}}{\alpha_k}U_i(k-1)$, for $i = 1, 2, .., l$,

$\widetilde{X}_i(k) = X_i(k-1) + \widetilde{\eta}_k \widetilde{U}_i(k)$,

$T_i(k) = \sum_{j=1}^{l}(A_{i,1,j}S_1(k)B_{i,1,j} + A_{i,2,j}S_2(k)B_{i,2,j} + ... + A_{i,l,j}S_l(k)B_{i,l,j})$, for $i = 1, 2, .., l$,

$\omega_k = \sum_{i=1}^{l}\langle S_i(k), T_i(k)\rangle / \sum_{i=1}^{l}\langle T_i(k), T_i(k)\rangle$,

$R_i(k) = S_i(k) - \omega_k T_i(k)$, for $i = 1, 2, .., l$, $\theta_k = (\sum_{i=1}^{l}||R_i(k)||^2)^{1/2}/\widetilde{\tau}$; $c = 1/\sqrt{1+\widetilde{\theta}_k^2}$, $\tau = \widetilde{\tau}\theta_k c$,

$\eta_k = c^2\omega_k$; $U_i(k) = S_i(k) + \frac{\widetilde{\theta}_k^2\widetilde{\eta}_k}{\omega_k}\widetilde{U}_i(k)$, for $i = 1, 2, .., l$,

$X_i(k) = \widetilde{X}_i(k) + \eta_k U_i(k)$, for $i = 1, 2, .., l$,
if $X_i(k)$ are accurate enough for $i = 1, 2, .., l$, then quit,
end.

The stopping criteria on Algorithm 24 can be used as

$$\sqrt{\sum_{i=1}^{l}||C_i - \sum_{j=1}^{l}(A_{i,1,j}X_1(k)B_{i,1,j} + A_{i,2,j}X_2(k)B_{i,2,j} + ... + A_{i,l,j}X_l(k)B_{i,l,j})||^2} \leq \varepsilon,$$

or

$$\sqrt{\sum_{i=1}^{l}||X_i(k) - X_i(k-1)||^2} \leq \varepsilon,$$

where $\varepsilon > 0$ is a small tolerance.

We can easily show that the periodic coupled matrix equations (83) are equivalent to the following generalized coupled Sylvester matrix equations

$$\begin{cases} A_1XB_1 + C_1XD_1 + E_1YF_1 = G_1, \\ A_2XB_2 + C_2XD_2 + E_2YF_2 = G_2, \end{cases} \tag{97}$$

where

$$A_j = \begin{pmatrix} 0 & \cdots & 0 & \mathcal{A}_{j,1} \\ \mathcal{A}_{j,2} & & & 0 \\ & \ddots & & \vdots \\ 0 & & \mathcal{A}_{j,\phi} & 0 \end{pmatrix}, \quad B_j = \begin{pmatrix} 0 & \mathcal{B}_{j,2} & & 0 \\ \vdots & & \ddots & \\ 0 & & & \mathcal{B}_{j,\phi} \\ \mathcal{B}_{j,1} & 0 & \cdots & 0 \end{pmatrix}, \quad C_j = \text{diag}\left(\mathcal{C}_{j,1}, \mathcal{C}_{j,2}, ..., \mathcal{C}_{j,\phi}\right),$$

$$D_j = \text{diag}\left(\mathcal{D}_{j,1}, \mathcal{D}_{j,2}, ..., \mathcal{D}_{j,\phi}\right), \quad E_j = \text{diag}\left(\mathcal{E}_{j,1}, \mathcal{E}_{j,2}, ..., \mathcal{E}_{j,\phi}\right), \quad F_j = \text{diag}\left(\mathcal{F}_{j,1}, \mathcal{F}_{j,2}, ..., \mathcal{F}_{j,\phi}\right),$$

$$G_j = \text{diag}\left(\mathcal{G}_{j,1}, \mathcal{G}_{j,2}, ..., \mathcal{G}_{j,\phi}\right), \quad X = \text{diag}\left(\mathcal{X}_2, \mathcal{X}_3, ..., \mathcal{X}_\phi, \mathcal{X}_1\right), \quad Y = \text{diag}\left(\mathcal{Y}_1, \mathcal{Y}_2, ..., \mathcal{Y}_\phi\right),$$

for $j = 1, 2$. It is obvious that the size of the matrix equations (97) is large. By considering (83), (97) and Algorithm 24, we extend the QMRCGSTAB algorithm for solving the periodic coupled matrix equations (83) as follows.

Algorithm 25. ***(The matrix form of QMRCGSTAB algorithm to solve (83))***

Choose the initial guesses $\mathcal{X}_t(0) \in \mathcal{R}^{n\times n}$ and $\mathcal{Y}_t(0) \in \mathcal{R}^{n\times n}$ for $t = 1, 2, .., \phi$,

set $\mathcal{X}_{\phi+1}(0) = \mathcal{X}_1(0)$, and $\mathcal{Y}_{\phi+1}(0) = \mathcal{Y}_1(0)$,

$R_{j,t}(0) = \mathcal{G}_{j,t} - \mathcal{A}_{j,t}\mathcal{X}_t(0)\mathcal{B}_{j,t} - \mathcal{C}_{j,t}\mathcal{X}_{t+1}(0)\mathcal{D}_{j,t} - \mathcal{E}_{j,t}\mathcal{Y}_t(0)\mathcal{F}_{j,t},$ for $t = 1, 2, .., \phi$ and $j = 1, 2,$
choose $R^*_{1,t}(0)$ and $R^*_{2,t}(0)$ (for example $R^*_{1,t}(0) = R_{1,t}(0)$ and $R^*_{2,t}(0) = R_{2,t}(0)$), for $t = 1, 2, .., \phi$,

$P_{j,t}(0) = V_{j,t}(0) = U_{j,t}(0) = 0,$ for $t = 1, 2, .., \phi$ and $j = 1, 2,$

$\rho_0 = \alpha_0 = \omega_0 = 1;\ \tau = (\sum_{t=1}^{\phi}[||R_{1,t}(0)||^2 + ||R_{2,t}(0)||^2])^{1/2},\ \theta_0 = 0,\ \eta_0 = 0,$
for $k = 1, 2, ...$ do

$\rho_k = \sum_{t=1}^{\phi}[\langle R^*_{1,t}(0), R_{1,t}(k-1)\rangle + \langle R^*_{2,t}(0), R_{2,t}(k-1)\rangle],\ \beta_k = (\rho_k\alpha_{k-1})/(\rho_{k-1}\omega_{k-1}),$

$P_{j,t}(k) = R_{j,t}(k-1) + \beta_k(P_{j,t}(k-1) - \omega_{k-1}V_{j,t}(k-1)),$ for $t = 1, 2, .., \phi$ and $j = 1, 2,$
set $P_{1,\phi+1}(k) = P_{1,1}(k)$ and $P_{2,\phi+1}(k) = P_{2,1}(k)$,

$V_{j,t}(k) = \mathcal{A}_{j,t}P_{1,t}(k)\mathcal{B}_{j,t} + \mathcal{C}_{j,t}P_{1,t+1}(k)(k)\mathcal{D}_{j,t} + \mathcal{E}_{j,t}P_{2,t}(k)\mathcal{F}_{j,t},$ for $t = 1, 2, .., \phi$ and $j = 1, 2,$

$\alpha_k = \rho_k / \sum_{t=1}^{\phi}[\langle R^*_{1,t}(0), V_{1,t}(k)\rangle + \langle R^*_{2,t}(0), V_{2,t}(k)\rangle],$

$S_{j,t}(k) = R_{j,t}(k-1) - \alpha_k V_{j,t}(k),$ for $t = 1, 2, .., \phi$ and $j = 1, 2,$
set $S_{1,\phi+1}(k) = S_{1,1}(k)$ and $S_{2,\phi+1}(k) = S_{2,1}(k)$,

$\widetilde{\theta}_k = (\sum_{t=1}^{\phi}[||S_{1,t}(k)||^2 + ||S_{2,t}(k)||^2])^{1/2}/\tau;\ c = 1/\sqrt{1+\widetilde{\theta}_k^2}, \widetilde{\tau} = \tau\widetilde{\theta}_k c,$

$\widetilde{\eta}_k = c^2\alpha_k, \widetilde{U}_{j,t}(k) = P_{j,t}(k) + \frac{\theta_{k-1}^2\eta_{k-1}}{\alpha_k}U_{j,t}(k-1),$ for $t = 1, 2, .., \phi$ and $j = 1, 2,$

$\widetilde{X}_t(k) = \mathcal{X}_t(k-1) + \widetilde{\eta}_k\widetilde{U}_{1,t}(k),$ for $t = 1, 2, .., \phi$,

$\widetilde{Y}_t(k) = \mathcal{Y}_t(k-1) + \widetilde{\eta}_k\widetilde{U}_{2,t}(k),$ for $t = 1, 2, .., \phi$,

$T_{j,t}(k) = \mathcal{A}_{j,t}S_{1,t}(k)\mathcal{B}_{j,t} + \mathcal{C}_{j,t}S_{1,t+1}(k)(k)\mathcal{D}_{j,t} + \mathcal{E}_{j,t}S_{2,t}(k)\mathcal{F}_{j,t},$ for $t = 1, 2, .., \phi$ and $j = 1, 2,$

$\omega_k = \sum_{t=1}^{\phi}[\langle S_{1,t}(k), T_{1,t}(k)\rangle + \langle S_{2,t}(k), T_{2,t}(k)\rangle] / \sum_{t=1}^{\phi}[\langle T_{1,t}(k), T_{1,t}(k)\rangle + \langle T_{2,t}(k), T_{2,t}(k)\rangle],$

$R_{j,t}(k) = S_{j,t}(k) - \omega_k T_{j,t}(k),$ for $t = 1, 2, .., \phi$ and $j = 1, 2,$

$\theta_k = (\sum_{t=1}^{\phi}[||R_{1,t}(k)||^2 + ||R_{2,t}(k)||^2])^{1/2}/\widetilde{\tau}, c = 1/\sqrt{1+\widetilde{\theta}_k^2}, \tau = \tilde{\tau}\theta_k c,$

$\eta_k = c^2\omega_k, U_{j,t}(k) = S_{j,t}(k) + \frac{\widetilde{\theta}_k^2\widetilde{\eta}_k}{\omega_k}\widetilde{U}_{j,t}(k),$ for $t = 1, 2, .., \phi$ and $j = 1, 2,$

$\mathcal{X}_t(k) = \widetilde{X}_t(k) + \eta_k U_{1,t}(k),$ for $t = 1, 2, .., \phi,$

$\mathcal{Y}_t(k) = \widetilde{Y}_t(k) + \eta_k U_{2,t}(k),$ for $t = 1, 2, .., \phi,$
set $\mathcal{X}_{\phi+1}(k) = \mathcal{X}_1(k)$ and $\mathcal{Y}_{\phi+1}(k) = \mathcal{Y}_1(k)$,
if $\mathcal{X}_t(k)$ and $\mathcal{Y}_t(k)$ are accurate enough for $t = 1, 2, .., \phi$, then quit,
end.

We can the stopping criteria on Algorithm 25 as follows

$$\sqrt{\sum_{j=1}^{2}\sum_{t=1}^{\phi} ||\mathcal{G}_{j,t} - \mathcal{A}_{j,t}\mathcal{X}_t(k)\mathcal{B}_{j,t} - \mathcal{C}_{j,t}\mathcal{X}_{t+1}(k)\mathcal{D}_{j,t} - \mathcal{E}_{j,t}\mathcal{Y}_t(k)\mathcal{F}_{j,t}||^2} \leq \varepsilon,$$

where $\varepsilon > 0$ is a small tolerance.

9. Numerical Results

In this section, some numerical examples are proposed for the validation of the proposed methods. We performed our computations using Matlab software on a Pentium IV.

Example 1. *Consider a system of matrix equations in the form of*

$$\begin{cases} AX_1B + CX_2D = M, \\ EX_1F + GX_2H = N, \end{cases}$$

with the following parameters

$$A = \begin{pmatrix} 1 & 1 & 2 & -3 & 4 \\ 3 & 4 & 2 & 2 & 1 \\ 0 & 4 & 7 & 2 & 4 \\ -1 & -1 & -1 & 2 & 4 \\ 4 & 4 & 3 & 2 & 1 \end{pmatrix}, \quad B = \begin{pmatrix} 2 & 2 & 3 & 1 & 1 \\ 0 & 5 & 4 & -2 & -2 \\ 2 & 3 & 4 & 1 & 1 \\ 2 & 0 & 2 & 0 & 1 \\ -3 & -3 & 1 & 2 & 2 \end{pmatrix}, \quad C = \begin{pmatrix} 4 & 3 & 4 & 4 & 1 \\ -2 & -2 & 3 & 4 & 4 \\ 5 & 6 & 5 & 0 & 1 \\ 5 & 4 & 5 & 3 & 3 \\ 1 & 2 & 0 & 0 & 1 \end{pmatrix},$$

$$D = \begin{pmatrix} -2 & -1 & -2 & 3 & 2 \\ 6 & 5 & 4 & 4 & 3 \\ 2 & 3 & 2 & 1 & 1 \\ 1 & 1 & 2 & 4 & 1 \\ 0 & 0 & 2 & 3 & 2 \end{pmatrix}, \quad E = \begin{pmatrix} 1 & 2 & 3 & 1 & 2 \\ 0 & 1 & 2 & 3 & 1 \\ 4 & 4 & 2 & 1 & 3 \\ 1 & 0 & 0 & -1 & 0 \\ 2 & 4 & 5 & 3 & 2 \end{pmatrix}, \quad F = \begin{pmatrix} 1 & 2 & 3 & -1 & -1 \\ 3 & 0 & 0 & 3 & 3 \\ 1 & 2 & 3 & 0 & -2 \\ -1 & -1 & 2 & 2 & 3 \\ 5 & 4 & 5 & 4 & 4 \end{pmatrix},$$

$$G = \begin{pmatrix} 1 & 2 & 1 & 2 & 1 \\ 3 & 3 & 3 & 1 & 2 \\ 1 & 2 & 3 & -4 & -4 \\ 5 & 5 & 5 & 4 & 4 \\ 2 & -2 & -2 & -2 & 1 \end{pmatrix}, \quad H = \begin{pmatrix} 2 & 4 & 3 & 2 & 1 \\ 0 & -3 & -3 & -3 & 2 \\ 1 & 1 & 1 & 0 & 0 \\ 1 & 1 & 2 & 3 & 4 \\ 2 & 3 & 4 & 5 & 2 \end{pmatrix}.$$

It can be verified that these matrix equations are consistent over generalized bisymmetric matrix pair (X_1, X_2) *and have a unique generalized bisymmetric solution pair* (X_1^*, X_2^*) *as follows*

$$X_1^* = \begin{pmatrix} 2 & 2 & 0 & 4 & 2 \\ 2 & -2 & 0 & 4 & 2 \\ 0 & 0 & 2 & 0 & 0 \\ 4 & 4 & 0 & 0 & 2 \\ 2 & 2 & 0 & 2 & 2 \end{pmatrix} \in \mathbb{BSR}_{P,P}^{5\times 5}, \qquad X_2^* = \begin{pmatrix} 2 & 0 & 0 & 4 & 0 \\ 0 & -2 & 4 & 0 & 2 \\ 0 & 4 & 2 & 0 & 2 \\ 4 & 0 & 0 & 0 & 0 \\ 0 & 2 & 2 & 0 & 2 \end{pmatrix} \in \mathbb{BSR}_{Q,Q}^{5\times 5},$$

with

$$P = \begin{pmatrix} 1 & 0 & 0 & 0 & 0 \\ 0 & 1 & 0 & 0 & 0 \\ 0 & 0 & -1 & 0 & 0 \\ 0 & 0 & 0 & 1 & 0 \\ 0 & 0 & 0 & 0 & 1 \end{pmatrix} \in \mathbb{SOR}^{5\times 5}, \qquad Q = \begin{pmatrix} 1 & 0 & 0 & 0 & 0 \\ 0 & -1 & 0 & 0 & 0 \\ 0 & 0 & -1 & 0 & 0 \\ 0 & 0 & 0 & 1 & 0 \\ 0 & 0 & 0 & 0 & -1 \end{pmatrix} \in \mathbb{SOR}^{5\times 5}.$$

By applying Algorithm 3 with the initial matrix pair $(X_1(1), X_2(1)) = 0$, *we obtain*

$$X_1(33) = \begin{pmatrix} 2.0000 & 2.0000 & -0.0000 & 4.0000 & 2.0000 \\ 2.0000 & -2.0000 & 0.0000 & 4.0000 & 2.0000 \\ 0.0000 & -0.0000 & 2.0000 & -0.0000 & -0.0000 \\ 4.0000 & 4.0000 & 0.0000 & 0.0000 & 2.0000 \\ 2.0000 & 2.0000 & 0.0000 & 2.0000 & 2.0000 \end{pmatrix},$$

$$X_2(33) = \begin{pmatrix} 2.0000 & -0.0000 & 0.0000 & 4.0000 & -0.0000 \\ 0.0000 & -2.0000 & 4.0000 & 0.0000 & 2.0000 \\ 0.0000 & 4.0000 & 2.0000 & -0.0000 & 2.0000 \\ 4.0000 & 0.0000 & -0.0000 & -0.0000 & 0.0000 \\ 0.0000 & 2.0000 & 2.0000 & 0.0000 & 2.0000 \end{pmatrix},$$

with corresponding residual

$$||R(33)|| = 1.0497 \times 10^{-11}.$$

The obtained results are presented in Figure 1, where

$$\delta(k) = \log_{10} \frac{||(X_1(k), X_2(k)) - (X_1^*, X_2^*)||}{||(X_1^*, X_2^*)||} \qquad \text{and} \qquad r(k) = \log_{10} ||R(k)||.$$

From Figure 1, we can see that Algorithm 3 is effective.

Example 2. *Consider the system of matrix equations*

$$\begin{cases} A_1 X D_1 = C_1, \\ A_2 X D_2 = C_2, \end{cases}$$

with the parameters

$$A_1 = \mathrm{tril}(\mathrm{rand}(10,10),1), \quad A_2 = \mathrm{tril}(\mathrm{rand}(10,10),1) + \mathrm{diag}(80 + \mathrm{diag}(\mathrm{rand}(10))),$$

$$D_1 = \mathrm{triu}(\mathrm{rand}(10,10),1), \quad D_2 = -\mathrm{tril}(\mathrm{rand}(10,10),1) + \mathrm{diag}(70 + \mathrm{diag}(\mathrm{rand}(10))),$$

$$C_1 = \begin{pmatrix} 0 & 0.4752 & 1.5819 & 1.2979 & 0.3225 & 2.3894 & 2.0078 & 1.7323 & 0.8166 & 0.8911 \\ 0 & 0.2283 & 1.2500 & 0.7226 & 0.5247 & 1.9341 & 1.9930 & 1.6817 & 1.0292 & 1.1276 \\ 0 & 0.5633 & 2.4746 & 1.7033 & 1.2994 & 4.5379 & 4.4119 & 3.9496 & 2.5569 & 2.3597 \\ 0 & 0.4243 & 2.4563 & 1.2918 & 1.1377 & 5.6673 & 3.7479 & 5.0162 & 3.1931 & 3.2798 \\ 0 & 0.8805 & 2.4842 & 2.3836 & 0.9075 & 5.8859 & 4.5445 & 4.6132 & 3.1371 & 1.9735 \\ 0 & 0.8405 & 2.8620 & 2.3549 & 1.2453 & 7.1221 & 5.5886 & 6.7545 & 4.9200 & 3.5456 \\ 0 & 0.7069 & 3.9678 & 2.1694 & 1.9116 & 9.5221 & 6.9670 & 9.7375 & 7.0484 & 6.7634 \\ 0 & 0.5821 & 3.5256 & 1.8747 & 1.9998 & 8.3373 & 6.8889 & 9.3980 & 7.1389 & 6.6456 \\ 0 & 1.1014 & 4.1475 & 3.1459 & 1.8569 & 9.3099 & 7.8557 & 10.2282 & 7.6441 & 6.8865 \\ 0 & 0.8221 & 5.2980 & 2.6205 & 2.7589 & 12.4282 & 9.3805 & 14.3536 & 10.6867 & 11.6303 \end{pmatrix},$$

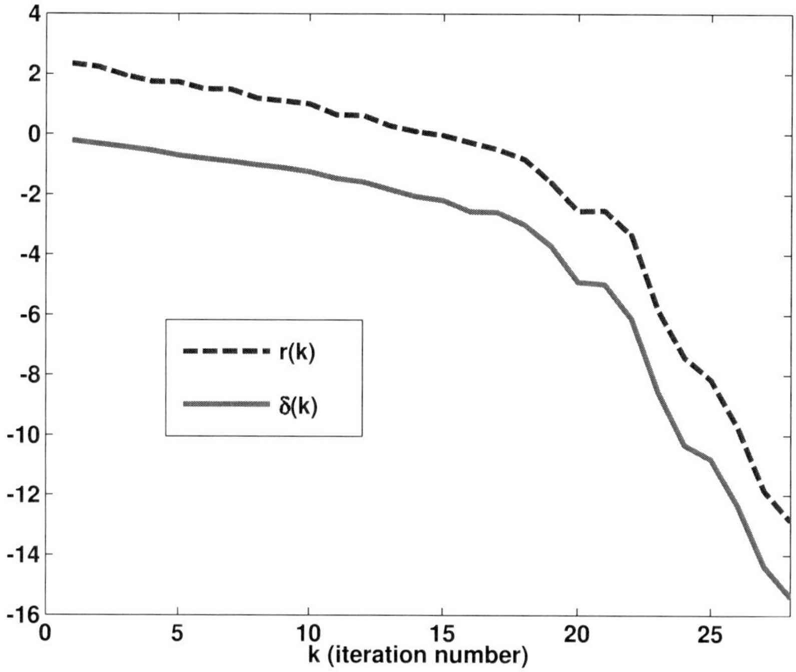

Figure 1. The obtained results for Example 1.

$$C_2 = 10^4 \begin{pmatrix} 1.0705 & 0.0016 & -0.0000 & 0 & 0 & 0 & 0 & 0 & 0 & 0 \\ 0.0054 & 0.8956 & 0.0009 & -0.0000 & 0 & 0 & 0 & 0 & 0 & 0 \\ 0.6863 & 0.0009 & 0.9202 & 0.0014 & -0.0001 & 0 & 0 & 0 & 0 & 0 \\ 0.0061 & 0.8296 & -0.0019 & 1.0586 & -0.0021 & -0.0000 & 0 & 0 & 0 & 0 \\ 1.0099 & 0.0089 & 0.1527 & 0.0009 & 0.9575 & 0.0057 & -0.0000 & 0 & 0 & 0 \\ 0.0054 & 0.4600 & -0.0088 & 0.4710 & -0.0022 & 0.7984 & 0.0012 & -0.0000 & 0 & 0 \\ 0.5210 & 0.0186 & 0.2178 & 0.0099 & 0.8071 & 0.0158 & 0.3886 & 0.0021 & -0.0000 & 0 \\ 0.0131 & 1.0395 & -0.0154 & 0.6018 & 0.0012 & 0.9916 & 0.0015 & 0.5023 & 0.0058 & -0.0000 \\ 0.9314 & 0.0064 & 0.2987 & 0.0029 & 0.3508 & 0.0071 & 0.3802 & -0.0064 & 0.9823 & 0.0018 \\ -0.0039 & 1.0096 & -0.0250 & 0.7476 & -0.0117 & 0.6680 & -0.0128 & 70.6905 & -0.0092 & 1.0990 \end{pmatrix},$$

It can be verified that this system is consistent over the generalized centro-symmetric matrices and has the generalized centro-symmetric solution X^*, that is

$$X^* = \begin{pmatrix} 1.9003 & 0 & 0 & 0 & 0 & 0 & 0 & 0 & 0 & 0 \\ 0 & 1.5839 & 0 & 0 & 0 & 0 & 0 & 0 & 0 & 0 \\ 1.2137 & 0 & 1.6263 & 0 & 0 & 0 & 0 & 0 & 0 & 0 \\ 0 & 1.4764 & 0 & 1.8636 & 0 & 0 & 0 & 0 & 0 & 0 \\ 1.7826 & 0 & 0.2778 & 0 & 1.6636 & 0 & 0 & 0 & 0 & 0 \\ 0 & 0.8114 & 0 & 0.8373 & 0 & 1.3958 & 0 & 0 & 0 & 0 \\ 0.9129 & 0 & 0.3974 & 0 & 1.4189 & 0 & 0.6839 & 0 & 0 & 0 \\ 0 & 1.8338 & 0 & 1.0503 & 0 & 1.7200 & 0 & 0.8898 & 0 & 0 \\ 1.6428 & 0 & 0.5444 & 0 & 0.6092 & 0 & 0.6824 & 0 & 1.7515 & 0 \\ 0 & 1.7873 & 0 & 1.3443 & 0 & 1.1871 & 0 & 1.2426 & 0 & 1.9767 \end{pmatrix}$$

$$\in \mathbb{SR}_{P,P}^{10\times 10},$$

$$P = \begin{pmatrix} -1 & 0 & 0 & 0 & 0 & 0 & 0 & 0 & 0 & 0 \\ 0 & 1 & 0 & 0 & 0 & 0 & 0 & 0 & 0 & 0 \\ 0 & 0 & -1 & 0 & 0 & 0 & 0 & 0 & 0 & 0 \\ 0 & 0 & 0 & 1 & 0 & 0 & 0 & 0 & 0 & 0 \\ 0 & 0 & 0 & 0 & -1 & 0 & 0 & 0 & 0 & 0 \\ 0 & 0 & 0 & 0 & 0 & 1 & 0 & 0 & 0 & 0 \\ 0 & 0 & 0 & 0 & 0 & 0 & -1 & 0 & 0 & 0 \\ 0 & 0 & 0 & 0 & 0 & 0 & 0 & 1 & 0 & 0 \\ 0 & 0 & 0 & 0 & 0 & 0 & 0 & 0 & -1 & 0 \\ 0 & 0 & 0 & 0 & 0 & 0 & 0 & 0 & 0 & 1 \end{pmatrix} \in \mathbb{SOR}^{10\times 10}.$$

For this problem, we apply Algorithm 7 to compute $X(k)$ with the initial matrix $X(1) = 0$. In Figure 2, we give the obtained results with several values of δ where

$$r(k) = \log_{10}(||C_1 - A_1X(k)D_1|| + ||C_2 - A_2X(k)D_2||), \quad \delta(k) = \log_{10}\frac{||X(k) - X^*||}{||X^*||}.$$

It can be observed from Figure 2 that Algorithm 7 is effective. The effect of changing the convergence factor ω is illustrated in Figure 2. We see that the larger the convergence factor ω is, the faster the convergence the algorithm.

Example 3. *We consider the Sylvester matrix equation $AX + XB = C$ with*

$$A = B = M + 2rN + \frac{100}{(n+1)^2}I \in \mathbb{R}^{n\times n},$$

where

$$M = \text{tridiag}(-1, 2, -1) \in \mathbb{R}^{n\times n} \ \textit{and} \ N = \text{tridiag}(0.5, 0, -0.5) \in \mathbb{R}^{n\times n}.$$

Now we obtain the sequence of solution $\{X(i)\}$ by Algorithms 11 and 12. The obtained results are presented in Figures 3 and 4 where

$$r_i = \log_{10}||C - AX(i) - X(i)B||, \quad \delta_i = \log_{10}(||X(i) - X^*||/||X^*||). \tag{98}$$

The results show that Algorithms 11 and 12 are quite efficient.

Example 4. *We study the periodic Sylvester matrix equation*

$$\widetilde{X}_l + \widetilde{C}_l\widetilde{X}_{l+1}\widetilde{D}_l = \widetilde{F}_l \quad \text{for} \quad l = 1, 2, \tag{99}$$

where

$$\begin{aligned} \widetilde{C}_1 &= \text{tril}(\text{rand}(n, n), n) + \text{diag}(1.75 + \text{diag}(\text{rand}(n))), \\ \widetilde{D}_1 &= \text{triu}(\text{rand}(n, n), n) - \text{diag}(2 - \text{diag}(\text{rand}(n))), \\ \widetilde{C}_2 &= -\text{tril}(\text{rand}(n, n), n) - \text{diag}(3 + \text{diag}(\text{rand}(n))), \\ \widetilde{D}_2 &= \text{tril}(\text{rand}(n, n), n) - \text{diag}(2 + \text{diag}(\text{rand}(n))), \end{aligned}$$

and

$$\widetilde{E}_1 = \widetilde{E}_2 = \text{rand}(n, n).$$

When $n = 30$ we apply Algorithms 13 and 14 for computing the solutions of the the periodic Sylvester matrix equation (99). The numerical results are depicted in Figure 5 where

$$r_{1,i} = \log_{10}||\widetilde{E}_1 - \widetilde{X}_1(i) - \widetilde{C}_1\widetilde{X}_2(i)\widetilde{D}_1|| \quad \text{and} \quad r_{2,i} = \log_{10}||\widetilde{E}_2 - \widetilde{X}_2(i) - \widetilde{C}_2\widetilde{X}_1(i)\widetilde{D}_2||.$$

From Figure 5, we can see that Algorithms 13 and 14 are effective.

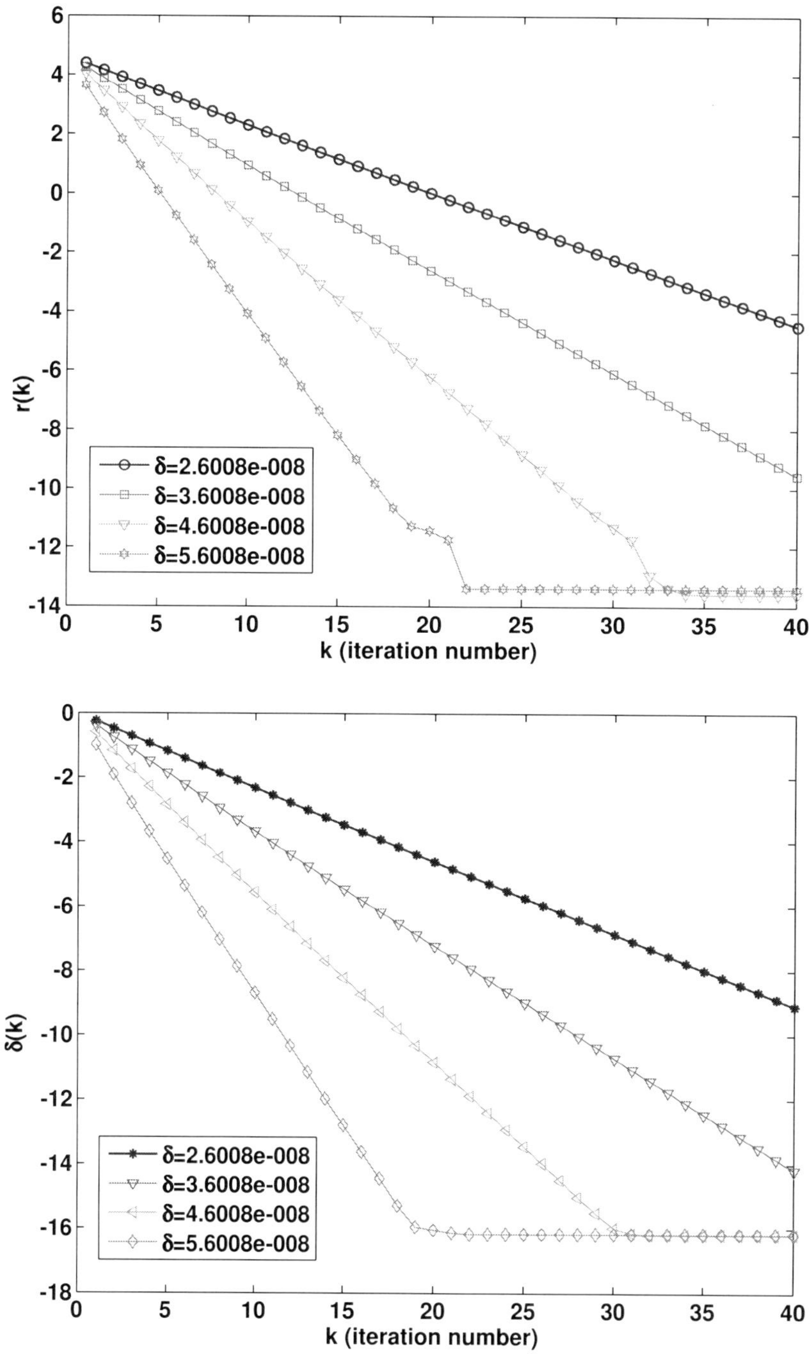

Figure 2. The obtained results for Example 2.

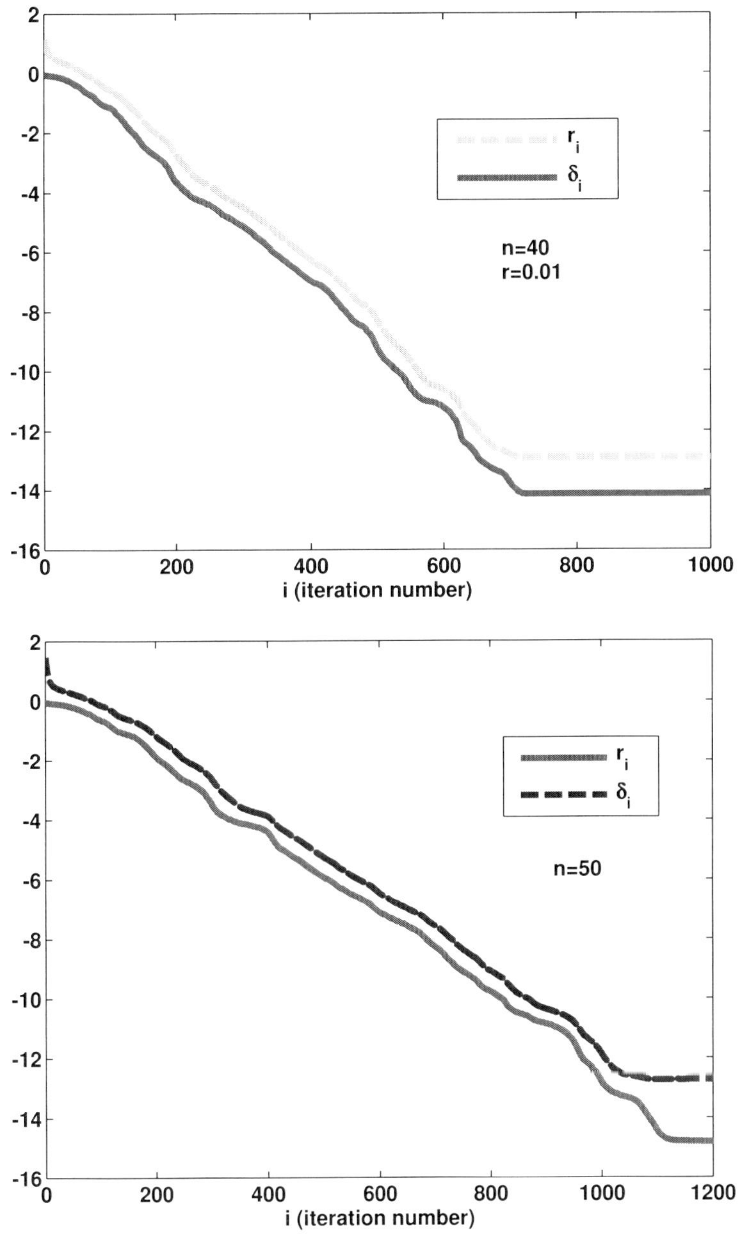

Figure 3. The obtained results for Example 3 with Algorithm 11.

Example 5. *As the final example, we consider the matrix equation*

$$AXB = C,$$

where

$$A = -\text{triu}(\text{rand}(n,n),1) + \text{diag}(1.5 + \text{diag}(\text{rand}(n))),$$

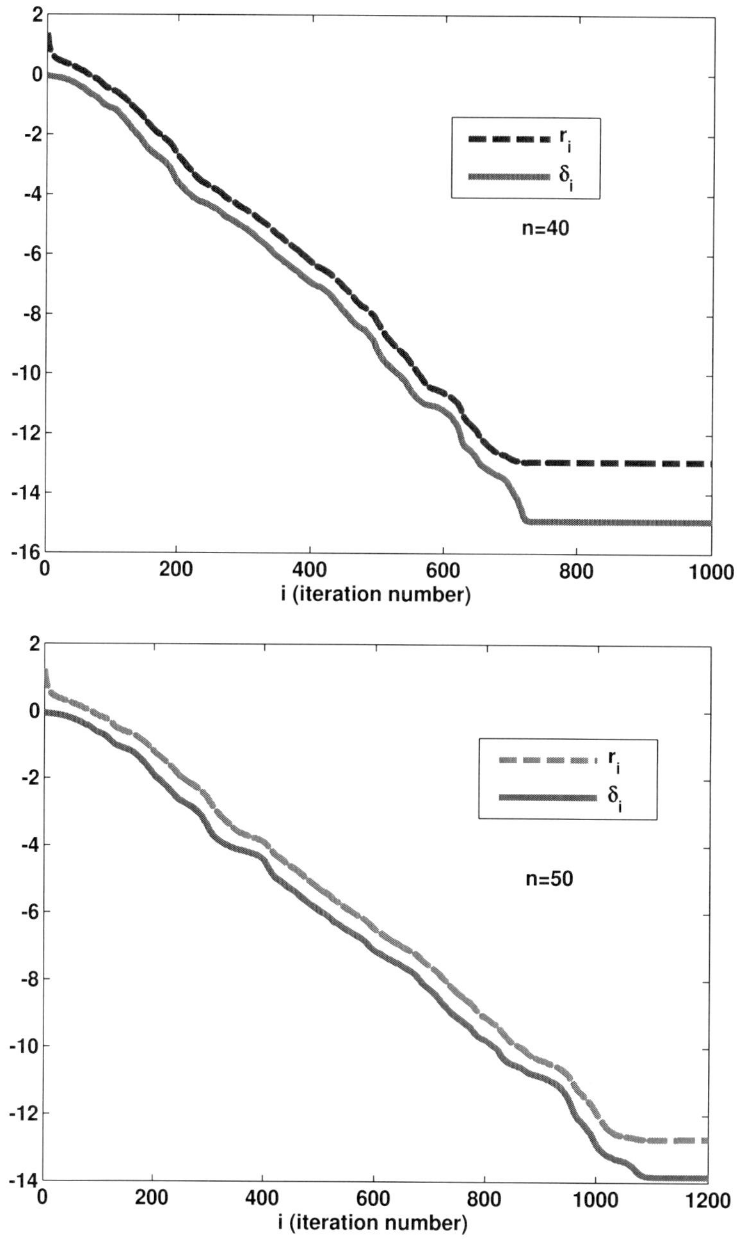

Figure 4. The obtained results for Example 3 with Algorithm 12.

$$B = \text{tril}(\text{rand}(n,n),1) + \text{diag}(1.25 + \text{diag}(\text{rand}(n))),$$

$$C = \text{rand}(n,n).$$

For $n = 50$ *we apply Algorithms 17, 18, 20, 22 and 24 for computing the solutions of this*

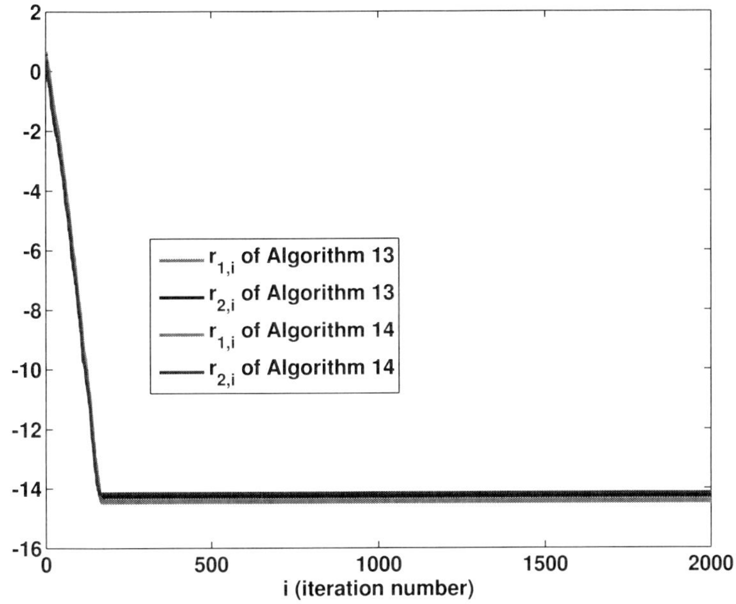

Figure 5. The obtained results for Example 5 with Algorithms 13 and 14.

matrix equation. The numerical results are presented in Figure 6 where

$$r(k) = \log_{10} ||C - AX(k)B||.$$

From Figure 6, we can see that Algorithms 17, 18, 20, 22 and 24 are efficient.

10. Conclusion

In summary, we have presented some of the latest developments in iterative algorithms for solving various matrix equations. The presented iterative algorithms were obtained by extending CGNE, LSQR, Bi-CG, Bi-CR, CGS, Bi-CGSTAB and QMRCGSTAB algorithms. Numerical results have indicated the effectiveness of the iterative algorithms.

References

[1] D. Bernstein, D. Hyland, The optimal projection equations for reduced-order state estimation, *IEEE Transactions on Automatic Control*, 30 (1985) 583-585.

[2] C. Brezinski, Projection methods for linear systems, *Journal of Computational and Applied Mathematics*, 77 (1997) 35-51.

[3] I. Borno, Parallel computation of the solutions of coupled algebraic Lyapunov equations, *Automatica*, 31 (1995) 1345-1347.

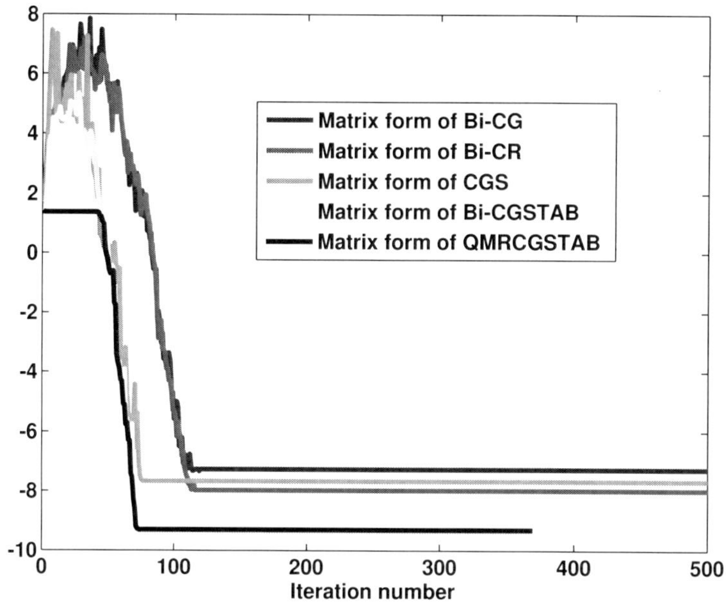

Figure 6. The obtained results for Example 5.

[4] T.F. Chan, E. Gallopoulos, V. Simoncini, T. Szeto, C.H. Tong, A Quasi-Minimal Residual Variant of the Bi-CGSTAB Algorithm for Nonsymmetric Systems, *SIAM Journal on Scientific Computing* 15 (1994) 338-347.

[5] H. Chan, J. Lam, D.W.C. Ho, Robust eigenvalue assignment in second-order systems: a gradient flow approach, *Optimal Control Applications and Methods*, 18 (1997) 283-296.

[6] H.C. Chen, Generalized reflexive matrices: special properties and applications, *SIAM Journal on Matrix Analysis and Applications*, 19 (1998) 140-153.

[7] E.J. Craig, The N-step iteration procedures, *Journal of Mathematical Physics*, 34 (1955) 64-73.

[8] M. Dehghan, M. Hajarian, The reflexive and anti-reflexive solutions of a linear matrix equation and systems of matrix equations, *Rocky Mountain Journal of Mathematics*, 40 (2010) 1-23.

[9] M. Dehghan, M. Hajarian, Two algorithms for the Hermitian reflexive and skew-Hermitian solutions of Sylvester matrix equations, *Applied Mathematics Letters*, 24 (2011) 444-449.

[10] M. Dehghan, M. Hajarian, The (R, S)-symmetric and (R, S)-skew symmetric solutions of the pair of matrix equations $A_1XB_1 = C_1$ and $A_2XB_2 = C_2$, *Bulletin of the Iranian Mathematical Society*, 37 (2011) 269-279.

[11] M. Dehghan, M. Hajarian, SSHI methods for solving general linear matrix equations, *Engineering Computations*, 28 (2012) 1028-1043.

[12] M. Dehghan, M. Hajarian, On the reflexive solutions of the matrix equation $AXB + CYD = E$, *Bulletin of the Korean Mathematical Society*, 46 (2009) 511–519.

[13] M. Dehghan and M. Hajarian, The general coupled matrix equations over generalized bisymmetric matrices, *Linear Algebra and its Applications*, 432, (2010) 1531-1552.

[14] M. Dehghan, M. Hajarian, On the generalized reflexive and anti-reflexive solutions to a system of matrix equations, *Linear Algebra and its Applications*, 437 (2012) 2793-2812.

[15] M. Dehghan, M. Hajarian, Analysis of an iterative algorithm to solve the generalized coupled Sylvester matrix equations, *Applied Mathematical Modelling*, 35 (2011) 3285-3300.

[16] M. Dehghan, M. Hajarian, On the generalized bisymmetric and skew-symmetric solutions of the system of generalized Sylvester matrix equations, *Linear and Multilinear Algebra*, 59 (2011) 1281-1309.

[17] M. Dehghan, M. Hajarian, Solving the generalized Sylvester matrix equation $\sum_{i=1}^{p} A_iXB_i + \sum_{j=1}^{q} C_jYD_j = E$ over reflexive and anti-reflexive matrices, *International Journal of Control, Automation and Systems*, 9 (2011) 118-124.

[18] M. Dehghan, M. Hajarian, The generalised Sylvester matrix equations over the generalised bisymmetric and skew-symmetric matrices, *International Journal of Systems Science*, 43 (2012) 1580-1590.

[19] M. Dehghan, M. Hajarian, An iterative algorithm for solving a pair of matrix equations $AYB = E$, $CYD = F$ over generalized centro-symmetric matrices, *Computers & Mathematics with Applications*, 56 (2008) 3246-3260.

[20] M. Dehghan, M. Hajarian, An iterative algorithm for the reflexive solutions of the generalized coupled Sylvester matrix equations and its optimal approximation, *Applied Mathematics and Computation*, 202 (2008) 571-588.

[21] M. Dehghan, M. Hajarian, Finite iterative algorithms for the reflexive and anti-reflexive solutions of the matrix equation $A_1X_1B_1 + A_2X_2B_2 = C$, *Mathematical and Computer Modelling*, 49 (2009) 1937-1959.

[22] M. Dehghan, M. Hajarian, Solving coupled matrix equations over generalized bisymmetric matrices, *International Journal of Control, Automation and Systems*, 10 (2012) 905-912.

[23] M. Dehghan, M. Hajarian, Iterative algorithms for the generalized centro-symmetric and central anti-symmetric solutions of general coupled matrix equations, *Engineering Computations*, 29 (2012) 528-560.

[24] M. Dehghan, M. Hajarian, Efficient iterative method for solving the second-order Sylvester matrix equation $EVF^2 - AVF - CV = BW$, *IET Control Theory & Applications*, 3 (2009) 1401-1408.

[25] F. Ding, P.X. Liu, J. Ding, Iterative solutions of the generalized Sylvester matrix equations by using the hierarchical identification principle, *Applied Mathematics and Computation*, 197 (2008) 41-50.

[26] F. Ding, T. Chen, Iterative least squares solutions of coupled Sylvester matrix equations, *Systems & Control Letters*, 54 (2005) 95-107.

[27] F. Ding, T. Chen, On iterative solutions of general coupled matrix equations, *SIAM Journal on Control and Optimization*, 44 (2006) 2269-2284.

[28] F. Ding, T. Chen, Gradient based iterative algorithms for solving a class of matrix equations, *IEEE Transactions on Automatic Control*, 50 (2005) 1216-1221.

[29] T.L. Doug, *Conjugate Gradient-Type Product Methods For Solving Nonsymmetric Linear Systems*, Ph.D. Thesis, 1994.

[30] G.R. Duan, Parametric eigenstructure assignment in high-order linear systems, *International Journal of Control, Automation and Systems*, 3 (2005) 419-429.

[31] G.R. Duan, R.J. Patton, Robust fault detection using Luenberger-type unknown input observers- a parametric approach, *International Journal of Systems Science*, 32 (2001) 533-540.

[32] G.R.Duan, B. Zhou, Solution to the second-order Sylvester matrix equation $MVF^2 - DVF - KV = BW$, *IEEE Transactions on Automatic Control*, 51 (2006) 805-809.

[33] L.R. Fletcher, J. Kuatsky, N.K. Nichols, Eigenstructure assignment in descriptor systems, *IEEE Transactions on Automatic Control*, 31 (1986) 1138-1141.

[34] R.W. Freund, A transpose-free quasi-minimal residual algorithm for non-Hermitian linear systems, *SIAM Journal on Scientific Computing*, 14 (1993) 470-482.

[35] M. Hajarian, A gradient-based iterative algorithm for generalized coupled Sylvester matrix equations over generalized centro-symmetric matrices, *Transactions of the Institute of Measurement and Control*, 36 (2014) 252-259.

[36] M. Hajarian, Efficient iterative solutions to general coupled matrix equations, *International Journal of Automation and Computing*, 10 (2013) 481-486.

[37] M. Hajarian, Developing Bi-CG and Bi-CR methods to solve generalized Sylvester-transpose matrix equations, *International Journal of Automation and Computing*, 11 (2014) 25-29.

[38] M. Hajarian, Matrix form of the CGS method for solving general coupled matrix equations, *Applied Mathematics Letters*, 34 (2014) 3742.

[39] M. Hajarian, Solving the general coupled and the periodic coupled matrix equations via the extended QMRCGSTAB algorithms, *Computational & Applied Mathematics*, 33 (2014) 349-362.

[40] M. Hajarian, Matrix form of the Bi-CGSTAB method for solving the coupled Sylvester matrix equations, *IET Control Theory & Applications*, 7 (2013) 1828-1833.

[41] M. Hajarian, Solving the system of linear operator equations over generalized bisymmetric matrices, *Transactions of the Institute of Measurement and Control*, 36 (2014) 541-550.

[42] M. Hajarian, Developing CGNE algorithm for the periodic discrete-time generalized coupled Sylvester matrix equations, *Computational & Applied Mathematics*, DOI: 10.1007/s40314-014-0138-7.

[43] M. Hajarian, M. Dehghan, The generalized centro-symmetric and least squares generalized centro-symmetric solutions of the matrix equation $AYB + CY^TD = E$, *Mathematical Methods in the Applied Sciences*, 34 (2011) 1562-1579.

[44] M. Hajarian, M. Dehghan, The reflexive and Hermitian reflexive solutions of the generalized Sylvester-conjugate matrix equation, *Bulletin of the Belgian Mathematical Society–Simon Stevin*, 20 (2013) 639-653.

[45] T. Hu, Z. Lin, J. Lam, Unified gradient approach to performance optimization under a pole assignment constrain, *Journal of Optimization Theory and Applications*, 121 (2004) 361-383.

[46] D.J. Inman, A. Kress, Eigenstructure assignment algorithm for second-order systems, *Journal of Guidance, Control, and Dynamics*, 22 (1999) 729-731.

[47] Y. Kim, H.S. Kim, Eigenstructure assignment algorithm for mechanical second-order systems, *Journal of Guidance, Control, and Dynamics*, 22 (1999) 729-731.

[48] I. Kyrchei, Determinantal representations of the Drazin inverse over the quaternion skew field with applications to some matrix equations, *Applied Mathematics and Computation*, 238 (2014) 193-207.

[49] I. Kyrchei, Explicit formulas for determinantal representations of the Drazin inverse solutions of some matrix and differential matrix equations, *Applied Mathematics and Computation*, 219 (2013) 7632-7644.

[50] I. Kyrchei, Explicit representation formulas for the minimum norm least squares solutions of some quaternion matrix equations, *Linear Algebra and its Applications*, 438 (2013) 136-152.

[51] I. Kyrchei, Analogs of Cramer's rule for the minimum norm least squares solutions of some matrix equations, *Applied Mathematics and Computation*, 218 (2012) 6375-6384.

[52] I.I. Kyrchei, Cramer's rule for some quaternion matrix equations, *Applied Mathematics and Computation*, 217 (2010) 2024-2030.

[53] L.P. LaSalle, S. Lefschetz, *Stability by Liapunov's Direct Method*, Academic Press, New York, 1961.

[54] A. Laub, A Schur method for solving algebraic Riccati equations, *IEEE Transactions on Automatic Control*, 24 (1979) 913-921.

[55] M. Mariton, *Jump Linear Systems in Automatic Control*, Marcel Dekker, New York, Basel, 1990.

[56] C.C. Paige, M.A. Saunders, LSQR: an algorithm for sparse linear equations and sparse least squares, *ACM Transactions on Mathematical Software*, 8 (1982) 43-47.

[57] Y.X. Peng, X.Y. Hu, L. Zhang, An iteration method for the symmetric solutions and the optimal approximation solution of the matrix equation $AXB = C$, *Applied Mathematics and Computation*, 160 (2005) 763-777.

[58] T. Sogabe, M. Sugihara, S.L. Zhang, An extension of the conjugate residual method to nonsymmetric linear systems, *Journal of Computational and Applied Mathematics*, 226 (2009) 103-113.

[59] P. Sonneveld, CGS, A fast Lanczos-type solver for nonsymmetric linear systems, *SIAM Journal on Scientific Computing*, 10 (1989) 3652.

[60] H.A. Van der Vorst, Bi-CGSTAB: a fast and smoothly converging variant of Bi-CG for the solution of nonsymmetric linear systems, *SIAM Journal on Scientific and Statistical Computing*, 13 (1992) 631-644.

[61] H.A. van der Vorst, *Iterative Krylov Methods for Large Linear Systems*, Cambridge University Press, Cambridge, 2003.

[62] Q.W. Wang, A system of matrix equations and a linear matrix equation over arbitrary regular rings with identity, *Linear Algebra and its Applications*, 384 (2004) 43-54.

[63] Q.W. Wang, C.K. Li, Ranks and the least-norm of the general solution to a system of quaternion matrix equations, *Linear Algebra and its Applications*, 430 (2009) 1626-1640.

[64] Q.W. Wang, H.X. Chang, Q.Ning, The common solution to six quaternion matrix equations with applications, *Applied Mathematics and Computation*, 198 (2008) 209-226.

[65] Q.W. Wang, F. Zhang, The reflexive re-nonnegative definite solution to a quaternion matrix equation, *Electronic Journal of Linear Algebra*, 17 (2008) 88-101.

[66] Q.W. Wang, H.S. Zhang, S.W. Yu, On solutions to the quaternion matrix equation $AXB + CYD = E$, *Electronic Journal of Linear Algebra*, 17 (2008) 343-358.

[67] Q.W. Wang, J.W. Woude, H.X. Chang, A system of real quaternion matrix equations with applications, *Linear Algebra and its Applications*, 431 (2009) 2291-2303.

[68] Q.W. Wang, C.K. Li, Ranks and the least-norm of the general solution to a system of quaternion matrix equations, *Linear Algebra and its Applications*, 430 (2009) 1626-1640.

[69] S.L. Zhang, A class of product-type Krylov-subspace methods for solving nonsymmetric linear systems, *Journal of Computational and Applied Mathematics*, 149 (2002) 297-305.

[70] B. Zhou, Z.Y. Li, G.R. Duan, Y. Wang, Solutions to a family of matrix equations by using the Kronecker matrix polynomials, *Applied Mathematics and Computation*, 212 (2009) 327-336.

[71] B. Zhou, J. Lam, G.R. Duan, On Smith-type iterative algorithms for the Stein matrix equation, *Applied Mathematics Letters*, 22 (2009) 1038-1044.

[72] B. Zhou, G.R. Duan, An explicit solution to the matrix equation $AX - XF = BY$, *Linear Algebra and its Applications*, 402 (2005) 345-366.

[73] B. Zhou, G.R. Duan, A new solution to the generalized Sylvester matrix equation $AV - EVF = BW$, *Systems & Control Letters*, 55 (2006) 193-198.

[74] B. Zhou, G.R. Duan, Solutions to generalized Sylvester matrix equation by Schur decomposition, *International Journal of Systems Science*, 38 (2007) 369-375.

[75] B. Zhou, Z.Y. Li, G.R. Duan, Y. Wang, Weighted least squares solutions to general coupled Sylvester matrix equations, *Journal of Computational and Applied Mathematics*, 224 (2009) 759-776.

[76] B. Zhou, Z.B. Yan, Solutions to right coprime factorizations and generalized Sylvester matrix equations, *Transactions of the Institute of Measurement and Control*, 30 (2008) 397-426.

[77] B. Zhou, G.R. Duan, On the generalized Sylvester mapping and matrix equations, *Systems & Control Letters*, 57 (2008) 200-208.

[78] B. Zhou, G.R. Duan, Z.Y. Li, Gradient based iterative algorithm for solving coupled matrix equations, *Systems & Control Letters*, 58 (2009) 327-333.

[79] B. Zhou, Z.Y. Li, G.R. Duan, Y. Wang, Solutions to a family of matrix equations by using the Kronecker matrix polynomials, *Applied Mathematics and Computation*, 212 (2009) 327-336.

[80] B. Zhou, J. Lamb, G.R. Duan. Toward solution of matrix equation $X = Af(X)B+C$, *Linear Algebra and its Applications*, 435 (2011) 1370-1398.

In: Advances in Linear Algebra Research
Editor: Ivan Kyrchei, pp. 287-297
ISBN: 978-1-63463-565-3

Chapter 8

Simultaneous Triangularization of a Pair of Matrices over a Principal Ideal Domain with Quadratic Minimal Polynomials

Volodymyr M. Prokip*
Pidstryhach Institute for Applied Problems
of Mechanics and Mathematics, Lviv, Ukraine

Abstract

The problems of existence of common eigenvectors and simultaneous triangularization of a pair of matrices over a principal ideal domain with quadratic minimal polynomials are investigated. The necessary and sufficient conditions of simultaneous triangularization of a pair of matrices with quadratic minimal polynomials are obtained. As a result, the approach offered provides the necessary and sufficient conditions of simultaneous triangularization of pairs of idempotent matrices and pairs of involutory matrices over a principal ideal domain.

Keywords: principal ideal domain; common eigenvector; simultaneous triangularization

AMS Subject Classification: 15A18, 15A21, 15A36

1. Introduction

Let R be a principal ideal domain with an identity element $e \neq 0$ and let $M_{m,n}(\mathrm{R})$ denote the set of $m \times n$ matrices over R. In what follows, I_n is the identity $n \times n$ matrix, $0_{m,k}$ is the zero $m \times k$ matrix, and O is a zero matrix whose order is obvious from the context. The symbol $[A, B]$ is the standard notation for the commutator $AB - BA$ of the matrices $A, B \in M_{n,n}(\mathrm{R})$.

*E-mail address: v.prokip@gmail.com

Matrices $A, B \in M_{n,n}(\mathrm{R})$ are said to be simultaneously triangularizable if there exists a matrix $U \in GL(n, \mathrm{R})$, such that

$$UAU^{-1} = \begin{bmatrix} \alpha_{11} & 0 & \dots & \dots & 0 \\ \alpha_{21} & \alpha_{22} & 0 & \dots & 0 \\ \dots & \dots & \dots & \dots & \dots \\ \alpha_{n1} & \alpha_{n,2} & \dots & \alpha_{n,n-1} & \alpha_{nn} \end{bmatrix}$$

and

$$UBU^{-1} = \begin{bmatrix} \beta_{11} & 0 & \dots & \dots & 0 \\ \beta_{21} & \beta_{22} & 0 & \dots & 0 \\ \dots & \dots & \dots & \dots & \dots \\ \beta_{n1} & \beta_{n,2} & \dots & \beta_{n,n-1} & \beta_{nn} \end{bmatrix}$$

are lower triangular matrices.

The question as to whether two complex matrices are simultaneously triangularizable is an old problem of linear algebra and has a long-standing history. In principle, it is solved by McCoy [1].

Theorem 1.1. *Let A and B be given $n \times n$ complex matrices. Matrices A and B are simultaneously triangularizable if and only if, for any polynomial $p(x, y)$ in noncommuting variables x and y, the matrix $p(A, B)(AB - BA)$ is nilpotent.*

The McCoy proof is complicated and contains no finite procedure for verifying the condition of the theorem. The more elementary proof is presented in [2] (see also [8, 10] and references therein).

Laffey [5] and Guralnick [6] have proved that if A and B are $n \times n$ matrices over any algebraically closed field and if $\operatorname{rank}(AB - BA) = 1$, then there exists an invertible matrix S such that SAS^{-1} and SBS^{-1} are both triangular. Laffey also gives the examples of pairs $\{A, B\}$ with $\operatorname{rank}(AB - BA) > 1$, where the conclusion does not hold. Several other interesting results were discussed in [9, 12, 14].

Simultaneous similarity and triangularization of matrices over integer domains were investigated in [15]. In this article a simple numerical criterion for the triangularization of sets of 2 by 2 matrices is proposed.

In this paper the problems of existence of common eigenvectors and simultaneous triangularization of a pair of matrices over a principal ideal domain with quadratic minimal polynomials are investigated. The paper is organized as follows. In Section 2 we present the necessary and sufficient conditions for diagonalizability of matrices of $M_{n,n}(\mathrm{R})$ with quadratic minimal polynomials. Using the results of the section 2, in Section 3 we indicate the conditions of existence of a common eigenvector for matrices over R with quadratic minimal polynomials. The necessary and sufficient conditions for the simultaneous triangularizability of two matrices over a principal ideal domain are established in section 4.

2. Preliminaries

If the matrix $A \in M_{n,n}(\mathrm{R})$ is similar to a diagonal matrix, then A is said to be diagonalizable. In other words, for A there exists a matrix $U \in GL(n, \mathrm{R})$, such that

$$UAU^{-1} = \operatorname{diag}(\alpha_1, \alpha_2, \dots, \alpha_n) \in M_{n,n}(\mathrm{R})$$

is a diagonal matrix. From this equality it follows, that the matrix $A \in M_{n,n}(\mathrm{R})$ is diagonalizable, then its characteristic polynomial $a(\lambda)$ admits the representation in the form of a product

$$a(\lambda) = \det(I_n\lambda - A) = (\lambda - \alpha_1)^{k_1}(\lambda - \alpha_2)^{k_2}\dots(\lambda - \alpha_r)^{k_r},$$

where $\alpha_i \in \mathrm{R}$, $i = 1, 2, \dots, r$ and $\alpha_i \neq \alpha_j$ for $i \neq j$. It is obvious that $A \in M_{n,n}(\mathrm{R})$ is a diagonalizable matrix, then its minimal polynomial $m(\lambda)$ has no multiple roots, i.e.,

$$m(\lambda) = (\lambda - \alpha_1)(\lambda - \alpha_2)\dots(\lambda - \alpha_r).$$

Let $\mathrm{R} = \mathbb{F}$ be a field. Then the last condition is necessary and sufficient for the diagonalizability of the matrix A over a field $\mathbb{F}$. It is easy to verify that this condition is not sufficient for the diagonalization of matrices over a commutative ring with an identity element (and over a principal ideal domain, in particular). Conditions under which a matrix with entries in a commutative ring can be reduced to a diagonal form by a similarity transformation were presented in [16, 17, 18]. In this section we present the necessary and sufficient conditions for the diagonalizability of matrices of $M_{n,n}(\mathrm{R})$ with quadratic minimal polynomials.

Let $A \in M_{n,n}(\mathrm{R})$ be an idempotent matrix, that is $A^2 = A$ and $1 \leq \operatorname{rank} A < n$. It is obvious that $m(\lambda) = \lambda(\lambda - e)$ is a minimal polynomial of the idempotent matrix A. It was proved in [3] that an idempotent matrix $A \in M_{n,n}(\mathrm{R})$ is diagonalized. In other words, for the idempotent matrix A, there exists a matrix $U \in GL(n, \mathrm{R})$ such that

$$UAU^{-1} = \left[\begin{array}{c|c} I_k & 0_{k,n-k} \\ \hline 0_{n-k,k} & 0_{n-k,n-k} \end{array}\right].$$

In what follows, we describe a structure of diagonalized matrices from $M_{n,n}(\mathrm{R})$ with minimal polynomial $m(\lambda) = (\lambda - \alpha)(\lambda - \beta)$, where $\alpha \neq \beta$.

Lemma 2.1. *Let $A \in M_{n,n}(\mathrm{R})$ be a matrix with characteristic polynomial $\det(I_n\lambda - A) = (\lambda - \alpha)^k(\lambda - \beta)^{n-k}$, where $1 \leq k < n$, $\alpha, \beta \in \mathrm{R}$ and $\alpha \neq \beta$. Then A is diagonalizable if and only if the two following conditions hold:*

(a) *$m(\lambda) = (\lambda - \alpha)(\lambda - \beta)$ is the minimal polynomial of the matrix A;*

(b) $(A - \alpha I_n) = 0_{n,n} \pmod{(\beta - \alpha)}$.

Proof. Let the matrix $A \in M_{n,n}(\mathrm{R})$ with characteristic polynomial

$$\det(I_n\lambda - A) = (\lambda - \alpha)_k(\lambda - \beta)_{n-k},$$

where $\alpha, \beta \in \mathrm{R}$, $\alpha \neq \beta$ and $1 \leq k < n$, be diagonalizable, i.e.,

$$UAU^{-1} = \left[\begin{array}{c|c} I_k\alpha & \mathrm{O} \\ \hline \mathrm{O} & \beta I_{n-k} \end{array}\right],$$

where $U \in GL(n, \mathrm{R})$. It is obvious that $m(\lambda) = (\lambda - \alpha)(\lambda - \beta)$ is the minimal polynomial of the matrix A. Then

$$A - \alpha I_n = U^{-1}\left[\begin{array}{c|c} \mathrm{O} & \mathrm{O} \\ \hline \mathrm{O} & (\beta - \alpha)I_{n-k} \end{array}\right]U =$$

$$U^{-1}\left[\begin{array}{c|c} \mathrm{O} & \mathrm{O} \\ \hline \mathrm{O} & I_{n-k} \end{array}\right]U(\beta - \alpha) = 0_{n,n} \pmod{(\beta - \alpha)}.$$

Conversely, let $A \in M_{n,n}(\mathrm{R})$ be a matrix with characteristic polynomial $\det(I_n\lambda - A) = (\lambda-\alpha)^k(\lambda-\beta)^{n-k}$, where $\alpha, \beta \in \mathrm{R}$, $\alpha \neq \beta, 1 \leq k < n$, and

$$(A - \alpha I_n) = 0_{n,n} \pmod{(\beta - \alpha)}. \tag{2.1}$$

From equality (2.1) it follows

$$A - I_n\alpha = (\beta - \alpha)P,$$

where $P \in M_{n,n}(\mathrm{R})$. Using the fact that $m(\lambda) = (\lambda-\alpha)(\lambda-\beta)$ is the minimal polynomial of the matrix A, we have

$$m(A) = (\alpha I_n - A)(\beta I_n - A) = (\beta - \alpha)^2(P^2 - P) = 0_{n,n}.$$

From the last equality it follows, that P is an idempotent matrix. Thus, the matrix A admits the representation in the form

$$A = I_n\alpha + (\beta - \alpha)P,$$

where P is the diagonalizable matrix. It is obvious that the matrix A is similar to the diagonal matrix $\left[\begin{array}{c|c} I_k\alpha & \mathrm{O} \\ \hline \mathrm{O} & \beta I_{n-k} \end{array}\right]$. The Lemma is proved. □

Corollary 2.1. *Let $A \in M_{n,n}(\mathrm{R})$ be a matrix with minimal polynomial $m(\lambda) = (\lambda - \alpha)(\lambda - \beta)$, where $\alpha, \beta \in \mathrm{R}$ and $\alpha \neq \beta$. If $(\alpha - \beta)$ is a divisor of unity in R, then the matrix A is diagonalizable.*

Corollary 2.2. *Let $A \in M_{n,n}(\mathrm{R})$ be a diagonalizable matrix with minimal polynomial $m(\lambda) = (\lambda - \alpha)(\lambda - \beta)$, where $\alpha, \beta \in \mathrm{R}$ and $\alpha \neq \beta$. Then for the matrix A there exists the unique pair of idempotent matrices $P_\alpha, P_\beta \in M_{n,n}(\mathrm{R})$ such that*

(a) $A = I_n\alpha + (\beta - \alpha)P_\beta$;

(b) $A = I_n\beta + (\alpha - \beta)P_\alpha$;

(c) $P_\alpha + P_\beta = I_n$;

(d) $A = \alpha P_\alpha + \beta P_\beta$.

Proof. Let $A \in M_{n,n}(\mathrm{R})$ be a diagonalized matrix with minimal polynomial $m(\lambda) = (\lambda-\alpha)(\lambda-\beta)$ and $\alpha \neq \beta$. From the proof of Lemma 2.1 we have that the matrix A admits a representation in the form

$$A = I_n\alpha + (\beta - \alpha)P_\beta, \tag{2.2}$$

where $P_\beta \in M_{n,n}(\mathrm{R})$ is an idempotent matrix. It is obvious that $I_n - P_\beta = P_\alpha$ is the idempotent matrix. We rewrite the equality (2.2) as

$$A = I_n\alpha + (\beta - \alpha)(I_n - P_\beta) = I_n\beta + (\alpha - \beta)P_\alpha.$$

By the relation $P_\alpha + P_\beta = I_n$, from (2.2) it follows that

$$A = (P_\alpha + P_\beta)\alpha + (\beta - \alpha)P_\beta = \alpha P_\alpha + \beta P_\beta.$$

For the matrix A we assume that, there exists another pair of idempotent matrices $Q_\alpha, Q_\beta \in M_{n,n}(\mathrm{R})$ that is different from the pair P_α, P_β, for which the conditions $Q_\alpha + Q_\beta = I_n$ and $A = \alpha Q_\alpha + \beta Q_\beta$ hold. Hence,

$$A = I_n\alpha + (\beta - \alpha)P_\beta = I_n\alpha + (\beta - \alpha)Q_\beta.$$

Since $\alpha \neq \beta$, the last equality yields $P_\beta = Q_\beta$. Analogously, we can show that $P_\alpha = Q_\alpha$. This completes the proof. □

3. Common Eigenvectors of Matrices with Quadratic Minimal Polynomials

It is said that the matrices $A, B \in M_{n,n}(\mathrm{R})$ have a common left eigenvector, if there exists a nonzero vector $\bar{u} \in M_{1,n}(\mathrm{R})$ such that $\bar{u}A = \bar{u}\alpha$ and $\bar{u}B = \bar{u}\beta$, where $\alpha, \beta \in \mathrm{R}$. Analogously, we introduce the notion of a common right eigenvector of the matrices A and B. It is obvious that if the matrices A and B have the common left eigenvector, then A and B have the common right eigenvector. Below, the term *common eigenvector* of the matrices A and B means that A and B have a common left eigenvector. It follows from the above consideration that the matrices A and B over the domain R can have a common eigenvector only in the case, if their characteristic polynomials $a(\lambda)$ and $b(\lambda)$ admit a representation in the forms $a(\lambda) = (\lambda - \alpha)c(\lambda)$ and $b(\lambda) = (\lambda - \beta)d(\lambda)$ respectively. We note that if R is a field, then the problem of the existence of a common eigenvector for the matrices A and B over a field was solved in [7] (see also [11], [13]). In this section we indicate the conditions of the existence of a common eigenvector for matrices over R with quadratic minimal polynomials.

Let A and B be $n \times n$ matrices over a principal ideal domain R chosen randomly. Then, with probability one, A and B do not have nontrivial common eigenvector. However, the situation can change if the choice of A and B is somehow restricted. In this respect, the following lemma, theorem and corollaries are of interest.

Lemma 3.1. *The idempotent matrices $A, B \in M_{n,n}(\mathrm{R})$ have a common eigenvector if and only if the commutator $[A, B]$ is a singular matrix.*

Proof. *Necessity.* Let $\bar{u} \in M_{1,n}(\mathrm{R})$ be a common eigenvector of the matrices $A, B \in M_{n,n}(\mathrm{R})$, that is

$$\bar{u}A = \bar{u}\alpha \qquad \text{and} \qquad \bar{u}B = \bar{u}\beta,$$

where $\alpha, \beta \in \mathrm{R}$. Then we have

$$\bar{u}\big(AB - BA\big) = \bar{u}\alpha B - \bar{u}\beta A = \bar{u}\big(\alpha\beta - \beta\alpha\big) = 0_{n,n}.$$

Since $\bar{u} \in M_{1,n}(\mathrm{R})$ is a nonzero vector, then from the last equality we have that the commutator $[A, B]$ is a singular matrix. The necessity is proved.

Sufficiency. Let for the idempotent matrices $A, B \in M_{n,n}(\mathrm{R})$ the commutator $[A, B]$ be a singular matrix. If one of the matrices A or B is the identity or zero matrix, then it is obvious that the matrices A and B have a common eigenvector.

We assume that $1 \leq \operatorname{rank} A = k < n$ and $1 \leq \operatorname{rank} B < n$. For the idempotent matrix A there exists a matrix $T \in GL(n, \mathrm{R})$ such that

$$TAT^{-1} = D = \begin{bmatrix} I_k & \mathrm{O} \\ \mathrm{O} & \mathrm{O} \end{bmatrix} \quad \text{and} \quad TBT^{-1} = C = \begin{bmatrix} C_{11} & C_{12} \\ C_{21} & C_{22} \end{bmatrix},$$

where $C_{11} \in M_k(\mathrm{R})$, $C_{12} \in M_{k,n-k}(\mathrm{R})$, $C_{21} \in M_{n-k,k}(\mathrm{R})$, $C_{22} \in M_{n-k}(\mathrm{R})$. It is obvious that the matrices A and B have a common eigenvector if and only if the matrices D and C have a common eigenvector.

Since $B^2 = B$, we have $C^2 = C$. The last equality yields

$$C_{11}C_{11} + C_{12}C_{21} = C_{11}, \tag{3.1}$$

$$C_{11}C_{12} + C_{12}C_{22} = C_{12}, \tag{3.2}$$

$$C_{21}C_{11} + C_{22}C_{21} = C_{21}, \tag{3.3}$$

$$C_{21}C_{12} + C_{22}C_{22} = C_{22}. \tag{3.4}$$

It is easily verified that $\operatorname{rank}\left[A, B\right] = \operatorname{rank}\left[D, C\right] < n$ and

$$[D, C] = \begin{bmatrix} \mathrm{O} & C_{12} \\ -C_{21} & \mathrm{O} \end{bmatrix}.$$

Since $\operatorname{rank}\left[D, C\right] < n$, we have that either $\operatorname{rank} C_{12} < k$ or $\operatorname{rank} C_{21} < n - k$.

A) Let $\operatorname{rank} C_{12} < k$. Then one of the following conditions is satisfied:

$$\operatorname{rank} \begin{bmatrix} C_{11} & C_{12} \end{bmatrix} < k \qquad \text{or} \qquad \operatorname{rank} \begin{bmatrix} C_{11} & C_{12} \end{bmatrix} = k.$$

If $\operatorname{rank} \begin{bmatrix} C_{11} & C_{12} \end{bmatrix} < k$, then there exists a nonzero vector $\bar{u} \in M_{1,k}(\mathrm{R})$ such that $\bar{u} \begin{bmatrix} C_{11} & C_{12} \end{bmatrix} = \bar{0}$ is the zero vector. Hence, the vector $\begin{bmatrix} \bar{u} & 0 & \dots & 0 \end{bmatrix} \in M_{1,n}(\mathrm{R})$ is the common left eigenvector of the matrices D and C.

Let $\operatorname{rank} \begin{bmatrix} C_{11} & C_{12} \end{bmatrix} = k$. Since $\operatorname{rank} C_{12} < k$, there exists a nonzero vector $\bar{u} \in M_{1,k}(\mathrm{R})$ such that $\bar{u}C_{12} = \bar{0}$ and $\bar{u}C_{11} \neq \bar{0}$. From equality (3.1) we obtain

$$(\bar{u}C_{11})\, C_{11} = \bar{u}C_{11}.$$

Similarly, from equality (3.2) it follows that

$$(\bar{u}C_{11})\, C_{12} = \bar{0}.$$

Hence, in this particular case the vector $\begin{bmatrix} \bar{u}C_{11} & 0 & \dots & 0 \end{bmatrix} \in M_{1,n}(\mathrm{R})$ is the common left eigenvector of the matrices D and C.

B) Let $\operatorname{rank} C_{21} < n - k$. Then one of the following conditions is satisfied:

$$\operatorname{rank} \begin{bmatrix} C_{21} & C_{22} \end{bmatrix} < n - k \qquad \text{or} \qquad \operatorname{rank} \begin{bmatrix} C_{21} & C_{22} \end{bmatrix} = n - k.$$

Let $\operatorname{rank} \begin{bmatrix} C_{21} & C_{22} \end{bmatrix} < n-k$. For the matrix $\begin{bmatrix} C_{21} & C_{22} \end{bmatrix}$ there exists a nonzero vector $\bar{u} \in M_{1,n-k}(\mathrm{R})$ such that

$$\bar{u} \begin{bmatrix} C_{21} & C_{22} \end{bmatrix} = O.$$

Hence, the vector $\begin{bmatrix} 0 & \dots & 0 & \bar{u} \end{bmatrix} \in M_{1,n}(\mathrm{R})$ is the common left eigenvector of the matrices D and C.

Let rank $\begin{bmatrix} C_{21} & C_{22} \end{bmatrix} = n - k$. Then there exists a nonzero vector $\bar{u} \in M_{1,n-k}(\mathrm{R})$ such that

$$\bar{u}C_{21} = \bar{0} \qquad \text{and} \qquad \bar{u}C_{22} \neq O.$$

From equality (3.4) we obtain $(\bar{u}C_{22})\, C_{22} = \bar{u}C_{22}$. Similarly, from equality (3.3) it follows that $(\bar{u}C_{22})\, C_{21} = \bar{0}$.

Hence, the vector $\begin{bmatrix} 0 & \dots & 0 & \bar{u}C_{22} \end{bmatrix} \in M_{1,n}(\mathrm{R})$ is the common left eigenvector of the matrices D and C. The Lemma is proved. □

Corollary 3.1. *Let $A, B \in M_{n,n}(\mathrm{R})$ be diagonalizable matrices with minimal polynomials $m_A(\lambda) = (\lambda - \alpha_1)(\lambda - \alpha_2)$ and $m_B(\lambda) = (\lambda - \beta_1)(\lambda - \beta_2)$ respectively, where $\alpha_i, \beta_i \in \mathrm{R}$, $\alpha_1 \neq \alpha_2$ and $\beta_1 \neq \beta_2$. The matrices A and B have a common eigenvector if and only if the commutator $[A, B]$ is a singular matrix.*

Proof. By Corollary 2.1, for diagonalizable matrices $A, B \in M_{n,n}(\mathrm{R})$ with minimal polynomials $m_A(\lambda) = (\lambda - \alpha_1)(\lambda - \alpha_2)$ and $m_B(\lambda) = (\lambda - \beta_1)(\lambda - \beta_2)$ respectively, there exist the idempotent matrices $P, Q \in M_{n,n}(\mathrm{R})$ such that

$$A = I_n\alpha_1 + (\alpha_2 - \alpha_1)P \qquad \text{and} \qquad B = I_n\beta_1 + (\beta_2 - \beta_1)Q.$$

Hence, the matrices A and B have a common eigenvector if and only if the idempotent matrices P and Q have a common eigenvector. Since $(\alpha_2 - \alpha_1)(\beta_2 - \beta_1) \neq 0$, it is easy to see that

$$\operatorname{rank} [A, B] = \operatorname{rank} [P, Q].$$

By Lemma 3.1, the matrices P and Q have a common eigenvector if and only if the commutator $[P, Q]$ is a singular matrix. This completes the proof. □

Theorem 3.1. *Let $A, B \in M_{n,n}(\mathrm{R})$ be the matrices with minimal polynomials $m_A(\lambda) = (\lambda - \alpha_1)(\lambda - \alpha_2)$ and $m_B(\lambda) = (\lambda - \beta_1)(\lambda - \beta_2)$ respectively, where $\alpha_i, \beta_i \in \mathrm{R}$, $\alpha_1 \neq \alpha_2$ and $\beta_1 \neq \beta_2$. The pair of matrices $A, B \in M_{n,n}(\mathrm{R})$ have a common eigenvector over R if and only if the commutator $[A, B]$ is a singular matrix.*

Proof. *Necessity* is evident.

Sufficiency. Let for the idempotent matrices $A, B \in M_{n,n}(\mathrm{R})$ the commutator $[A, B]$ be a singular matrix. Further, let $\mathbb{F}$ be a quotient field of the domain R, that is $\mathrm{R} \subset \mathbb{F}$. It is clear, that $m_A(\lambda)$ and $m_B(\lambda)$ are minimal polynomials of the matrices A and B over $\mathbb{F}$, respectively. Thus, matrices A and B over $\mathbb{F}$ are diagonalizable. By Corollary 3.1 from inequality $\operatorname{rank} [A, B] < n$ it follows that for the matrices A and B there exists a common eigenvector $\overline{u} \in M_{1,n}(\mathbb{F})$. Now we write the vector $\overline{u}$ in the form

$$\overline{u} = \begin{bmatrix} \dfrac{u_1}{v_1} & \dfrac{u_2}{v_2} & \dots & \dfrac{u_n}{v_n} \end{bmatrix},$$

where $u_i \in \mathrm{R}$ and $v_i \in \mathrm{R}\backslash\{0\}$, $i = 1, 2, \dots, n$. Put $v = \prod_{i=1}^{n} v_i$. It is obvious that the vector $\overline{u}_0 = v \cdot \overline{u} \in M_{1,n}(\mathrm{R})$ is the common eigenvector of the matrices A and B over R. The proof of Theorem 3.1 is completed. □

Let $A \in M_{n,n}(\mathrm{R})$ be an involutory matrix, that is $A^2 = I_n$. If $A \neq \pm I_n$, then $m(\lambda) = (\lambda - e)(\lambda + e)$ is the minimal polynomial of an involutory matrix A. The impotent application of Theorem 3.1 is presented in the following corollary.

Corollary 3.2. *The involutory matrices $A, B \in M_{n,n}(\mathrm{R})$ have a common eigenvector if and only if the commutator $\left[A, B\right]$ is a singular matrix.*

Let $\overline{v} = \left[\begin{array}{cccc} v_1 & v_2 & \ldots & v_n \end{array}\right] \in M_{1,n}(\mathrm{R})$ be a vector. If $e = \gcd\,(v_1\,,\, v_2\,,\, \ldots\,,\, v_n)$ is the greatest common divisor of elements $v_1\,,\, v_2\,,\, \ldots\,,\, v_n$, then $\overline{v}$ is called an unimodular vector.

Suppose that $\overline{w} = \left[\begin{array}{cccc} w_1 & w_2 & \ldots & w_n \end{array}\right] \in M_{1,n}(\mathrm{R})$ is a common eigenvector of matrices $A, B \in M_{n,n}(\mathrm{R})$ and $\gcd\,(w_1\,,\, w_2\,,\, \ldots\,,\, w_n) = d \neq e$. Then $w_i = du_i$, where $u_i \in \mathrm{R}$ for all $i = 1, 2, \ldots, n$, and $\gcd\,(u_1\,,\, u_2\,,\, \ldots\,,\, u_n) = e$. Obviously, that $\overline{u} = \left[\begin{array}{cccc} u_1 & u_2 & \ldots & u_n \end{array}\right]$ is an unimodular vector and $\overline{w} = d \cdot \overline{u}$. This means that the unimodular vector $\overline{u}$ is a common eigenvector of matrices $A, B \in M_{n,n}(\mathrm{R})$ also. We record this fact as a remark to Theorem 3.1.

Remark 1. *Let a pair of matrices $A, B \in M_{n,n}(\mathrm{R})$ have a common eigenvector over* R. *Then for matrices A and B there exists an unimodular common eigenvector over* R.

4. Simultaneous Triangularization of a Pair of Matrices with Quadratic Minimal Polynomials

In this section, we present new, relatively simple necessary and sufficient conditions for the simultaneous triangularization of a pair of matrices over a principal ideal domain with quadratic minimal polynomials. Let $\mathbb{F}$ be an algebraically closed field. If matrices A and B over a field $\mathbb{F}$ commute, then their simultaneous triangulability is a known fact in linear algebra (see [8], Chapter 2). It is easy to prove the following propositions.

Proposition 4.1. *Let $A, B \in M_{n,n}(\mathrm{R})$ be matrices with characteristic polynomials*

$$a(\lambda) = \det(I_n\lambda - A) = (\lambda - \alpha_1)^{k_1}(\lambda - \alpha_2)^{k_2}\ldots(\lambda - \alpha_r)^{k_r},\ \alpha_i \in \mathrm{R}, i = 1, 2, \ldots, r;$$

and

$$b(\lambda) = \det(I_n\lambda - B) = (\lambda - \beta_1)^{k_1}(\lambda - \beta_2)^{k_2}\ldots(\lambda - \beta_r)^{k_l},\ \beta_j \in \mathrm{R}, j = 1, 2, \ldots, l,$$

respectively. If matrices A and B commute, then A and B are simultaneously triangularizable.

Proposition 4.2. *Let matrices $A, B \in M_{n,n}(\mathrm{R})$ be simultaneously triangularizable. Then the commutator $\left[A, B\right]$ is a nilpotent matrix.*

The following question is natural: for which classes of matrices $A, B \in M_{n,n}(\mathrm{R})$ the condition of Proposition 4.2 is sufficient as well. The following theorem and its corollaries specify the classes of such pair of matrices.

Theorem 4.1. *Let $A, B \in M_{n,n}(\mathrm{R})$ be the matrices with minimal polynomials $m_A(\lambda) = (\lambda-\alpha_1)(\lambda-\alpha_2)$ and $m_B(\lambda) = (\lambda-\beta_1)(\lambda-\beta_2)$ respectively, where $\alpha_i, \beta_i \in \mathrm{R}$, $\alpha_1 \neq \alpha_2$ and $\beta_1 \neq \beta_2$. The pair of matrices $A, B \in M_{n,n}(\mathrm{R})$ are simultaneously triangularizable if and only if the commutator $\left[A, B\right]$ is a nilpotent matrix.*

Proof. The *necessity* follows from Proposition 4.2.

Sufficiency. Put $\Lambda_A = \{\alpha_1, \alpha_2\}$ and $\Lambda_B = \{\beta_1, \beta_2\}$. Since $[A, B]$ is a nilpotent matrix, then by Theorem 3.1 and Remark 1, for the matrices A and B there exists an unimodular common eigenvector $\overline{u}_1 \in M_{1,n}(\mathrm{R})$, that is

$$\bar{u}_1 A = \bar{u}_1 \alpha_{11} \qquad \text{and} \qquad \bar{u}_1 B = \bar{u}_1 \beta_{11},$$

where $\alpha_{11} \in \Lambda_A$, $\beta_{11} \in \Lambda_B$. For the vector $\overline{u}_1$ there exists a matrix $U_1 \in GL(n, \mathrm{R})$ with first row $\overline{u}_1$ (see [4], Chapter II), that is $U_1 = \begin{bmatrix} \overline{u}_1 \\ * \end{bmatrix}$. For the matrix U_1 the following relations holds:

$$U_1 A U_1^{-1} = \left[\begin{array}{c|ccc} \alpha_{11} & 0 & \cdots & 0 \\ \hline * & & A_1 & \end{array}\right] \quad \text{and} \quad U_1 B U_1^{-1} = \left[\begin{array}{c|ccc} \beta_{11} & 0 & \cdots & 0 \\ \hline * & & B_1 & \end{array}\right],$$

where $A, B \in M_{n-1,n-1}(\mathrm{R})$. From equality

$$U_1 \left[A, B\right] U_1^{-1} = \left[\begin{array}{c|ccc} 0 & 0 & \cdots & 0 \\ \hline * & & \left[A_1, B_1\right] & \end{array}\right]$$

it follows that $\left[A_1, B_1\right]$ is a nilpotent matrix.

By Theorem 3.1 and Remark 1, for the matrices A_1 and B_1 there exists an unimodular common eigenvector $\overline{u}_2 \in M_{1,n-1}(\mathrm{R})$, that is

$$\bar{u}_2 A_1 = \bar{u}_2 \alpha_{22} \qquad \text{and} \qquad \bar{u}_2 B_1 = \bar{u}_2 \beta_{22},$$

where $\alpha_{22} \in \Lambda_A$, $\beta_{22} \in \Lambda_B$. Similarly, for the vector $\overline{u}_2$ there exists a matrix $U_2 = \begin{bmatrix} \overline{u}_2 \\ * \end{bmatrix} \in GL(n-1, \mathrm{R})$ and for the matrix U_2 the following relations holds:

$$U_2 A_1 U_2^{-1} = \left[\begin{array}{c|ccc} \alpha_{22} & 0 & \cdots & 0 \\ \hline * & & A_2 & \end{array}\right] \quad \text{and} \quad U_2 B_1 U_2^{-1} = \left[\begin{array}{c|ccc} \beta_{22} & 0 & \cdots & 0 \\ \hline * & & B_2 & \end{array}\right],$$

where $\alpha_{22} \in \Lambda_A$, $\beta_{22} \in \Lambda_B$ and $A_2, B_2 \in M_{n-2,n-2}(\mathrm{R})$. It is clear that $\left[A_2, B_2\right]$ is a nilpotent matrix.

Put $U_{21} = \mathrm{diag}\left(e, \;\; U_2\right) U_1 \in GL(n, \mathrm{R})$. It is easy to see that

$$U_{21} A U_{21}^{-1} = \left[\begin{array}{cc|ccc} \alpha_{11} & 0 & 0 & \cdots & 0 \\ \alpha_{21} & \alpha_{22} & 0 & \cdots & 0 \\ \hline * & * & & A_2 & \end{array}\right]$$

and

$$U_{21}BU_{21}^{-1} = \left[\begin{array}{cc|ccc} \beta_{11} & 0 & 0 & \dots & 0 \\ \beta_{21} & \beta_{22} & 0 & \dots & 0 \\ \hline * & * & & B_2 & \end{array}\right].$$

Continuing these procedure further, after a finite number of steps we obtain that for the matrices A and B there exists a matrix $U \in GL(n, \mathrm{R})$ such that UAU^{-1} and UBU^{-1} are lower triangular matrices. This completes the proof of Theorem 4.1. □

The following results are immediate consequences of Theorem 4.1.

Corollary 4.1. *The idempotent matrices $A, B \in M_{n,n}(\mathrm{R})$ are simultaneously triangularizable if and only if the commutator $[A, B]$ is a nilpotent matrix.*

Corollary 4.2. *The involutory matrices $A, B \in M_{n,n}(\mathrm{R})$ are simultaneously triangularizable if and only if the commutator $[A, B]$ is a nilpotent matrix.*

Conclusion

We note that the obtained results are true for matrices over the domains of elementary divisors. In addition, some of them can be extended for matrices over the ID-rings [3], i.e., over the commutative rings with an identity element, over which an idempotent matrix is diagonalized. It is noted that the simultaneous triangularization of the set of matrices over commutative rings with an identity element (and over the principal ideal domain R, in particular) is an open problem.

References

[1] N.H. McCoy, On the characteristic roots of matrix polynomials, *Bull. Amer. Math. Soc.* 42 (1936) 592–600.

[2] M. P. Drazin, J. W. Dungey, and K. W. Gruenberg, Some theorems on commutative matrices, *J. London Math. Soc.* 26 (1951) 221–228.

[3] A. Steger, Diagonability of idempotent matrices, *Pacific J. Math.* 19 (1966) 535–542.

[4] M.Newman, *Integer Matrices*, Acad. Press, New York (1972).

[5] T.J. Laffey. Simultaneous triangularization of matrices – low rank cases and the non-derogatory case, *Linear and Multilinear Algebra* 6 (1978) 269–305.

[6] R. M. Guralnick, A note on pairs of matrices with rank one commutator, *Linear and Multilinear Algebra* 8 (1979) 97–99.

[7] D.Shemesh, Common eigenvectors of two matrices, *Linear Algebra and Appl.* 62 (1984) 11–18.

[8] R. A. Horn, C. R. Johnson, *Matrix Analysis*, Cambridge University Press, Cambridge (1985).

[9] G.Szep, Simultaneous triangularization of projector matrices, *Acta Math. Hung.* 48 (1986) 285–288.

[10] V. V. Prasolov, Problems and Theorems in Linear Algebra, *Amer. Math. Soc.*, Providence, R.I. (1994).

[11] M. Tsatsomeros, A criterion for the existence of common invariant subspaces of matrices, *Linear Algebra and Appl.* 322 (1999) 51–59.

[12] Yu. A. Alpin, N. A. Koreshkov, On the Simultaneous Triangulability of Matrices, *Mathematical Notes* 431 (2000) 552–555.

[13] A. George and Kh.D. Ikramov, Common invariant subspaces of two matrices, *Linear Algebra Appl.* 287 (2001) 171–179.

[14] C. Dubi, An algorithmic approach to simultaneous triangularization, *Linear Algebra Appl.* 430 (2009) 2975–2981.

[15] Carlos A.A. Florentino, Simultaneous similarity and triangularization of sets of 2 by 2 matrices, *Linear Algebra Appl.* 431 (2009) 1652–1674.

[16] V. M. Prokip, Diagonalization of matrices over the domain of principal ideals with minimal polynomial $m(\lambda) = (\lambda - \alpha)(\lambda - \beta)$, $\alpha \neq \beta$, *J. Math. Sciences* 174 (2011) 481–485.

[17] V. M. Prokip, Diagonalizability of matrices over a principal ideal domain, *Ukrainian Math. J.* 64 (2012) 316–323.

[18] Dan Laksov, Diagonalization of matrices over rings, *J. Algebra* 376 (2013) 123–138.

In: Advances in Linear Algebra Research
Editor: Ivan Kyrchei, pp. 299-324
ISBN: 978-1-63463-565-3

Chapter 9

Relation of Row-Column Determinants with Quasideterminants of Matrices over a Quaternion Algebra

Aleks Kleyn[1,*] ***and Ivan I. Kyrchei***[2,†]
[1]American Mathematical Society, USA
[2]Pidstryhach Institute for Applied Problems
of Mechanics and Mathematics, Lviv, Ukraine

Abstract

Since product of quaternions is noncommutative, there is a problem how to determine a determinant of a matrix with noncommutative elements (it's called a noncommutative determinant). We consider two approaches to define a noncommutative determinant. Primarily, there are row – column determinants that are an extension of the classical definition of the determinant; however we assume predetermined order of elements in each of the terms of the determinant. In the chapter we extend the concept of an immanant (permanent, determinant) to a split quaternion algebra using methods of the theory of the row and column determinants.

Properties of the determinant of a Hermitian matrix are established. Based on these properties, analogs of the classical adjont matrix over a quaternion skew field have been obtained. As a result we have a solution of a system of linear equations over a quaternion division algebra according to Cramer's rule by using row–column determinants.

Quasideterminants appeared from the analysis of the procedure of a matrix inversion. By using quasideterminants, solving of a system of linear equations over a quaternion division algebra is similar to the Gauss elimination method.

The common feature in definition of row and column determinants and quasideterminants is that we have not one determinant of a quadratic matrix of order n with noncommutative entries, but certain set (there are n^2 quasideterminants, n row determinants, and n column determinants). We have obtained a relation of row-column determinants with quasideterminants of a matrix over a quaternion division algebra.

*E-mail address: Aleks_Kleyn@MailAPS.org
†E-mail address: kyrchei@online.ua

Keywords: quaternion algebra, immanant, permanent, determinant, quasideterminant, system of linear equations, Cramer's rule

AMS abs Subject Classification: 16K20, 15A15, 15A06

1. Introduction

Linear algebra is a powerful tool that we use in different areas of mathematics, including the calculus, the analytic and differential geometry, the theory of differential equations, and the optimal control theory. Linear algebra has accumulated a rich set of different methods. Since some methods have a common final result, this gives us the opportunity to choose the most effective method, depending on the nature of calculations.

At transition from linear algebra over a field to linear algebra over a division ring, we want to save as much as possible tools that we regularly use. Already in the early XX century, shortly after Hamilton created a quaternion algebra, mathematicians began to search the answer how looks like the algebra with noncommutative multiplication. In particular, there is a problem how to determine a determinant of a matrix with elements belonging to a noncommutative ring. Such determinant is also called a noncommutative determinant.

There were a lot of approaches to the definition of the noncommutative determinant. However none of the introduced noncommutative determinants maintained all those properties that determinant possessed for matrices over a field. Moreover, in paper [1], J. Fan proved that there is no unique definition of determinant which would expands the definition of determinant of real matrices for matrices over the division ring of quaternions. Therefore, search for a solution of the problem to define a noncommutative determinant is still going on.

In this chapter, we consider two approaches to define noncommutative determinant. Namely, we explore row-column determinants and quasideterminant.

Row-column determinants are an extension of the classical definition of the determinant, however we assume predetermined order of elements in each of the terms of the determinant. Using row-column determinants, we obtain a solution of a system of linear equations over a quaternion division algebra according to Cramer's rule.

Quasideterminant appeared from the analysis of the procedure of a matrix inversion. Using quasideterminant, solving of a system of linear equations over a quaternion division algebra is similar to the Gauss elimination method.

There is common in definition of row and column determinants and quasideterminant. In both cases, we have not one determinant in correspondence to quadratic matrix of order n with noncommutative entries, but certain set (there are n^2 quasideterminant, n row determinants, and n column determinants).

Today there is wide application of quasideterminants in linear algebra ([2, 3]), and in physics ([4, 5, 6]). Row and column determinants ([7, 8]) introduced relatively recently are less well known. Purpose of the chapter is establishment of a relation of row-column determinants with quasideterminants of a matrix over a quaternion algebra. The authors are hopeful that the establishment of this relation can provide mutual development of both the theory of quasideterminants and the theory of row-column determinants.

1.1. Convention about Notations

There are different forms to write elements of a matrix. In this paper, we denote a_{ij} an element of the matrix $\mathbf{A}$. The index i labels rows, and the index j labels columns.

We use the following notation for different minors of the matrix $\mathbf{A}$.

$\mathbf{a}_{i\,.}$ the i-th row

$\mathbf{A}_{S\,.}$ the minor obtained from $\mathbf{A}$ by selecting rows with index from the set S

$\mathbf{A}^{i\,.}$ the minor obtained from $\mathbf{A}$ by deleting row $\mathbf{a}_{i\,.}$

$\mathbf{A}^{S\,.}$ the minor obtained from $\mathbf{A}$ by deleting rows with index from the set S

$\mathbf{a}_{.\,j}$ the j-th column

$\mathbf{A}_{.\,T}$ the minor obtained from $\mathbf{A}$ by selecting columns with index from the set T

$\mathbf{A}^{.\,j}$ the minor obtained from $\mathbf{A}$ by deleting column $\mathbf{a}_{.\,j}$

$\mathbf{A}^{.\,T}$ the minor obtained from $\mathbf{A}$ by deleting columns with index from the set T

$\mathbf{A}_{.j}\left(\mathbf{b}\right)$ the matrix obtained from $\mathbf{A}$ by replacing its j-th column by the column $\mathbf{b}$

$\mathbf{A}_{i.}\left(\mathbf{b}\right)$ the matrix obtained from $\mathbf{A}$ by replacing its i-th row by the row $\mathbf{b}$

Considered notations can be combined. For instance, the record

$$\mathbf{A}^{ii}_{k.}\left(\mathbf{b}\right)$$

means replacing of the k-th row by the vector $\mathbf{b}$ followed by removal of both the i-th row and the i-th column.

As was noted in section 2.2 of the paper [9], we can define two types of matrix products: either product of rows of first matrix over columns of second one, or product of columns of first matrix over rows of second one. However, according to the theorem 2.2.5 in the paper [9], this product is symmetric relative operation of transposition. Hence in the chapter, we will restrict ourselves by traditional product of rows of first matrix over columns of second one; and we do not indicate clearly the operation like it was done in [9].

1.2. Preliminaries. A Brief Overview of the Theory of Noncommutative Determinants

Theory of determinants of matrices with noncommutative elements can be divided into three groups regarding their methods of definition. Denote $\mathrm{M}(n, \mathbf{K})$ the ring of matrices with elements from the ring $\mathbf{K}$. One of the ways to determine determinant of a matrix of $\mathrm{M}\left(n, \mathbf{K}\right)$ is following ([11, 12, 13]).

Definition 1.1. *Let the functional*

$$\mathrm{d} : \mathrm{M}\left(n, \mathbf{K}\right) \to \mathbf{K}$$

satisfy the following axioms.

Axiom 1. $\mathrm{d}(\mathbf{A}) = 0$ *iff* $\mathbf{A}$ *is singular (irreversible).*

Axiom 2. $\forall \mathbf{A}, \mathbf{B} \in \mathrm{M}(n, \mathbf{K})$, $\mathrm{d}(\mathbf{A} \cdot \mathbf{B}) = \mathrm{d}(\mathbf{A}) \cdot \mathrm{d}(\mathbf{B})$.

Axiom 3. *If we obtain a matrix* $\mathbf{A}'$ *from matrix* $\mathbf{A}$ *either by adding of an arbitrary row multiplied on the left with its another row or by adding of an arbitrary column multiplied on the right with its another column, then*

$$\mathrm{d}(\mathbf{A}') = \mathrm{d}(\mathbf{A})$$

Then the value of the functional d *is called determinant of* $\mathbf{A} \in \mathrm{M}(n, \mathbf{K})$. □

The known determinants of Dieudonné and Study are examples of such functionals. Aslaksen [11] proved that determinants which satisfy Axioms 1, 2 and 3 take their value in some commutative subset of the ring. It makes no sense for them such property of conventional determinants as the expansion along an arbitrary row or column. Therefore a determinantal representation of an inverse matrix using only these determinants is impossible. This is the reason that causes to introduce determinant functionals that do not satisfy all Axioms. Dyson [13] considers Axiom 1 as necessary to determine a determinant.

In another approach, a determinant of a square matrix over a noncommutative ring is considered as a rational function of entries of a matrix. The greatest success is achieved by Gelfand and Retakh [14, 15, 16, 17] in the theory of quasideterminants. We present introduction to the theory of quasideterminants in the section 5.

In third approach, a determinant of a square matrix over a noncommutative ring is considered as an alternating sum of $n!$ products of entries of a matrix. However, it assumed certain fixed order of factors in each term. E. H. Moore was first who achieved implementation of the key Axiom 1 using such definition of a noncommutative determinant. Moore had done this not for all square matrices, but only for Hermitian. He defined the determinant of a Hermitian matrix[1] $\mathbf{A} = (a_{ij})_{n \times n}$ over a division ring with involution by induction over n following way (see [13])

$$\mathrm{Mdet}\mathbf{A} = \begin{cases} a_{11}, & n = 1 \\ \sum\limits_{j=1}^{n} \varepsilon_{ij} a_{ij} \mathrm{Mdet}\left(\mathbf{A}(i \to j)\right), & n > 1 \end{cases} \tag{1.1}$$

Here $\varepsilon_{kj} = \begin{cases} 1, & i = j \\ -1, & i \neq j \end{cases}$, and $\mathbf{A}(i \to j)$ denotes the matrix obtained from $\mathbf{A}$ by replacing its j-th column with the i-th column and then by deleting both the i-th row and column. Another definition of this determinant is presented in [11] by using permutations,

$$\mathrm{Mdet}\,\mathbf{A} = \sum_{\sigma \in S_n} |\sigma| a_{n_{11} n_{12}} \cdot \ldots \cdot a_{n_{1l_1} n_{11}} \cdot a_{n_{21} n_{22}} \cdot \ldots \cdot a_{n_{rl_1} n_{r1}} .$$

Here S_n is symmetric group of n elements. A cycle decomposition of a permutation σ has form,

$$\sigma = (n_{11} \ldots n_{1l_1})(n_{21} \ldots n_{2l_2}) \ldots (n_{r1} \ldots n_{rl_r}).$$

[1]Hermitian matrix is such matrix $\mathbf{A} = (a_{ij})$ that $a_{ij} = \overline{a_{ji}}$.

However, there was no any generalization of the definition of Moore's determinant to arbitrary square matrices. Freeman J. Dyson [13] pointed out the importance of this problem.

L. Chen [18, 19] offered the following definition of determinant of a square matrix over the quaternion skew field $\mathbf{H}$, by putting for $\mathbf{A} = (a_{ij}) \in \mathrm{M}(n, \mathbf{H})$,

$$\det \mathbf{A} = \sum_{\sigma \in S_n} \varepsilon(\sigma)\, a_{n_1 i_2} \cdot a_{i_2 i_3} \cdot \ldots \cdot a_{i_s n_1} \cdot \ldots \cdot a_{n_r k_2} \cdot \ldots \cdot a_{k_l n_r},$$
$$\sigma = (n_1 i_2 \ldots i_s) \ldots (n_r k_2 \ldots k_r),$$
$$n_1 > i_2, i_3, \ldots, i_s; \ldots, n_r > k_2, k_3, \ldots, k_l,$$
$$n = n_1 > n_2 > \ldots > n_r \geq 1.$$

Despite the fact that this determinant does not satisfy Axiom 1, L. Chen got a determinantal representation of an inverse matrix. However it can not been expanded along arbitrary rows and columns (except for n-th row). Therefore, L. Chen did not obtain a classical adjoint matrix as well. For $\mathbf{A} = (\alpha_1, \ldots, \alpha_m)$ over the quaternion skew field $\mathbf{H}$, if $\|\mathbf{A}\| := \det(\mathbf{A}^* \mathbf{A}) \neq 0$, then $\exists \mathbf{A}^{-1} = (b_{jk})$, where

$$\overline{b_{jk}} = \frac{1}{\|\mathbf{A}\|} \omega_{kj}, \quad \left(j, k = \overline{1, n}\right),$$

$$\omega_{kj} = \det\left(\alpha_1 \ldots \alpha_{j-1} \alpha_n \alpha_{j+1} \ldots \alpha_{n-1} \delta_k\right)^* \left(\alpha_1 \ldots \alpha_{j-1} \alpha_n \alpha_{j+1} \ldots \alpha_{n-1} \alpha_j\right).$$

Here α_i is the i-th column of $\mathbf{A}$, δ_k is the n-dimensional column with 1 in the k-th entry and 0 in other ones. L. Chen defined $\|\mathbf{A}\| := \det(\mathbf{A}^* \mathbf{A})$ as the double determinant. If $\|\mathbf{A}\| \neq 0$, then the solution of a right system of linear equations

$$\sum\nolimits_{j=1}^{n} \alpha_j x_j = \beta$$

over $\mathbf{H}$ is represented by the following formula, which the author calls Cramer's rule

$$x_j = \|\mathbf{A}\|^{-1} \overline{\mathbf{D}_j},$$

for all $j = \overline{1, n}$, where

$$\mathbf{D}_j = \det \begin{pmatrix} \alpha_1^* \\ \vdots \\ \alpha_{j-1}^* \\ \alpha_n^* \\ \alpha_{j+1}^* \\ \vdots \\ \alpha_{n-1}^* \\ \beta^* \end{pmatrix} \begin{pmatrix} \alpha_1 & \ldots & \alpha_{j-1} & \alpha_n & \alpha_{j+1} & \ldots & \alpha_{n-1} & \alpha_j \end{pmatrix}.$$

Here α_i^* is the i-th row of $\mathbf{A}^*$ and β^* is the n-dimensional vector-row conjugated with β.

In this chapter we explore the theory of row and column determinants which develops the classical approach to the definition of determinant of a square matrix, as an alternating sum of products of entries of a matrix but with a predetermined order of factors in each of the terms of the determinant.

2. Quaternion Algebra

A quaternion algebra $\mathbb{H}(a,b)$ (we also use notation $\left(\frac{a,b}{\mathbb{F}}\right)$) is a four-dimensional vector space over a field $\mathbb{F}$ with basis $\{1,i,j,k\}$ and the following multiplication rules:

$$\begin{aligned} i^2 &= a, \\ j^2 &= b, \\ ij &= k, \\ ji &= -k. \end{aligned}$$

The field $\mathbb{F}$ is the center of the quaternion algebra $\mathbb{H}(a,b)$.

In the algebra $\mathbb{H}(a,b)$ there are following mappings.

- A quadratic form

  $$\mathrm{n} : x \in \mathbb{H} \to \mathrm{n}(x) \in \mathbb{F}$$

  such that

  $$\mathrm{n}(x \cdot y) = \mathrm{n}(x)\mathrm{n}(y) \quad x, y \in \mathbb{H}$$

  is called the norm on a quaternion algebra $\mathbb{H}$.

- The linear mapping

  $$\mathrm{t} : x = x^0 + x^1 i + x^2 j + x^3 k \in \mathbb{H} \to \mathrm{t}(x) = 2x^0 \in \mathbb{F}$$

  is called the trace of a quaternion. The trace satisfies permutability property of the trace,

  $$\mathrm{t}\,(q \cdot p) = \mathrm{t}\,(p \cdot q)\,.$$

  From the theorem 10.3.3 in the paper [9], it follows

  $$\mathrm{t}(x) = \frac{1}{2}(x - ixi - jxj - kxk). \tag{2.1}$$

- A linear mapping

  $$x \to \overline{x} = \mathrm{t}(x) - x \tag{2.2}$$

  is an involution. The involution has following properties

  $$\begin{aligned} \overline{\overline{x}} &= x, \\ \overline{x + y} &= \overline{x} + \overline{y}, \\ \overline{x \cdot y} &= \overline{y} \cdot \overline{x}. \end{aligned}$$

  A quaternion $\overline{x}$ is called the conjugate of $x \in \mathbb{H}$. The norm and the involution satisfy the following condition:

  $$\mathrm{n}\left(\overline{q}\right) = \mathrm{n}(q).$$

  The trace and the involution satisfy the following condition,

  $$\mathrm{t}\left(\overline{x}\right) = \mathrm{t}(x).$$

From equations (2.1), (2.2), it follows that

$$\overline{x} = -\frac{1}{2}(x + ixi + jxj + kxk).$$

Depending on the choice of the field $\mathbb{F}$, a and b, on the set of quaternion algebras there are only two possibilities [20]:

1. $\left(\frac{a,b}{\mathbb{F}}\right)$ is a division algebra.
2. $\left(\frac{a,b}{\mathbb{F}}\right)$ is isomorphic to the algebra of all 2×2 matrices with entries from the field $\mathbb{F}$. In this case, quaternion algebra is splittable.

The most famous example of a non-split quaternion algebra is Hamilton's quaternions $\mathbf{H} = (\frac{-1,-1}{\mathbb{R}})$, where $\mathbb{R}$ is real field. The set of quaternions can be represented as

$$\mathbf{H} = \{q = q_0 + q_1 i + q_2 j + q_3 k;\ q_0, q_1, q_2, q_3 \in \mathbb{R}\},$$

where $i^2 = j^2 = k^2 = -1$ and $ijk = -1$. Consider some non-isomorphic quaternion algebra with division.

1. $\left(\frac{a,b}{\mathbb{R}}\right)$ is isomorphic to the Hamilton quaternion skew field $\mathbf{H}$ whenever $a < 0$ and $b < 0$. Otherwise $\left(\frac{a,b}{\mathbb{R}}\right)$ is splittable.
2. If $\mathbb{F}$ is the rational field $\mathbb{Q}$, then there exist infinitely many nonisomorphic division quaternion algebras $\left(\frac{a,b}{\mathbb{Q}}\right)$ depending on choice of $a < 0$ and $b < 0$.
3. Let $\mathbb{Q}_p$ be the p-adic field where p is a prime number. For each prime number p there is a unique division quaternion algebra.

The famous example of a split quaternion algebra is split quaternions of James Cockle $\mathbf{H_S}(\frac{-1,1}{\mathbb{R}})$, which can be represented as

$$\mathbf{H_S} = \{q = q_0 + q_1 i + q_2 j + q_3 k;\ q_0, q_1, q_2, q_3 \in \mathbb{R}\},$$

where $i^2 = -1$, $j^2 = k^2 = 1$ and $ijk = 1$. Unlike quaternion division algebra, the set of split quaternions is a noncommutative ring with zero divisors, nilpotent elements and nontrivial idempotents. Recently there was conducted a number of studies in split quaternion matrices (see, for ex. [21, 22, 23, 24]).

3. Introduction to the Theory of the Row and Column Determinants over a Quaternion Algebra

The theory of the row and column determinants was introduced [7, 8] for matrices over a quaternion division algebra. Now this theory is in development for matrices over a split quaternion algebra. In the following two subsections we extend the concept of immanant (permanent, determinant) to a split quaternion algebra using methods of the theory of the row and column determinants.

3.1. Definitions and Properties of the Column and Row Immanants

The immanant of a matrix is a generalization of the concepts of determinant and permanent. The immanant of a complex matrix was defined by Dudley E. Littlewood and Archibald Read Richardson [25] as follows.

Definition 3.1. *Let $\sigma \in S_n$ denote the symmetric group on n elements. Let $\chi : S_n \to \mathbb{C}$ be a complex character. For any $n \times n$ matrix $\mathbf{A} = (a_{ij}) \in \mathbb{C}^{n\times n}$ define the immanant of $\mathbf{A}$ as*

$$\mathrm{Imm}_\chi(\mathbf{A}) = \sum_{\sigma\in S_n} \chi(\sigma) \prod_{i=1}^{n} a_{i\,\sigma(i)}$$

Special cases of immanants are determinants and permanents. In the case where χ is the constant character ($\chi(x) = 1$ for all $x \in S_n$), $\mathrm{Imm}_\chi(\mathbf{A})$ is the permanent of $\mathbf{A}$. In the case where χ is the sign of the permutation (which is the character of the permutation group associated to the (non-trivial) one-dimensional representation), $\mathrm{Imm}_\chi(\mathbf{A})$ is the determinant of $\mathbf{A}$.

Denote by $\mathbb{H}^{n\times m}$ a set of $n \times m$ matrices with entries in an arbitrary (split) quaternion algebra $\mathbb{H}$ and $\mathrm{M}\,(n, \mathbb{H})$ a ring of matrices with entries in $\mathbb{H}$. For $\mathbf{A} = (a_{ij}) \in \mathrm{M}\,(n, \mathbb{H})$ we define n row immanants as follows.

Definition 3.2. *The i-th row immanant of $\mathbf{A} = (a_{ij}) \in \mathrm{M}\,(n, \mathbb{H})$ is defined by putting*

$$\mathrm{rImm}_i\mathbf{A} = \sum_{\sigma\in S_n} \chi(\sigma) a_{i\,i_{k_1}} a_{i_{k_1}\,i_{k_1+1}} \ldots a_{i_{k_1+l_1}\,i} \ldots a_{i_{k_r}\,i_{k_r+1}} \ldots a_{i_{k_r+l_r}\,i_{k_r}},$$

where left-ordered cycle notation of the permutation σ is written as follows

$$\sigma = \left(i\, i_{k_1} i_{k_1+1} \ldots i_{k_1+l_1}\right) \left(i_{k_2} i_{k_2+1} \ldots i_{k_2+l_2}\right) \ldots \left(i_{k_r} i_{k_r+1} \ldots i_{k_r+l_r}\right). \tag{3.1}$$

Here the index i starts the first cycle from the left and other cycles satisfy the following conditions

$$i_{k_2} < i_{k_3} < \ldots < i_{k_r}, \quad i_{k_t} < i_{k_t+s}. \tag{3.2}$$

for all $t = \overline{2, r}$ and $s = \overline{1, l_t}$.

Consequently we have the following definitions.

Definition 3.3. *The i-th row permanent of $\mathbf{A} = (a_{ij}) \in \mathrm{M}\,(n, \mathbb{H})$ is defined as*

$$\mathrm{rper}_i\mathbf{A} = \sum_{\sigma\in S_n} a_{i\,i_{k_1}} a_{i_{k_1}\,i_{k_1+1}} \ldots a_{i_{k_1+l_1}\,i} \ldots a_{i_{k_r}\,i_{k_r+1}} \ldots a_{i_{k_r+l_r}\,i_{k_r}},$$

where left-ordered cycle notation of the permutation σ satisfies the conditions (3.1) and (3.2).

Definition 3.4. *The i-th row determinant of $\mathbf{A} = (a_{ij}) \in \mathrm{M}\,(n, \mathbb{H})$ is defined as*

$$\mathrm{rdet}_i\mathbf{A} = \sum_{\sigma\in S_n} (-1)^{n-r}\, a_{i\,i_{k_1}} a_{i_{k_1}\,i_{k_1+1}} \ldots a_{i_{k_1+l_1}\,i} \ldots a_{i_{k_r}\,i_{k_r+1}} \ldots a_{i_{k_r+l_r}\,i_{k_r}},$$

where left-ordered cycle notation of the permutation σ satisfies the conditions (3.1) and (3.2), (since $\mathrm{sign}(\sigma) = (-1)^{n-r}$).

For $\mathbf{A} = (a_{ij}) \in \mathrm{M}(n, \mathbb{H})$ we define n column immanants as well.

Definition 3.5. *The j-th column immanant of $\mathbf{A} = (a_{ij}) \in \mathrm{M}(n, \mathbb{H})$ is defined as*

$$\mathrm{cImm}_j \mathbf{A} = \sum_{\tau \in S_n} \chi(\tau) a_{j_{k_r} j_{k_r+l_r}} \cdots a_{j_{k_r+1} j_{k_r}} \cdots a_{j\, j_{k_1+l_1}} \cdots a_{j_{k_1+1} j_{k_1}} a_{j_{k_1} j},$$

where right-ordered cycle notation of the permutation $\tau \in S_n$ is written as follows

$$\tau = \left(j_{k_r+l_r} \cdots j_{k_r+1} j_{k_r}\right) \cdots \left(j_{k_2+l_2} \cdots j_{k_2+1} j_{k_2}\right) \left(j_{k_1+l_1} \cdots j_{k_1+1} j_{k_1} j\right). \tag{3.3}$$

Here the first cycle from the right begins with the index j and other cycles satisfy the following conditions

$$j_{k_2} < j_{k_3} < \ldots < j_{k_r}, \quad j_{k_t} < j_{k_t+s}, \tag{3.4}$$

for all $t = \overline{2, r}$ and $s = \overline{1, l_t}$.

Consequently we have the following definitions as well.

Definition 3.6. *The j-th column permanent of $\mathbf{A} = (a_{ij}) \in \mathrm{M}(n, \mathbb{H})$ is defined as*

$$\mathrm{rper}_j \mathbf{A} = \sum_{\tau \in S_n} a_{j_{k_r} j_{k_r+l_r}} \cdots a_{j_{k_r+1} j_{k_r}} \cdots a_{j\, j_{k_1+l_1}} \cdots a_{j_{k_1+1} j_{k_1}} a_{j_{k_1} j},$$

where right-ordered cycle notation of the permutation σ satisfies the conditions (3.3) and (3.4).

Definition 3.7. *The j-th column determinant of $\mathbf{A} = (a_{ij}) \in \mathrm{M}(n, \mathbb{H})$ is defined as*

$$\mathrm{rdet}_j \mathbf{A} = \sum_{\tau \in S_n} (-1)^{n-r} a_{j_{k_r} j_{k_r+l_r}} \cdots a_{j_{k_r+1} j_{k_r}} \cdots a_{j\, j_{k_1+l_1}} \cdots a_{j_{k_1+1} j_{k_1}} a_{j_{k_1} j},$$

where right-ordered cycle notation of the permutation σ satisfies the conditions (3.3) and (3.4).

Consider the basic properties of the column and row immanants over $\mathbb{H}$.

Proposition 3.8. *(The first theorem about zero of an immanant) If one of the rows (columns) of $\mathbf{A} \in \mathrm{M}(n, \mathbb{H})$ consists of zeros only, then $\mathrm{rImm}_i \mathbf{A} = 0$ and $\mathrm{cImm}_i \mathbf{A} = 0$ for all $i = \overline{1, n}$.*

Proof. The proof immediately follows from the definitions. □

Denote by $\mathbb{H}a$ and $a\mathbb{H}$ left and right principal ideals of $\mathbb{H}$, respectively.

Proposition 3.9. *(The second theorem about zero of an row immanant) Let $\mathbf{A} = (a_{ij}) \in \mathrm{M}(n, \mathbb{H})$ and $a_{ki} \in \mathbb{H}a_i$ and $a_{ij} \in \overline{a_i}\mathbb{H}$, where $n(a_i) = 0$ for $k, j = \overline{1, n}$ and for all $i \neq k$. Let $a_{11} \in \mathbb{H}a_1$ and $a_{22} \in a_1\mathbb{H}$ if $k = 1$, and $a_{kk} \in \mathbb{H}a_k$ and $a_{11} \in \overline{a_k}\mathbb{H}$ if $k = i > 1$, where $n(a_k) = 0$. Then $\mathrm{rImm}_k \mathbf{A} = 0$.*

Proof. Let $i \neq k$. Consider an arbitrary monomial of $\mathrm{rImm}_k\mathbf{A}$, if $i \neq k$,

$$d = \chi(\sigma)a_{ki}a_{ij}\ldots a_{lm}$$

where $\{l, m\} \subset \{1, ..., n\}$. Since there exists $a_i \in \mathbb{H}$ such that $n(a_i) = 0$, and $a_{ki} \in \mathbb{H}a_i$, $a_{ij} \in \overline{a_i}\mathbb{H}$, than $a_{ki}a_{ij} = 0$ and $d = 0$.

Let $i = k = 1$. Then an arbitrary monomial of $\mathrm{rImm}_1\mathbf{A}$,

$$d = \chi(\sigma)a_{11}a_{22}\ldots a_{lm}.$$

Since there exists $a_1 \in \mathbb{H}$ such that $n(a_1) = 0$, and $a_{11} \in \mathbb{H}a_1$, $a_{22} \in \overline{a_1}\mathbb{H}$, then $a_{11}a_{22} = 0$ and $d = 0$.

If $k = i > 1$, then an arbitrary monomial of $\mathrm{rImm}_k\mathbf{A}$,

$$d = \chi(\sigma)a_{kk}a_{11}\ldots a_{lm}.$$

Since there exists $a_k \in \mathbb{H}$ such that $n(a_k) = 0$, and $a_{kk} \in \mathbb{H}a_k$, $a_{11} \in \overline{a_k}\mathbb{H}$, then $a_{kk}a_{11} = 0$ and $d = 0$. □

Proposition 3.10. *(The second theorem about zero of an column immanant) Let* $\mathbf{A} = (a_{ij}) \in \mathrm{M}(n, \mathbb{H})$ *and* $a_{ik} \in a_i\mathbb{H}$ *and* $a_{ji} \in \mathbb{H}\overline{a_i}$, *where* $n(a_i) = 0$ *for* $k, j = \overline{1, n}$ *and for all* $i \neq k$. *Let* $a_{11} \in a_1\mathbb{H}$ *and* $a_{22} \in \mathbb{H}\overline{a_1}$ *if* $k = 1$, *and* $a_{kk} \in a_k\mathbb{H}$ *and* $a_{11} \in \mathbb{H}\overline{a_k}$ *if* $k = i > 1$, *where* $n(a_k) = 0$. *Then* $\mathrm{cImm}_k\mathbf{A} = 0$.

Proof. The proof is similar to the proof of the Proposition 3.9. □

The proofs of the next theorems immediately follow from the definitions.

Proposition 3.11. *If the* i*-th row of* $\mathbf{A} = (a_{ij}) \in \mathrm{M}(n, \mathbb{H})$ *is left-multiplied by* $b \in \mathbb{H}$, *then* $\mathrm{rImm}_i\,\mathbf{A}_{i.}(b \cdot \mathbf{a}_{i.}) = b \cdot \mathrm{rImm}_i\,\mathbf{A}$ *for all* $i = \overline{1, n}$.

Proposition 3.12. *If the* j*-th column of* $\mathbf{A} = (a_{ij}) \in \mathrm{M}(n, \mathbb{H})$ *is right-multiplied by* $b \in \mathbb{H}$, *then* $\mathrm{cImm}_j\,\mathbf{A}_{.j}(\mathbf{a}_{.j} \cdot b) = \mathrm{cImm}_j\,\mathbf{A} \cdot b$ *for all* $j = \overline{1, n}$.

Proposition 3.13. *If for* $\mathbf{A} = (a_{ij}) \in \mathrm{M}(n, \mathbb{H})$ *there exists* $t \in \{1, ..., n\}$ *such that* $a_{tj} = b_j + c_j$ *for all* $j = \overline{1, n}$, *then for all* $i = \overline{1, n}$

$$\begin{aligned}
\mathrm{rImm}_i\,\mathbf{A} &= \mathrm{rImm}_i\,\mathbf{A}_{t.}(\mathbf{b}) + \mathrm{rImm}_i\,\mathbf{A}_{t.}(\mathbf{c}),\\
\mathrm{cImm}_i\,\mathbf{A} &= \mathrm{cImm}_i\,\mathbf{A}_{t.}(\mathbf{b}) + \mathrm{cImm}_i\,\mathbf{A}_{t.}(\mathbf{c}),
\end{aligned}$$

where $\mathbf{b} = (b_1, \ldots, b_n)$, $\mathbf{c} = (c_1, \ldots, c_n)$.

Proposition 3.14. *If for* $\mathbf{A} = (a_{ij}) \in \mathrm{M}(n, \mathbb{H})$ *there exists* $t \in \{1, ..., n\}$ *such that* $a_{i\,t} = b_i + c_i$ *for all* $i = \overline{1, n}$, *then for all* $j = \overline{1, n}$

$$\begin{aligned}
\mathrm{rImm}_j\,\mathbf{A} &= \mathrm{rImm}_j\,\mathbf{A}_{.\,t}(\mathbf{b}) + \mathrm{rImm}_j\,\mathbf{A}_{.\,t}(\mathbf{c}),\\
\mathrm{cImm}_j\,\mathbf{A} &= \mathrm{cImm}_j\,\mathbf{A}_{.\,t}(\mathbf{b}) + \mathrm{cImm}_j\mathbf{A}_{.\,t}(\mathbf{c}),
\end{aligned}$$

where $\mathbf{b} = (b_1, \ldots, b_n)^T$, $\mathbf{c} = (c_1, \ldots, c_n)^T$.

Proposition 3.15. *If* $\mathbf{A}^*$ *is the Hermitian adjoint matrix (conjugate and transpose) of* $\mathbf{A} \in \mathrm{M}(n, \mathbb{H})$, *then* $\mathrm{rImm}_i\, \mathbf{A}^* = \overline{\mathrm{cImm}_i\, \mathbf{A}}$ *for all* $i = \overline{1, n}$.

Particular cases of these properties for the row-column determinants and permanents are evident.

Remark 3.16. *The peculiarity of the column immanant (permanent, determinant) is that, at the direct calculation, factors of each of the monomials are written from right to left.* □

In Lemmas 3.17 and 3.18, we consider the recursive definition of the column and row determinants. This definition is an analogue of the expansion of a determinant along a row and a column in commutative case.

Lemma 3.17. *Let* R_{ij} *be the right* ij*-th cofactor of* $\mathbf{A} = (a_{ij}) \in \mathrm{M}(n, \mathbb{H})$, *namely*

$$\mathrm{rdet}_i\, \mathbf{A} = \sum_{j=1}^{n} a_{ij} \cdot R_{ij}$$

for all $i = \overline{1, n}$. *Then*

$$R_{ij} = \begin{cases} -\mathrm{rdet}_j\, (\mathbf{A}^{ii}_{.j}(\mathbf{a}_{.i})), & i \neq j \\ \mathrm{rdet}_k\, \mathbf{A}^{ii}, & i = j \end{cases}$$

$$k = \begin{cases} 2, & i = 1 \\ 1, & i > 1 \end{cases}$$

where the matrix $(\mathbf{A}^{ii}_{.j}(\mathbf{a}_{.i}))$ *is obtained from* $\mathbf{A}$ *by replacing its* j*-th column with the* i*-th column and then by deleting both the* i*-th row and column.* □

Lemma 3.18. *Let* L_{ij} *be the left* ij*th cofactor of entry* a_{ij} *of matrix* $\mathbf{A} = (a_{ij}) \in \mathrm{M}(n, \mathbb{H})$, *namely*

$$\mathrm{cdet}_j\, \mathbf{A} = \sum_{i=1}^{n} L_{ij} \cdot a_{ij}$$

for all $j = \overline{1, n}$. *Then*

$$L_{ij} = \begin{cases} -\mathrm{cdet}_i\, (\mathbf{A}^{jj}_{i.}(\mathbf{a}_{j.})), & i \neq j \\ \mathrm{cdet}_k\, \mathbf{A}^{jj}, & i = j \end{cases}$$

$$k = \begin{cases} 2, & j = 1 \\ 1, & j > 1 \end{cases}$$

where the matrix $(\mathbf{A}^{jj}_{i.}(\mathbf{a}_{j.}))$ *is obtained from* $\mathbf{A}$ *by replacing its* i*th row with the* j*th and then by deleting both the* j*th row and column.* □

Remark 3.19. *Clearly, an arbitrary monomial of each row or column determinant corresponds to a certain monomial of another row or column determinant such that both of them have the same sign, consist of the same factors and differ only in their ordering. If the entries of* $\mathbf{A}$ *are commutative, then* $\mathrm{rdet}_1\, \mathbf{A} = \ldots = \mathrm{rdet}_n \mathbf{A} = \mathrm{cdet}_1\, \mathbf{A} = \ldots = \mathrm{cdet}_n \mathbf{A}$. □

4. An Immanant of a Hermitian Matrix

If $\mathbf{A}^* = \mathbf{A}$ then $\mathbf{A} \in \mathbb{H}^{n\times n}$ is called a Hermitian matrix. In this section we consider the key theorem about row-column immanants of a Hermitian matrix.

The following lemma is needed for the sequel.

Lemma 4.1. *Let T_n be the sum of all possible products of n factors, each of their are either $h_i \in \mathbb{H}$ or $\overline{h_i}$ for all $i = \overline{1, n}$, by specifying the ordering in the terms, $T_n = h_1 \cdot h_2 \cdot \ldots \cdot h_n + \overline{h_1} \cdot h_2 \cdot \ldots \cdot h_n + \ldots + \overline{h_1} \cdot \overline{h_2} \cdot \ldots \cdot \overline{h_n}$. Then T_n consists of the 2^n terms and $T_n = \mathrm{t}\,(h_1)\ \mathrm{t}\,(h_2)\ \ldots\ \mathrm{t}\,(h_n)$.*

Theorem 4.2. *If $\mathbf{A} \in \mathrm{M}\,(n, \mathbb{H})$ is a Hermitian matrix, then*

$$\mathrm{rImm}_1\mathbf{A} = \ldots = \mathrm{rImm}_n\mathbf{A} = \mathrm{cImm}_1\mathbf{A} = \ldots = \mathrm{cImm}_n\mathbf{A} \in \mathbb{F}.$$

Proof. At first we note that if $\mathbf{A} = (a_{ij}) \in \mathbb{H}^{n\times n}$ is Hermitian, then we have $a_{ii} \in \mathbb{F}$ and $a_{ij} = \overline{a_{ji}}$ for all $i, j = \overline{1, n}$.

We divide the set of monomials of $\mathrm{rImm}_i\mathbf{A}$ for some $i \in \{1, ..., n\}$ into two subsets. If indices of coefficients of monomials form permutations as products of disjoint cycles of length 1 and 2, then we include these monomials to the first subset. Other monomials belong to the second subset. If indices of coefficients form a disjoint cycle of length 1, then these coefficients are a_{jj} for $j \in \{1, ..., n\}$ and $a_{jj} \in \mathbb{F}$.

If indices of coefficients form a disjoint cycle of length 2, then these entries are conjugated, $a_{i_k i_{k+1}} = \overline{a_{i_{k+1} i_k}}$, and

$$a_{i_k i_{k+1}} \cdot a_{i_{k+1} i_k} = \overline{a_{i_{k+1} i_k}} \cdot a_{i_{k+1} i_k} = \mathrm{n}(a_{i_{k+1} i_k}) \in \mathbb{F}.$$

So, all monomials of the first subset take on values in $\mathbb{F}$.

Now we consider some monomial d of the second subset. Assume that its index permutation σ forms a direct product of r disjoint cycles. Denote $i_{k_1} := i$, then

$$\begin{aligned} d = \chi(\sigma) a_{i_{k_1} i_{k_1+1}} \ldots a_{i_{k_1+l_1} i_{k_1}} a_{i_{k_2} i_{k_2+1}} \ldots a_{i_{k_2+l_2} i_{k_2}} \ldots a_{i_{k_m} i_{k_m+1}} \ldots \times \\ \times a_{i_{k_m+l_m} i_{k_m}} \ldots a_{i_{k_r} i_{k_r+1}} \ldots a_{i_{k_r+l_r} i_{k_r}} = \chi(\sigma) h_1 h_2 \ldots h_m \ldots h_r, \end{aligned} \tag{4.1}$$

where $h_s = a_{i_{k_s} i_{k_s+1}} \cdot \ldots \cdot a_{i_{k_s+l_s} i_{k_s}}$ for all $s = \overline{1, r}$, and $m \in \{1, \ldots, r\}$. If $l_s = 1$, then $h_s = a_{i_{k_s} i_{k_s+1}} a_{i_{k_s+1} i_{k_s}} = \mathrm{n}(a_{i_{k_s} i_{k_s+1}}) \in \mathbb{F}$. If $l_s = 0$, then $h_s = a_{i_{k_s} i_{k_s}} \in \mathbb{F}$. If $l_s = 0$ or $l_s = 1$ for all $s = \overline{1, r}$ in (4.1), then d belongs to the first subset. Let there exists $s \in I_n$ such that $l_s \geq 2$. Then

$$\overline{h_s} = \overline{a_{i_{k_s} i_{k_s+1}} \ldots a_{i_{k_s+l_s} i_{k_s}}} = \overline{a_{i_{k_s+l_s} i_{k_s}}} \ldots \overline{a_{i_{k_s} i_{k_s+1}}} = a_{i_{k_s} i_{k_s+l_s}} \ldots a_{i_{k_s+1} i_{k_s}}.$$

Denote by $\sigma_s\,(i_{k_s}) := (i_{k_s} i_{k_s+1} \ldots i_{k_s+l_s})$ a disjoint cycle of indices of d for some $s \in \{1, ..., r\}$, then $\sigma = \sigma_1\,(i_{k_1})\,\sigma_2\,(i_{k_2}) \ldots \sigma_r\,(i_{k_r})$. The disjoint cycle $\sigma_s\,(i_{k_s})$ corresponds to the factor h_s. Then $\sigma_s^{-1}\,(i_{k_s}) = (i_{k_s} i_{k_s+l_s} i_{k_s+1} \ldots i_{k_s+1})$ is the inverse disjoint cycle and $\sigma_s^{-1}\,(i_{k_s})$ corresponds to the factor $\overline{h_s}$. By the Lemma 4.1, there exist another $2^p - 1$ monomials for d, (where $p = r - \rho$ and ρ is the number of disjoint cycles of length 1 and 2), such that their index permutations form the direct products of r disjoint cycles either $\sigma_s\,(i_{k_s})$

or $\sigma_s^{-1}(i_{k_s})$ by specifying their ordering by s from 1 to r. Their cycle notations are left-ordered according the to the Definition 3.2. These permutations are unique decomposition of the permutation σ including their ordering by s from 1 to r. Suppose C_1 is the sum of these $2^p - 1$ monomials and d, then, by the Lemma 4.1, we obtain

$$C_1 = \chi(\sigma)\alpha\, \mathrm{t}(h_{\nu_1}) \ \ldots\ \mathrm{t}(h_{\nu_p}) \in \mathbb{F}.$$

Here $\alpha \in \mathbb{F}$ is the product of coefficients whose indices form disjoint cycles of length 1 and 2, $\nu_k \in \{1, \ldots, r\}$ for all $k = \overline{1, p}$.

Thus for an arbitrary monomial of the second subset of $\mathrm{rImm}_i\, \mathbf{A}$, we can find the 2^p monomials such that their sum takes on a value in $\mathbb{F}$. Therefore, $\mathrm{rImm}_i\, \mathbf{A} \in \mathbb{F}$.

Now we prove the equality of all row immanants of $\mathbf{A}$. Consider an arbitrary $\mathrm{rImm}_j\, \mathbf{A}$ such that $j \neq i$ for all $j = \overline{1, n}$. We divide the set of monomials of $\mathrm{rImm}_j\, \mathbf{A}$ into two subsets using the same rule as for $\mathrm{rImm}_i\, \mathbf{A}$. Monomials of the first subset are products of entries of the principal diagonal or norms of entries of $\mathbf{A}$. Therefore they take on a value in $\mathbb{F}$ and each monomial of the first subset of $\mathrm{rImm}_i\, \mathbf{A}$ is equal to a corresponding monomial of the first subset of $\mathrm{rImm}_j\, \mathbf{A}$.

Now consider the monomial d_1 of the second subset of monomials of $\mathrm{rImm}_j\, \mathbf{A}$ consisting of coefficients that are equal to the coefficients of d but they are in another order. Consider all possibilities of the arrangement of coefficients in d_1.

(i) Suppose that the index permutation σ' of its coefficients form a direct product of r disjoint cycles and these cycles coincide with the r disjoint cycles of d but differ by their ordering. Then $\sigma' = \sigma$ and we have

$$d_1 = \chi(\sigma)\alpha h_\mu \ldots h_\lambda,$$

where $\{\mu, \ldots, \lambda\} = \{\nu_1, \ldots, \nu_p\}$. By the Lemma 4.1, there exist $2^p - 1$ monomials of the second subset of $\mathrm{rImm}_j\, \mathbf{A}$ such that each of them is equal to a product of p factors either h_s or $\overline{h_s}$ for all $s \in \{\mu, \ldots, \lambda\}$. Hence by the Lemma 4.1, we obtain

$$C_2 = \chi(\sigma)\alpha\, t(h_\mu) \ \ldots\ t(h_\lambda) = \chi(\sigma)\ \alpha\ t(h_{\nu_1}) \ldots\ t(h_{\nu_p}) = C_1.$$

(ii) Now suppose that in addition to the case (i) the index j is placed inside some disjoint cycle of the index permutation σ of d, e.g., $j \in \{i_{k_m+1}, \ldots, i_{k_m+l_m}\}$. Denote $j = i_{k_m+q}$. Considering the above said and $\sigma_{k_m+1}(i_{k_m+1}) = \sigma_{k_m+q}(i_{k_m+q})$, we have $\sigma' = \sigma$. Then d_1 is represented as follows:

$$\begin{aligned} d_1 = \chi(\sigma) a_{i_{k_m+q} i_{k_m+q+1}} \cdots \quad a_{i_{k_m+l_m} i_{k_m}}\, a_{i_{k_m} i_{k_m+1}} \cdots \times \\ \times a_{i_{k_m+q-1} i_{k_m+q}} a_{i_{k_\mu} i_{k_\mu+1}} \cdots a_{i_{k_\mu+l_\mu} i_{k_\mu}} \cdots a_{i_{k_\lambda} i_{k_\lambda+1}} \cdots a_{i_{k_\lambda+l_\lambda} i_{k_\lambda}} = \\ = \chi(\sigma)\alpha \tilde{h}_m h_\mu \ldots h_\lambda, \end{aligned} \tag{4.2}$$

where $\{m, \mu, \ldots, \lambda\} = \{\nu_1, \ldots, \nu_p\}$. Except for $\tilde{h}_m$, each factor of d_1 in (4.2) corresponds to the equal factor of d in (4.1). By the rearrangement property of the trace, we have $t(\tilde{h}_m) = t(h_m)$. Hence by the Lemma 4.1 and by analogy to the previous case, we obtain,

$$\begin{aligned} C_2 = \chi(\sigma)\alpha\, t(\tilde{h}_m)\, t(h_\mu) \ \ldots\ t(h_\lambda) = \\ = \chi(\sigma)\ \alpha\ t(h_{\nu_1}) \ \ldots\ t(h_m) \ \ldots\ t(h_{\nu_p}) = C_1. \end{aligned}$$

(iii) If in addition to the case (i) the index i is placed inside some disjoint cycles of the index permutation of d_1, then we apply the rearrangement property of the trace to this cycle. As in the previous cases we find 2^p monomials of the second subset of $\mathrm{rImm}_j\,\mathbf{A}$ such that by Lemma 4.1 their sum is equal to the sum of the corresponding 2^p monomials of $\mathrm{rImm}_i\mathbf{A}$. Clearly, we obtain the same conclusion at association of all previous cases, then we apply twice the rearrangement property of the trace.

Thus, in any case each sum of 2^p corresponding monomials of the second subset of $\mathrm{rImm}_j\,\mathbf{A}$ is equal to the sum of 2^p monomials of $\mathrm{rImm}_i\,\mathbf{A}$. Here p is the number of disjoint cycles of length more than 2. Therefore, for all $i, j = \overline{1,n}$ we have

$$\mathrm{rImm}_i\,\mathbf{A} = \mathrm{rImm}_j\,\mathbf{A} \in \mathbb{F}.$$

The equality $\mathrm{cImm}_i\,\mathbf{A} = \mathrm{rImm}_i\,\mathbf{A}$ for all $i = \overline{1,n}$ is proved similarly. □

Remark 4.3. *If $\mathbf{A} \in \mathbb{H}^{n\times n}$ is skew-hermitian ($\mathbf{A} = -\mathbf{A}^*$), then the Theorem 4.2 is not meaningful. It follows from the next example.*

Example 4.4. *Consider the following skew-hermitian matrix over the split quaternions of James Cockle $\mathbf{H_S}(\frac{-1,1}{\mathbb{R}})$,*

$$\mathbf{A} = \begin{pmatrix} j & 2+i \\ -2+i & -k \end{pmatrix}.$$

Since

$$\begin{aligned} \mathrm{rImm}_1\,\mathbf{A} &= -jk - (2+i)(-2+i) = 5+i, \\ \mathrm{rImm}_2\,\mathbf{A} &= -(-2+i)(2+i) - kj = 5-i, \end{aligned}$$

then $\mathrm{rImm}_1\,\mathbf{A} \neq \mathrm{rImm}_2\,\mathbf{A}$.

Since the Theorem 4.2, we have the following definition.

Definition 4.5. *Since all column and row immanants of a Hermitian matrix over $\mathbb{H}$ are equal, we can define the immanant (permanent, determinant) of a Hermitian matrix $\mathbf{A} \in \mathbb{H}^{n\times n}$. By definition, we put for all $i = \overline{1,n}$*

$$\begin{aligned} \mathrm{Imm}\,\mathbf{A} &:= \mathrm{rImm}_i\,\mathbf{A} = \mathrm{cImm}_i\,\mathbf{A}, \\ \mathrm{per}\,\mathbf{A} &:= \mathrm{rper}_i\,\mathbf{A} = \mathrm{cper}_i\,\mathbf{A}, \\ \det\mathbf{A} &:= \mathrm{rdet}_i\,\mathbf{A} = \mathrm{cdet}_i\,\mathbf{A}. \end{aligned}$$

4.1. Cramer's Rule for System of Linear Equations over a Quaternion Division Algebra

In this subsection we shall be consider $\mathbb{H}$ as a quaternion division algebra especially since quasideterminants are defined over the skew field as well.

Properties of the determinant of a Hermitian matrix is completely explored in [7, 8] by its row and column determinants. Among all, consider the following.

Theorem 4.6. *If the i-th row of the Hermitian matrix $\mathbf{A} \in \mathrm{M}\,(n, \mathbb{H})$ is replaced with a left linear combination of its other rows*

$$\mathbf{a}_{i\,.} = c_1\mathbf{a}_{i_1\,.} + \ldots + c_k\mathbf{a}_{i_k\,.}$$

where $c_l \in \mathbb{H}$ *for all* $l = \overline{1,k}$ *and* $\{i, i_l\} \subset \{1, \dots, n\}$, *then for all* $i = \overline{1,n}$

$$\mathrm{cdet}_i \mathbf{A}_{i.}\left(c_1 \cdot \mathbf{a}_{i_1.} + \dots + c_k \cdot \mathbf{a}_{i_k.}\right) = \mathrm{rdet}_i \mathbf{A}_{i.}\left(c_1 \cdot \mathbf{a}_{i_1.} + \dots + c_k \cdot \mathbf{a}_{i_k.}\right) = 0.$$

Theorem 4.7. *If the* j*-th column of a Hermitian matrix* $\mathbf{A} \in \mathrm{M}(n, \mathbb{H})$ *is replaced with a right linear combination of its other columns*

$$\mathbf{a}_{.j} = \mathbf{a}_{.j_1} c_1 + \dots + \mathbf{a}_{.j_k} c_k$$

where $c_l \in \mathbb{H}$ *for all* $l = \overline{1,k}$ *and* $\{j, j_l\} \subset \{1, \dots, n\}$, *then for all* $j = \overline{1,n}$

$$\mathrm{cdet}_j \mathbf{A}_{.j}\left(\mathbf{a}_{.j_1} \cdot c_1 + \dots + \mathbf{a}_{.j_k} \cdot c_k\right) = \mathrm{rdet}_j \mathbf{A}_{.j}\left(\mathbf{a}_{.j_1} \cdot c_1 + \dots + \mathbf{a}_{.j_k} \cdot c_k\right) = 0.$$

The following theorem on the determinantal representation of an inverse matrix of Hermitian follows immediately from these properties.

Theorem 4.8. *There exist a unique right inverse matrix* $(R\mathbf{A})^{-1}$ *and a unique left inverse matrix* $(L\mathbf{A})^{-1}$ *of a nonsingular Hermitian matrix* $\mathbf{A} \in \mathrm{M}(n, \mathbb{H})$, *(*$\det \mathbf{A} \neq 0$*), where* $(R\mathbf{A})^{-1} = (L\mathbf{A})^{-1} =: \mathbf{A}^{-1}$. *Right inverse and left inverse matrices has following determinantal representation*

$$(R\mathbf{A})^{-1} = \frac{1}{\det \mathbf{A}} \begin{pmatrix} R_{11} & R_{21} & \cdots & R_{n1} \\ R_{12} & R_{22} & \cdots & R_{n2} \\ \cdots & \cdots & \cdots & \cdots \\ R_{1n} & R_{2n} & \cdots & R_{nn} \end{pmatrix},$$

$$(L\mathbf{A})^{-1} = \frac{1}{\det \mathbf{A}} \begin{pmatrix} L_{11} & L_{21} & \cdots & L_{n1} \\ L_{12} & L_{22} & \cdots & L_{n2} \\ \cdots & \cdots & \cdots & \cdots \\ L_{1n} & L_{2n} & \cdots & L_{nn} \end{pmatrix},$$

where R_{ij}, L_{ij} *are right and left* ij*-th cofactors of* $\mathbf{A}$, *respectively, for all* $i, j = \overline{1,n}$.

To obtain the determinantal representation for an arbitrary inverse matrix over a quaternion division algebra $\mathbb{H}$, we consider the right $\mathbf{A}\mathbf{A}^*$ and left $\mathbf{A}^*\mathbf{A}$ corresponding Hermitian matrices.

Theorem 4.9 ([7]). *If an arbitrary column of* $\mathbf{A} \in \mathbb{H}^{m \times n}$ *is a right linear combination of its other columns, or an arbitrary row of* $\mathbf{A}^*$ *is a left linear combination of its other rows, then* $\det \mathbf{A}^*\mathbf{A} = 0$.

Since principal submatrices of a Hermitian matrix are also Hermitian, then the basis principal minor may be defined in this noncommutative case as a principal nonzero minor of a maximal order. We also can introduce the notion of the rank of a Hermitian matrix by principal minors, as a maximal order of a principal nonzero minor. The following theorem establishes the correspondence between the rank by principal minors of a Hermitian matrix and the rank of the corresponding matrix that are defined as a maximum number of right-linearly independent columns or left-linearly independent rows, which form a basis.

Theorem 4.10 ([7]). *A rank by principal minors of a Hermitian matrix* $\mathbf{A}^*\mathbf{A}$ *is equal to its rank and a rank of* $\mathbf{A} \in \mathbb{H}^{m \times n}$.

Theorem 4.11 ([7]). *If $\mathbf{A} \in \mathbb{H}^{m\times n}$, then an arbitrary column of $\mathbf{A}$ is a right linear combination of its basic columns or arbitrary row of $\mathbf{A}$ is a left linear combination of its basic rows.*

It implies a criterion for the singularity of a corresponding Hermitian matrix.

Theorem 4.12 ([7]). *The right linearly independence of columns of $\mathbf{A} \in \mathbb{H}^{m\times n}$ or the left linearly independence of rows of $\mathbf{A}^*$ is the necessary and sufficient condition for*

$$\det \mathbf{A}^*\mathbf{A} \neq 0$$

Theorem 4.13 ([7]). *If $\mathbf{A} \in \mathrm{M}(n, \mathbb{H})$, then $\det \mathbf{A}\mathbf{A}^* = \det \mathbf{A}^*\mathbf{A}$.*

In the following example, we shall prove the Theorem 4.13 for the case $n = 2$.

Example 4.14. *Consider the matrix $\mathbf{A} = \begin{pmatrix} a_{11} & a_{12} \\ a_{21} & a_{22} \end{pmatrix}$, then $\mathbf{A}^* = \begin{pmatrix} \overline{a_{11}} & \overline{a_{21}} \\ \overline{a_{12}} & \overline{a_{22}} \end{pmatrix}$. Respectively, we have*

$$\mathbf{A}\mathbf{A}^* = \begin{pmatrix} a_{11}\overline{a_{11}} + a_{12}\overline{a_{12}} & a_{11}\overline{a_{21}} + a_{12}\overline{a_{22}} \\ a_{21}\overline{a_{11}} + a_{22}\overline{a_{12}} & a_{21}\overline{a_{21}} + a_{22}\overline{a_{22}} \end{pmatrix},$$

$$\mathbf{A}^*\mathbf{A} = \begin{pmatrix} \overline{a_{11}}a_{11} + \overline{a_{21}}a_{21} & \overline{a_{11}}a_{12} + \overline{a_{21}}a_{22} \\ \overline{a_{12}}a_{11} + \overline{a_{22}}a_{21} & \overline{a_{12}}a_{12} + \overline{a_{22}}a_{22} \end{pmatrix}.$$

According to thw Theorem 4.2 and the Definition 4.5, we have

$$\begin{aligned} \det \mathbf{A}\mathbf{A}^* &= \mathrm{rdet}_1 \mathbf{A}\mathbf{A}^*, \\ \det \mathbf{A}^*\mathbf{A} &= \mathrm{rdet}_1 \mathbf{A}^*\mathbf{A} \end{aligned}$$

According to the Lemma 3.17

$$\begin{aligned} \det \mathbf{A}\mathbf{A}^* &= (\mathbf{A}\mathbf{A}^*)_{11}(\mathbf{A}\mathbf{A}^*)_{22} - (\mathbf{A}\mathbf{A}^*)_{12}(\mathbf{A}\mathbf{A}^*)_{21} \\ &= (a_{11}\overline{a_{11}} + a_{12}\overline{a_{12}})(a_{21}\overline{a_{21}} + a_{22}\overline{a_{22}}) \\ &-(a_{11}\overline{a_{21}} + a_{12}\overline{a_{22}})(a_{21}\overline{a_{11}} + a_{22}\overline{a_{12}}) \\ &= a_{11}\overline{a_{11}}a_{21}\overline{a_{21}} + a_{12}\overline{a_{12}}a_{21}\overline{a_{21}} \\ &+a_{11}\overline{a_{11}}a_{22}\overline{a_{22}} + a_{12}\overline{a_{12}}a_{22}\overline{a_{22}} \\ &-a_{11}\overline{a_{21}}a_{21}\overline{a_{11}} - a_{12}\overline{a_{22}}a_{21}\overline{a_{11}} \\ &-a_{11}\overline{a_{21}}a_{22}\overline{a_{12}} - a_{12}\overline{a_{22}}a_{22}\overline{a_{12}} \\ &= a_{12}\overline{a_{12}}a_{21}\overline{a_{21}} + a_{11}\overline{a_{11}}a_{22}\overline{a_{22}} \\ &-a_{12}\overline{a_{22}}a_{21}\overline{a_{11}} - a_{11}\overline{a_{21}}a_{22}\overline{a_{12}} \end{aligned}, \tag{4.3}$$

$$\begin{aligned} \det \mathbf{A}^*\mathbf{A} &= (\mathbf{A}^*\mathbf{A})_{11}(\mathbf{A}^*\mathbf{A})_{22} - (\mathbf{A}^*\mathbf{A})_{12}(\mathbf{A}^*\mathbf{A})_{21} \\ &= (\overline{a_{11}}a_{11} + \overline{a_{21}}a_{21})(\overline{a_{12}}a_{12} + \overline{a_{22}}a_{22}) \\ &-(\overline{a_{11}}a_{12} + \overline{a_{21}}a_{22})(\overline{a_{12}}a_{11} + \overline{a_{22}}a_{21}) \\ &= \overline{a_{11}}a_{11}\overline{a_{12}}a_{12} + \overline{a_{21}}a_{21}\overline{a_{12}}a_{12} \\ &+\overline{a_{11}}a_{11}\overline{a_{22}}a_{22} + \overline{a_{21}}a_{21}\overline{a_{22}}a_{22} \\ &-\overline{a_{11}}a_{12}\overline{a_{12}}a_{11} - \overline{a_{21}}a_{22}\overline{a_{12}}a_{11} \\ &-\overline{a_{11}}a_{12}\overline{a_{22}}a_{21} - \overline{a_{21}}a_{22}\overline{a_{22}}a_{21} \\ &= \overline{a_{21}}a_{21}\overline{a_{12}}a_{12} + \overline{a_{11}}a_{11}\overline{a_{22}}a_{22} \\ &-\overline{a_{21}}a_{22}\overline{a_{12}}a_{11} - \overline{a_{11}}a_{12}\overline{a_{22}}a_{21} \end{aligned}. \tag{4.4}$$

Positive terms in equations (4.3), (4.4) *are real numbers and they obviously coincide. To prove equation*

$$a_{12}\overline{a_{22}}a_{21}\overline{a_{11}} + a_{11}\overline{a_{21}}a_{22}\overline{a_{12}} = \overline{a_{21}}a_{22}\overline{a_{12}}a_{11} + \overline{a_{11}}a_{12}\overline{a_{22}}a_{21} \tag{4.5}$$

we use the rearrangement property of the trace of elements of the quaternion algebra, $\mathrm{t}(pq) = \mathrm{t}(qp)$*. Indeed,*

$$a_{12}\overline{a_{22}}a_{21}\overline{a_{11}} + a_{11}\overline{a_{21}}a_{22}\overline{a_{12}} = a_{12}\overline{a_{22}}a_{21}\overline{a_{11}} + \overline{a_{12}\overline{a_{22}}a_{21}\overline{a_{11}}} = \mathrm{t}(a_{12}\overline{a_{22}}a_{21}\overline{a_{11}}),$$

$$\overline{a_{21}}a_{22}\overline{a_{12}}a_{11} + \overline{a_{11}}a_{12}\overline{a_{22}}a_{21} = \overline{\overline{a_{11}}a_{12}\overline{a_{22}}a_{21}} + \overline{a_{11}}a_{12}\overline{a_{22}}a_{21} = \mathrm{t}(\overline{a_{11}}a_{12}\overline{a_{22}}a_{21})$$

Then by the rearrangement property of the trace, we obtain (4.5).

According to the Theorem 4.13, we introduce the concept of double determinant. For the first time this concept was introduced by L. Chen ([18]).

Definition 4.15. *The determinant of corresponding Hermitian matrices is called the double determinant of* $\mathbf{A} \in \mathrm{M}(n, \mathbb{H})$*, i.e.,* $\mathrm{ddet}\mathbf{A} := \det(\mathbf{A}^*\mathbf{A}) = \det(\mathbf{A}\mathbf{A}^*)$.

If $\mathbb{H}$ is the Hamilton's quaternion skew field $\mathbf{H}$, then the following theorem establishes the validity of Axiom 1 for the double determinant.

Theorem 4.16. *If* $\{\mathbf{A}, \mathbf{B}\} \subset \mathrm{M}(n, \mathbf{H})$*, then* $\mathrm{ddet}(\mathbf{A} \cdot \mathbf{B}) = \mathrm{ddet}\mathbf{A} \cdot \mathrm{ddet}\mathbf{B}$.

Unfortunately, if a non-Hermitian matrix is not full rank, then nothing can be said about singularity of its row and column determinant. We show it in the following example.

Example 4.17. *Consider the matrix*

$$\mathbf{A} = \begin{pmatrix} i & j \\ j & -i \end{pmatrix}.$$

Its second row is obtained from the first row by left-multiplying by k*. Then, by the Theorem 4.12,* $\mathrm{ddet}A = 0$*. Indeed,*

$$\mathbf{A}^*\mathbf{A} = \begin{pmatrix} -i & -j \\ -j & i \end{pmatrix} \cdot \begin{pmatrix} i & j \\ j & -i \end{pmatrix} = \begin{pmatrix} 2 & -2k \\ 2k & 2 \end{pmatrix}.$$

Then $\mathrm{ddet}\mathbf{A} = 4 + 4k^2 = 0$*. However*

$$\mathrm{cdet}_1\mathbf{A} = \mathrm{cdet}_2\mathbf{A} = \mathrm{rdet}_1\mathbf{A} = \mathrm{rdet}_2\mathbf{A} = -i^2 - j^2 = 2.$$

At the same time $\mathrm{rank}\,\mathbf{A} = 1$*, that corresponds to the Theorem 4.10.*

The correspondence between the double determinant and the noncommutative determinants of Moore, Stady and Dieudonné are as follows,

$$\mathrm{ddet}\mathbf{A} = \mathrm{Mdet}(\mathbf{A}^*\mathbf{A}) = \mathrm{Sdet}\mathbf{A} = \mathrm{Ddet}^2\mathbf{A}.$$

Definition 4.18. *Let* $\mathrm{ddet}\mathbf{A} = \mathrm{cdet}_j(\mathbf{A}^*\mathbf{A}) = \sum\limits_i \mathbb{L}_{ij} \cdot a_{ij}$ *for* $j = \overline{1, n}$*. Then* $\mathbb{L}_{ij}$ *is called the left double* ij*-th cofactor of* $\mathbf{A} \in \mathrm{M}(n, \mathbb{H})$.

Definition 4.19. *Let* $\mathrm{ddet}\mathbf{A} = \mathrm{rdet}_i\left(\mathbf{A}\mathbf{A}^*\right) = \sum\limits_j a_{ij} \cdot \mathbb{R}_{ij}$ *for* $i = \overline{1,n}$. *Then* $\mathbb{R}_{ij}$ *is called the right double* ij*-th cofactor of* $\mathbf{A} \in \mathrm{M}\left(n, \mathbb{H}\right)$.

Theorem 4.20. *The necessary and sufficient condition of invertibility of a matrix* $\mathbf{A} = (a_{ij}) \in \mathrm{M}(n, \mathbb{H})$ *is* $\mathrm{ddet}\mathbf{A} \neq 0$. *Then* $\exists \mathbf{A}^{-1} = (L\mathbf{A})^{-1} = (R\mathbf{A})^{-1}$, *where*

$$(L\mathbf{A})^{-1} = \left(\mathbf{A}^*\mathbf{A}\right)^{-1}\mathbf{A}^* = \frac{1}{\mathrm{ddet}\mathbf{A}} \begin{pmatrix} \mathbb{L}_{11} & \mathbb{L}_{21} & \dots & \mathbb{L}_{n1} \\ \mathbb{L}_{12} & \mathbb{L}_{22} & \dots & \mathbb{L}_{n2} \\ \dots & \dots & \dots & \dots \\ \mathbb{L}_{1n} & \mathbb{L}_{2n} & \dots & \mathbb{L}_{nn} \end{pmatrix} \tag{4.6}$$

$$(R\mathbf{A})^{-1} = \mathbf{A}^*\left(\mathbf{A}\mathbf{A}^*\right)^{-1} = \frac{1}{\mathrm{ddet}\mathbf{A}^*} \begin{pmatrix} \mathbb{R}_{11} & \mathbb{R}_{21} & \dots & \mathbb{R}_{n1} \\ \mathbb{R}_{12} & \mathbb{R}_{22} & \dots & \mathbb{R}_{n2} \\ \dots & \dots & \dots & \dots \\ \mathbb{R}_{1n} & \mathbb{R}_{2n} & \dots & \mathbb{R}_{nn} \end{pmatrix} \tag{4.7}$$

and $\mathbb{L}_{ij} = \mathrm{cdet}_j(\mathbf{A}^*\mathbf{A})_{.j}\left(\mathbf{a}^*_{.i}\right)$, $\mathbb{R}_{ij} = \mathrm{rdet}_i(\mathbf{A}\mathbf{A}^*)_{i.}\left(\mathbf{a}^*_{j.}\right)$ *for all* $i, j = \overline{1,n}$.

Remark 4.21. *In the Theorem 4.20, the inverse matrix* $\mathbf{A}^{-1}$ *of an arbitrary matrix* $\mathbf{A} \in \mathrm{M}(n, \mathbb{H})$ *under the assumption of* $\mathrm{ddet}\mathbf{A} \neq 0$ *is represented by the analog of the classical adjoint matrix. If we denote this analog of the adjoint matrix over* $\mathbb{H}$ *by* $\mathrm{Adj}[[\mathbf{A}]]$, *then the next formula is valid over* $\mathbb{H}$:

$$\mathbf{A}^{-1} = \frac{\mathrm{Adj}[[\mathbf{A}]]}{\mathrm{ddet}\mathbf{A}}.$$

An obvious consequence of a determinantal representation of the inverse matrix by the classical adjoint matrix is Cramer's rule.

Theorem 4.22. *Let*

$$\mathbf{A} \cdot \mathbf{x} = \mathbf{y} \tag{4.8}$$

be a right system of linear equations with a matrix of coefficients $\mathbf{A} \in \mathrm{M}(n, \mathbb{H})$, *a column of constants* $\mathbf{y} = (y_1, \dots, y_n)^T \in \mathbb{H}^{n\times 1}$, *and a column of unknowns* $\mathbf{x} = (x_1, \dots, x_n)^T$. *If* $\mathrm{ddet}\mathbf{A} \neq 0$, *then* (4.8) *has a unique solution that has represented as follows,*

$$x_j = \frac{\mathrm{cdet}_j(\mathbf{A}^*\mathbf{A})_{.j}\left(\mathbf{f}\right)}{\mathrm{ddet}\mathbf{A}}, \qquad \forall j = \overline{1,n} \tag{4.9}$$

where $\mathbf{f} = \mathbf{A}^*\mathbf{y}$.

Theorem 4.23. *Let*

$$\mathbf{x} \cdot \mathbf{A} = \mathbf{y} \tag{4.10}$$

be a left system of linear equations with a matrix of coefficients $\mathbf{A} \in \mathrm{M}(n, \mathbb{H})$, *a column of constants* $\mathbf{y} = (y_1, \dots, y_n) \in \mathbb{H}^{1\times n}$ *and a column of unknowns* $\mathbf{x} = (x_1, \dots, x_n)$. *If* $\mathrm{ddet}\mathbf{A} \neq 0$, *then* (4.10) *has a unique solution that has represented as follows,*

$$x_i = \frac{\mathrm{rdet}_i\left(\mathbf{A}\mathbf{A}^*\right)_{i.}\left(\mathbf{z}\right)}{\mathrm{ddet}\mathbf{A}}, \qquad \forall i = \overline{1,n} \tag{4.11}$$

where $\mathbf{z} = \mathbf{y}\mathbf{A}^*$.

Equations (4.9) and (4.11) are the obvious and natural generalizations of Cramer's rule for systems of linear equations over a quaternion division algebra. As follows from the Theorem 4.8, the closer analog to Cramer's rule can be obtained in the following specific cases.

Theorem 4.24. *Let* $\mathbf{A} \in \mathrm{M}(n, \mathbb{H})$ *be Hermitian in* (4.8). *Then the solution of* (4.8) *has represented by the equation,*

$$x_j = \frac{\mathrm{cdet}_j \mathbf{A}_{.j}(\mathbf{y})}{\det \mathbf{A}}, \quad \forall j = \overline{1, n}.$$

Theorem 4.25. *Let* $\mathbf{A} \in \mathrm{M}(n, \mathbb{H})$ *be Hermitian in* (4.10). *Then the solution of* (4.10) *has represented as follows,*

$$x_i = \frac{\mathrm{rdet}_i \mathbf{A}_{i.}(\mathbf{y})}{\det \mathbf{A}}, \quad \forall i = \overline{1, n}.$$

An application of the column-row determinants in the theory of generalized inverse matrices over the quaternion skew field recently has been received in [26, 27, 28, 29, 30, 31, 32, 33, 34, 35, 36, 37, 38].

5. Quasideterminants over a Quaternion Division Algebra

Theorem 5.1. *Suppose a matrix*

$$\mathbf{A} = \begin{pmatrix} a_{11} & \dots & a_{1n} \\ \dots & \dots & \dots \\ a_{n1} & \dots & a_{nn} \end{pmatrix}$$

with entries from a quaternion division algebra has an inverse $\mathbf{A}^{-1}$.[2] *Then a minor of the inverse matrix satisfies the following equation, provided that the inverse matrices exist*

$$((\mathbf{A}^{-1})_{IJ})^{-1} = \mathbf{A}_{JI} - \mathbf{A}_{J.}^{\cdot I}(\mathbf{A}^{JI})^{-1}\mathbf{A}_{\cdot I}^{J.} \tag{5.1}$$

Proof. Definition of an inverse matrix leads to the system of linear equations

$$\mathbf{A}^{JI}(\mathbf{A}^{-1})_{\cdot J}^{I.} + \mathbf{A}_{\cdot I}^{J.}(\mathbf{A}^{-1})_{IJ} = 0 \tag{5.2}$$

$$\mathbf{A}_{J.}^{\cdot I}(\mathbf{A}^{-1})_{\cdot J}^{I.} + \mathbf{A}_{JI}(\mathbf{A}^{-1})_{IJ} = \mathbf{I} \tag{5.3}$$

where $\mathbf{I}$ is a unit matrix. We multiply (5.2) by $\left(\mathbf{A}^{JI}\right)^{-1}$

$$(\mathbf{A}^{-1})_{\cdot J}^{I.} + (\mathbf{A}^{JI})^{-1}\mathbf{A}_{\cdot I}^{J.}(\mathbf{A}^{-1})_{IJ} = 0 \tag{5.4}$$

Now we can substitute (5.4) into (5.3)

$$\mathbf{A}_{JI}(\mathbf{A}^{-1})_{IJ} - \mathbf{A}_{J.}^{\cdot I}(\mathbf{A}^{JI})^{-1}\mathbf{A}_{\cdot I}^{J.}(\mathbf{A}^{-1})_{IJ} = \mathbf{I} \tag{5.5}$$

(5.1) follows from (5.5). □

[2]This statement and its proof are based on statement 1.2.1 from [17] (page 8) for matrix over free division ring.

Corollary 5.2. *Suppose a matrix* $\mathbf{A}$ *has the inverse matrix. Then elements of the inverse matrix satisfy to the equation*

$$((\mathbf{A}^{-1})_{ij})^{-1} = a_{ji} - \mathbf{A}_{j.}^{.i}(\mathbf{A}^{ji})^{-1}\mathbf{A}_{.i}^{j.} \tag{5.6}$$

Example 5.3. *Consider a matrix*

$$\mathbf{A} = \begin{pmatrix} a_{11} & a_{12} \\ a_{21} & a_{22} \end{pmatrix}$$

According to (5.6)

$$(\mathbf{A}^{-1})_{11} = (a_{11} - a_{12}(a_{22})^{-1}\, a_{21})^{-1} \tag{5.7}$$

$$(\mathbf{A}^{-1})_{21} = (a_{21} - a_{22}(a_{12})^{-1}\, a_{11})^{-1} \tag{5.8}$$

$$(\mathbf{A}^{-1})_{12} = (a_{12} - a_{11}(a_{21})^{-1}\, a_{22})^{-1} \tag{5.9}$$

$$(\mathbf{A}^{-1})_{22} = (a_{22} - a_{21}(a_{11})^{-1}\, a_{12})^{-1} \tag{5.10}$$

We call a matrix

$$\mathcal{H}\mathbf{A} = ((\mathcal{H}\mathbf{A})_{ij}) = ((a_{ji})^{-1}) \tag{5.11}$$

a Hadamard inverse of[3] $\mathbf{A}$.

Definition 5.4. *The* (ji)*-quasideterminant of* $\mathbf{A}$ *is formal expression*

$$|\mathbf{A}|_{ji} = (\mathcal{H}\mathbf{A}^{-1})_{ji} = ((\mathbf{A}^{-1})_{ij})^{-1} \tag{5.12}$$

We consider the (ji)*-quasideterminant as an element of the matrix* $|\mathbf{A}|$*, which is called a quasideterminant.*

Theorem 5.5. *Expression for the* (ji)*-quasideterminant has form*

$$|\mathbf{A}|_{ji} = a_{ji} - \mathbf{A}_{j.}^{.i}(\mathbf{A}^{ji})^{-1}\mathbf{A}_{.i}^{j.} \tag{5.13}$$

$$|\mathbf{A}|_{ji} = a_{ji} - \mathbf{A}_{j.}^{.i}\, \mathcal{H}|\mathbf{A}^{ji}|\, \mathbf{A}_{.i}^{j.} \tag{5.14}$$

Proof. The statement follows from (5.6) and (5.12). □

Example 5.6. *Let*

$$\mathbf{A} = \begin{pmatrix} 1 & 0 \\ 0 & 1 \end{pmatrix} \tag{5.15}$$

It is clear from (5.7) *and* (5.10) *that* $(\mathbf{A}^{-1})_{11} = 1$ *and* $(\mathbf{A}^{-1})_{22} = 1$. *However expression for* $(\mathbf{A}^{-1})_{21}$ *and* $(\mathbf{A}^{-1})_{12}$ *cannot be defined from* (5.8) *and* (5.9) *since* $(a_{21} - a_{22}(a_{12})^{-1}\, a_{11})^{-1} = (a_{12} - a_{11}(a_{21})^{-1}\, a_{22})^{-1} = 0$. *We can transform these expressions. For instance*

$$\begin{aligned}(\mathbf{A}^{-1})_{21} &= (a_{21} - a_{22}(a_{12})^{-1}\, a_{11})^{-1} \\ &= (a_{11}((a_{11})^{-1}\, a_{12} - (a_{21})^{-1}\, a_{22}))^{-1} \\ &= ((a_{21})^{-1}\, a_{11}(a_{21}(a_{11})^{-1}\, a_{12} - a_{22}))^{-1} \\ &= (a_{11}(a_{21}(a_{11})^{-1}\, a_{12} - a_{22}))^{-1}\, a_{21}\end{aligned}$$

[3] See also page 4 in paper [16].

It follows immediately that $(\mathbf{A}^{-1})_{21} = 0$. *In the same manner we can find that* $(\mathbf{A}^{-1})_{12} = 0$. *Therefore,*

$$\mathbf{A}^{-1} = \begin{pmatrix} 1 & 0 \\ 0 & 1 \end{pmatrix} \tag{5.16}$$

□

From the Example 5.6 we see that we cannot always use Equation (5.6) to find elements of the inverse matrix and we need more transformations to solve this problem. From the theorem 4.6.3 in the paper [9], it follows that if

$$\operatorname{rank}\begin{pmatrix} a_{11} & \dots & a_{1n} \\ \dots & \dots & \dots \\ a_{n1} & \dots & a_{nn} \end{pmatrix} \leq n-2$$

then $|\mathbf{A}|_{ij}$ for all $i, j = \overline{1, n}$ is not defined. From this, it follows that although a quasideterminant is a powerful tool, use of a determinant is a major advantage.

Theorem 5.7. *Let a matrix* $\mathbf{A}$ *have an inverse. Then for any matrices* $\mathbf{B}$ *and* $\mathbf{C}$ *equation*

$$\mathbf{B} = \mathbf{C} \tag{5.17}$$

follows from the equation

$$\mathbf{BA} = \mathbf{CA} \tag{5.18}$$

Proof. Equation (5.17) follows from (5.18) if we multiply both parts of (5.18) over $\mathbf{A}^{-1}$. □

Theorem 5.8. *The solution of a nonsingular system of linear equations*

$$\mathbf{A}x = b \tag{5.19}$$

is determined uniquely and can be presented in either form[4]

$$x = \mathbf{A}^{-1}b \tag{5.20}$$

$$x = \mathcal{H}|\mathbf{A}|\,b \tag{5.21}$$

Proof. Multiplying both sides of (5.19) from left by $\mathbf{A}^{-1}$ we get (5.20). Using the Definition 5.4, we get (5.21). Since the Theorem 5.7, the solution is unique. □

6. Relation of Row-Column Determinants with Quasideterminants

Theorem 6.1. *If* $\mathbf{A} \in \mathrm{M}(n, \mathbb{H})$ *is an invertible matrix, then, for arbitrary* $p, q = \overline{1, n}$, *we have the following representation of the pq-quasideterminant*

$$|\,\mathbf{A}\,|_{pq} = \frac{\operatorname{ddet}\mathbf{A} \cdot \overline{\operatorname{cdet}_q(\mathbf{A}^*\mathbf{A})_{.q}\left(\mathbf{a}^*_{.p}\right)}}{\mathrm{n}(\operatorname{cdet}_q(\mathbf{A}^*\mathbf{A})_{.q}\left(\mathbf{a}^*_{.p}\right))}, \tag{6.1}$$

[4]See similar statement in the theorem 1.6.1 in the paper [17] on pagen 19.

$$\mid \mathbf{A} \mid_{pq} = \frac{\mathrm{ddet}\mathbf{A} \cdot \overline{\mathrm{rdet}_p(\mathbf{A}\mathbf{A}^*)_{p.}\left(\mathbf{a}^*_{q.}\right)}}{\mathrm{n}(\mathrm{rdet}_p(\mathbf{A}\mathbf{A}^*)_{p.}\left(\mathbf{a}^*_{q.}\right))}. \tag{6.2}$$

Proof. Let $\mathbf{A}^{-1} = (b_{ij})$ to $\mathbf{A} \in \mathrm{M}(n, \mathbb{H})$. Equation (5.12) reveals the relationship between a quasideterminant $\mid \mathbf{A} \mid_{p,q}$ of $\mathbf{A} \in \mathrm{M}(n, \mathbb{H})$ and elements of the inverse matrix $\mathbf{A}^{-1} = (b_{ij})$, namely

$$\mid \mathbf{A} \mid_{pq} = b_{qp}^{-1}$$

for all $p, q = \overline{1, n}$. At the same time, the theory of row and column determinants (the theorem 4.20) gives us representation of the inverse matrix through its left (4.6) and right (4.7) double cofactors. Thus, accordingly, we obtain

$$\mid \mathbf{A} \mid_{pq} = b_{qp}^{-1} = \left(\frac{\mathbb{L}_{pq}}{\mathrm{ddet}\mathbf{A}}\right)^{-1} = \left(\frac{\mathrm{cdet}_q(\mathbf{A}^*\mathbf{A})_{.q}\left(\mathbf{A}^*_{.p}\right)}{\mathrm{ddet}\mathbf{A}}\right)^{-1}, \tag{6.3}$$

$$\mid \mathbf{A} \mid_{pq} = b_{qp}^{-1} = \left(\frac{\mathbb{R}_{pq}}{\mathrm{ddet}\mathbf{A}}\right)^{-1} = \left(\frac{\mathrm{rdet}_p(\mathbf{A}\mathbf{A}^*)_{p.}\left(\mathbf{A}^*_{q.}\right)}{\mathrm{ddet}\mathbf{A}}\right)^{-1}. \tag{6.4}$$

Since $\mathrm{ddet}\mathbf{A} \neq 0 \in \mathbb{F}$, then $\exists(\mathrm{ddet}\mathbf{A})^{-1} \in \mathbb{F}$. It follows that

$$\mathrm{cdet}_q(\mathbf{A}^*\mathbf{A})_{.q}\left(\mathbf{A}^*_{.p}\right)^{-1} = \frac{\overline{\mathrm{cdet}_q(\mathbf{A}^*\mathbf{A})_{.q}\left(\mathbf{A}^*_{.p}\right)}}{\mathrm{n}(\mathrm{cdet}_q(\mathbf{A}^*\mathbf{A})_{.q}\left(\mathbf{A}^*_{.p}\right))}, \tag{6.5}$$

$$\mathrm{rdet}_p(\mathbf{A}\mathbf{A}^*)_{p.}\left(\mathbf{A}^*_{q.}\right)^{-1} = \frac{\overline{\mathrm{rdet}_p(\mathbf{A}\mathbf{A}^*)_{p.}\left(\mathbf{A}^*_{q.}\right)}}{\mathrm{n}(\mathrm{rdet}_p(\mathbf{A}\mathbf{A}^*)_{p.}\left(\mathbf{A}^*_{q.}\right))}. \tag{6.6}$$

Substituting (6.5) into (6.3), and (6.6) into (6.4), we accordingly obtain (6.1) and (6.2).

We proved the theorem. □

Equation (6.1) gives an explicit representation of a quasideterminant $\mid \mathbf{A} \mid_{p,q}$ of $\mathbf{A} \in \mathrm{M}(n, \mathbb{H})$ for all $p, q = \overline{1, n}$ by the column determinant of its corresponding left Hermitian matrix $\mathbf{A}^*\mathbf{A}$, and (6.2) does by the row determinant of its corresponding right Hermitian matrix $\mathbf{A}\mathbf{A}^*$.

Example 6.2. *Consider a matrix*

$$\mathbf{A} = \begin{pmatrix} a_{11} & a_{12} \\ a_{21} & a_{22} \end{pmatrix}$$

According to (5.13)

$$|\mathbf{A}| = \begin{pmatrix} a_{11} - a_{12}(a_{22})^{-1}a_{21} & a_{12} - a_{11}(a_{21})^{-1}a_{22} \\ a_{21} - a_{22}(a_{12})^{-1}a_{11} & a_{22} - a_{21}(a_{11})^{-1}a_{12} \end{pmatrix} \tag{6.7}$$

Our goal is to find this quasideterminant using the Theorem 6.1. It is evident that

$$\mathbf{A}^* = \begin{pmatrix} \overline{a_{11}} & \overline{a_{21}} \\ \overline{a_{12}} & \overline{a_{22}} \end{pmatrix} \quad \mathbf{A}^*\mathbf{A} = \begin{pmatrix} \mathrm{n}(a_{11}) + \mathrm{n}(a_{21}) & \overline{a_{11}}a_{12} + \overline{a_{21}}a_{22} \\ \overline{a_{12}}a_{11} + \overline{a_{22}}a_{21} & \mathrm{n}(a_{12}) + \mathrm{n}(a_{22}) \end{pmatrix}.$$

Calculate the necessary determinants

$$
\begin{aligned}
\mathrm{ddet}\mathbf{A} &= \mathrm{rdet}_1(\mathbf{A}^*\mathbf{A})\\
&= (\mathrm{n}(a_{11}) + \mathrm{n}(a_{21})) \cdot (\mathrm{n}(a_{12}) + \mathrm{n}(a_{22}))\\
&-(\overline{a_{11}}a_{12} + \overline{a_{21}}a_{22}) \cdot (\overline{a_{12}}a_{11} + \overline{a_{22}}a_{21})\\
&= \mathrm{n}(a_{11})\mathrm{n}(a_{12}) + \mathrm{n}(a_{11})\mathrm{n}(a_{22}) + \mathrm{n}(a_{21})\mathrm{n}(a_{12}) + \mathrm{n}(a_{21})\mathrm{n}(a_{22})\\
&-\overline{a_{11}}a_{12}\overline{a_{12}}a_{11} - \overline{a_{11}}a_{12}\overline{a_{22}}a_{21} - \overline{a_{21}}a_{22}\overline{a_{12}}a_{11} - \overline{a_{21}}a_{22}\overline{a_{22}}a_{21}\\
&= \mathrm{n}(a_{11})\mathrm{n}(a_{22}) + \mathrm{n}(a_{21})\mathrm{n}(a_{12}) - (\overline{a_{11}}a_{12}\overline{a_{22}}a_{21} + \overline{\overline{a_{11}}a_{12}\overline{a_{22}}a_{21}})\\
&= \mathrm{n}(a_{11})\mathrm{n}(a_{22}) + \mathrm{n}(a_{21})\mathrm{n}(a_{12}) - \mathrm{t}(\overline{a_{11}}a_{12}\overline{a_{22}}a_{21})
\end{aligned}
$$

$$
\begin{aligned}
\mathrm{cdet}_1(\mathbf{A}^*\mathbf{A})_{.1}(\mathbf{a}^*_{.2}) &= \mathrm{cdet}_1\begin{pmatrix} \overline{a_{21}} & \overline{a_{11}}a_{12} + \overline{a_{21}}a_{22}\\ \overline{a_{22}} & \mathrm{n}(a_{12}) + \mathrm{n}(a_{22}) \end{pmatrix}\\
&= \mathrm{n}(a_{12})\overline{a_{21}} + \mathrm{n}(a_{22})\overline{a_{21}} - \overline{a_{11}}a_{12}\overline{a_{22}} - \overline{a_{21}}a_{22}\overline{a_{22}}\\
&= \mathrm{n}(a_{12})\overline{a_{21}} - \overline{a_{11}}a_{12}\overline{a_{22}}.
\end{aligned}
$$

Then

$$
\overline{\mathrm{cdet}_1(\mathbf{A}^*\mathbf{A})_{.1}(\mathbf{a}^*_{.2})} = \mathrm{n}(a_{12})a_{21} - a_{22}\overline{a_{12}}a_{11},
$$

$$
\begin{aligned}
\mathrm{n}(\mathrm{cdet}_1(\mathbf{A}^*\mathbf{A})_{.1}(\mathbf{a}^*_{.2})) &= \overline{\mathrm{cdet}_1(\mathbf{A}^*\mathbf{A})_{.1}(\mathbf{a}^*_{.2})} \cdot \mathrm{cdet}_1(\mathbf{A}^*\mathbf{A})_{.1}(\mathbf{a}^*_{.2})\\
&= (\mathrm{n}(a_{12})a_{21} - a_{22}\overline{a_{12}}a_{11}) \cdot (\mathrm{n}(a_{12})\overline{a_{21}} - \overline{a_{11}}a_{12}\overline{a_{22}})\\
&= \mathrm{n}^2(a_{12})\mathrm{n}(a_{21}) - \mathrm{n}(a_{12})a_{21}\overline{a_{11}}a_{12}\overline{a_{22}}\\
&-\mathrm{n}(a_{12})a_{22}\overline{a_{12}}a_{11}\overline{a_{21}} + a_{22}\overline{a_{12}}a_{11}\overline{a_{11}}a_{12}\overline{a_{22}}\\
&= \mathrm{n}(a_{12})(\mathrm{n}(a_{12})\mathrm{n}(a_{21}) - \mathrm{t}(\overline{a_{11}}a_{12}\overline{a_{22}}a_{21}) + \mathrm{n}(a_{21})\mathrm{n}(a_{12}))\\
&= \mathrm{n}(a_{12})\mathrm{ddet}\mathbf{A}.
\end{aligned}
$$

Following (6.1)*, we obtain*

$$
\begin{aligned}
|\mathbf{A}|_{21} &= \frac{\mathrm{ddet}\mathbf{A}}{\mathrm{n}(\mathrm{cdet}_1(\mathbf{A}^*\mathbf{A})_{.1}(\mathbf{a}^*_{.2}))}\overline{\mathrm{cdet}_1(\mathbf{A}^*\mathbf{A})_{.1}(\mathbf{a}^*_{.2})}\\
&= \frac{\mathrm{ddet}\mathbf{A}}{\mathrm{n}(a_{12})\mathrm{ddet}\mathbf{A}}\overline{\mathrm{cdet}_1(\mathbf{A}^*\mathbf{A})_{.1}(\mathbf{a}^*_{.2})}\\
&= \frac{1}{\mathrm{n}(a_{12})} \cdot \overline{\mathrm{cdet}_1(\mathbf{A}^*\mathbf{A})_{.1}(\mathbf{a}^*_{.2})}\\
&= \frac{1}{\mathrm{n}(a_{12})} \cdot (\mathrm{n}(a_{12})a_{21} - a_{22}\overline{a_{12}}a_{11})\\
&= a_{21} - a_{22}(a_{12})^{-1}a_{11}.
\end{aligned} \tag{6.8}
$$

The last expression in (6.8) *coincides with the expression* $|\mathbf{A}|_{21}$ *in* (6.7). □

7. Conclusion

In the chapter we consider two approaches to define a noncommutative determinant, row-column determinants and quasideterminants. These approaches of studying of a matrix with entryes from non commutative division ring have their own field of applications.

The theory of the row and column determinants as an extension of the classical definition of determinant has been elaborated for matrices over a quaternion division algebra. It has applications in the theories of matrix equations and of generalized inverse matrices over the quaternion skew field. Now it is in development for matrices over a split quaternion algebra. In the chapter we have extended the concepts of an immanant, a permanent and a determinant to a split quaternion algebra and have established their basic properties.

Quasideterminants of Gelfand-Retax are rational matrix functions that requires the invertibility of certain submatrices. Now they are widely used. Though we can use quasideterminant in any division ring,[5] row-column determinant is more attractive to find solution of system of linear equations when division ring has conjugation.

In the chapter we have derived relations of row-column determinants with quasideterminants of a matrix over a quaternion division algebra. The use of equations (6.1) and (6.2) allows us direct calculation of quasideterminants. It already gives significance in establishing these relations.

References

[1] J. Fan, Determinants and multiplicative functionals on quaternion matrices, *Linear Algebra and Its Applications*, 369 (2003) 193-201.

[2] A. Lauve, Quantum- and quasi-Plucker coordinates, *Journal of Algebra*, 296(2) (2006) 440-461.

[3] T. Suzuki, Noncommutative spectral decomposition with quasideterminant, *Advances in Mathematics*, 217(5) (2008) 2141-2158.

[4] C.R.Gilson, J.J.C.Nimmo, Y.Ohta, Quasideterminant solutions of a non-Abelian Hirota-Miwa equation, *Journal of Physics A: Mathematical and Theoretical*, 40(42) (2007) 12607-12617.

[5] B. Haider, M. Hassan, Quasideterminant solutions of an integrable chiral model in two dimensions, *Journal of Physics A: Mathematical and Theoretical*, 42(35) (2009) 355211-355229.

[6] C.X.Li, J.J.C. Nimmo, Darboux transformations for a twisted derivation and quasideterminant solutions to the super KdV equation, Proceedings of the Royal Society A: Mathematical, *Physical and Engineering Sciences*, 466(2120) (2010) 2471-2493.

[7] I.I. Kyrchei, Cramer's rule for quaternion systems of linear equations, *Journal of Mathematical Sciences*, 155(6) (2008) 839-858.
eprint http://arxiv.org/abs/math/0702447arXiv:math.RA/0702447 (2007)

[8] Ivan I. Kyrchei, The theory of the column and row determinants in a quaternion linear algebra. In: Albert R. Baswell (Eds.), *Advances in Mathematics Research*, 15, pp. 301-359, Nova Sci. Publ., New York, 2012.

[9] Aleks Kleyn, *Lectures on Linear Algebra over Division Ring*,
eprint http://arxiv.org/abs/math.GM/0701238arXiv:math.GM/0701238 (2010).

[5]See for instance sections 2.3, 2.4, 2.5 in the paper [10].

[10] Aleks Kleyn, *Linear Maps of Free Algebra*, eprint http://arxiv.org/abs/1003.1544 arXiv:1003.1544 (2010).

[11] H. Aslaksen. Quaternionic determinants, *Math. Intelligencer*, 18(3) (1996) 57-65.

[12] N. Cohen, S. De Leo, The quaternionic determinant, *The Electronic Journal Linear Algebra*, 7 (2000) 100-111.

[13] F. J. Dyson, Quaternion determinants, *Helvetica Phys. Acta*, 45 (1972) 289-302.

[14] I. Gelfand and V. Retakh, Determinants of Matrices over Noncommutative Rings, *Funct. Anal. Appl.*, 25(2) (1991) 91-102.

[15] I. Gelfand and V. Retakh, A Theory of Noncommutative Determinants and Characteristic Functions of Graphs, *Funct. Anal. Appl.*, 26(4) (1992) 1-20.

[16] I. Gelfand, V. Retakh, *Quasideterminants*, I, eprint http://arxiv.org/abs/q-alg/9705026 arXiv:q-alg/9705026 (1997).

[17] I. Gelfand, S. Gelfand, V. Retakh, R. Wilson, *Quasideterminants*, eprint http://arxiv.org/abs/math.QA/0208146arXiv:math.QA/0208146 (2002).

[18] L. Chen, Definition of determinant and Cramer solutions over quaternion field, *Acta Math. Sinica* (N.S.) 7 (1991) 171-180.

[19] L. Chen, Inverse matrix and properties of double determinant over quaternion field, *Sci. China, Ser. A*, 34 (1991) 528-540.

[20] Lewis D. W. Quaternion algebras and the algebraic legacy of Hamilton's quaternions, *Irish Math. Soc. Bulletin*, 57 (2006) 41-64.

[21] M. Erdoğdu, M. Ożdemir, On complex split quaternion matrices, *Advances in Applied Clifford Algebras*, 23 (2013) 625-638.

[22] M. Erdoğdu, M. Ożdemir, On eigenvalues of split quaternion matrices, *Advances in Applied Clifford Algebras*, 23 (2013) 615-623.

[23] L. Kula, Y. Yayli, Split quaternions and rotations in semi euclidean space E_2^4, *J. Korean Math. Soc.*, 44 (2007) 1313-1327.

[24] Y. Alagöz, K. H. Oral, S.Yüce, Split quaternion matrices, *Miskolc Mathematical Notes*, 13 (2012) 223-232.

[25] D.E. Littlewood, A.R. Richardson, Group characters and algebras, *Proc. London Math. Soe.*, 39 (1935), 150-199.

[26] I.I. Kyrchei, Cramer's rule for some quaternion matrix equations, *Applied Mathematics and Computation*, 217(5) (2010) 2024-2030. eprint http://arxiv.org/abs/1004.4380 arXiv:math.RA/arXiv:1004.4380 (2010).

[27] I.I. Kyrchei, Determinantal representation of the Moore-Penrose inverse matrix over the quaternion skew field, *Journal of Mathematical Sciences*, 180(1) (2012) 23-33.

[28] I.I. Kyrchei, Determinantal representations of the Moore-Penrose inverse over the quaternion skew field and corresponding Cramer's rules, *Linear Multilinear Algebra*, 59 (2011) 413-431. eprint http://arxiv.org/abs/1005.0736arXiv:math.RA/1005.0736 (2010).

[29] Ivan Kyrchei, Explicit representation formulas for the minimum norm least squares solutions of some quaternion matrix equations, *Linear Algebra and Its Applications*, 438(1) (2013) 136-152.

[30] Ivan Kyrchei, Determinantal representations of the Drazin inverse over the quaternion skew field with applications to some matrix equations, *Applied Mathematics and Computation*, 238 (2014) 193-207.

[31] G. Song, Q. Wang, H. Chang. Cramer rule for the unique solution of restricted matrix equations over the quaternion skew field, *Comput. Math. Appl.*, (61) (2011) 1576–1589.

[32] G. Song, Q. Wang, Condensed Cramer rule for some restricted quaternion linear equations, *Applied Mathematics and Computation*, 218 (2011) 3110–3121.

[33] G. Song, Determinantal representation of the generalized inverses over the quaternion skew field with applications, *Applied Mathematics and Computation*, 219 (2012) 656–667.

[34] G. Song, Determinantal representations of the generalized inverses $A^{(2)}_{T,S}$ over the quaternion skew field with applications, *Journal of Applied Mathematics and Computing*, 39 (2012) 201–220.

[35] G. Song, Bott-Duffin inverse over the quaternion skew field with applications, *Journal of Applied Mathematics and Computing*, 41 (2013) 377–392.

[36] G. Song, X. Wang, X. Zhang, On solutions of the generalized Stein quaternion matrix equation, *Journal of Applied Mathematics and Computing*, 43 (2013) 115–131.

[37] G. Song, Characterization of the W-weighted Drazin inverse over the quaternion skew field with applications, *Electronic Journal of Linear Algebra*, 26 (2013) 1–14.

[38] G. Song, H. Chang, Z. Wu, Cramer's rules for various solutions to some restricted quaternionic linear systems, *Journal of Applied Mathematics and Computing*, (2014) in press.

In: Advances in Linear Algebra Research
Editor: Ivan Kyrchei, pp. 325-343

ISBN: 978-1-63463-565-3

Chapter 10

FIRST ORDER CHEMICAL KINETICS MATRICES AND STABILITY OF O.D.E. SYSTEMS

Victor Martinez-Luaces*
Electrochemistry Engineering Multidisciplinary Research Group,
Udelar, Uruguay

Abstract

First order chemical reaction mechanisms are modeled through Ordinary Differential Equations (O.D.E.) systems of the form: $\dot{X} = AX$, X being the chemical species concentrations vector, $\dot{X}$ its time derivative, and A the associated system matrix.

A typical example of these reactions, which involves two species, is the Mutarotation of Glucose [1], which has a corresponding matrix with a null eigenvalue whereas the other one is negative.

A very simple example with three chemical compoundsis grape juice, when it is converted into wine and then transformed into vinegar [2]. A more complicated example,also involving three species, is the adsorption of Carbon Dioxide over Platinum surfaces [3]. Although, in these examples the chemical mechanisms are very different, in both cases the O.D.E. system matrix has two negative eigenvalues and the other one is zero. Consequently, in all these cases that involve two or three chemical species, solutions show a weak stability (i.e., they are stable but not asymptotically). This fact implies that small errors due to measurements in the initial concentrations will remain bounded, but they do not tend to vanish as the reaction proceeds.

In order to know if these results can be extended or not to other chemical mechanisms, a possible general result is studied through an inverse modeling approach, like in previous papers ([3, 4, 5]). For this purpose, theoretical mechanisms involving two or more species are proposed and a general type of matrices – so-called First Order Chemical Kinetics Mechanisms (F.O.C.K.M.) matrices – is studied from the eigenvalues and eigenvectors view point.

This chapter shows that in an F.O.C.K.M. matrix all columns add to zero, all the diagonal elements are non-positive and all the other matrix entries are non-negative. Because of this particular structure, the Gershgorin Circles Theorem [6] can be applied to show that all the eigenvalues are negative or zero. Moreover, it can be proved that in the case of the null

** E-mail address: victorml@fing.edu.uy

eigenvalues – under certain conditions – algebraic and geometric multiplicities give the same number.

As an application of these results, several conclusions about the stability of the O.D.E. solutions are obtained for these chemical reactions, and its consequences on the propagation of concentrations and/or surface concentration measurement errors, are analyzed.

Introduction

A typical example of first order O.D.E. is given by the mathematical model corresponding to the unimolecular chemical reaction. This separable variables O.D.E. problem is usually included in classical mathematics textbooks like Courant [7].

The original problem comes from an important paper published by L.E. Wilhelmy in 1850 [8]. This German physicist studied the inversion of sugar with acids, using a new technique (Polarimetry) for evaluating the dependency of reaction velocity on the quantity of reactants and temperature. As S. Zambelli mentioned, "in this paper probably appeared the first differential equation in chemistry" and "although written in a prestigious journalthe paper passed unnoticed by contemporary scholars. It will be rediscovered only in 1884 by Ostwald" [9].

The O.D.E. proposed by Wilhelmy was:

$$-\frac{dZ}{dt} = MZS \tag{1}$$

where the reaction velocity is the negative derivative of , the sugar quantity in time , is the acid quantity and finally, represents the quantity of inverted sugar in the differential time dt. If there is an excess of acid, then can be considered as a constant.Wilhelmy also verified that remains almost constant in time at a certain temperature. Then, the O.D.E. solution under these conditions is:

$$Z = Z_0 \exp(-MSt) \tag{2}$$

The O.D.E. problem and the corresponding solution is the same as Courant's textbook example, but with a different notation.

This unimolecular first order chemical reaction can be schematized more easily as follows:

$$A \xrightarrow{k} B \tag{3}$$

where A represents sugar, is time, is the kinetic constant and is the inverted sugar. The corresponding O.D.E. mathematical model is:

$$\frac{d[A]}{dt} = -k[A] \tag{4}$$

where $[A]$ represents the concentration of substance , is time and the minus sign expresses that the sugar is being transformed and its concentration diminishes with time.

The O.D.E. solution is:

$$[A]=[A]_0 \exp(-kt) \tag{5}$$

being $[A]_0$ the initial concentration of the reactant.

In both cases (Courant's textbook and Wilhelmy's paper), the authors were concerned about the sugar concentration and its variation against time. Now, if the inverted sugar concentration is also considered, then we will have the following O.D.E. system:

$$\begin{cases} \dfrac{d[A]}{dt} = -k[A] \\ \dfrac{d[B]}{dt} = k[B] \end{cases} \tag{6}$$

that corresponds to the following vectorial O.D.E. problem: $\dot{X} = AX$, where $X = \begin{pmatrix} [A] \\ [B] \end{pmatrix}$ is the concentrations vector, $\dot{X}$ is its time derivative and is the associated system matrix.

One more interesting problem takes place when the reaction between and is reversible. In this case, we will have:

$$A \xrightarrow{K} B \text{ and } B \xrightarrow{k} A \tag{7}$$

where K and are the corresponding kinetic constants.

This situation can be schematized as:

$$A \underset{k}{\overset{K}{\rightleftarrows}} B \tag{8}$$

And the O.D.E. system will be:

$$\begin{cases} \dfrac{d[A]}{dt} = -K[A]+k[B] \\ \dfrac{d[B]}{dt} = K[A]-k[B] \end{cases} \tag{9}$$

Once again, this system can be written more easily as $\dot{X} = AX$, being in this case:

$$A = \begin{pmatrix} -K & k \\ K & -k \end{pmatrix} \tag{10}$$

A real-life example of this chemical reaction is given by the Mutarotation of Glucose [1].

It is important to note that in both cases (i.e., the irreversible reaction and the reversible one), there is a negative eigenvalue and the other one is zero. In fact, as the kinetic constants are always positive, the matrix eigenvalues are and for the irreversible reaction (Eq. 3), while in the reversible one (Eq. 8), the corresponding eigenvalues are and (see [3] for this result).

An example involving three different species takes place when a chemical substance reacts giving a chemical compound which reacts again to give C, the final product of the whole reactions sequence [2]. This is a very common situation in real life, for example, when grape juice is converted into wine and then, it is transformed into vinegar.

This sequence of reactions is a first order chemical kinetics mechanism (F.O.C.K.M), which can be represented as follows:

$$E_1 \xrightarrow{k_1} E_2 \xrightarrow{k_2} E_3 \tag{11}$$

where k_1 and are the corresponding kinetic constants.

In the production of wine and vinegar, is a carbohydrate, is ethylic alcohol, and is acetic acid. The mathematical model for this F.O.C.K.M. is the following:

$$\begin{cases} \frac{d[E_1]}{dt} = -k_1[E_1] \\ \frac{d[E_2]}{dt} = k_1[E_1] - k_2[E_2] \\ \frac{d[E_3]}{dt} = k_2[E_2] \end{cases} \tag{12}$$

and the associated matrix is:

$$A = \begin{pmatrix} -k_1 & 0 & 0 \\ k_1 & -k_2 & 0 \\ 0 & k_2 & 0 \end{pmatrix} \tag{13}$$

whose eigenvalues are obvious , and (note that the system matrix is a triangular one).

Another interesting case involving three species comes from the study of the adsorption of Carbon Dioxide (CO_2) on Platinum (Pt) surfaces [3-10-11]. The research found three different adsorbates, E_1 , E_2 and E_3, and their surface concentrations were measured by conventional electrochemical techniques. Several mechanisms were proposed and the theoretical curves were compared with the experimental ones. The best fit was obtained by the following mechanism:

$$\begin{aligned} &\mathrm{E}_1 \xrightarrow{k_1} \mathrm{E}_2 \\ &\mathrm{E}_1 \xrightarrow{k_2} \mathrm{E}_3 \\ &\mathrm{E}_2 \underset{k_{-3}}{\overset{k_3}{\rightleftarrows}} \mathrm{E}_3 \end{aligned} \tag{14}$$

If represents the adsorbate E_i surface concentration and k_1, k_2, k_3 and are the kinetic constants, the corresponding mathematical model [3] is:

$$\begin{cases} \frac{d[E_1]}{dt} = -(k_1 + k_2)[E_1] \\ \frac{d[E_2]}{dt} = k_1[E_1] - k_3[E_2] + k_{-3}[E_3] \\ \frac{d[E_3]}{dt} = k_2[E_1] + k_3[E_2] - k_{-3}[E_3] \end{cases} \tag{15}$$

The associated system matrix is:

$$A = \begin{pmatrix} -k_1 - k_2 & 0 & 0 \\ k_1 & -k_3 & k_{-3} \\ k_2 & k_3 & -k_{-3} \end{pmatrix} \tag{16}$$

which has the following eigenvalues: $\lambda_1 = -k_1 - k_2 < 0$, and $\lambda_3 = 0$, as it was proved in [3].

It can be noted that once again – like in all the other cases – all the eigenvalues are negative, except one, which is zero.

In all the previous examples, the chemical or electrochemical processes considered were quite different; however in all of them the mathematical models showed a certain regularity that poses the following questions:

- Is there any general form for the associated matrices corresponding to the F.O.C.K.M. proposed in the examples? If so, could this form be generalized to any F.O.C.K.M.?
- Can it be proved that all the eigenvalues of the F.O.C.K.M. matrices are negative except one, which is always zero?
- In all the previous examples the O.D.E. system solutions are stable but not asymptotically [3-12]. Is it a general result valid for any F.O.C.K.M.?
- Which would be the practical consequences of the previous statements if they were true?

These questions, their answers, and possible generalizations among other issues will be considered in this book chapter, following an inverse modeling approach [3-4-5-13-14], i.e., proposing different theoretical F.O.C.K.M. and analyzing the corresponding O.D.E. systems and their associated matrices, in order to get general results and/or interesting counterexamples.

A General Form for F.O.C.K.M. Matrices

Let us consider again two chemical substances and B. If all the possible first order chemical reactions take place, we will have a mechanism like: and $B \xrightarrow{k} A$, previously analyzed in (Eq. 7). For this mechanism, we refer back to Eq. 9 and 10, the corresponding O.D.E. system is:

$$\begin{cases} \dfrac{d[A]}{dt} = -K[A] + k[B] \\ \dfrac{d[B]}{dt} = K[A] - k[B] \end{cases}$$

and its associated matrix is:

$$A = \begin{pmatrix} -K & k \\ K & -k \end{pmatrix}$$

It is easy to note that in any other simpler F.O.C.K.M., like (Eq. 3), the O.D.E. system (refers back to Eq. 6):

$$\begin{cases} \dfrac{d[A]}{dt} = -k[A] \\ \dfrac{d[B]}{dt} = k[B] \end{cases}$$

and its associated matrix are just particular cases of (Eq. 9) and (Eq. 10), with a null constant and a different notation.

In a similar way, if three chemical species E_1 , and E_3 are involved, and all the possible first order reactions take place, we havethe following mechanism:

$$E_1 \underset{K_{-1}}{\overset{K_1}{\rightleftarrows}} E_2 ,\ E_2 \underset{K_{-2}}{\overset{K_2}{\rightleftarrows}} E_3 \text{ and } E_1 \underset{K_{-3}}{\overset{K_3}{\rightleftarrows}} E_3 \tag{17}$$

using the notation of a previous article [3].

For such mechanism, the mathematical model can be written as:

$$\begin{pmatrix} x \\ y \\ z \end{pmatrix}' = \begin{pmatrix} -k_1 - k_3 & k_{-1} & k_{-3} \\ k_1 & -k_{-1} - k_2 & k_{-2} \\ k_3 & k_2 & -k_{-2} - k_{-3} \end{pmatrix} \begin{pmatrix} x \\ y \\ z \end{pmatrix} \tag{18}$$

where the concentrations $[E_1], [E_2]$ and were replaced by x, y and so as to simplify the notation.

Once again, it can be observed that the systems given by (Eq.12) and (Eq.15) and their corresponding associated matrices – see (Eq. 13) and (Eq. 16) – are just particular cases if we substitute several matrix entries by zero.

$$\text{So, } \begin{pmatrix} -K & k \\ K & -k \end{pmatrix} \text{(Eq. 10) and } \begin{pmatrix} -k_1 - k_3 & k_{-1} & k_{-3} \\ k_1 & -k_{-1} - k_2 & k_{-2} \\ k_3 & k_2 & -k_{-2} - k_{-3} \end{pmatrix} \tag{19}$$

can be considered as the general forms for and F.O.C.K.M matrices.

In order to get an general form, let us consider chemical species , E_2 ,... E_n and suppose that all the possible first order chemical reactions take place. If any of these reactions does not occur, then, the corresponding constant will be considered null. For instance, (Eq. 14) can be rewritten as:

$$E_1 \xrightarrow{k_{12}} E_2,\ E_1 \xrightarrow{k_{13}} E_3,\ E_2 \xrightarrow{k_{23}} E_3$$

and

$$E_3 \xrightarrow{k_{32}} E_2 \text{, with } k_{21} = k_{31} = 0 \tag{20}$$

Then, in a general form, corresponding to a F.O.C.K.M., direct reactions involving species are:

$$E_1 \xrightarrow{k_{12}} E_2,\ E_1 \xrightarrow{k_{13}} E_3,\ \dots,\ E_1 \xrightarrow{k_{1n}} E_n \tag{21}$$

and the opposed reactions are:

$$E_2 \xrightarrow{k_{21}} E_1,\ E_3 \xrightarrow{k_{31}} E_1,\ \dots,\ E_n \xrightarrow{k_{n1}} E_1 \tag{22}$$

Consequently, the corresponding O.D.E. for the variation of concentration with time is:

$$\frac{d[E_1]}{dt} = -k_{12}[E_1] - k_{13}[E_1] - \dots - k_{1n}[E_1] + k_{21}[E_2] + k_{31}[E_3] + \dots + k_{n1}[E_n] \tag{23}$$

or:

$$\frac{d[E_1]}{dt} = -s_1[E_1] + k_{21}[E_2] + k_{31}[E_3] + \dots + k_{n1}[E_n] \tag{24}$$

being

$$s_1 = k_{12} + k_{13} + \dots + k_{1n} = \sum_{j \neq 1} k_{1j} \tag{25}$$

Following a similar reasoning for species we have:

$$\frac{d[E_i]}{dt} = -s_i[E_i] + k_{1i}[E_1] + \dots + k_{i-1,i}[E_{i-1}] + k_{i+1,i}[E_{i+1}] \dots + k_{ni}[E_n] \tag{26}$$

being

$$s_i = k_{i1} + \dots + k_{i,i-1} + k_{i,i+1} \dots + k_{in} = \sum_{j \neq i} k_{ij} \tag{27}$$

Then, the general F.O.C.K.M. mathematical model is:

$$\begin{pmatrix} [E_1] \\ [E_2] \\ \vdots \\ [E_n] \end{pmatrix}' = \begin{pmatrix} -s_1 & k_{21} \cdots & k_{n1} \\ k_{12} & -s_2 \cdots & k_{n2} \\ \vdots & \vdots & \vdots \\ k_{1n} & k_{2n} \cdots & -s_n \end{pmatrix} \begin{pmatrix} [E_1] \\ [E_2] \\ \vdots \\ [E_n] \end{pmatrix} \tag{28}$$

and its associated matrix is:

$$A = \begin{pmatrix} -s_1 & k_{21} \cdots & k_{n1} \\ k_{12} & -s_2 \cdots & k_{n2} \\ \vdots & \vdots & \vdots \\ k_{1n} & k_{2n} \cdots & -s_n \end{pmatrix} \tag{29}$$

where all the non-diagonal entries are non-negative and the diagonal elements are $-s_i = -\sum_{j \neq i} k_{ij}$, so all the matrix columns add to zero.

To summarize, the characteristics that define these F.O.C.K.M. matrices, are the following:

- A is a $n \times n$ matrix with real entries a_{ij}.
- The non-diagonal entries are always non-negative numbers, i.e., $a_{ij} \geq 0 \quad \forall i \neq j$.
- The diagonal elements are $-s_i$, s_i is the sum of the non-diagonal entries in the i th column.

Two obvious corollaries can be easily obtained for this general form:

- $\det(A) = 0$.
- $\lambda = 0$ is an eigenvalue of A.

The first one is a straightforward conclusion from $row_1 + row_2 + \ldots + row_n = \vec{0}$ (Eq. 30), and the second one is an obvious consequence of having a null determinant.

To end this section, two more comments can be made. The first one is that if all the reactions involved are reversible ones, then:

$$E_i \xrightarrow{k_{ij}} E_j \text{, with } k_{ij} \neq 0 \tag{31}$$

$$E_j \xrightarrow{k_{ji}} E_i \text{, with } k_{ji} \neq 0 \tag{32}$$

So, in this particular case, all the non-diagonal entries are positive and all the diagonal ones are negative. These mechanisms where all the reactions are reversible were already studied in another paper [12].

The second comment is about the previous corollaries, i.e., and $\lambda = 0$ is an eigenvalue of A. Both of them can be obtained because of Lavoisier's Law of Conservation of Mass. This different approach [3] establishes a strong connection with the original modeling problem, but in the opposite way, getting a mathematical result because of a chemical law.

This inverse modeling approach was widely used in previous books and articles like [3], [4], [5], [12], [13] and [14] and it will be utilized one more time in this chapter in order to get other general results.

The Gershgorin Circle Theorem and Its Application to F.O.C.K.M. Problems

The Gershgorin circle theorem was first published by S. A. Gershgorin in 1931 and may be used to bind the spectrum of a complex $n \times n$ matrix [6]. If with $i, j \in \{1,\ldots,n\}$ are the matrix entries and $R_i = -\sum_{j \neq i} |a_{ij}|$ is the sum of the non-diagonal entries modules in the i th row, then $\overline{D}(a_{ii}, R_i)$, the closed disc centered at with radius R_i, is called a Gershgorin disc. The theorem states that every eigenvalue of lies within at least one of the Gershgorin discs $\overline{D}(a_{ii}, R_i)$.

A corollary of this theorem can be obtained by applying the Gershgorin circle theorem to A^T. The straightforward conclusion is that all the eigenvalues of A lie within the Gershgorin discs corresponding to the columns of A.

For a general F.O.C.K.M. involving species E_1, … the O.D.E. system is like (Eq. 28) and the associated matrix is given by (Eq. 29). For instance, for the first column the Gershgorin disc is where $R_1 = \sum_{j \neq 1} |k_{1j}| = \sum_{j \neq 1} k_{1j} = s_1$. Then, the first column Gershgorin disc is and the same happens with all the other columns of A (Eq. 29), so all the eigenvalues lie in $\bigcup_{i=1}^{n} \overline{D}(-s_i, s_i)$

It is important to note that every disc is centered in a non-positive number $-s_i \leq 0$ and the circle radius is the absolute value of this number (i.e., $|-s_i| = s_i$), then all the Gershgorin discs are tangent to the imaginary axis.

One of these circles can be schematized like in Figure 1.

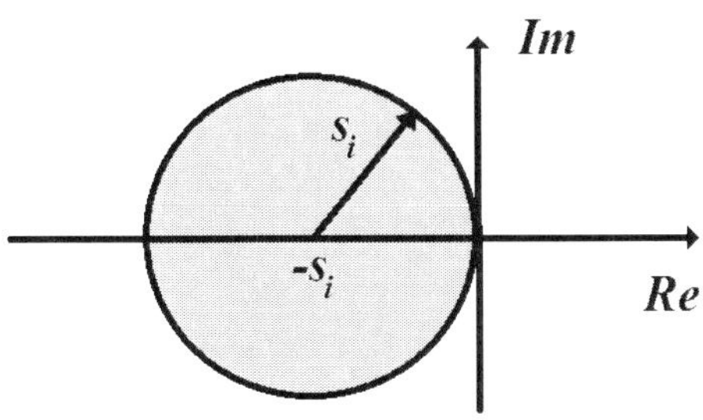

Figure 1. The i thGershgorin circle $\overline{D}(-s_i, s_i)$.

Then, all the Gershgorin circles are contained in the closure of the left semi-plane, $\overline{C}_{(-)}$, i.e.,

$$\bigcup_{i=1}^{n} \overline{D}(-s_i, s_i) \subset \{z \in C / \mathrm{Re}(z) \leq 0\} = \overline{C}_{(-)} \tag{33}$$

This situation is shown in the next figure:

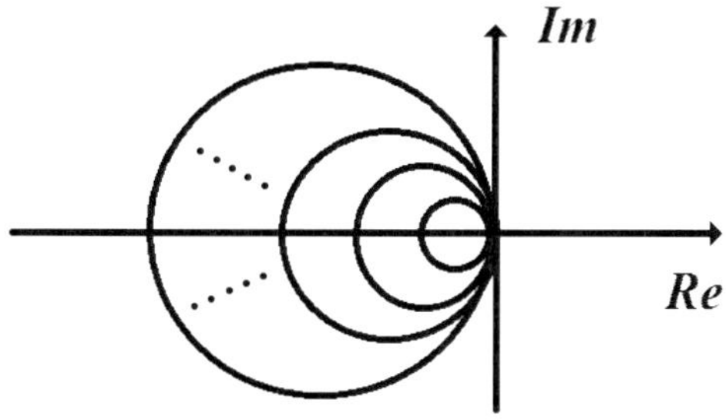

Figure 2. The Gershgorin circles.

As a consequence of (Eq. 33), the spectrum of A is also included in $\overline{C}_{(-)}$, or in other words:

$$\lambda_i \in \overline{C}_{(-)} \quad \forall i = 1,2,\ldots,n \tag{34}$$

According to the previous results, it follows that:

$$\mathrm{Re}(\lambda_i) \leq 0 \ \ \forall i = 1,2,\ldots,n \tag{35}$$

and

$$\mathrm{Re}(\lambda_k) = 0 \Leftrightarrow \lambda_k = 0 \tag{36}$$

From the O.D.E. solutions point of view, the non-zero eigenvalues give linear combinations of functions like:

$$\exp(\lambda t), t\exp(\lambda t), t^2 \exp(\lambda t), \ldots, t^p \exp(\lambda t) \tag{37}$$

depending on the algebraic multiplicity (A.M.) and the corresponding geometric multiplicity (G.M.) of the eigenvalue λ.

An interesting example of this kind of analysis was included in a previous article [3], where a F.O.C.K.M. involving three species was studied and the corresponding O.D.E. system was analyzed. In that mechanism, three possible cases were described: three different eigenvalues, a double eigenvalue and a simple one, and finally, a unique triple eigenvalue. In all those cases the O.D.E. system solutions showed a weak stability, i.e., they were stable solutions but not asymptotically stable.

In order to generalize the previous result already mentioned, to any other F.O.C.K.M., a general matrix like (Eq. 29) will be considered. For this general form, above in this chapter it was proved that: (Eq. 35) and (Eq. 36), so there are two different cases to analyze: the eigenvalues with a negative real part and the null eigenvalue.

Case 1. is an eigenvalue with $\mathrm{Re}(\lambda)<0$

In this case, if is an eigenvalue with $a<0$, the O.D.E. solutions are a linear combination of $\exp(\lambda t), t\exp(\lambda t), t^2\exp(\lambda t), \ldots, t^p\exp(\lambda t)$, where depends on the A.M. and G.M. corresponding to λ. Taking into account that $\exp(\lambda t)=e^{at}(\cos bt + i\sin bt)$, being $a<0$, it follows that and the same happens with all the other functions: $t\exp(\lambda t), t^2\exp(\lambda t), \ldots, t^p\exp(\lambda t)$.

Then, all the O.D.E. solutions associated with the eigenvalue tend to vanish with time, independently of corresponding A.M. and/or G.M.

Case 2. $\lambda=0$

The null eigenvalue is always present in F.O.C.K.M. problems as it was proved above in this chapter. As it was mentioned before, the same result can be obtained from a different approach, because of Lavoisier's Law of Conservation of Mass [3].

The O.D.E. system solutions associated with the null eigenvalue are linear combinations of the following functions: $\left\{e^{0t}, te^{0t}, \ldots, t^q e^{0t}\right\}$, or the equivalent: $\left\{1, t, \ldots, t^q\right\}$. Then, the solutions due to the null eigenvalue are polynomial functions, which grade depends on both the A.M. and the G.M., corresponding to $\lambda=0$.

It follows straightforward that only if $q=0$ will the polynomial solutions remain bound when tends to infinity.

To sum up the previous results, it can be stated that only the null eigenvalue – and particularly, its A.M. and G.M. – is relevant to make predictions about the stability of the O.D.E. system solutions. The study for the different cases, corresponding to the eigenvalue $\lambda=0$, will be carried out in the following section of this chapter, through an inverse modeling approach.

The Multiplicities of the Null Eigenvalue

In the previous sections, a general F.O.C.K.M. was analyzed. For this mechanism, the mathematical model is an O.D.E. linear system whose associated matrix has its spectrum in the closure of the left semi-plane in the complex numbers. With the purpose of studying the stability of the O.D.E. solutions, only the A.M. and the G.M. corresponding to are relevant.

Taking into account all these facts, the following questions must be considered:

- Is it possible to find a F.O.C.K.M. with a multiple (double, triple, etc.) null eigenvalue?
- If so, is it possible to have a F.O.C.K.M. such that $AM_{\lambda=0} \neq GM_{\lambda=0}$?

In order to answer these questions an inverse modeling approach will be put into practice, like in other papers [4-12-13-14] and books [3-5].

As W. Blum et al. mentioned, modeling is a process that goes from the real world towards mathematics [15]. Examples of this process were presented above when (Eq. 9) was proposed as the mathematical model for the F.O.C.K.M. considered in (Eq. 7), or when (Eq. 12) was obtained from the chemical reactions of (Eq. 11) in the wine/vinegar example, and finally when (Eq. 15) was derived from (Eq. 14) in the adsorption of CO_2 on surfaces.

In an inverse modeling approach, O.D.E. systems and/or their associated matrices are the inputs and the main objective is to find a chemical reaction or a chemical mechanism that fits with the proposed O.D.E. system or its associated matrix. Then, only matrices such as those described in the second section of this chapter, e.g., (Eq.29), will be considered,since these are the unique matrices that are able to fit with a F.O.C.K.M., as it was proved before.

In order to answer the first question proposed – about double, triple, etc.,and null eigenvalues – the following matrix can be analyzed:

$$A = \begin{pmatrix} -5 & 0 & 0 \\ 3 & 0 & 0 \\ 2 & 0 & 0 \end{pmatrix} \tag{38}$$

It is easy to observe that A has two different eigenvalues: (simple) and (double). Therefore, the inverse modeling question is, if it exists, a F.O.C.K.M. that fits with this matrix and the answer is affirmative, since the F.O.C.K.M.:

$$E_1 \xrightarrow{k_{12}} E_2 \, , E_1 \xrightarrow{k_{13}} E_3 \tag{39}$$

gives an O.D.E. like:

$$\begin{cases} \dfrac{d[E_1]}{dt} = -(k_{12} + k_{13})[E_1] \\ \dfrac{d[E_2]}{dt} = k_{12}[E_1] \\ \dfrac{d[E_3]}{dt} = k_{13}[E_1] \end{cases} \tag{40}$$

and its associated matrix is:

$$\begin{pmatrix} -k_{12} - k_{13} & 0 & 0 \\ k_{12} & 0 & 0 \\ k_{13} & 0 & 0 \end{pmatrix} \tag{41}$$

This matrix corresponds with the matrix given in (Eq. 38) if and $k_{13} = 2$. It is interesting to note that this mechanism is like the F.O.C.K.M. corresponding to the adsorption of on surfaces (Eq. 14), without the reversible reaction between and E_3 .

The mechanism can be easily generalized for n chemical species E_1, E_2 ... in the following way: $E_1 \xrightarrow{k_{12}} E_2$, $E_1 \xrightarrow{k_{13}} E_3$, ..., (Eq. 42), then:

$$\begin{cases} \frac{d[E_1]}{dt} = -(k_{12} + k_{13} + \ldots + k_{1n})[E_1] \\ \frac{d[E_2]}{dt} = k_{12}[E_1] \\ \vdots \\ \frac{d[E_n]}{dt} = k_{1n}[E_1] \end{cases} \tag{43}$$

and its associated matrix is:

$$\begin{pmatrix} -s_1 & 0 & \cdots & 0 \\ k_{12} & 0 & \cdots & 0 \\ \vdots & \vdots & & \vdots \\ k_{1n} & 0 & \cdots & 0 \end{pmatrix} \tag{44}$$

being $s_1 = k_{12} + k_{13} + \ldots + k_{1n} = \sum_{j \neq 1} k_{1j}$

It is obvious that has two different eigenvalues: $\lambda_1 = -s_1 < 0$ (simple) and $\lambda_2 = 0$ with

$AM_{\lambda=0} = n-1$.

In the previous examples was found in (Eq. 10), (Eq. 13) and (Eq. 16), and was found in (Eq. 41) and (Eq. 44).

Now, an interesting question arises: Is it possible to get an intermediate for the null eigenvalue?

Once again the answer is affirmative, for instance if

$$A = \begin{pmatrix} -s_1 & k_{21} & 0 & 0 \\ k_{12} & -s_2 & 0 & 0 \\ k_{13} & k_{23} & 0 & 0 \\ k_{14} & k_{24} & 0 & 0 \end{pmatrix} \tag{45}$$

Then, algebraic manipulations give:

$p(\lambda) = \det(A - \lambda I) = \lambda^2[(s_1 + \lambda)(s_2 + \lambda) - k_{12}k_{21}]$ and if we take $k_{12} = k_{13} = k_{14} = k_{21} = k_{23} = k_{24} = 1$ the characteristic polynomial will be $p(\lambda) = \lambda^2(2+\lambda)(4+\lambda)$, which has two negative eigenvalues: $\lambda_1 = -2 < 0$ (simple), $\lambda_2 = -4 < 0$ (simple) and $\lambda_3 = \lambda_4 = 0$ (double).

An inverse modeling approach suggests considering a mechanism with one reversible reaction: $E_1 \underset{k_{21}}{\overset{k_{12}}{\rightleftarrows}} E_2$ and four irreversible ones: $E_1 \xrightarrow{k_{13}} E_3$, $E_1 \xrightarrow{k_{14}} E_4$, and .

This F.O.C.K.M can be easily generalized for n species E_1, ... E_n. If we consider the reversible reactions: $E_i \underset{k_{ji}}{\overset{k_{ij}}{\rightleftarrows}} E_j$ being and the irreversible ones: $E_i \xrightarrow{k_{ij}} E_j$ being and then, the corresponding matrix is:

$$A = \begin{pmatrix} -s_1 & \cdots & k_{n-p,1} & 0 & \cdots & 0 \\ \vdots & \ddots & \vdots & \vdots & & \vdots \\ k_{1,n-p} & \cdots & -s_{n-p} & 0 & \cdots & 0 \\ \vdots & & \vdots & 0 & \cdots & 0 \\ \vdots & & \vdots & \vdots & \ddots & \vdots \\ k_{1n} & \cdots & k_{n-p,n} & 0 & \cdots & 0 \end{pmatrix} \quad (46)$$

In this matrix, if the kinetic constants are well chosen, the first $n-p$ columns will be linear independent and $AM_{\lambda=0} = p$.

The previous examples show that the A.M. corresponding to the null eigenvalue can be so, the next question is: can be equal to n?

In this case, the answer is negative as can be proved by analyzing the trace of A. It is well known that is by definition the sum of the diagonal entries of and it also equalsthe sum of the eigenvalues [16]. So, on one hand we have: (Eq.47), and on the other hand:

$$tr(A) = -s_1 - s_2 - \ldots - s_n = -\sum_{i,j=1}^{n} k_{ij} \quad (48)$$

If then and so must be zero. In this theoretical situation and there are no chemical reactions or mechanism to be studied.

Summarizing all these results, (Eq. 49), and can take any of these values.

This result gives an affirmative answer to the first question posed at the beginning of the section.

In order to respond to the second question, the G.M. corresponding to the eigenvalue must be analyzed.

If only two species are involved, and the must be 1, due to (Eq. 49) and in this case is a simple eigenvalue and so, $GM_{\lambda=0} = AM_{\lambda=0} = 1$.

If three chemical compounds are involved in the F.O.C.K.M., the analysis is not as simple as in the previous case. In this new situation, and taking into account the result of (Eq. 49).

In the first sub-case, i.e., $AM_{\lambda=0} = 1$, once again is a simple eigenvalue and the result is: $GM_{\lambda=0} = AM_{\lambda=0} = 1$.

In the second sub-case (i.e., $AM_{\lambda=0} = 2$) a priori the can be 1 or 2. We will see that in F.O.C.K.M. matrices it is always 2. For this purpose, let us consider the general 3×3 matrix:

$$A = \begin{pmatrix} -s_1 & k_{21} & k_{31} \\ k_{12} & -s_2 & k_{32} \\ k_{13} & k_{23} & -s_3 \end{pmatrix} \tag{50}$$

where

$$s_1 = k_{12} + k_{13} \text{, and } s_3 = k_{31} + k_{32} \tag{51}$$

The characteristic polynomial is:

$$p(\lambda) = \det(A - \lambda I) = \det \begin{pmatrix} -s_1 - \lambda & k_{21} & k_{31} \\ k_{12} & -s_2 - \lambda & k_{32} \\ k_{13} & k_{23} & -s_3 - \lambda \end{pmatrix} \tag{52}$$

which can be written as:

$$p(\lambda) = \det(A - \lambda I) = c_3 \lambda^3 + c_2 \lambda^2 + c_1 \lambda + c_0 \tag{53}$$

In this formula, several coefficients can be easily determined [16]. In fact, it is well known that: , $c_2 = tr(A)$ and , so in this case:

$$\begin{cases} c_3 = (-1)^n = (-1)^3 = -1 \\ c_2 = tr(A) = -\sum_{i=1}^{n} s_i = -\sum_{i,j=1}^{3} k_{ij} < 0 \\ c_0 = \det(A) = 0 \end{cases} \tag{54}$$

then:

$$p(\lambda) = -\lambda^3 + [tr(A)]\lambda^2 + c_1 \lambda \tag{55}$$

In this polynomial, c_1 must be zero in order to have $AM_{\lambda=0} = 2$. Developing the determinant in (Eq. 52) it is easy to obtain:

$$c_1 = -s_1 s_2 - s_1 s_3 - s_2 s_3 + k_{12}k_{21} + k_{13}k_{31} + k_{23}k_{32} \tag{56}$$

and if c_1 must be zero, then:

$$s_1 s_2 + s_1 s_3 + s_2 s_3 = k_{12}k_{21} + k_{13}k_{31} + k_{23}k_{32} \tag{57}$$

Finally, combining (Eq. 51) and (Eq. 57) the result is:

$$k_{12}k_{23}+k_{13}k_{21}+k_{13}k_{23}+k_{12}k_{31}+k_{12}k_{32}+k_{13}k_{32}+k_{21}k_{31}+k_{21}k_{32}+k_{23}k_{31}=0 \qquad (58)$$

The constants k_{ij} are non-negative $\forall i, j$, therefore, all the products included in (Eq. 58) must be zero. In order to analyze all the possibilities, the tree diagram of figure 3 will be followed.

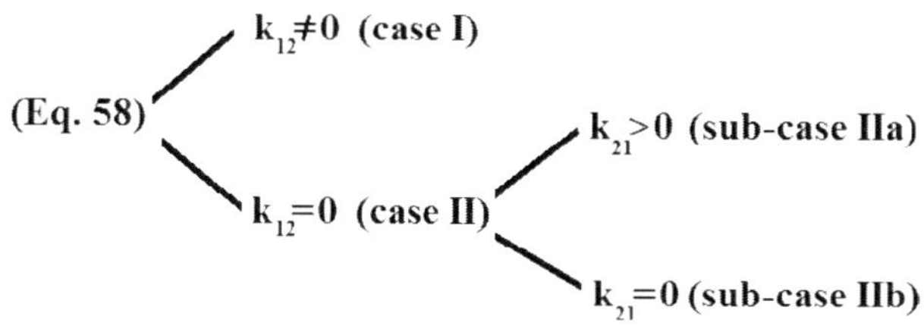

Figure 3. Tree diagram for the analysis of possible cases in (Eq. 58).

Case I. $k_{12} \neq 0$

If and all the products in (Eq. 58) must be zero, then: $k_{23}=0$, and $k_{32}=0$, and this result implies that $k_{13} \neq 0$. In fact, if then, the whole mechanism will be only: $E_1 \rightleftarrows E_2$ and species is not involved in the F.O.C.K.M.

So, here we have: $k_{12} \neq 0$, and $k_{23}=k_{31}=k_{32}=0$. Moreover, being the second product in (Eq. 58) $k_{13}k_{21}=0$ and $k_{13} \neq 0$, then $k_{21}=0$ and the F.O.C.K.M. is: $E_1 \xrightarrow{k_{12}} E_2$, $E_1 \xrightarrow{k_{13}} E_3$. This case was already considered (Eq. 39-40) and the associated matrix was:

$$\begin{pmatrix} -k_{12}-k_{13} & 0 & 0 \\ k_{12} & 0 & 0 \\ k_{13} & 0 & 0 \end{pmatrix} \qquad (41)$$

This matrix has a null double eigenvalue, such that the canonic vectors: $\vec{e}_2=(0,1,0)$ and are associated eigenvectors and so, $GM_{\lambda=0}=AM_{\lambda=0}=2$.

Case II. $k_{12}=0$

If then k_{21} may be positive (sub-case IIa) or (sub-case IIb). Both sub-cases will be analyzed in the following paragraphs.

Sub-case IIa. and $k_{21}>0$

If the constant is positive, then since all products in (Eq. 58) must be zero. The remaining mechanism is given only by: $E_2 \xrightarrow{k_{21}} E_1$, $E_2 \xrightarrow{k_{23}} E_3$ (Eq. 59) and the associated matrix is:

$$\begin{pmatrix} 0 & k_{21} & 0 \\ 0 & -(k_{21}+k_{23}) & 0 \\ 0 & k_{23} & 0 \end{pmatrix} \tag{60}$$

It is easy to observe that the canonic vectors: $\vec{e}_1 = (1,0,0)$ and are eigenvectors associated with the null eigenvalue and so, once again: $GM_{\lambda=0} = AM_{\lambda=0} = 2$.

Sub-case IIb. and $k_{21} = 0$

If then (Eq. 58) is converted into: (Eq. 61), where as always, $k_{ij} \geq 0 \ \forall i, j$.

It is important to note that if then $k_{13} = k_{31} = 0$, since all products in (Eq. 61) – particularly the first and the last products – must be zero. In this situation, $k_{12} = k_{21} = 0$ and $k_{13} = k_{31} = 0$, so the species E_1 is not involved in the mechanism. Then, if three chemical substances are considered, must be zero and (Eq. 61) is converted into:

$$k_{13}k_{32} = 0 \tag{62}$$

We already have $k_{12} = k_{21} = 0$ and $k_{23} = 0$, so must be positive (if not, the species E_2 is not part of the F.O.C.K.M.). Then, it follows from (Eq. 62) that $k_{13} = 0$.

To summarize this sub-case, we have $k_{12} = k_{21} = 0$, $, k_{23} = 0$ and the remaining mechanism is $E_3 \xrightarrow{k_{31}} E_1$, $E_3 \xrightarrow{k_{32}} E_2$ (Eq. 63), and the associated matrix is:

$$\begin{pmatrix} 0 & 0 & k_{31} \\ 0 & 0 & k_{32} \\ 0 & 0 & -k_{31}-k_{32} \end{pmatrix} \tag{64}$$

For this matrix, it is easy to observe that the canonic vectors $\vec{e}_1 = (1,0,0)$ and are eigenvectors associated with the null eigenvalue and we have again: $GM_{\lambda=0} = AM_{\lambda=0} = 2$, like in the previous sub-case.

As a summary of this section, in all mechanisms involving two or three species, the A.M. and the G.M., corresponding to the null eigenvalue are the same. The consequences of this result on the stability of the O.D.E. solutions and its possible generalizations, among other conclusions, will be the core of the next section.

Conclusion

In the preceding sections, a general form for matrices associated to F.O.C.K.M. problems was obtained.

Because of this structure, several properties were proved. Particularly, for a general $n \times n$ matrix A, corresponding to a given F.O.C.K.M., the following statements were demonstrated:

- $\det(A) = 0$
- If is an eigenvalue of A, then $\mathrm{Re}(\lambda) \leq 0$
- $\mathrm{Re}(\lambda) = 0$ if and only if $\lambda = 0$
- For the null eigenvalue is and can take any of these possible values.

If two or three chemical substances are considered, the matrix that corresponds to this F.O.C.K.M. verifies that $GM_{\lambda=0} = AM_{\lambda=0}$.

This algebraic result has an analytical corollary: the O.D.E. solutions for F.O.C.K.M. involving two or three species are always stable, but not asymptotically. This weak stability has an important chemical consequence, since it implies that small errors in the initial concentration measurements will remain bound as the reactions take place, but they will not tend to disappear when $t \to +\infty$.

If more than three substances are involved in the F.O.C.K.M., this weak stability result can easily be generalized in the particular case where only reversible reactions are considered [12].

Other qualitative results can be obtained by analyzing the form of the solutions for the O.D.E. linear system. For instance, the existence and number of inflexion points in curves of vs. were previously obtained in [3], among other conclusions.

It is important to note that the cases studied in this chapter – i.e., F.O.C.K.M. involving two or three species – are especially important since they are the most common situations in chemical kinetics problems and they appear regularly in the corresponding mathematical models.

Finally, the study of other stability properties and qualitative results, for any number of reactants and for any kind of chemical reactions (reversible, irreversible, second and third order reactions, etc.), represents a challenging problem and an opportunity for further research in this area.

Acknowledgments

The author wishes to thank Marjorie Chaves for useful discussions relating to this chapter.

References

[1] Guerasimov, Y.A. et al. (1995), *Physical Chemistry,* 2nd Ed., Houghton-Mifflin, Boston.

[2] Martinez-Luaces, V. (2005), Engaging secondary school and university teachers in modelling: some experiences in South American countries, *International Journal of Mathematical Education in Science and Technology*,Vol. 36, N° 2–3, 193–205.

[3] Martinez-Luaces, V. (2012), Chemical Kinetics and Inverse Modelling Problems, in *Chemical Kinetics*, In Tech Open Science Eds., Rijeka, Croatia.

[4] Martinez-Luaces, V. (2009), Modelling and Inverse Modelling: Experiences with O.D.E. linear systems in engineering courses, *International Journal of Mathematical Education in Science and Technology*,Vol. 40, N° 2, 259-268.

[5] Martinez-Luaces, V. (2012), *Problemas inversos y de modelado inverso en Matemática Educativa*, Editorial Académica Española, Saarbrücken, Germany.

[6] Varga, R.S. (2004),*Geršgorin and His Circles*. Springer-Verlag, Berlin.

[7] Courant, R. (1937). *Differential and Integral Calculus*. Volume I, 2nd Edition. Blackie & Son Limited. London and Glasgow.

[8] Wilhelmy, L. (1850), Über das Gesetz, nachwelchem die Einwirkung der Säuren auf den Rohrzuckerstattfindet, *Pogg. Ann.* Vol. 81, pp. 413-433.Available from http://gallica.bnf.fr/ark:/12148/bpt6k15166k/f427.table

[9] Zambelli, S.(2012), Chemical Kinetics, a Historical Introduction, in *Chemical Kinetics*, In Tech Open Science Eds., Rijeka, Croatia.

[10] Zinola, F., Méndez, E. & Martínez Luaces, V., (1997). Modificación de estados adsorbidos de Anhídrido Carbónico reducido por labilización electroquímica en superficies facetadas de platino. *Proceedings of X Congreso Argentino de Fisicoquímica*, Tucumán, Argentina.

[11] Martínez- Luaces, V. & Guineo Cobs, G., (2002). Un problema de Electroquímica y su Modelación Matemática, *Anuario Latinoamericano de Educación Química.* Año 2002, pp. 272 – 276.

[12] Martinez-Luaces, V. (2014), *Stability of O.D.E. solutions corresponding to chemical mechanisms based-on unimolecular first order reactions,* 3rd International Eurasian Conference on Mathematical Sciences and Applications(IECMSA-2014), Vienna, Austria, in press.

[13] Martínez-Luaces, V., (2007). Inverse-modelling problems in Chemical Engineering courses. Vision and change for a new century. *Proceedings of Calafate Delta '07*.El Calafate, Argentina.

[14] Martínez-Luaces,V., (2009). Modelling, applications and Inverse Modelling: Innovations in Differential Equations courses. *Proceedings of Southern Right Gordon's Bay Delta '09*, Gordon's Bay, South Africa.

[15] Blum, W. et al. (2002). ICMI Study 14: applications and modelling in mathematics education discussion document. *Educational Studies in Mathematics*, Vol. 51, No. 1 2, 149-171.

[16] Lebanon, G. (2012). *Probability. The Analysis of Data*, Volume 1, 1st Edition, CreateSpace Publishing.

ABOUT THE EDITOR

Dr. Ivan Kyrchei
Department of Differential Equations and Theory Of Functions
Pidstryhach Institute for Applied Problems of Mechanics and Mathematics
National Academy of Sciences of Ukraine
3-b, Naukova Str., Lviv, 79060, Ukraine
Email: kyrchei@online.ua

INDEX

#

A

B

C

G

H

I

J

K

L

M

N

O

P

Q

R

S

T

U

V

W

Z